Digital Image Processing: Advanced Technologies and Applications

Digital Image Processing: Advanced Technologies and Applications

Editors

Zahid Mehmood Jehangiri
Mohsin Shahzad
Uzair Khan

MDPI

Basel • Beijing • Wuhan • Barcelona • Belgrade • Novi Sad • Cluj • Manchester

Editors

Zahid Mehmood Jehangiri	Mohsin Shahzad	Uzair Khan
COMSATS University	COMSATS University	COMSATS University
Islamabad	Islamabad	Islamabad
Abbottabad	Abbottabad	Abbottabad
Pakistan	Pakistan	Pakistan

Editorial Office
MDPI AG
Grosspeteranlage 5
4052 Basel, Switzerland

This is a reprint of articles from the Special Issue published online in the open access journal *Applied Sciences* (ISSN 2076-3417) (available at: https://www.mdpi.com/journal/applsci/special_issues/XER023WESS).

For citation purposes, cite each article independently as indicated on the article page online and as indicated below:

Lastname, A.A.; Lastname, B.B. Article Title. *Journal Name* **Year**, *Volume Number*, Page Range.

ISBN 978-3-7258-1825-9 (Hbk)
ISBN 978-3-7258-1826-6 (PDF)
doi.org/10.3390/books978-3-7258-1826-6

Contents

About the Editors

Zahid Mehmood Jehangiri

Zahid Mahmood Jehangiri received his B.S. degree in Computer Engineering from COMSATS University Abbottabad, Pakistan, in 2007, his MS degree in Electrical Engineering from Hanyang University, South Korea, in 2011, and his Ph.D. in Electrical Engineering from North Dakota State University, USA, in 2015. Currently, he serves as an Associate Professor in the Electrical and Computer Engineering Department at COMSATS Abbottabad. Zahid Mahmood's research expertise encompass topics such as Digital Image Processing, Machine Learning, and Signal Processing. His work has appeared in over 65 publications. He is the recipient of the Government of Pakistan scholarship award for MS and Ph.D. studies.

Mohsin Shahzad

Mohsin Shahzad received his B.E degree in Industrial Electronics Engineering from N.E.D University of Engineering and Technology, Karachi, Pakistan, in 2007, his MSc (ENG) in Avionic Systems from the University of Sheffield, UK, in 2009, and his Ph.D. in Electrical Engineering (January 2017) from the Vienna University of Technology, Austria. Currently, he serves as an Assistant Professor in the Electrical and Computer Engineering Department at COMSATS Abbottabad. Mohsin Shahzad's research expertise encompass the topic of Electrical Power System Planning.

Uzair Khan

Uzair Khan received his B.S. degree in Electronics Engineering from COMSATS University Abbottabad, Pakistan, in 2008, his MS degree in Electrical Engineering from NUST (CEME), Rawalpindi, Pakistan, in 2010, and his Ph.D. in Electronic Systems Engineering from Hanyang University, South Korea, in 2015. Currently, he serves as an Associate Professor in the Electrical and Computer Engineering Department at COMSATS Abbottabad. Uzair Khan's research expertise encompass topics such as Target Tracking, Sensor Fusion, Estimation Theory, Control Systems, and Guidance and Navigation Systems.

Preface

In an era where digital technology is the backbone of numerous innovations, digital image processing stands at the forefront of technological advancements. It is a field that produces meaningful information, enabling us to interpret and manipulate visual content in a variety of ways that were once the realm of science fiction.

"Digital Image Processing: Advanced Technologies and Applications" is a comprehensive exploration of the state-of-the-art methodologies and practical applications that define this dynamic discipline. This reprint is crafted to serve as both a foundational text for students entering the field and for professionals who seek to stay updated on the latest developments. Throughout the chapters, we dig into the advanced algorithms and techniques that power modern image processing systems. From fundamental concepts such as object detection and recognition to advanced topics such as deep learning-based image analysis, this book covers a broad spectrum of technologies that are driving the future of digital imaging.

What sets this reprint apart is its focus on real-world applications. We examine how digital image processing is revolutionizing industries such as surveillance, satellite imaging, and medicine. Each application is discussed with detailed case studies and provides readers with insights into how these technologies are implemented and the challenges they address. The journey through this book begins with a solid grounding in the basic principles of digital image processing, ensuring that readers have a firm grasp of the essential concepts.

I am deeply grateful to the numerous researchers, practitioners, and educators whose work and insights have significantly contributed to the compilation of this reprint. Their dedication to advancing the field of digital image processing has been a source of inspiration throughout this writing process.

I am confident that "Digital Image Processing: Advanced Technologies and Applications" will not only educate and inform but also inspire innovations and further research. Whether you are a student, a beginner, or a professional, this reprint is designed to be a valuable companion in your exploration of the interesting world of digital image processing.

Welcome to a journey of discovery, where pixels and algorithms converge to create endless possibilities.

Zahid Mehmood Jehangiri, Mohsin Shahzad, and Uzair Khan
Editors

applied sciences

MDPI

Editorial

Digital Image Processing: Advanced Technologies and Applications

Zahid Mahmood

Department of Electrical and Computer Engineering, COMSATS University Islamabad, Abbottabad Campus, Tobe Camp, Abbottabad 22060, Pakistan; zahid0987@cuiatd.edu.pk

1. Introduction

A few decades ago, conventional image processing methods mostly focused on basic tasks such as image enhancement, registration, or edge detection. Early attempts to achieve these tasks mostly utilized grayscale images. Over time, simple methods to process grayscale images resulted in performance degradation for RGB images [1]. Ultimately, RGB image processing received more attention and subsequent advancements were made, including color preservation and fusion-based processing [2]. Currently, deep learning is used extensively in various fields, such as speech recognition and healthcare domains, with encouraging outcomes in image processing, such as image classification and segmentation [3]. A recent study showed that deep learning-based approaches significantly improve the performance of many image-related tasks, such as object detection, recognition, or segmentation compared, compared to conventional methods.

With the evolution of convolutional neural networks (CNNs), supervised learning techniques were used to train CNNs, which aimed to extract efficient features to meet their gold label requirements [4]. The performance of these methods strictly relied on the available training data. Subsequently, the limited annotated training data failed to acquire the particulars of the image details. Since the supervised learning approaches learned nonlinear mapping, they tended to primarily focus on the limited training data. As a result, the trained model struggled to yield encouraging results on unseen image data [5].

The domain of digital image processing has experienced amazing advancements, particularly through the evolution of deep learning-based algorithms, which have enhanced capabilities in many real-life applications, such as image object detection [6], recognition [7], segmentation [8], edge detection [9], and restoration [10]. Despite these advances, critical gaps remain in research and knowledge, especially in the applications and exploration of deep learning models' robustness in several challenging situations. Deep learning models also have great ease and efficiency in processing high-dimensional data [11].

This Special Issue entitled "Digital Image Processing: Advanced Technologies and Applications" addresses these challenges by collecting 15 state-of-the-art research contributions that reinforce current methodologies and offer inventive solutions and novel perspectives. Looking ahead, future research will likely focus on developing more robust and explainable AI models to enhance the feasibility of image processing systems.

Future research can also focus on exploring the potential of quantum computing to process progressively complex image data, in addition to current deep learning models. These directions will not only satisfy existing knowledge gaps but also open new possibilities for advanced applications and technologies in digital image processing.

2. An Overview of Published Articles

During the past three decades, a large number of diverse methods have appeared in computer vision and machine learning. Many of them utilize conventional machine learning. However, the recent trend in deep learning has yielded encouraging results. This section gives a brief overview of the works collected in this Special Issue.

Citation: Mahmood, Z. Digital Image Processing: Advanced Technologies and Applications. *Appl. Sci.* **2024**, *14*, 6051. https://doi.org/10.3390/app14146051

Received: 6 July 2024
Accepted: 8 July 2024
Published: 11 July 2024

In contribution one, researchers proposed an AI-enabled setup to analyze animal behavior with the objective of providing better flexibility and scalability to make the proposed setup more feasible. One of the interesting aspects of this work is that users can compliantly extend different behavior recognition algorithms to recognize animal behaviors and enjoy convenient human–computer interaction through natural language descriptions. A case study is discussed that evaluates behavioral variations between sick and healthy animals in a medical laboratory.

License plate recognition (LPR) is a key part of current intelligent systems that locate and identify varying license plates. LPR is a challenging task due to the various designs of LPs, a lack of standard LP templates, unconventional outlines, and angle dissimilarities/occlusion. These factors influence the appearance of the LP and degrade the detection and recognition abilities of algorithms. However, recent rising trends in the development of machine learning algorithms have prompted authors to solve this problem, which is the second contribution in this Special Issue. Particularly, this contribution presents a novel LPR algorithm to solve the aforedescribed challenges. This method is composed of three interconnected steps: Initially, a vehicle is detected using the Faster-RCNN algorithm. Next, the LP is spotted by applying the morphological operations of imaging. Lastly, LPR is accomplished using a deep learning network. Experiments conducted on several datasets indicate a mean LPR accuracy of over 96% on three different datasets.

The third contribution in this manuscript is about Urdu numeral classification and recognition. Urdu is one of the most complex languages, as it is a combination of several languages. Therefore, its character recognition is a difficult task. It is a bidirectional language that induces complexities during the recognition procedure. This contribution uses CNN and its variants to extract features, which are used by the Softmax activation function and SVM classifier. The obtained results are compared with GoogLeNet and the residual network (ResNet). This contribution reports 98.41% accuracy with the Softmax classifier and 99.0% with the SVM classifier. For GoogLeNet, the obtained accuracies are 95.61% and 96.4%, respectively, on ResNet.

Unmanned aerial vehicle (UAV) image capture is a promising means for acquiring geospatial data. Securing even and consistent quality in UAV images is hard due to the use of low-cost steering devices and non-surveying cameras. In addition, no specific procedures exist to perform quantitative tests on UAV images. Hence, in the fourth contribution, the authors conducted a modulation transfer function (MTF) investigation using a slanted-edge target and a ground sample distance (GSD) analysis to verify the basics of MTF analysis. This was used to verify the basics of MTF analysis in assessing UAV image values.

The accurate extraction of individual features in multi-view and multi-modal datasets is a difficult topic. In the fifth contribution, researchers present PhotoMatch, an open-source tool for multi-view and multi-modal feature-based image matching. The software contains several recently developed methods to process, extract, and match features. It also offers tools for a thorough assessment and judgement of the numerous methods and allows the user to select the top combination of methods for every modality in the dataset. A set of thirteen case studies, which included six multi-view and six multi-modal image datasets, were processed by following different methodologies.

In recognition of the importance of the video classification task and to summarize the success of deep learning models, contribution six is a concise review of the said topic. Particularly, this work highlights several major findings that are based on existing deep learning algorithms. This review emphasizes the type of architectures used, the evaluation criteria, and the experimented datasets. Moreover, a fair insight into the recently reported deep learning methods and traditional approaches is also provided. Furthermore, the important tasks based on the targets are highlighted to calculate the technical advancement of these systems.

In the seventh contribution, researchers addressed the task of multiple-object tracking (MOT) in complex scenarios, such as instances of missed detections, false alarms, and frequent target switching. This contribution has explicit potential applications in security applications, which include public safety and fire prevention, to track crucial targets. Therefore, researchers proposed an approach to multi-object tracking and an identity validity discrimination module. The authors raised the KC-YOLO detection model for tracking, optimized detection frames, and implemented adaptive feature refinement to solve challenges, for instance, incomplete pedestrian features, which are caused by occlusion. The method proposed in this work improves pedestrian tracking accuracy along with pedestrian characteristics. In experiments on the MOT16, MOT17, and MOT20 datasets, this method resulted in substantial findings and encouraging results.

The eighth contribution in this Special Issue is related to the study of recognizing handwritten Arabic characters. Given the fundamental complexities of the Arabic characters that encompass semi-cursive styles, apparent character models, and the insertion of diacritical spots, this area of research has great potential. Highlights in this work are on children's handwritten Arabic writing. This area is recognized for its apparent challenges, for example, variations in writing and distortions. The researchers also collected a dataset, referred to as "Dhad". Their investigation employs a tri-fold experimental approach, covering the investigation of pre-trained deep learning algorithms, custom-designed ConvNets architectures, and established classifiers. These findings sort out the efficiency of fine-tuned models, the potential of custom ConvNets designs, and the details associated with several classification paradigms. The pre-trained model yields the best test accuracy, at 93.59%, with the authors' collected dataset. Moreover, researchers also proposed the idea of a novel application specifically for children younger than 13, with the aim of improving their handwriting skills.

The ninth contribution in this Special Issue is related to the analysis of mammography images using multi-branch attentional ConvNets. In this work, a research team proposed a method based on the multi-label classification of two-view mammography images. It influences the correlation between lesion type and its different states. It then classifies mammograms into density, anomaly type, and difficulty level. It takes two-view mammograms as input, analyzes them using ConvNeXt and the channel attention mechanism, and integrates this information. Finally, the combined information is fed into multi-branches, which learn pattern representations to predict the appropriate state. This algorithm was evaluated on two public domain benchmark datasets, INBreast and the Curated Breast Imaging Subset of DDSM. The developed CAD method discusses the holistic performance of a patient's state. It guides radiologists in the analysis of mammograms with a facility to prepare a complete report of a patient's condition with high confidence.

The tenth contribution in this Special Issue is about the detection and classification of vehicles from publicly available datasets through YOLO-v5. The authors use a transfer learning method on the packed traffic patterns. The datasets were made thorough by introducing various aspects, for example, high- and low-density traffic images and distinct weather environments. Eventually, the improved YOLO-v5 algorithm becomes familiar to any traffic examples. Through fine-tuning the pre-trained system, the authors validated that the proposed YOLO-v5 has surpassed various traditional vehicle detection algorithms in terms of accuracy and complexity. The experiments were conducted on three different datasets to demonstrate its effectiveness in varying real-life conditions.

The eleventh contribution in this Special Issue discusses segmentation in X-ray computed tomography (CT) data for non-destructive testing (NDT) by combining the segment anything model (SAM) with tile-based flood-filling networks (FFN). This method evaluates the performance of the SAM on volumetric NDT datasets and demonstrates its effectiveness to segment instances in challenging imaging scenarios. The authors implemented different methods to analyze the image-based SAM algorithm for use with volumetric datasets. This investigation enables the segmentation of 3D objects using FFN's spatial flexibility. The piecewise method for SAM influences FFN's abilities to segment various sized objects. This research has huge potential for merging SAM with FFN for volumetric instance segmentation, particularly for large objects.

The twelfth contribution in this Special Issue discusses a novel methodology that combines bidirectional feature learning and generative networks to innovatively approach the domain adaptation problem. This study proves that merging bidirectional feature learning and generative networks is an effective solution for domain adaptation. Through various evaluations, authors verify that merging outperforms the existing works.

The thirteenth contribution in this study proposes a fruit freshness classification method through deep learning. After the fruit data was gathered, the data was pre-processed, including augmentation and labeling. Later, the AlexNet model was used. Meanwhile, transfer learning and fine-tuning of the CNN was accomplished. Lastly, the Softmax classifier was used for classification. Experiments were performed using three commonly available datasets. The proposed model achieved highly favorable results in all three datasets by yielding an over 98% classification accuracy. In addition, this method is also computationally efficient and works in real-time to yield the final classification result.

The fourteenth contribution is about a survey of optical character recognition (OCR). OCR is a process of extracting handwritten or printed text from a scanned or printed image and converting it to a machine-readable form for further data processing. OCR technology helps digitize documents for improved productivity and accessibility. Currently, the OCR is useful for preserving historical documents. The authors briefly discuss the recent OCR methods and identify the best-performing approach that researchers could utilize in their developed applications. This contribution also covers research gaps and presents future directions for Arabic language OCR in a systematic way.

In the fifteenth contribution, the authors present a method of transfer style patterns while fusing the confined style construction with the local contented arrangement. In this contribution, numerous levels of coarse stylized features are reconstructed at low resolution using a coarse network. While achieving this, the color distribution is transferred, and the content structure is integrated with the initial style structure. Then, both the reconstructed and the content features are embraced to produce high-quality, structure-aware stylized images that have a high resolution. This is obtained through a fine network that has three structural selective fusion (SSF) sections. This method has proven to be robust by generating high-quality stylization outcomes.

3. Conclusions

The contributions listed in this Special Issue can be combined into three major groups with the following key attributes.

Group 1: Object detection: In this category, contributions 7 and 10 inspect various object detection methods. In particular, contribution 7 addresses multi-pedestrian detection and tracking. Whereas contribution 10 addresses license plate detection in real-life images in an open environment.

Group 2: Object recognition: In this category, several state-of-the-art contributions were accepted, which include contributions 1, 2, 3, 6, 8, 13, and 14. The afore-listed contributions either use AI methods or use deep learning methods to inspect various objects. This group is most prominent in this Special Issue and gathers significant scientific findings in the object recognition domain.

Group 3: Image Manipulations: This group gathers contributions 4, 5, 9, 11, 12, and 15. Specifically, contribution 4 evaluates the quality of aerial images. Whereas 5, 9, 11, 12, and 15 perform image manipulations through various methods listed therein. In these collections, contribution 12 is particularly related to image segmentation, which is currently a challenging task in various real-life scenarios.

After thoroughly analyzing the gathered contributions, the following important points are highlighted:

- With the rapid advancements in AI and machine learning, the use of deep learning in various applications has become obvious in many industries due to its ability to process complex patterns and make reliable predictions. Therefore, deep learning algorithms have found their place in crucial fields, including object detection and

recognition, natural language processing, and medical imaging. For instance, CNNs are extensively employed for tasks such as image classification, object detection, segmentation, and medical imaging. Similarly, recurrent neural networks (RNNs) and transformers have advanced the capabilities of many applications. For instance, real-time translation, sentiment analysis, and conversational agents. The development of GPUs and large-scale datasets has further driven deep learning's adoption for solving complex problems with exceptional accuracy and efficiency.
- With the evolution of RGB images, several state-of-the-art algorithms have also appeared in the literature. Most of the images related papers collected in this Special Issue address RGB images using an intelligent combination of machine learning-based methods to achieve desired outcomes.

Final Remarks: *Digital Image Processing: Advanced Technologies and Applications* will serve as a fundamental resource for researchers and practitioners. It will also assist students who aim to orient their career in machine learning and deep learning. It not only imparts basic knowledge but also stimulates advanced thinking and exploration in recent technological advancements. As the digital imaging domain continues to grow, the insights and methodologies collected in this Special Issue will provide resources and applications for newcomers. The following are a few major takeaways from the collections presented in this Special Issue:

Technological Integration: The collection presents a combination of digital image processing with advanced technologies, such as machine learning and deep learning, and demonstrates their potential for solving complex, real-world problems.

Algorithmic Development: This collection emphasizes the development and optimization of recent algorithms, which process images, extract features, and report efficient processed images results.

Innovative Applications: A variety of applications in several domains are collected, which include traffic images, medical imaging, and aerial images. Each manuscript gathered here underscores their practical relevance to modern-day technology.

Future Directions: This Special Issue also hints towards future directions in several domains, such as colored image processing, image analysis, and the development of more robust procedures, which are capable of handling a variety of datasets. Finally, the conventional object detection, recognition, and segmentation methods [12] can be integrated with recent deep learning algorithms to build a more accurate and feasible system to be deployed for various scenarios.

Conflicts of Interest: The authors declare no conflicts of interest.

List of Contributions:

1. Chen, Y.; Jiao, T.; Song, J.; He, G.; Jin, Z. AI-Enabled Animal Behavior Analysis with High Usability: A Case Study on Open-Field Experiments. *Appl. Sci.* **2024**, *14*, 4583. https://doi.org/10.3390/app14114583.
2. Sultan, F.; Khan, K.; Shah, Y.A.; Shahzad, M.; Khan, U.; Mahmood, Z. Towards Automatic License Plate Recognition in Challenging Conditions. *Appl. Sci.* **2023**, *13*, 3956. https://doi.org/10.3390/app13063956.
3. Bhatti, A.; Arif, A.; Khalid, W.; Khan, B.; Ali, A.; Khalid, S.; Rehman, A.U. Recognition and classification of handwritten urdu numerals using deep learning techniques. *Appl. Sci.* **2023**, *13*, 1624. https://doi.org/10.3390/app13031624.
4. Kim, J.H.; Sung, S.M. Quality Analysis of Unmanned Aerial Vehicle Images Using a Resolution Target. *Appl. Sci.* **2024**, *14*, 2154. https://doi.org/10.3390/app14052154.
5. Ruiz de Oña, E.; Barbero-García, I.; González-Aguilera, D.; Remondino, F.; Rodríguez-Gonzálvez, P.; Hernández-López, D. PhotoMatch: An Open-Source Tool for Multi-View and Multi-Modal Feature-Based Image Matching. *Appl. Sci.* **2023**, *13*, 5467. https://doi.org/10.3390/app13095467.
6. Belhaouari, A.R.S.B.; Kabir, M.A.; Khan, A. On the Use of Deep Learning for Video Classification. *Appl. Sci.* **2022**, *13*, 2007. https://doi.org/10.3390/app13032007.

7. Li, J.; Wu, W.; Zhang, D.; Fan, D.; Jiang, J.; Lu, Y.; Gao, E.; Yue, T. Multi-Pedestrian Tracking Based on KC-YOLO Detection and Identity Validity Discrimination Module. *Appl. Sci.* **2023**, *10*, 2228. https://doi.org/10.3390/app132212228.
8. AlMuhaideb, S.; Altwaijry, N.; AlGhamdy, A.D.; AlKhulaiwi, D.; AlHassan, R.; AlOmran, H.; AlSalem, A.M. Dhad—A Children's Handwritten Arabic Characters Dataset for Automated Recognition. *Appl. Sci.* **2024**, *10*, 2332. https://doi.org/10.3390/app14062332.
9. Al-Mansour, E.; Hussain, M.; Aboalsamh, H.A.; Al-Ahmadi, S.A. Comprehensive Analysis of Mammography Images Using Multi-Branch Attention Convolutional Neural Network. *Appl. Sci.* **2024**, *5*, 12995. https://doi.org/10.3390/app132412995.
10. Farid, A.; Hussain, F.; Khan, K.; Shahzad, M.; Khan, U.; Mahmood, Z. A Fast and Accurate Real-time Vehicle Detection Method Using Deep Learning for Unconstrained Environments. *Appl. Sci.* **2023**, *30*, 3059. https://doi.org/10.3390/app13053059.
11. Gruber, R.; Rüger, S.; Wittenberg, T. Adapting the Segment Anything Model for Volumetric X-ray Data-Sets of Arbitrary Sizes. *Appl. Sci.* **2024**, *17*, 3391. https://doi.org/10.3390/app14083391.
12. Han, C.; Choo, H.; Jeong, J. Bidirectional-Feature-Learning-Based Adversarial Domain Adaptation with Generative Network. *Appl. Sci.* **2023**, *13*, 11825. https://doi.org/10.3390/app132111825.
13. Amin, U.; Shahzad, M.I.; Shahzad, A.; Shahzad, M.; Khan, U.; Mahmood, Z. Automatic fruits freshness classification using CNN and transfer learning. *Appl. Sci.* **2023**, *11*, 8087. https://doi.org/10.3390/app13148087.
14. Faizullah, S.; Ayub, M.S.; Hussain, S.; Khan, M.A. A survey of OCR in Arabic language: Applications, techniques, and challenges. *Appl. Sci.* **2023**, *13*, 4584. https://doi.org/10.3390/app13074584.
15. Liu, K.; Yuan, G.; Wu, H.; Qian, W. Coarse-to-Fine Structure-Aware Artistic Style Transfer. *Appl. Sci.* **2023**, *13*, 952. https://doi.org/10.3390/app13020952.

References

1. Zhang, X.; Wang, X.; Yan, C.; Sun, Q. EV-fusion: A novel infrared and low-light color visible image fusion network integrating unsupervised visible image enhancement. *IEEE Sens. J.* **2024**, *73*, 5020911. [CrossRef]
2. Yin, M.; Du, X.; Liu, W.; Yu, L.; Xing, Y. Multiscale fusion algorithm for underwater image enhancement based on color preservation. *IEEE Sens. J.* **2023**, *23*, 7728–7740. [CrossRef]
3. Qi, Y.; Guo, Y.; Wang, Y. Image Quality Enhancement Using a Deep Neural Network for Plane Wave Medical Ultrasound Imaging. *IEEE Trans. Ultrason. Ferroelectr. Freq. Control* **2021**, *68*, 926–934. [CrossRef] [PubMed]
4. Ye, T.; Qin, W.; Zhao, Z.; Gao, X.; Deng, X.; Ouyang, Y. Real-time object detection network in UAV-vision based on CNN and transformer. *IEEE Trans. Instrum. Meas.* **2023**, *72*, 2505713. [CrossRef]
5. Mahmood, Z.; Ullah, A.; Khan, T.; Zahir, A. Miscellaneous Objects Detection Using Machine Learning Under Diverse Environments. *Adv. Deep Gener. Models Med. Artif. Intell. Stud. Comput. Intell.* **2023**, *1124*, 201–223.
6. Alassafi, M.O.; Ibrahim, M.S.; Naseem, I.; AlGhamdi, R.; Alotaibi, R.; Kateb, F.A.; Oqaibi, H.M.; Alshdadi, A.A.; Yusuf, S.A. A novel deep learning architecture with image diffusion for robust face presentation attack detection. *IEEE Access* **2023**, *11*, 59204–59216. [CrossRef]
7. Tan, Z.; Liu, A.; Wan, J.; Liu, H.; Lei, Z.; Guo, G.; Li, S.Z. Cross-batch hard example mining with pseudo large batch for id vs. spot face recognition. *IEEE Trans. Image Process.* **2022**, *31*, 3224–3235. [CrossRef]
8. Sheikhjafari, A.; Krishnaswamy, D.; Noga, M.; Ray, N.; Punithakumar, K. Deep learning based parameterization of diffeomorphic image registration for cardiac image segmentation. *IEEE Trans. NanoBiosci.* **2023**, *22*, 800–807. [CrossRef] [PubMed]
9. Felt, V.; Kacker, S.; Kusters, J.; Pendergrast, J.; Cahoy, K. Fast ocean front detection using deep learning edge detection models. *IEEE Trans. Geosci. Remote Sens.* **2023**, *61*, 4204812. [CrossRef]
10. Zhang, Q.; Dong, Y.; Yuan, Q.; Song, M.; Yu, H. Combined deep priors with low-rank tensor factorization for hyperspectral image restoration. *IEEE Geosci. Remote Sens. Lett.* **2023**, *20*, 5500205. [CrossRef]
11. Wu, R.; Zheng, F.; Li, M.; Huang, S.; Ge, X.; Liu, L.; Liu, Y.; Ni, G. Toward ground-truth optical coherence tomography via three-dimensional unsupervised deep learning processing and data. *IEEE Trans. Med. Imaging* **2024**, *43*, 2395–2407.
12. Mahmood, Z.; Muhammad, N.; Bibi, N.; Ali, T. A Review on state-of-the-art Face Recognition Approaches. *Fractals Complex Geom. Patterns Scaling Nat. Soc.* **2017**, *25*, 1750025. [CrossRef]

applied sciences

MDPI

Article

AI-Enabled Animal Behavior Analysis with High Usability: A Case Study on Open-Field Experiments

Yuming Chen [1], Tianzhe Jiao [1], Jie Song [1,*], Guangyu He [2] and Zhu Jin [2]

[1] Software College, Northeastern University, Shenyang 112000, China; 2290126@stu.neu.edu.cn (Y.C.); jiaotianzhe@stumail.neu.edu.cn (T.J.)
[2] Technology Strategy and Development Department, Neusoft Group, Shenyang 110002, China; hegy@neusoft.com (G.H.); jin_zh@neusoft.com (Z.J.)
* Correspondence: songjie@mail.neu.edu.cn

Featured Application: In this study, we designed a highly available animal behavior analysis platform that can help researchers significantly improve their work efficiency. In addition, the platform has good flexibility, scalability, and human-machine interaction. Researchers can easily configure and use the platform for behavioral observation experiments with minimal learning costs.

Abstract: In recent years, with the rapid development of medicine, pathology, toxicology, and neuroscience technology, animal behavior research has become essential in modern life science research. However, the current mainstream commercial animal behavior recognition tools only provide a single behavior recognition method, limiting the expansion of algorithms and how researchers interact with experimental data. To address this issue, we propose an AI-enabled, highly usable platform for analyzing experimental animal behavior, which aims to provide better flexibility, scalability, and interactivity to make the platform more usable. Researchers can flexibly select or extend different behavior recognition algorithms for automated recognition of animal behaviors or experience more convenient human-computer interaction through natural language descriptions only. A case study at a medical laboratory where the platform was used to evaluate behavioral differences between sick and healthy animals demonstrated the high usability of the platform.

Keywords: analysis platform; behavior recognition; human-computer interaction

Citation: Chen, Y.; Jiao, T.; Song, J.; He, G.; Jin, Z. AI-Enabled Animal Behavior Analysis with High Usability: A Case Study on Open-Field Experiments. *Appl. Sci.* **2024**, *14*, 4583. https://doi.org/10.3390/app14114583

Academic Editors: Zahid Mehmood Jehangiri, Mohsin Shahzad and Uzair Khan

Received: 10 April 2024
Revised: 21 May 2024
Accepted: 23 May 2024
Published: 27 May 2024

1. Introduction

Animal behavior is the body language by which an animal expresses its psychological and physiological state and its overall function. Typical model animals, such as mice, rabbits, and goats, are widely used to analyze different behaviors in the open field to measure the effectiveness of experiments in biology, toxicology, neuroscience, pharmacology, animal husbandry, and genetics [1–3].

Due to the advancement of embedded technology and automation technology, animal behavior recognition has progressed rapidly. For example, Arablouei et al. [4] utilize embedded devices and corresponding behavior recognition methods for efficient behavior recognition of livestock, such as cows. Roughan et al. [5] proposd automation technology to predict the behavioral changes of mice undergoing surgery and observe the effects of painkillers. With the development of deep learning techniques, the performance of animal behavior recognition has been significantly improved. Natarajan et al. [6] achieve high-accuracy detection of wild-animal behavior using deep learning models. In order to ensure real-time performance, Fuentes et al. [7] proposed a behavior recognition algorithm for cattle based on a spatial-and-temporal information framework. Despite the success of these studies in specific scenarios, these studies generally need to rely on commercialized tools to provide convenient human–computer interaction and data retrieval.

Although commercialized tools can simplify human–computer interaction and reduce labor costs for behavior recognition, they still need more flexibility, scalability, and interactivity. Specifically, these tools usually can only detect specific behaviors, making it difficult to adapt to different experimental needs. In addition, designing the underlying algorithms of commercialized tools is often scenario-specific. It cannot generate intermediate results such as single-frame pose data, limiting its application scope. The limitations of hardware and architecture also make it difficult for these tools to change or expand the underlying algorithms. It affects the accuracy and speed of the experimental results. Finally, commercialized tools have a limited scope of human-computer interaction for data retrieval and analysis, and researchers need to resort to other specialized tools or programming languages, such as structured query language (SQL) or Python, which increases the learning cost and reduces the usability of the tools. Therefore, designing and implementing a laboratory animal behavior analysis platform that can efficiently identify animal behavioral actions, effectively manage animal behavioral data, support changes in the underlying algorithms, and provide convenient human-computer interaction capabilities is of great significance in reducing the workload of related researchers.

In order to compensate for the shortcomings of existing tools and to design a highly usable platform for animal behavior analysis, we established three goals that the platform should achieve:

1. The platform should be flexible enough to support researchers in selecting different behavior recognition methods and behavior detection categories.
2. The platform should be scalable to support researchers in upgrading or expanding the underlying algorithms.
3. The platform should have flexible and convenient interactivity so that researchers can use the platform's preset human-computer interaction functions when using experimental data systems or commercial tools for data querying and statistical analysis or use more adaptive interaction methods to meet the changing functional needs of researchers.

When designing the platform, we integrated various architecture design methods, such as microservices and plug-in design, to support researchers in flexibly configuring the detection methods and behavioral categories and having good flexibility and scalability. In order to improve the interactivity of the platform and enable researchers to retrieve and analyze data more flexibly, we have introduced natural language processing algorithms into the platform, through which we analyze the intent of the user's natural language query and convert the command to generate database execution statements that can be executed. Although the fusion architecture and natural language processing technology can bring higher usability to the platform, the effective integration, replacement, extension, and management of different behavior recognition algorithms, the adaptability of natural language processing algorithms in the field of animal experiments, and the cross-language problem of natural language processing algorithms still pose significant challenges to the implementation of the platform.

We propose a high-availability animal behavior analysis platform that combines good architectural design practices and natural language processing techniques, and Figure 1 shows the overall architecture of the platform. We build the platform ecosystem as a hybrid architecture, where behavior recognition services can be flexibly configured or extended to efficiently recognize multiple behavioral categories, including fine-grained movements, and produce intermediate results that meet specific experimental requirements. The platform is highly integrated with multiple business modules, which can automatically identify the behavioral actions based on the input data and store the identified behavioral data information directly in the database without manual recording or inputting information, thus effectively reducing the labor cost. The platform provides natural language query interface services. In addition to predefined platform functions, it can also use natural language descriptions for more flexible data retrieval and analysis.

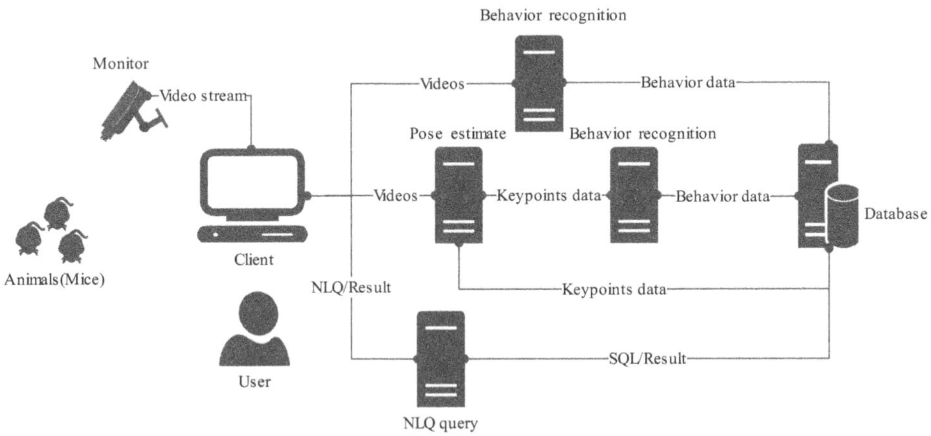

Figure 1. Overall structure of the platform.

In summary, our main contributions are as follows:

1. We have developed an AI-enabled, highly available platform that centralizes necessary functions for researchers, streamlining their workflow, reducing costs, and enhancing efficiency.
2. We have enhanced the platform's architecture with multiple design patterns to boost its flexibility and scalability, allowing for easy selection and extension of various algorithms and integration of posture estimation and behavior recognition for diverse experimental needs.
3. We have incorporated natural language processing to improve user interaction, eliminating the need for additional programming or complex database operations for data analysis.
4. We have validated the platform's effectiveness through a case study on UBE3A gene deletion, highlighting its practical utility in real-world scenarios.

The paper is organized as follows: Section 2 presents the related work. Section 3 describes the overall system architecture. Section 4 outlines methods to improve the usability of behavior recognition. Section 5 outlines methods to improve the usability of human-computer interaction. Section 6 discusses the case study. Section 7 provides the conclusion.

2. Related Work

In recent years, the field of animal behavior analysis has made remarkable progress because of the application of commercial tools and advanced behavior recognition algorithms. These techniques not only improve research efficiency but also provide strong support for animal welfare and disease research.

Commercial tools. EthoWatcher is an open-source software designed to record and analyze animal behavior [8]. It can process video files and offers rich features to label and quantify animal movement information. EthoWatcher provides a user-friendly interface for various experimental setups, which makes it easy for researchers to perform behavioral analyses. The ToxTrac software utilizes a second-order Kalman filter to estimate a detected object's trajectory and can fuse existing trajectory segments to generate a complete trajectory [9]. ToxTrac also provides various tools and features for analyzing animal behavior, such as path length, average velocity, and dwell time. These features make ToxTrac a powerful tool in animal behavior research. ANY-maze is an animal behavior analysis system developed by Stoelting, Inc., Kiel, WI, USA [10]. By marking a point on the back of a mouse, ANY-maze can calculate the distance the mouse moves in the open field, thus determining the mouse's locomotor ability. Although various parameters can be generated automatically, statistical analysis software such as GraphPad is required to analyze the differences between disease and normal mice [11].

Animal behavior recognition algorithm. With the development of machine learning and deep learning technology, many scholars have started to apply methods based on machine learning and deep learning to animal behavior recognition, and they have achieved good results. Fang et al. [12] proposed an animal behavior classification method based on six features (keypoint location, depression, skeleton, shape feature, skeleton angle, and elongation) and a naive Bayes model (NBM), which can effectively identify and classify the daily behaviors of animals. In order to improve detection accuracy, Nasiri et al. [13] fully utilized the advantage of long short-term memory (LSTM) in processing time series data. They accurately assessed the lameness status of broilers by successively extracting keypoints into the LSTM model and classifying the lameness degree of broilers according to the six-point assessment method. To further improve detection accuracy Lin et al. [14] first estimated bird keypoints using HRNet to generate global and local features [15]. After that, the excitation region was localized by keypoint clustering. Finally, bird behavior recognition achieved significant results by combining ResNet [16]. In order to compensate for the lack of single morphological features, Li et al. [17] fused multi-features to realize efficient lameness classification, including red–green–blue (RGB), optical flow, and skeleton. They utilized VGG-19 to extract skeleton joint point features and analyze spatiotemporal features by ST-GCN [18,19]. Chen et al. [20] ameliorated the deep learning method for pig aggression behavior recognition, using video data as input and extracting temporal and spatial features based on VGG-16 and LSTM models [18]. Its recognition accuracy reached 98.4% and significantly improves prediction efficiency. In order to meet the demand for real-time monitoring in the production environment, Zhang et al. [21] designed a real-time sow behavior detection algorithm (SBDA-DL), based on MobileNet and a single-shot multi-box detector (SSD). They trained and predicted sow behaviors, including watering, urinating, and crawling, and obtained satisfactory results. Moreover, several methods focus on behavioral recognition detection in complex wild environments. For example, Schindler et al. [22] used an infrared camera to capture the activity of deer, wild boar, fox, and hare in the wild environment. It can recognize the feeding, moving, and gaze behaviors based on the ResNet variant and SlowFast framework [23].

Table 1 compares our platform with other results. Our platform has flexible architecture and algorithm service management. It can quickly adapt to different behavior recognition requirements and support the training of proprietary models. Our platform incorporates natural language processing for a natural language query interface. It improves data retrieval and analysis processes and reduces reliance on traditional statistical tools.

Table 1. Comparison of our platform with commercial tools and behavior recognition algorithms.

Source	Research Object	Behavior Type	Behavior Recognition Method	Data Retrieval and Analysis	Scalability	Interactivity
EthoWatcher	Animals	Extracts only activity-related parameters	Digital image processing techniques	Contains certain functions, further analysis requires reliance on other tools	Not supporting algorithm replacement	Graphical interface
ToxTrac	Animals	Extracts only activity-related parameters	Digital image processing techniques and second-order Kalman filter	Contains certain functions, further analysis requires reliance on other tools	Not supporting algorithm replacement	Graphical interface
ANY-maze	Animals	Hairdressing, stiffness, movement, stillness, activity related parameter	Experimental animal trajectory prediction algorithm	Contains certain functions, further analysis requires reliance on other tools	Not supporting algorithm replacement	Graphical interface

Table 1. *Cont.*

Source	Research Object	Behavior Type	Behavior Recognition Method	Data Retrieval and Analysis	Scalability	Interactivity
Fang	Broiler chicken	Standing, walking, running, feeding, resting carding	Naive Bayes	Data retrieval and analysis rely on other tools	-	Command line
Nasiri	Broiler chicken	Limp	LSTM	Data retrieval and analysis rely on other tools	-	Command line
Lin	Bird	Swimming, flapping wings, standing, shoot winging, feeding, squatting	ResNet18	Data retrieval and analysis rely on other tools	-	Command line
Li	Dairy cattle	Limp	ST-GCN	Data retrieval and analysis rely on other tools	-	Command line
Ours	Animals	Customizable	Replaceable	Contains certain functions and natural language query interface	Supporting algorithm replacement	Graphical interface and natural language query interface

3. Overall System Architecture

The software system we designed is an animal behavior analysis platform that can be deployed in medical laboratories. It automatically acquires the behavioral information of animals in open-field videos and generates corresponding data reports to be stored in a database. The platform allows researchers to flexibly select existing behavioral recognition algorithms or extend new ones into the platform for use according to different experimental needs. In addition, the platform utilizes an integrated natural language query interface service to provide high usability, as researchers are no longer limited by preset functions when conducting data retrieval or analysis. We selected relational databases for the platform because the relational database can effectively manage and maintain relationships between data, ensure data integrity and consistency, and support complex query operations in SQL language. We used a model–template–view (MTV) pattern similar to the model–view–controller (MVC) pattern, which can effectively separate data, business logic, and user interface. The pattern can improve the maintainability and scalability of the system and make the platform easier to develop and maintain. The model layer handles the application's data logic and database interaction. The template layer is responsible for building the structure and style of the page, typically using hypertext markup language (HTML) and template languages. As a traditional controller, the view layer receives and processes user requests and passes the model data to the template for display. In addition, we selected JavaScript object notation (JSON) as the structural form for data exchange to achieve lightweight data transmission and parsing, which can improve system performance and efficiency. We developed the algorithm library using Python. We used Flask, Pytorch frameworks, and open-source libraries (such as OpenCV-python, Numpy, Scikit-video, and others) in the development project. We deployed the platform, pose estimation service, behavior recognition service, and natural language query interface service on four identical devices. Each device adopted the Ubuntu 18.04 system with a CPU model of 2vCPU Intel (R) Xeon (R) Platinum 8352V, Intel Corporation and NVIDIA Corporation, Santa Clara, California, USA and had 90 GB of memory. The device that was deployed for the pose estimation, behavior recognition, or natural language query interface service had anRTX4090 (24 GB) GPU, Santa Clara, California, USA. Figure 2 shows the overall architecture of the platform, where the gray part indicates the main service modules, and Table 2 describes the relevant information of each module.

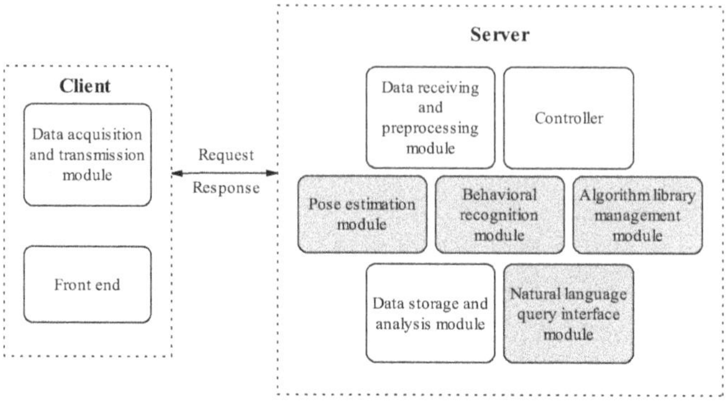

Figure 2. The main architecture of the platform, in which the pose estimation module, the behavior recognition module and the natural language processing model are the main modules.

Table 2. Relevant information of each module.

Module Name	Functionality
Data acquisition and transmission module	Capture video data and pass it on to the server.
Data receiving and preprocessing module	Receive and convert data into standardized form.
Controller	Receive and process the requests sent by the client.
Algorithm library management module	Manage plug-ins and interface services within the algorithm library.
Pose estimation module	Training or execution of different pose estimation algorithms.
Behavior recognition module	Training or execution different behavior recognition algorithms.
Natural language query interface module	Converting natural language queries entered by researchers into computer instructions.

The platform has two working modes—training and inference—to adapt to different practical use cases. In the training mode, researchers first need to input different training information, including animal category, number of key points, number of behavioral categories, etc., through the front-end page according to the different algorithms selected, then import and label the training data. Specifically, the labeled key point data will be used to train within the pose estimation model, while the labeled behavioral category data will be used to train the behavioral recognition model. Once the training starts, the training information and progress will be fed back to the researcher in real time through the front-end page. In the inference mode, the video capture and sending module will first send the captured video data to the receiving and preprocessing module on the server side. Then, after the preprocessing module processes the video data, there will be two different kinds of subsequent processing depending on the requirements. One is to input the preprocessed data into the pose estimation module, and the obtained pose estimation result will be temporarily stored as an intermediate result at the service end. Then, the posture estimation results are input into the behavior recognition model to generate specific behavioral category information, which is then stored in the database. The other is to directly input the preprocessed data into the behavior recognition model for analysis and keep the analysis results. Once reasoning is complete, researchers can describe their data retrieval or analysis needs in natural language on the front-end page. These requirements are sent, received, processed, and then fed into the natural speech query interface module, which converts the requirements into commands understandable by the platform and executes them. Finally, the query results are displayed to the researchers through the front-end page.

4. High Availability of Animal Behavior Recognition

Automatic recognition of animal behavior is the core function of the platform. It is of great significance for medical animal behavior experiments to be able to select behavior recognition methods flexibly and effectively recognize the behavior of animals in experimental videos. In order to improve the usability of the platform and enable it to flexibly select and extend the behavior recognition algorithm according to the actual needs of researchers, we integrated the design idea of plug-in management, unified interface, and algorithm library when designing the platform architecture.

Plugin design is a software architecture design pattern that modularizes the functionality of software, and each module can be developed and used as a separate plugin. In the laboratory animal behavior analysis platform, we design each behavior recognition algorithm as a plugin and deploy it locally. Each plugin contains all the code and resources to implement a certain behavior recognition algorithm. These behavior recognition algorithm plugins can be developed and tested independently of the platform and only need to follow certain interface specifications and data formats. At runtime, the platform can dynamically load and unload plugins to execute the algorithms in the plugins. To realize this design, we define a local plugin interface and require all plugins to implement this interface. This interface includes the initialization of the plugin, setting and obtaining parameters, executing analysis, obtaining results, and other operations. The plugin design allows the platform to have higher flexibility and extensibility. Researchers can choose and combine plugins according to their needs and even develop their own plugins.

Although plug-in local deployment can bring good response speed, local deployment is sometimes limited by hardware performance. Therefore, in addition to supporting plug-in algorithm access for local deployment, the platform also supports access to non-locally deployed behavior recognition algorithm services in the form of a unified interface. The platform requires all behavior recognition algorithms to have a unified remote web interface. We define the basic operations of this interface about behavior recognition algorithms, such as initialization, setting parameters, executing analysis, obtaining results, and so on. The unified network interface approach allows the platform to use behavior recognition services deployed on other computing resources. In addition, the unified interface service removes the platform's focus on the specific details of the service and is not affected by upgrades or replacements of the algorithms.

An algorithm library is a software library that stores and manages algorithms. The platform stores all available behavior recognition algorithm plug-ins and remote service interfaces in an algorithm library. The algorithm library contains service resources for all available behavior recognition algorithms. Researchers can use the search and filter functions to select appropriate algorithms for behavioral analysis experiments. To realize this design, we define the structure of an algorithm library containing storage paths of algorithms, service interface paths, metadata formats, etc. The platform provides a module to manage the algorithm library, which contains functions such as adding, deleting, searching, and loading algorithms. The application of the algorithm library enables the platform to centrally manage decentralized algorithm plug-ins and remote services, which is convenient for researchers to find and select and improves the usability of the platform.

Unlike commercial tools, the platform uses a more flexible architecture that allows it to change different behavior recognition methods according to user needs, such as skeleton keypoint-based methods, optical flow information-based methods, depth image-based methods, and appearance contour-based methods. Among them, the skeleton keypoint-based method needs to rely on a posture estimation algorithm to obtain information about the key points of the animal's skeleton. The timing information of these key points can reliably describe the subtle changes in the animal's posture and serve as the basic data for analyzing other motion indicators. The behavior recognition algorithm based on skeleton key points can identify specific categories of behaviors based on the key point sequence information. Combining posture estimation with key point-based behavior recognition algorithms satisfies the reliability of behavior recognition in animal experiments and increases the flexibility of adapting to different experimental needs. The key points must be accurately mapped onto the animal limbs to recognize the animal-generated pose data during the experiment. The accuracy of the key point information of experimental animals has an extremely important impact on subsequent behavior recognition and other experimental tasks.

The platform currently provides two pose estimation algorithm plug-ins, one of which is the DeepLabCut pose estimation algorithm, which combines target detection, target tracking, and semantic segmentation algorithms to accurately locate key points on the limbs of experimental animals without the need for labeling [24]. DeepLabCut reduces the computational cost of the pose estimation algorithm by transforming the complex pose estimation task into key point detection and tracking, significantly reducing the computational cost. The network architecture of DeepLabCut is based on a convolutional neural network, as shown in Figure 3. DeepLabCut consists of the following main components: a feature extraction layer for extracting features from the input image, a fully connected layer for key point regression, a loss function that measures the difference between the predicted key point position and the actual calibrated position, and the optimizer that adjusts the network parameters to minimize the prediction error. Overall, the network structure of DeepLabCut is a convolutional neural network based on a backbone network such as ResNet, which implements key point localization and tracking through a feature extraction layer and a key point regression layer.

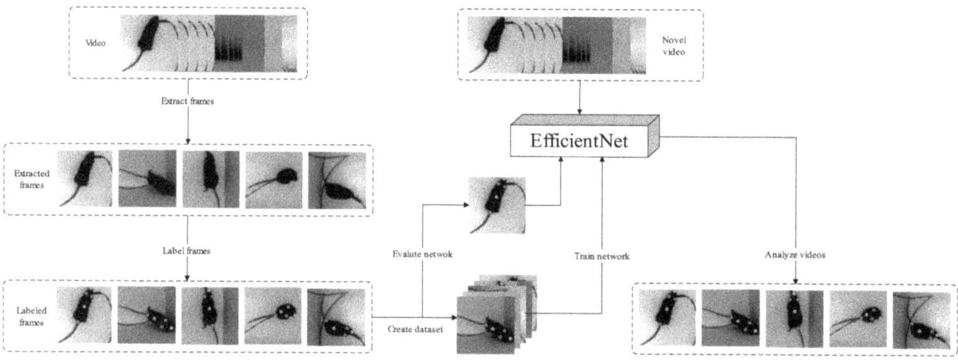

Figure 3. DeepLabCut network architecture, where EfficientNet is the replaceable feature extraction network.

Another plug-in for various pose estimation algorithms provided by the platform is YOLOX-Pose [25], an algorithm for multi-person pose estimation based on the popular YOLO target detection framework [26]. The algorithm combines the advantages of top-down and bottom-up approaches by simultaneously detecting the bounding boxes and corresponding 2D poses of multiple people through a forward propagation process. Unlike traditional heatmap-based two-stage approaches, YOLO-Pose is end-to-end trainable and optimized to evaluate the metric of object key point similarity (OKS) instead of using L1 loss as a proxy for training. In addition, YOLO-Pose does not require the post-processing step of the underlying method to group the detected key points into skeletons, as each bounding box has an associated pose, which enables the natural grouping of key points. YOLO-Pose achieved new optimal results on the COCO validation set and the test set (90.2% AP50 and 90.3% AP50) and outperform all existing key points in a single forward propagation process, outperforming all existing bottom-up methods without the need for flip tests, multi-scale tests, or other test time enhancements. While the original YOLO pose implements single-shot pose estimation based on the YOLOv5 target detection framework, the platform extends it based on the better-performing YOLOX framework. It provides network structures with different parameter scales such as YOLOX-tiny-Pose, YOLOX-s-Pose, YOLOX-m Pose, YOLOX-s-Pose, YOLOX-m-Pose, and YOLOX-l-Pose. The network structures with different parameter scales can meet the performance constraints of different hardware resources.

Behavior recognition based on skeleton keypoints is usually done in two ways: one is based on keypoint coordinate information by manual design of matching rules for behavioral categories, such as linear motion behaviors that can be matched by the linear change of the coordinates of the center point of the animal's body between consecutive frames; the other method based on deep learning by the model autonomously learns the keypoint change characteristics of different behavioral categories.

The platform provides a behavior recognition algorithm plug-in based on skeleton key points, which selects the ST-GCN network as the underlying algorithm, as shown in Figure 4. For the first time, this network combines the graph convolution operation, which captures spatial dimensional information, and the temporal convolution operation, which captures temporal dimensional information, to form spatiotemporal convolution modules. These modules can extract high-level features of skeleton graph sequences through multiple layers. The ST-GCN network mainly consists of nine layers of basic modules, with the output channels of the first three layers being 64, the middle three layers being 128, and the output channels of the last three layers being 256. In addition, the size of the temporal convolution kernel of each layer is 9. To reduce the feature loss of the network during feature extraction and to improve the feature extraction capability of the model, residual concatenation is used in each base unit to realize the cross-region feature fusion. Meanwhile, to avoid overfitting during the training process and improve the robustness of the model, a dropout layer is added to each base unit. After these processes, the feature vectors generated from the skeleton sequences will finally be fed into the SoftMax classifier for behavioral action classification.

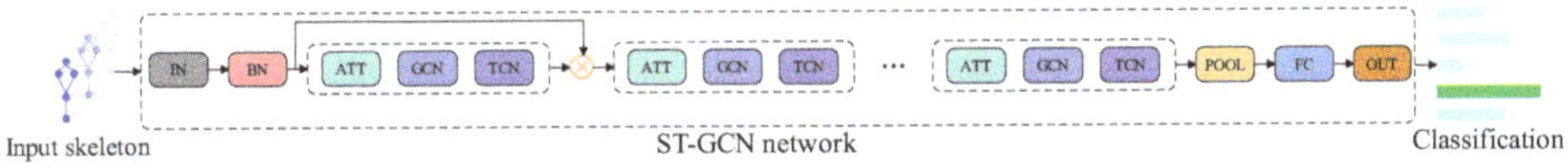

Figure 4. Principle of ST-GCN algorithm, where GCN is spatial graph convolution and TCN is temporal graph convolution.

Another behavior recognition algorithm plug-in provided by the platform is the SlowFast algorithm based on optical flow features. The algorithm uses hand-designed optical flow features to characterize the movement information of the target between two frames. SlowFast is biologically inspired by a two-pathway structural model, Slow Pathway and Fast Pathway, concerning the characteristics of P-cells, which are used to capture spatial information, and M-cells, which are used to capture fast-moving information, in the retinal cells of primates. Slow Pathway is used to capture spatial semantic information reflected by sparse frames, and it uses a very low frame frequency; Fast Pathway is used to capture rapidly changing running information, and it uses a very high frame frequency. In addition, Slow Pathway has a larger model volume, like 80% P-cells; Fast Pathway is lightweight, like 20% M-cells. In the middle of the two pathways is a Fast to Slow passthrough connection, i.e., the fusion of motion information to spatial semantics. Finally, the two-pathway information is fused for classification.

5. High Availability of Human–Computer Interaction

With the changing needs of animal behavior experiments, commercial tools or independent experimental data management systems have gradually become unable to meet researchers' data retrieval and analysis needs. The main reason is that the interactivity of commercial tools or independent experimental data management systems could be better, and there are problems such as limited query syntax, pre-written queries, lack of context understanding, and strict format requirements. In order to improve the usability of human–computer interaction in data retrieval and analysis, the platform combines text-to-SQL-related algorithm models into natural language query interface services in the form of plug-ins and integrates them into the platform.

Natural language query interface service can bring much convenience to the human-computer interaction of the platform. First, the natural language query interface service can provide researchers with more free query methods. Traditional database queries need to write SQL statements according to specific syntax and structure, which limits users' query methods. Researchers can use their familiar vocabulary and expressions for querying without being restricted to a specific query syntax. Secondly, the natural language query interface service can provide dynamic query functions. Researchers can adjust the query conditions according to real-time needs without writing fixed SQL statements in advance. For example, the user can say, "Show the experimental mice that have been assisted to stand more than five times in the past three days"; the natural language query interface service can understand the user's intent and generate and execute the corresponding SQL statement.

The natural language query interface service makes human–computer interaction more flexible and natural through free querying methods, dynamic querying, and contextual understanding. Users can query in a way they are familiar with and make flexible adjustments according to real-time needs, thus improving the flexibility and adaptability of the interaction. In addition, the natural language query interface service brings significant advantages to human–computer interaction by improving query accuracy, lowering the use threshold, and providing a better user experience.

Currently, the platform provides two algorithms as natural language query interface services for researchers: namely, RAT-SQL and RYANSQL [27,28]. RAT-SQL encodes schema links and table structures based on Transformer by adding a relation-aware self-attention mechanism; Figure 5 illustrates the model structure of RAT-SQL. RAT-SQL transforms a database schema into a directed graph G_q, describes known relationships by adding biases, and encodes associations between natural language questions and database schemas using name-based and value-based strategies. Eventually, these encoding results are fed into a tree Decoder and decoded according to the syntax rules of SQL to generate SQL statements. Due to the cross-linguistic issues, the platform also incorporates a cross-linguistic common sense knowledge graph and a cross-domain common sense knowledge graph (ConceptNet) [29] into the schema-concatenation phase of RAT-SQL, which results in improved accuracy of RAT-SQL execution for medical animal experiment information retrieval.

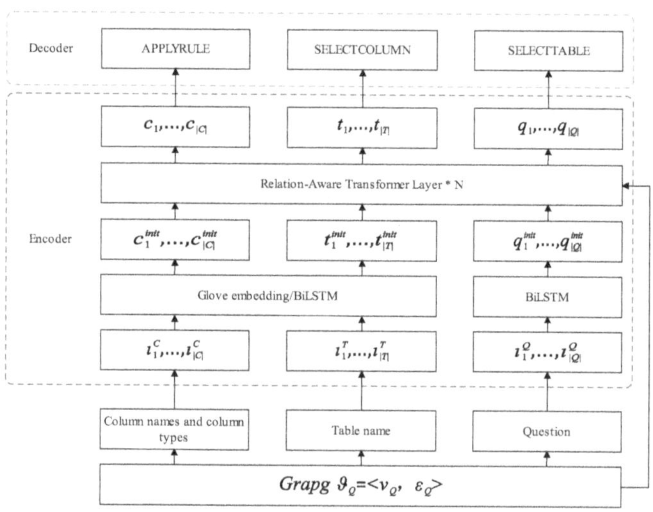

Figure 5. The model structure of RAT-SQL. * N represents the number of Tansformer layers.

RYANSQL model mainly uses a sketch-based slot-filling method and marks the complex structure of SQL statements by SQL statement position code (SPC). RYANSQL divides the generation of SQL statements into two phases: sketch generation and slot filling. For nested statements, RYANSQL first splits the SQL statement into non-nested SELECT statement blocks and represents the relationship between the blocks by SPC. Then, the final SQL statement is generated by recursively predicting the SPC and the corresponding SELECT statement blocks.

6. Application Case

The platform was deployed in a medical laboratory in Liaoning Province, China, with a wide variety of experimental animals and sufficient experimental video resources. In this facility, we chose "A study on the pathological mechanism of motor defects due to UBE3A gene deletion" as a validation case for the platform. Figure 6 shows the detailed process of this case study and the external environmental dependencies. In this study, the researchers used AS mice to simulate clinical Angelman Syndrome patients, observed the behavioral differences between the disease model mice and normal mice, and investigated the specific locomotor differences by combining postural keypoint locomotor information with calcium signaling information. Since postural and behavioral data were needed for this case, the researchers selected a skeleton keypoint-based behavioral recognition method. In deploying and using the platform, we instructed the researchers to train the corresponding DeepLabCut posture estimation model and ST-GCN behavior recognition model according to the actual needs. We retrained the RAT-SQL model in the natural language query interface service for data retrieval and analysis needs.

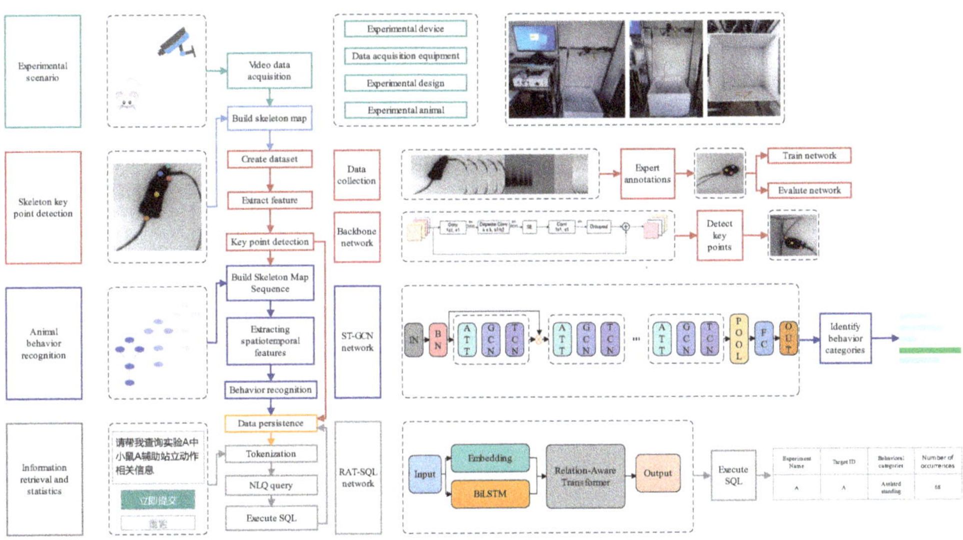

Figure 6. The detailed process and the external environmental dependencies. The Chinese meaning in the picture is to help me search for relevant information on the auxiliary standing movement of mouse A in Experiment A.

6.1. Dataset

6.1.1. Mouse Behavioral Dataset

To train and validate the actual effects of the pose estimation model and the behavior recognition model, we randomly selected 3000 experimental mouse behavioral videos. Each video was about 150 frames, mainly containing behavioral actions, such as stationary, standing, curling up, rectilinear movement, and steering movement. We divided these videos according to the ratio of 8:2, which constituted the training set and test set of the behavior recognition model. At the same time, we randomly selected 500 frames of images from these 3000 videos, and the experimenter labeled five key points on each mouse according to the experimental needs and the characteristics of the mouse skeleton. Then, we divided these images according to the ratio of 9:1, constituting the training set and test set of the pose estimation model.

6.1.2. Text-to-SQL Medical Animal Experiment Chinese Dataset

In order to train and verify the actual effect of the natural language query interface module in converting natural language into SQL statements, we have collected about 1500 SQL statement scripts used by researchers in the past and supplemented the natural language descriptions of these SQL statements and the corresponding table structure information according to the relevant information. The types of query statements include queries with keywords such as group by, order by, and having but also multi-table join queries, Nested queries, and more comprehensive calculation queries. Compared with CSpider [30], TableQA [31], and other datasets, the Chinese dataset of medical animal experiments is relatively simple. However, it is more domain-specific and in line with the actual needs of medical animal experiments' information retrieval. We divided the dataset into a training set and a testing set according to the ratio of 2:1.

6.2. Behavioral Recognition

DeepLabCut is fully pre-trained on the ImageNet dataset. In addition, DeepLabCut has been tested and calibrated on behavioral data generated by different species of organisms, such as mice and fruit flies. These diverse data make the model robust. In this case, we trained DeepLabCut specifically using the mouse mentioned above behavioral dataset mentioned above, due to the differences in key point locations. We evaluated the model's accuracy by comparing the deviation between the pixel coordinates of the key points predicted by the model and the coordinates labeled by the expert. We used the change in the mean value of the deviation for each key point to objectively assess the stability of the model. The root mean square error (*RMSE*) measures the root mean square difference between the predicted and true values, thus indicating the average degree of deviation between the predicted and true values. The formula for the RMSE is shown below:

$$RMSE = \sqrt{\frac{\sum_{i=1}^{n} \left[x_{t,i} - x_{p,i}\right]^2}{n}} \quad (1)$$

where n is the number of observations, x_t is the true value, and x_p is the predicted value. Since DeepLabCut supports a variety of feature extraction networks, we selected six different feature extraction network models to obtain the best performance, including ResNet-50 and ResNet-101 for training [32,33]. We evaluated the detection performance of these networks on mouse skeletal key points and selected the most suitable feature extraction network for this case.

As shown in Figure 7, the RMSEs of different feature extraction networks for the three key points of the mice varied. The EfficientNet-b6 network has the smallest error on the test set [34], and the difference between its predicted coordinates and the true pixel coordinates was close to 5.9 pixels at the tail. Considering the higher pixels occupied by the nose part of the mouse in the high-resolution image in this experiment, such a coordinate deviation is acceptable.

In addition to this, we also evaluated the processing speed of different feature extraction networks. For example, when using MobileNet-V2-0.35 [35], the model can reach a processing speed of 16.5 frames/sed but with a relatively high error rate. The slowest detection speed is EfficientNet-b6, about 3.8 frames/sec. In medical animal experiments, the need for accuracy is usually higher than the processing speed. Therefore, we finally chose EfficientNet-b6 as the feature extraction network for DeepLabCut.

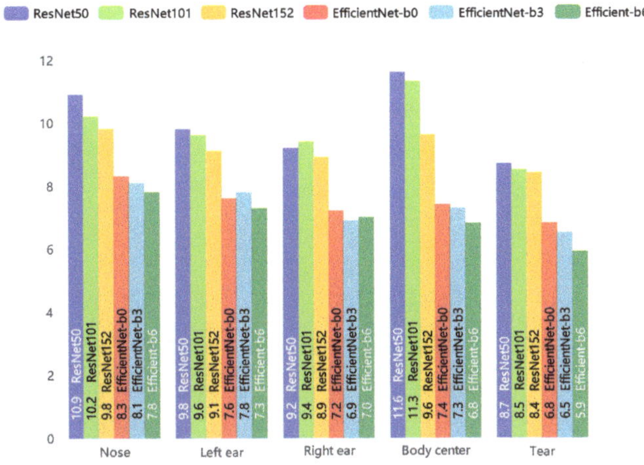

Figure 7. *RMSE* of five critical points in mice extracted by different feature extraction networks.

In the study of the pathological mechanism of motor deficits caused by UBE3A gene deletion, in addition to capturing the relationship between motor function and calcium signals in mice through changes in skeleton key point coordinates, the main focus was on whether the frequency of occurrence of the five movements of stationary, stand, rectilinear movement, steering movement and curl up in diseased mice and normal mice has changed. To analyze the changes in the frequency of these actions, the experimentalists, in constructing the behavioral dataset of the mice, selected only the videos that contained these actions. Figure 8 demonstrates the different behavioral actions in a single pose frame. Then, we input this dataset into DeepLabCut, which led to the corresponding skeleton key point data. After data preprocessing, we obtained valid inputs applicable to the ST-GCN network, and using these valid inputs, we retrained the ST-GCN network. Table 3 shows the detection accuracy of the ST-GCN network for different classes of actions. For actions with significant pose changes, ST-GCN has high detection accuracy. However, for steering movement, which is similar to rectilinear movement, the model may not be able to learn enough detailed information due to the small number of key points, resulting in average detection accuracy. Therefore, we suggest the experimenter mark more key points in the video to track the detailed changes in the pose more accurately. Figure 9 shows the detection results of ST-GCN for different behavioral categories in detail. Overall, ST-GCN offers high accuracy in mouse behavioral action detection, which meets the experimenter's accuracy requirements.

Table 3. Detection results of ST-GCN for different behavioral categories.

Behavioral Categories	Amount	Correct Amount	Accuracy
Stationary	528	496	94.12
Stand	545	502	92.11
Curl up	478	453	94.76
Rectilinear movement	1015	987	97.24
Steering movement	434	382	88.01
Aggregation	3000	2821	94.03

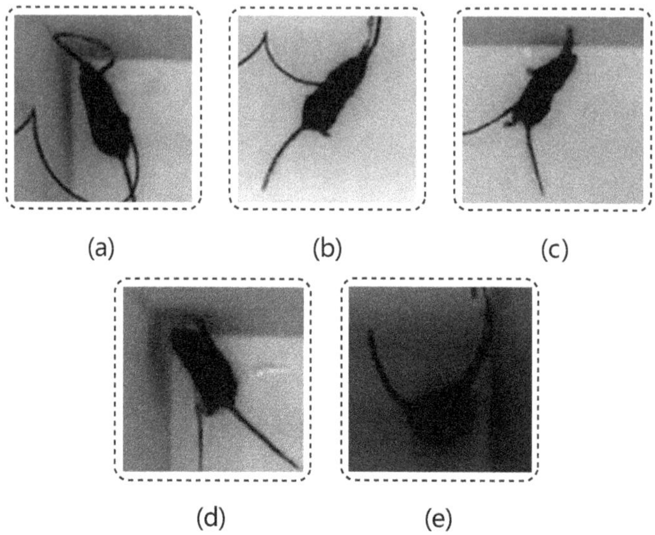

Figure 8. Pose in mice. (**a**) Stationary. (**b**) Rectilinear movement. (**c**) Steering movement. (**d**) Stand. (**e**) Curl up.

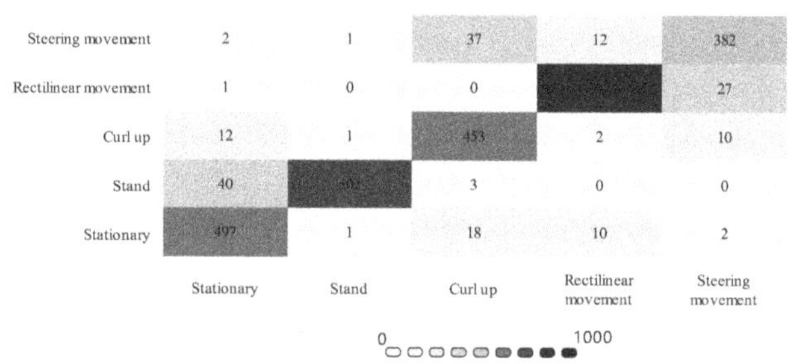

Figure 9. Confusion Matrix for Behavioral Identification Results.

6.3. *Natural Language Query Interface*

To capture the alignment relationship between natural language issues raised by users and database patterns, we needed to perform simultaneous semantic encoding on both in the RAT-SQL algorithm of the natural language query interface service. Considering the excellent performance of the BERT pre-trained model in natural language processing tasks, we choose to use its multi-language version to solve the cross-language problem in the Chinese Text-to-SQL task.

For Chinese natural language problems, we first needed to perform the divide words operation. In this case, we chose to use a Chinese word-splitter tool with high accuracy, i.e., Jieba, which can process Chinese natural language problems and return the combination of words with the highest probability. However, since natural language problems often contain Arabic numerals, unit symbols, and punctuation marks, in addition to Chinese characters, we further processed the output of the Jieba lexer tool by combining the substrings separated by the above cases to keep their original meanings unmodified.

For the column and table names of the database, we adopted the method consisting of English words with underlined separators according to the needs of engineering practice. Therefore, we only needed to divide words based on the underline. After completing the divide-words work, we spliced the obtained natural language questions, data tables, and data columns, and we connected each data column with its corresponding type.

Since the input contains Chinese and English, we encode it using a multilingual BERT pre-trained model (Multilingual-Bert) [36]. In the original RAT-SQL model, the schema-linking operation utilizes strings for matching, and therefore, the matching mechanism will not work correctly when multiple languages are involved. To address this problem, we introduced a multilingual, cross-domain common sense knowledge graph (ConceptNet [37]) and optimized the schema-linking process using its tautological edges. ConceptNet is a directed graph structure whose vertices are natural language words and phrases, and its edges are labeled 'types' and 'weights'. Figure 10 shows its seven commonly used relation types.

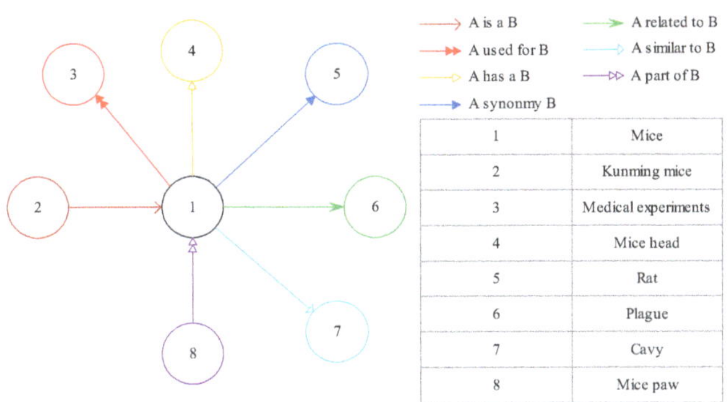

Figure 10. Seven Common Relationship Types in ConceptNet.

The evaluation metrics for the natural language query interface task consist of two main aspects: first, the exact matching rate of the structure of the generated SQL statements to the standard SQL statements, and second, the execution accuracy of the SQL query statements in the given database. In this case, we realized that experimenters are usually only concerned with whether the actual output meets their needs. Therefore, we focused more on how to make the SQL statements generated by the text-to-SQL model obtain the correct results after execution. First, we used the improved RAT-SQL model for pre-training on Chinese datasets such as CSpider, DuSQL [38] , and TableQA [31] . Then, we performed special training on the Chinese dataset of medical animal experiments. Table 4 shows the accuracy performance of RAT-SQL on the test set after introducing ConceptNet.

Table 4. The accuracy performance of RAT-SQL.

Type	Total Sample Size	Correct Amount	Execution Accuracy
Easy	272	267	0.98
Medium	146	137	0.93
Hard	93	81	0.87
Aggregation	511	485	0.95

As shown in the above table, RAT-SQL performed excellently on the medical experimental animal dataset, although its conversion ability was weak when dealing with difficult samples. Highly difficult samples only accounted for a small portion of the actual demand. After introducing ConceptNet, RAT-SQL was able to meet the retrieval needs of the experimentalists sufficiently. For the problem of low accuracy of highly difficult samples, we plan to use multi-round quizzing to improve its conversion ability in the subsequent work.

7. Conclusions

We propose an AI-enabled and highly available animal behavior analysis platform, which has been applied to a medical experimental institution in China to evaluate the behavioral differences between disease model mice and normal mice in a specific case.

In this case, the platform obtained the experimental video through the video capture device. Then, it used the pose estimation model to extract the single-frame pose features of experimental animals. Then, the platform used the behavior recognition model to process the continuous single-frame pose features to capture and store experimental animals' behavior information automatically.

In the platform's design, we mainly focused on how to improve the flexibility and scalability of behavior recognition and the interactivity of the platform to reduce the learning cost for researchers to use the platform and improve its usability. Therefore, the platform architecture integrates plug-in management, unified interface, and algorithm library design ideas so that it can be flexibly configured and extended. Whether high-precision behavior recognition can be achieved in animal behavior experiments is often one of many evaluation standards. The intermediate result output in the recognition process, such as the pose data of each frame, is also of great significance. Flexible algorithm replacement and expansion can enable researchers to choose more suitable identification methods according to actual needs and make the integration of algorithm upgrading and other algorithms more convenient, making the platform more competitive in the flexibility and scalability of algorithms.

The platform mainly relies on three core services. The pose estimation service performs pose estimation on experimental animals in the same experimental environment to obtain each animal's key point coordinate information. The behavior recognition service extracts and classifies the behavior feature vector of each animal in the video frame in different ways according to the selected algorithm. The natural language query interface service can convert the researchers' natural language query requirements into executable SQL statements and obtain the corresponding results from the database. The natural language query interface service provides more flexible and efficient information retrieval and improves the platform's interaction. Based on these three core services, we have built a highly available animal behavior analysis platform, which not only realizes the automatic identification of animal behavior but also enables researchers to flexibly select behavior recognition algorithms through the algorithm library, eliminating technical barriers and reducing researchers' dependence on experts. By utilizing natural language query interface services, the platform can open data access to all researchers and provide higher usability behavioral interactions.

The platform relies on computer vision technology in deep learning. However, the recognition effect may be affected when the video quality is too low, or the limbs between animals are blocked too much. In addition, in the natural language query interface service, researchers' inaccurate language description may also affect the effect of information retrieval. In order to further improve the availability of the platform, we plan to reduce the dependence of the platform on video quality and improve the accuracy of behavior recognition. Therefore, we will consider making the platform compatible with multimodal behavior recognition algorithms, reducing the dependence on a single video image by fusing data information such as voice, electroencephalogram signal, or pressure sensor signal, and improving the detection effect. For the inaccurate description of researchers in information retrieval, we plan to use multiple rounds of questions and answers to guide researchers in expressing their needs more accurately to improve the accuracy and availability of platform data retrieval and analysis.

Author Contributions: Conceptualization, J.S. and G.H.; methodology, Y.C., T.J. and J.S.; software, T.J. and Z.J.; validation, J.S., Y.C. and G.H.; formal analysis, Y.C.; resources, G.H. and Z.J.; data curation, Y.C.; writing—original draft preparation, Y.C.; writing—review and editing, J.S.; visualization, Y.C.; supervision, J.S. and G.H.; project administration, G.H. All authors have read and agreed to the published version of the manuscript.

Funding: This research was funded by the National Natural Science Foundation of China (Grant No. 62302086) and the Natural Science Foundation of Liaoning Province (Grant No. 2023-MSBA-070).

Institutional Review Board Statement: Not applicable.

Informed Consent Statement: Not applicable.

Data Availability Statement: The data provided in this study may be provided at the request of the corresponding author due to ethical and privacy protection restrictions.

Conflicts of Interest: The authors declare no conflicts of interest.

References

1. Broomé, S.; Feighelstein, M.; Zamansky, A.; Carreira Lencioni, G.; Haubro Andersen, P.; Pessanha, F.; Mahmoud, M.; Kjellström, H.; Salah, A.A. Going Deeper than Tracking: A Survey of Computer-Vision Based Recognition of Animal Pain and Emotions. *Int. J. Comput. Vis.* **2023**, *131*, 572–590. [CrossRef]
2. Chen, J.; Hu, M.; Coker, D.J.; Berumen, M.L.; Costelloe, B.; Beery, S.; Rohrbach, A.; Elhoseiny, M. MammalNet: A Large-Scale Video Benchmark for Mammal Recognition and Behavior Understanding. In Proceedings of the 2023 IEEE/CVF Conference on Computer Vision and Pattern Recognition (CVPR), Vancouver, BC, Canada, 17–24 June 2023; IEEE: New York, NY, USA, 2023; pp. 13052–13061. [CrossRef]
3. Da Silva Santos, A.; De Medeiros, V.W.C.; Gonçalves, G.E. Monitoring and Classification of Cattle Behavior: A Survey. *Smart Agric. Technol.* **2023**, *3*, 100091. [CrossRef]
4. Arablouei, R.; Wang, L.; Currie, L.; Yates, J.; Alvarenga, F.A.; Bishop-Hurley, G.J. Animal Behavior Classification via Deep Learning on Embedded Systems. *Comput. Electron. Agric.* **2023**, *207*, 107707. [CrossRef]
5. Roughan, J.V.; Wright-Williams, S.L.; Flecknell, P.A. Automated Analysis of Postoperative Behaviour: Assessment of HomeCageScan as a Novel Method to Rapidly Identify Pain and Analgesic Effects in Mice. *Lab. Anim.* **2009**, *43*, 17–26. [CrossRef]
6. Natarajan, B.; Elakkiya, R.; Bhuvaneswari, R.; Saleem, K.; Chaudhary, D.; Samsudeen, S.H. Creating Alert Messages Based on Wild Animal Activity Detection Using Hybrid Deep Neural Networks. *IEEE Access* **2023**, *11*, 67308–67321. [CrossRef]
7. Fuentes, A.; Yoon, S.; Park, J.; Park, D.S. Deep Learning-Based Hierarchical Cattle Behavior Recognition with Spatio-Temporal Information. *Comput. Electron. Agric.* **2020**, *177*, 105627. [CrossRef]
8. Crispim Junior, C.F.; Pederiva, C.N.; Bose, R.C.; Garcia, V.A.; Lino-de-Oliveira, C.; Marino-Neto, J. ETHOWATCHER: Validation of a Tool for Behavioral and Video-Tracking Analysis in Laboratory Animals. *Comput. Biol. Med.* **2012**, *42*, 257–264. [CrossRef] [PubMed]
9. Rodriguez, A.; Zhang, H.; Klaminder, J.; Brodin, T.; Andersson, P.L.; Andersson, M. *ToxTrac*: A Fast and Robust Software for Tracking Organisms. *Methods Ecol. Evol.* **2018**, *9*, 460–464. [CrossRef]
10. Lim, C.J.; Platt, B.; Janhunen, S.K.; Riedel, G. Comparison of Automated Video Tracking Systems in the Open Field Test: ANY-Maze versus EthoVision XT. *J. Neurosci. Methods* **2023**, *397*, 109940. [CrossRef]
11. Meade, M.J. Medication-Related Osteonecrosis of the Jaw: A Cross-Sectional Survey Assessing the Quality of Information on the Internet. *Oral Surg. Oral Med. Oral Pathol. Oral Radiol.* **2022**, *133*, e83–e90. [CrossRef]
12. Fang, C.; Zhang, T.; Zheng, H.; Huang, J.; Cuan, K. Pose Estimation and Behavior Classification of Broiler Chickens Based on Deep Neural Networks. *Comput. Electron. Agric.* **2020**, *180*, 105863. [CrossRef]
13. Nasiri, A.; Yoder, J.; Zhao, Y.; Hawkins, S.; Prado, M.; Gan, H. Pose Estimation-Based Lameness Recognition in Broiler Using CNN-LSTM Network. *Comput. Electron. Agric.* **2022**, *197*, 106931. [CrossRef]
14. Lin, C.W.; Hong, S.; Lin, M.; Huang, X.; Liu, J. Bird Posture Recognition Based on Target Keypoints Estimation in Dual-Task Convolutional Neural Networks. *Ecol. Indic.* **2021**, *135*, 108506. [CrossRef]
15. Ren, Q.; Lu, Z.; Wu, H.; Zhang, J.; Dong, Z. HR-Net: A Landmark Based High Realistic Face Reenactment Network. *IEEE Trans. Circuits Syst. Video Technol.* **2023**, *33*, 6347–6359. [CrossRef]
16. Asma-Ull, H.; Yun, I.D.; Yun, B.L. Regression to Classification: Ordinal Prediction of Calcified Vessels Using Customized ResNet50. *IEEE Access* **2023**, *11*, 48783–48796. [CrossRef]
17. Li, Z.; Zhang, Q.; Lv, S.; Han, M.; Jiang, M.; Song, H. Fusion of RGB, Optical Flow and Skeleton Features for the Detection of Lameness in Dairy Cows. *Biosyst. Eng.* **2022**, *218*, 62–77. [CrossRef]
18. Shah, S.R.; Qadri, S.; Bibi, H.; Shah, S.M.W.; Sharif, M.I.; Marinello, F. Comparing Inception V3, VGG 16, VGG 19, CNN, and ResNet 50: A Case Study on Early Detection of a Rice Disease. *Agronomy* **2023**, *13*, 1633. [CrossRef]
19. Yan, S.; Xiong, Y.; Lin, D. Spatial Temporal Graph Convolutional Networks for Skeleton-Based Action Recognition. *Proc. AAAI Conf. Artif. Intell.* **2018**, *32*, 7444–7452. [CrossRef]
20. Chen, C.; Zhu, W.; Steibel, J.; Siegford, J.; Wurtz, K.; Han, J.; Norton, T. Recognition of Aggressive Episodes of Pigs Based on Convolutional Neural Network and Long Short-Term Memory. *Comput. Electron. Agric.* **2020**, *169*, 105166. [CrossRef]
21. Zhang, Y.; Cai, J.; Xiao, D.; Li, Z.; Xiong, B. Real-Time Sow Behavior Detection Based on Deep Learning. *Comput. Electron. Agric.* **2019**, *163*, 104884. [CrossRef]
22. Schindler, F.; Steinhage, V. Identification of Animals and Recognition of Their Actions in Wildlife Videos Using Deep Learning Techniques. *Ecol. Inform.* **2021**, *61*, 101215. [CrossRef]
23. Sun, G.; Liu, T.; Zhang, H.; Tan, B.; Li, Y. Basic behavior recognition of yaks based on improved SlowFast network. *Ecol. Inform.* **2023**, *78*, 102313. [CrossRef]

24. Lauer, J.; Zhou, M.; Ye, S.; Menegas, W.; Schneider, S.; Nath, T.; Rahman, M.M.; Di Santo, V.; Soberanes, D.; Feng, G.; et al. Multi-Animal Pose Estimation, Identification and Tracking with DeepLabCut. *Nat. Methods* **2022**, *19*, 496–504. [CrossRef] [PubMed]
25. Hua, Z.; Wang, Z.; Xu, X.; Kong, X.; Song, H. An effective PoseC3D model for typical action recognition of dairy cows based on skeleton features. *Comput. Electron. Agric.* **2023**, *212*, 108152. [CrossRef]
26. Sriharipriya, K.C. Enhanced Pothole Detection System Using YOLOX Algorithm. *Auton. Intell. Syst.* **2022**, *2*, 22. [CrossRef]
27. Wang, B.; Shin, R.; Liu, X.; Polozov, O.; Richardson, M. RAT-SQL: Relation-Aware Schema Encoding and Linking for Text-to-SQL Parsers. In Proceedings of the 58th Annual Meeting of the Association for Computational Linguistics, Online, 5–10 July 2020; pp. 7567–7578. [CrossRef]
28. Katsogiannis-Meimarakis, G.; Koutrika, G. A Survey on Deep Learning Approaches for Text-to-SQL. *VLDB J.* **2023**, *32*, 905–936. [CrossRef]
29. Liu, H.; Singh, P. ConceptNet — A Practical Commonsense Reasoning Tool-Kit. *BT Technol. J.* **2004**, *22*, 211–226. [CrossRef]
30. Min, Q.; Shi, Y.; Zhang, Y. A Pilot Study for Chinese SQL Semantic Parsing. *arXiv* **2019**, arXiv:1909.13293.
31. Sun, N.; Yang, X.; Liu, Y. TableQA: A Large-Scale Chinese Text-to-SQL Dataset for Table-Aware SQL Generation. *arXiv* **2020**, arXiv:2006.06434.
32. Hien, P.T.; Hong, I.P. Millimeter Wave SAR Imaging Denoising and Classification by Combining Image-to-Image Translation with ResNet. *IEEE Access* **2023**, *11*, 70203–70215. [CrossRef]
33. Nijaguna, G.; Babu, J.A.; Parameshachari, B.; De Prado, R.P.; Frnda, J. Quantum Fruit Fly Algorithm and ResNet50-VGG16 for Medical Diagnosis. *Appl. Soft Comput.* **2023**, *136*, 110055. [CrossRef]
34. Bharadwaj, G.V.; Sree, Y.R.; Varshita, J.L.; Chebrolu, S. Ensemble Model of U-Net EfficientNet-B3, U-Net EfficientNet B6, CoaT, SegFormer for Segmenting Functional Tissue Units in Various Human Organs. In Proceedings of the 2023 14th International Conference on Computing Communication and Networking Technologies (ICCCNT), Delhi, India, 6–8 July 2023; pp. 1–8. [CrossRef]
35. Kumar Shukla, R.; Kumar Tiwari, A. Masked Face Recognition Using MobileNet V2 with Transfer Learning. *Comput. Syst. Sci. Eng.* **2023**, *45*, 293–309. [CrossRef]
36. Devlin, J.; Chang, M.W.; Lee, K.; Toutanova, K. BERT: Pre-training of Deep Bidirectional Transformers for Language Understanding. *arXiv* **2018**, arXiv:1810.04805
37. Pan, M.; Pei, Q.; Liu, Y.; Li, T.; Huang, E.A.; Wang, J.; Huang, J.X. SPRF: A Semantic Pseudo-relevance Feedback Enhancement for Information Retrieval via ConceptNet. *Knowl.-Based Syst.* **2023**, *274*, 110602. [CrossRef]
38. Wang, L.; Zhang, A.; Wu, K.; Sun, K.; Li, Z.; Wu, H.; Zhang, M.; Wang, H. DuSQL: A Large-Scale and Pragmatic Chinese Text-to-SQL Dataset. In Proceedings of the 2020 Conference on Empirical Methods in Natural Language Processing (EMNLP). Association for Computational Linguistics, Online, 16–20 November 2020; pp. 6923–6935. [CrossRef]

applied sciences

MDPI

Article

Adapting the Segment Anything Model for Volumetric X-ray Data-Sets of Arbitrary Sizes

Roland Gruber [1,2,*], **Steffen Rüger** [1] **and Thomas Wittenberg** [1,2]

[1] Fraunhofer IIS, Fraunhofer Institute for Integrated Circuits IIS, Division Development Center X-ray Technology, 90768 Fürth, Germany
[2] Chair for Visual Computing, Friedrich-Alexander-Universität Erlangen-Nürnberg, 91058 Erlangen, Germany
* Correspondence: roland.gruber@iis.fraunhofer.de

Abstract: We propose a new approach for volumetric instance segmentation in X-ray Computed Tomography (CT) data for Non-Destructive Testing (NDT) by combining the Segment Anything Model (SAM) with tile-based Flood Filling Networks (FFN). Our work evaluates the performance of SAM on volumetric NDT data-sets and demonstrates its effectiveness to segment instances in challenging imaging scenarios. We implemented and evaluated techniques to extend the image-based SAM algorithm for the use with volumetric data-sets, enabling the segmentation of three-dimensional objects using FFN's spatial adaptability. The tile-based approach for SAM leverages FFN's capabilities to segment objects of any size. We also explore the use of dense prompts to guide SAM in combining segmented tiles for improved segmentation accuracy. Our research indicates the potential of combining SAM with FFN for volumetric instance segmentation tasks, particularly in NDT scenarios and segmenting large entities and objects. While acknowledging remaining limitations, our study provides insights and establishes a foundation for advancements in instance segmentation in NDT scenarios.

Keywords: instance segmentation; Segment Anything Model; computed tomography; non-destructive testing; neural networks; machine learning

Citation: Gruber, R.; Rüger, S.; Wittenberg, T. Adapting the Segment Anything Model for Volumetric X-ray Data-Sets of Arbitrary Sizes. *Appl. Sci.* **2024**, *14*, 3391. https://doi.org/10.3390/app14083391

Academic Editors: Zahid Mehmood Jehangiri, Mohsin Shahzad and Uzair Khan

Received: 6 March 2024
Revised: 5 April 2024
Accepted: 8 April 2024
Published: 17 April 2024

1. Introduction

In the field of Non-Destructive Testing (NDT) of large-scale components and assemblies, cars [1], shipping containers [2,3], or even airplanes [4,5] are often captured using large-scale 3D X-ray computed tomography (CT) and are subsequently subjected to automated analysis and evaluation. In this context, an important step of the analysis process consists of instance segmentation, where an attempt is made to assign a unique semantic identifier or label to each entity in a data-set. For example, all voxels belonging to a specific screw are hereby assigned the same unique identifier, while voxels belonging to another component are assigned a different unique identifier.

The complexity of computing accurate instance segmentation varies significantly across different problem domains and data-sets. While simple threshold- or flood-filling-based methods from classical image processing suffice for data-sets from many fields, it remains uncertain as to whether an adequate solution for segmentation is feasible for others. Recent efforts, such as those in a challenge [6], tested multiple techniques to segment the data-set of a Me 163 [7], a historic German airplane with a rocket engine during the Second World War, with mixed success. This contribution aims to evaluate the suitability of an approach based on the currently highly appraised Segment Anything Model (SAM) [8], a foundational model for instance segmentation of such complex data-sets.

The task of instance segmentation shown in Figure 1 exemplifies this attempt using the XXL-CT data-set of the historic airplane. It begins with acquisition of data from the specimen, in this case the airplane, and proceeds with the reconstruction of a volumetric voxel data-set (Figure 1a). Figure 1b,c shows a sub-volume of size 512 × 512 × 512 voxels of the reconstruction and the instance segmentation. In Figure 1c, each semantic entity within the sub-volume is assigned a unique identifier. The classes of these entities (primarily screws and metal plates) are not considered, as the classification of the entities is not performed and is the focus of future work.

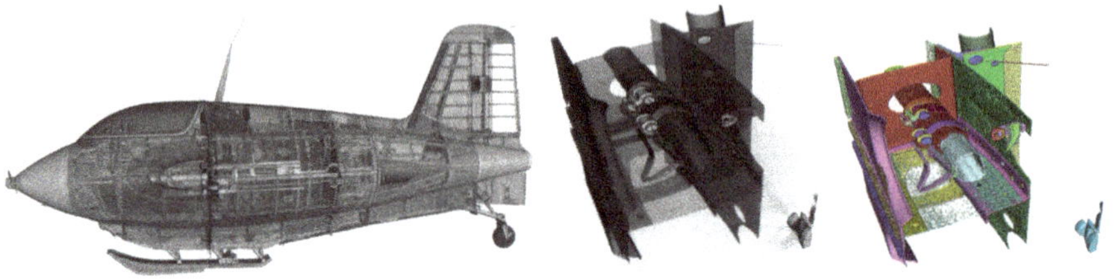

(**a**) Reconstruction (**b**) Input sub-volume (**c**) Reference sub-volume

Figure 1. Rendered example of instance segmentation (**c**), of a sub-volume of size 512 × 512 × 512 voxels (**b**), from the XXL-CT Me 163 data-set with a data resolution of 10,000 × 10,000 × 8000 voxels (**a**). The objective of instance segmentation is to generate a plausible segmentation of individual objects or instances, as depicted in (**c**), from an input sub-volume such as that shown in (**b**), applicable to data-sets of any size, akin to the one demonstrated in (**a**).

Instance segmentation is essential for automated image processing and data exploration in NDT and medical [9] applications. By segmenting a large-scale volumetric image data-set into its semantic instances, it becomes easier to extract valuable information and to analyse complex component geometries. This is particularly important in cases where the data-set contains various acquisition and reconstruction that can make interpretation difficult for both experts and non-experts.

Instance segmentation is a critical task in computer vision, leading to the proposal and development of numerous methods that leverage both classical image processing and neural networks. These approaches, however, are not without their limitations. Some methods necessitate manual intervention and corrections [10,11]; others are specifically tailored to predefined component classes [12]. Challenges associated with data quality, particularly in data-sets with a high incidence of artefacts, can significantly hinder the effectiveness of segmentation algorithms.

1.1. Segment Anything Model

The Segment Anything Model (SAM) [8] is an instance segmentation model based on the vision transformer architecture [13]. It is an advanced model for segmenting arbitrary entities out of photographs. It stands out primarily for its high quality, robustness, and minimal required user input. One of its notable features is the ability to be queried using a variety of prompts, allowing it to segment a RGB input image with a spatial resolution up to 1024 × 1024 pixels into multiple segments in one inference call. SAM supports prompts in various forms such as seed points (point prompts), bounding boxes, brush masks (dense prompts), and text prompts.

Furthermore, SAM allows the generation of multiple output masks for each input prompt, hence enabling image segmentations at varying hierarchical levels of granularity. Another advancement presented by the SAM is the extensive training data-set SA-1B, which has been iteratively collected and refined through prior versions of SAM during its own training process.

A multitude of studies and publications are currently emerging, which aim to apply SAM as a foundation model across a diverse range of fields, testing its segmentation quality. The application domains are varied. For instance, Li et al. [14] assess SAM for GeoAI vision tasks particularly in permafrost mapping. Alternatively, Noe et al. [15] utilise SAM to introduce a new approach for tracking black cattle on photographs. Another application within the domain of so-called "Precision Agriculture" is investigated by Carraro et al. [16], where mapping of crop features by automated mechanisms is conducted. In the field of NDT, the work by Weinberger et al. [17] examines how SAM can distinguish various segments in CT volume slices through unsupervised learning techniques. However, the direct application of SAM for instance segmentation is not the only focus of resent research. For example, Xu et al. [18] explored how an expanded data-set computed via SAM can be used to train an object detection network to improve license plate detection under severe weather conditions. Similarly, Liu [19] employed SAM to optimise road sign detection by using the model for background pixel exclusion in the data-set. In all these named studies, SAM exhibits a performance ranging from high quality to mixed results, which are strongly influenced by the data-set and specific problem domain under investigation.

1.2. Combination with Tile-Based FFN

This work aims to evaluate the applicability of SAM for segmenting volumetric NDT data-sets and to examine its potential enhancement through the integration of Flood Filling Networks (FFN), initially proposed by Januszewski et al. [20]. FFNs are instance segmentation methods originally based on convolutional networks [21,22], which are able to segment arbitrarily large data-sets based on tiles. Originally, FFN was developed for the segmentation of organic objects but in the past, was extended to other applications, including the delineation of large-scale XXL-CT data [4].

The FFN approach maintains the current state of segmentation within an accumulator volume, which is sized to match the dimensions of the input volume. During each segmentation step, a sub-volume or tile of the input volume and the corresponding partially computed tile of the accumulator is passed to the model (in our case, a volumetric variant of SAM). The segmentation proposal of the tile is then updated and written back to the corresponding tile position within the accumulator.

Candidates for neighbouring tile positions with significant overlap, which could extend the current segment, are determined using the updated accumulator state and added to a queue of tiles pending processing. In the subsequent iteration, the next unprocessed tile is removed from the front of the queue for processing. Starting from a seed point, the FFN then processes all of the tiles that potentially belong to the current segment. The processing of the current segment is completed when the queue of potentially belonging tiles is depleted. The algorithm then proceeds with the next segment starting from another seed point.

The seed points of the segments can be manually specified or computed automatically by a reasonable algorithm.

1.3. Contributions

In this work, we propose a novel approach for volumetric instance segmentation in NDT by combining SAM with FFN. Our contributions include the following:

1. Evaluation of SAM on NDT data-sets
We assess the performance of SAM on data-sets from the field of non-destructive testing and demonstrate its effectiveness in accurately segmenting instances in challenging CT imaging scenarios.

2. Implementation and evaluation of various methods to combine image-based SAM for the application with volumetric data-sets
We implement and evaluate different techniques to integrate and fuse the output of the image-based SAM approach for the application of volumetric data-sets, hence enabling the segmentation of three-dimensional objects using FFN's spatially adaptive capabilities.

3. Extending SAM for objects of arbitrary size through tile-based approaches
We propose a tile-based approach that leverages FFN's capabilities to segment objects of arbitrary size. By initially dividing the input volumes into tiles and then applying SAM on each tile individually, we achieve accurate and efficient segmentation results for objects of any size.

4. Utilizing dense prompts for SAM to combine tiles in an accumulator
To further improve the accuracy of the proposed tiled-based approach of SAM, we use dense prompts to guide SAM in combining the segmented tiles into a cohesive instance segmentation result. By leveraging the accumulated information from neighbouring tiles, we try to achieve more robust and accurate instance segmentation results.

2. Materials and Methods

This section presents the methodology and the experimental setup used, including the introduction of the data-sets (Section 2.1) used for the evaluation of the proposed methods. Furthermore, we describe a technique to improve the image segmentation performance of SAM with respect to the Me 163 airplane XXL-CT data-set by fine-tuning it specifically for this task (Section 2.2). Additionally, we detail our inference workflow in Section 2.3, which adapts the top-performing SAM model for volumetric data-sets. This process includes tile-based segmentation, accumulator-based dense prompts, and post-processing. The workflow aims to integrate the best model into a cohesive volumetric inference approach.

2.1. Data-Sets and Data Processing

To demonstrate, exemplify, and evaluate our achievements, we make use of three distinct data-sets. A specific sub-volume of the Me 163 data-set of a Second World War fighter airplane [7] as well as two bulk material data-sets depicting entities of glass marbles and corn kernels [4]. Figure 2 shows a photograph of each specimen, along with one typical slice from the reconstructed volume and a corresponding reference segmentation.

The Me 163 data-set utilized in this study consists of a volumetric subset and a manually obtained reference segmentation XXL-CT data-set from a historic airplane [5], which itself was extracted from an XXL-CT reconstruction. The reference segmentation sub-volumes of the Me 163 data-set were manually annotated and underwent morphological post-processing to clean up the edges. The acquisition process involved addressing challenging aspects such as noisy data, low contrast, and limited spatial resolution. A detailed description of the data-set creation, including the annotation and post-processing process, can be found in [7].

The data-set consists of eight sets of sub-volume pairs, each sub-volume having the spatial dimensions of $512 \times 512 \times 512$ voxels. For training, six sub-volume pairs of the data-set are used, while one sub-volume pair is used for validation and one for testing, respectively. Each sub-volume pair consists of a reconstructed sub-volume (see Figure 2b) and its corresponding reference segmentation sub-volume (see Figure 2c).

The reconstruction sub-volume is a small volumetric region that is extracted from the reconstructed Me 163 XXL-CT data. To ensure compatibility with SAM, both the reconstruction or input sub-volumes and the corresponding reference segmentation sub-volumes are extended with zero-padded 512 voxels in every direction. This results in an embedded version of the sub-volumes with working dimensions of $1536 \times 1536 \times 1536$ voxels. This arrangement allows for the extraction of a slice, centred on any arbitrary voxel within the original sub-volume, with the resolution of $1024 \times 1024 \times 1$ voxels, matching the native input dimensions required by SAM.

The first row of Figure 3 illustrates the described enframing process for the Me 163 data-set. The green rectangles in the first two columns indicate the unembedded region with $512 \times 512 \times 512$ voxels and their manually annotated references. Due to the fact that the input sub-volumes of this data-set are located directly at the edge of the XXL-CT volume, it was not possible to fill the border of the sub-volumes with actual reconstruction values. Instead, we decided to use a border with a constant value of zero in all directions. The last two columns of Figure 3 display the prepared input and reference slices used in the subsequent processing.

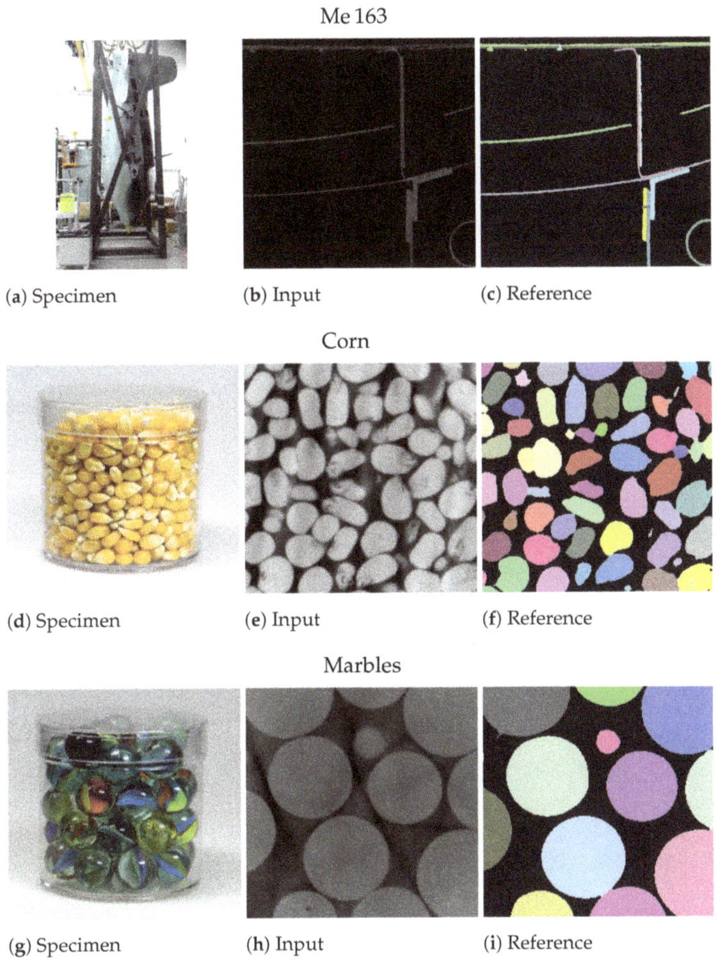

Figure 2. Photographs, exemplary CT slices, and reference segmentation of the Me 163 (**a–c**), corn (**d–f**), and marbles (**g–i**) data-sets, respectively.

The other two data-sets, which consist of CT scans of jars filled with marbles and corn, also contain two sub-volumes each: one for the input CT reconstruction sub-volume and one for its reference segmentation sub-volume. The segmentation process to yield the reference volumes of the bulk material data-set involved semi-automatic segmentation using threshold binarization with a threshold obtained from Otsu's method [23], followed by a distance transform, watershed transform, and label-wise morphological closing, as described in more detail in [4]. As this traditional computer-vision process resulted in some erroneous segmentations in the contact regions between the jar and the bulk material, we only used a correctly segmented sub-volume in the centre of the jar, having a spatial dimension of $256 \times 256 \times 256$ voxels (denoted by the green rectangle in Figure 3). Also, the sub-volumes of the bulk material were enframed by a border of 512 voxels thickness with a constant value of zero.

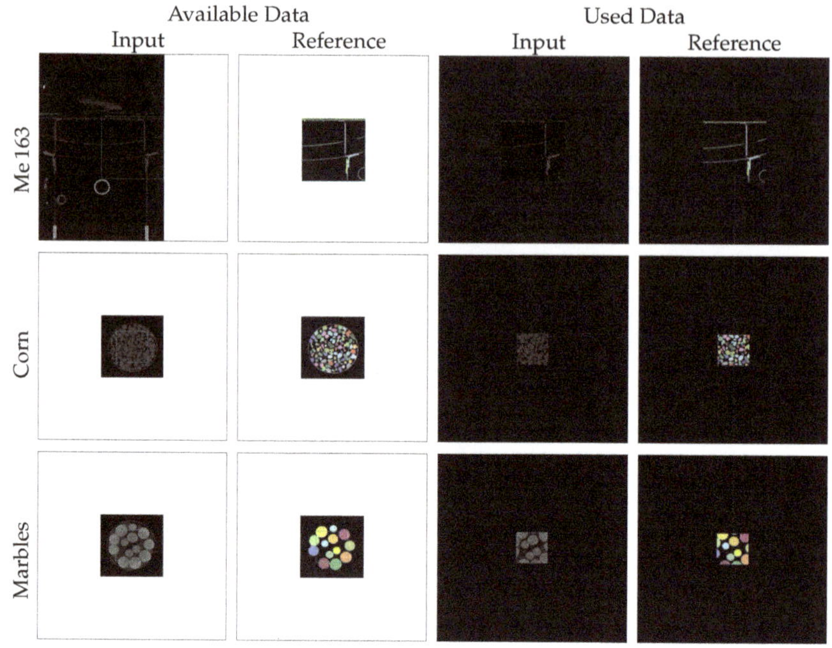

Figure 3. Zero-padding preparation steps were performed on the input and reference slices of the different data-sets to create slices of size 1024 × 1024 pixels centred around each possible seed point. The white border regions in the available input and reference slices were filled with constant values of zero.

2.2. Fine-Tuning on the NDT Data-Set

The SA-1B training data-set published by the authors of the SAM [8] contains predominantly coloured natural photographs, such as street scenes or still life compositions of semantically well-known objects from daily life. In contrast, volumetric data-sets obtained from the NDT field and particularly the slices extracted from the volumes are frequently of a rather abstract nature and do not depict recognizable objects. Hence, these NDT images deviate from the familiar photographic data-set used by SAM and this deviation poses several challenges in achieving sufficient segmentation quality (see Section 3.1). This, within the CT imaging domain, means that even familiar objects can be difficult to recognize for non-experts, as they exhibit unusual structures or non-orthogonal sections due to the specimen's imaging geometry; or, they may contain strong imaging and reconstruction artefacts.

Ma et al. [24] showcased a potential improvement in segmentation quality by fine-tuning SAM on the problem domain, which inspired us to adopt a similar fine-tuning approach.

In this study, we opted to perform fine-tuning on a certain part of the SAM, specifically the Mask Decoder. For this purpose, we utilized, extracted, and pre-processed slices from the Me 163 training data-set. Our approach adhered to the guidelines outlined in [24], which have previously been employed for fine-tuning on medical volume CT data-sets.

The Me 163 data-set was chosen due to its distinct level of complexity, setting it apart from the bulk material data-sets also being investigated. In contrast, the marble and corn data-sets can be segmented relatively easily using conventional image processing techniques.

For the fine-tuning process, we randomly selected voxel positions from the Me 163 training data-set. If the chosen voxel was a foreground voxel belonging to a known labeled entity, three orthogonal slices centred around its position were extracted. These slices were used as training examples, with the data range of the input slice normalised to [0.0, 255.0]. For the target slice, all voxels of the entities belonging to the centre voxel were one-hot encoded.

The original SAM operates on images, while our attempted input is a single slice from a volumetric data-set. To ensure that a three-dimensional connected object was represented by a single segment in the two-dimensional slices, a connected component analysis (CCA) was performed on the one-hot encoded target slice. This issue is depicted in Figure 4. Specifically, in the one-hot encoded a foreground target after the CCA (Figure 4d), where only the central component is visible, as we isolated the segment connected to the centre of the target slice, marked by a green cross. This central segment was then selected as the training target. The surrounding image does not provide sufficient information to distinguish if neighbouring non-touching segments belong to the same segment. Thus, we performed a CCA and treated the parts of segments not connected in the current slice as separate segments.

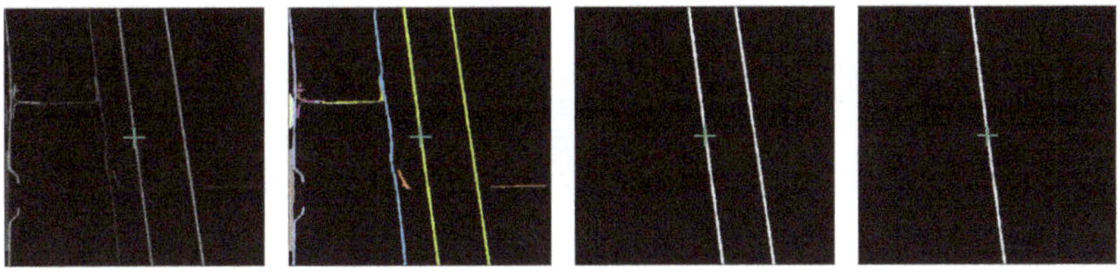

(**a**) Input (**b**) Reference (**c**) One-hot encoded foreground target (**d**) Connected component target

Figure 4. Processing of an example foreground slice used for fine-tuning SAM. Consisting of reconstruction slice (**a**), reference slice (**b**), one-hot encoded slice (**c**), and connected component training target slice (**d**). The green cross marks the centre of the slice.

If the voxel at the centre of a slice represented the background, we generated three orthogonal background examples, each containing a normalised input slice and a target slice. We evaluated three versions: *ForegroundOnly*, which included only foreground input slices; *ConstantValueBackground*, where we provided both background and foreground input and target slices for training but expected SAM to produce a completely empty response for background slices; and *ConnectedComponentBackground*, where we identified all background voxels connected to the centre voxel of the slice as the target segment. This was achieved through CCA on the data-set's background, formed by also enframing the reference segmentation with a zero-padded boundary. Consequently, the network was prompted to consider all voxels connected to the air space in the slice's centre as part of that segment. Figure 5 provides an illustrative example of the different target versions.

Due to the significantly lower count of foreground voxels (0.1–9.4%) compared to background voxels in the Me 163 data-set, we included all foreground examples while randomly selecting a subset of background examples of the same size. This approach ensured a balanced representation of both classes. To prevent batches from containing closely located examples, the selected examples were shuffled and grouped into batches, with each batch containing 16 foreground examples and 16 background examples. Additionally, to further diversify the examples within each batch, we employed a relatively large stride during the example extraction process. This ensured that the examples originated from different sub-volumes within the data-set. In each iteration over the data-sets, a new random initial position offset was chosen, employing a non-repetitive selection process to extract different examples.

We chose a single point prompt in the exact centre of each slice as the input for SAM during training. This choice aligns with the input for our validation application as well as the tile-based SAM integration for volume data-sets (see Section 2.3).

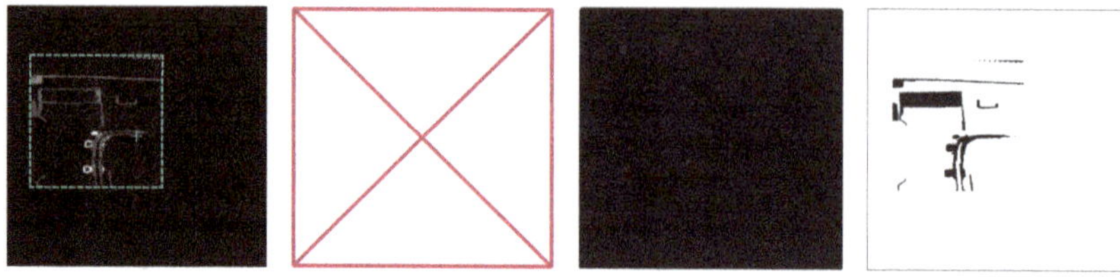

(**a**) Input (**b**) No background (**c**) Constant value background (**d**) Connected component background

Figure 5. Processing of an example background slice used for fine tuning SAM. The green cross marks the centre of the slice, which is located in the background of the reconstruction. The green border around the reconstruction slice in (**a**) depicts the original volume size, which was then enframed with a constant value border. The other sub-figures show the tested possibilities for target slices for the fine-tuning: *ForegroundOnly* (**b**), *ConstantValueBackground* (**c**), and *ConnectedComponentBackground* (**d**).

The batch size was set to 64. We initiated the training with a learning rate of 8×10^{-4}, which was linearly increased over the first 250 iterations. For optimization, we utilized the AdamW optimizer [25] with $\beta_1 = 0.9$ and $\beta_2 = 0.999$, along with a weight decay of 0.1. Our loss function consisted of a combination of dice loss (sigmoid = true, squared-pred = true, and mean reduction) and binary cross-entropy loss (mean reduction). We let the training run until overfitting for 10 to 25 days. We selected the model with the lowest validation loss, determined at moving window intervals of 128 iterations.

2.3. Inference Workflow for Volumetric Data-Sets

Since SAM works only on RGB image data-sets but we wanted to segment volumetric data-sets, we had to incorporate an adequate workflow to translate between these two spatial domains. Since our goal was to evaluate SAM for volumetric data-sets and not necessarily to implement a complete new volumetric version, we referred to simple operators. Figure 6 shows an overview of the approximate workflow for a volumetric data inference of SAM. In short, we extract a sub-volume tile from the input volume and pass it to the volumetric SAM adaption, which transforms it into three orthogonal slice stacks.

For each slice stack, we perform slice preparations (such as normalization and zero-padding), a forward pass through SAM, selection of the corresponding outputs, and slice post-processing. The output slice stacks are then merged and undergo further volumetric post-processing to generate segmentation proposals, which are returned from the volumetric SAM adaption into the inference algorithm. The evaluated algorithms are listed and compared in Table 1.

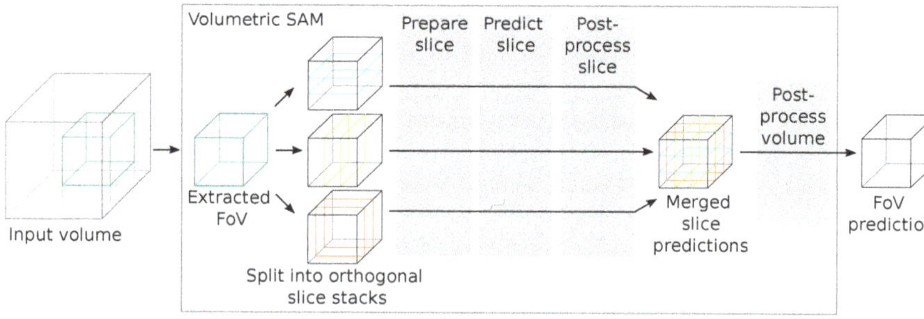

Figure 6. Schematic workflow of the volumetric data inference segmentation using SAM. Algorithm options and steps for the configurable stages (grey boxes) are listed in Table 1.

Table 1. Overview of algorithm choices and options for different stages of the volumetric SAM adaption seen in Figure 6.

Stage	**Algorithm**	**Description (*Options*)**
Preprocess Slice Algorithms		
	Slice Normalization	Normalization of pixel values in each slice to the minimum and maximum range of the slice.
	Outlier and Empty Slice Detection	Identification and handling of outlier and empty slices.
	RGB Conversion	Conversion of grey values to RGB colour in order to comply with SAM interface requirements.
	Enframing	Adds a zero-padded border to each slice to centre the seed point to comply with SAM interface requirements.
	Estimated Foreground Volume	Utilizes different *binarization strategies* and *thresholds* to estimate the foreground volume.
Predict Slice Algorithms		
	Prompt Type	Type of prompt is used for invoking SAM: point prompt for tile centre and dense prompt from accumulator.
	Multimask Output Selector	Select mask from multiple disambiguating instance output channels predicted by SAM: maximum predicted IoU, fixed index of channel; maximum IoU with estimated foreground to avoid segmenting background; and minimum count of voxels to reduce under segmentation.
	Mask Output Selector	Selected output format of SAM: binary full resolution mask and quarter resolution logits with subsequent *threshold* and *upscaling algorithm.*
Postprocess Slice Algorithms		
	Seed Point Filter	Aborts or continues prediction based on the seed point's classification as background or foreground (*count of slices*).
	Merge Slice Rule	Rule that should be used to decide if and how to merge slices to stacks and when to abort an computation stack: *Always*; *BreakOnEmptySlice*; *MinimumIOUToLastSlice* (*threshold*); *MinimumIOUToForeground* (*threshold*).
	Slice Median Connected Component Analysis and analyse connected components and keep only segment connected to seed point (*enabled or disabled*).	Apply median filter to each slice (*enabled or disabled*).
Postprocess Volume Algorithms		
	Merge Slice Predictions	Merge orthogonal slice stack predictions based on *count* of foreground voxels.
	Volume Median	Apply median filter to merged volume (*enabled or disabled*).

2.3.1. Adapting SAM for Volumetric Data-Sets

Adapting SAM, which was originally designed for segmenting image data-sets, to our volumetric CT data-sets required certain modifications and the implementation of appropriate post-processing steps. In this section, we explore various possibilities for this transition and subsequently outline the approach we finally selected.

Several 2D to 3D techniques can be utilized to facilitate this transformation [26]. For example, in [27], a Volumetric Fusion Net (VFN) was employed to merge multiple 2D segmentation predictions into a comprehensive 3D prediction volume. In a related work, Ref. [28] adopted a similar methodology for pancreas segmentation, albeit utilizing a different VFN. According to [26], other approaches involve incorporating neighbouring 2D slices as additional channel information or utilizing specialized topologies to extract and merge features in both the 2D and 3D domains. However, the effectiveness of these methods for improving segmentation results heavily depends on the specific data-sets at hand.

Due to reports on the segmentation performance of SAM on volumetric medical data-sets, such as those in [29] and our own preliminary experiments, which suggested that the segmentation quality of SAM was likely to be mixed, we opted for a simple majority voting approach to merge the 2D predictions into 3D volumes.

During the slice merging process, we experimented with different rules to determine when to terminate the slice-wise merging. We either combined all slice within the current field of view regardless of their content or stopped at the first empty slice, i.e., a slice without foreground voxels. We also tested various rules based on different thresholds of overlap or Intersection over Union (IoU) between the proposed segmentation of the current slice and the preceding slice or a foreground volume obtained through global Otsu thresholding followed by a morphological closing step.

As an optimization strategy, slice-wise prediction was performed in an alternating manner, starting from the centre of the current sub-volume and moving outward slice-wise in both directions. This approach was implemented to save computational time and prevent the segmentation of unconnected segments, ensuring that only cohesive regions were accurately identified.

In situations where the segmentation results in an identification of unconnected segments, the algorithm may inadvertently continue segmenting entire regions composed of non-cohesive segments. This phenomenon occurs when the segmentation quality is significantly compromised. During the subsequent hyperparameter search, we also permitted segmentations without applying these rules. However, it appears that these deviations have only minimal impact on the output quality.

Subsequently, a new target volume is constructed. Voxels are included in the output volume if they are segmented as the foreground in at least one and depending on the configuration, up to three slice-wise predictions.

Additionally, we employed post-processing techniques such as slice-wise and volume-based median filtering and CCA prior to and after merging the slices into volumes to smooth scattered and miss segmented voxels.

We also conducted experiments with different variants of SAM's outputs. Since SAM has the ability to generate multiple outputs per prompt, such as separating a backpack from a person wearing it, we investigated whether selecting any of these outputs could improve the segmentation quality. Specifically, we examined whether it is better for volumetric segmentation to use the segmentation proposal provided by SAM with the highest probable IoU or the one with the maximum IoU of the approximated foreground volume. Additionally, as SAM often tends to under-segment and include background or neighbouring segments as part of the foreground, we investigated whether selecting the output with the smallest count of voxels among the multiple outputs would improve the segmentation quality.

In this context, experiments were conducted using both the binarized output of SAM and the raw probability values, which are available at a lower resolution than the binary mask. After upscaling, different threshold values can be applied to the probability outputs for further processing and experimentation.

2.3.2. Tile-Based Segmentation for Data-Sets of Arbitrary Size

Due to SAM's image-based nature, we encounter segmentation challenges when dealing with topologically complex objects depicted by volumetric CT NDT data-sets. These volumes may contain holes or inclusions; complex folds are spatially sparse or may extend beyond the boundaries of the currently processed tile.

To clarify this, Figure 7 offers a visual exposition of several schematically depicted objects of varying complexity. The figure serves to illustrate how, in a volumetric context, such complex segments are easier to understand but when segmenting them slice by slice there is a risk of mistakenly delineating them as multiple segments. This effect also occurs when the tile is smaller than the entity's size.

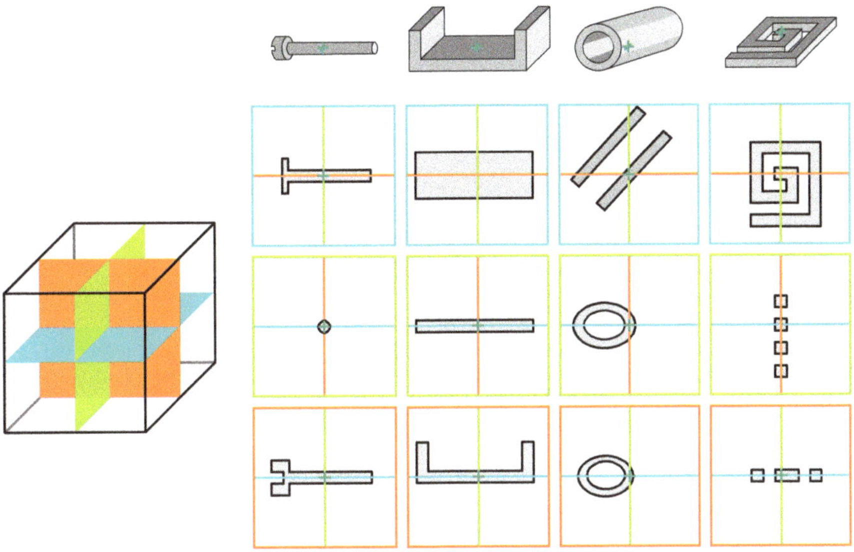

Figure 7. Schematic views of multiple simple volumetric objects (bolt, U-profile, pipe, and spiral spring) and cross-sectional slices along their central axes in three orthogonal directions marked by three respective colours (■, ■, ■). The disjunction of simple objects into multiple components if processed slice-wise poses a challenge as there are no straightforward rules for merging them without a step-by-step traversal of the object.

To overcome these challenges, we utilize volume-based SAM inference (see Section 2.3.1) within the FFN framework (see Section 1.2). The inference process starts with a single seed point and is applied to a small sub-volume tile. The resulting segmentation proposal is then stored in a result buffer, the accumulator volume. If a segment intersects the outer boundaries of a tile, the intersection position is added to a queue. In subsequent iterations, corresponding slightly shifted tiles aimed at these intersection points are processed by the volume-based SAM inference. This iterative process generates segmentation proposals, which are incorporated into the accumulator. This process repeats until the intersection points queue is empty and the segmentation proposal in the accumulator is no longer constrained by the boundary of the processed tiles.

As an optimization step, the proposed additional intersection positions are filtered based on the approximated foreground volume. They are added to the intersection points queue only if the corresponding voxels have a high probability to be foreground voxels.

The proposed combination of SAM and FFN allows us to compute segments and input volumes of arbitrary size by combining multiple overlapping tiles using a temporary accumulator volume. Nevertheless, this approach also increases the runtime due to the recomputation of the overlapping tiles.

The choice of using 48 voxels per tile side was made heuristically based on the original FFN algorithm, which also uses this tile size. However, the algorithm can be adjusted by changing the tile size up to 1024 voxels in each dimension; the maximum dimension SAM can handle without resizing the input. When the tile size is below this threshold, no resizing of tiles is required as we add a constant value border around the tile. Additionally, the step width between tiles and the overlap of the tiles can be adjusted to mitigate artefacts caused by the tile-based algorithm. Tile-based algorithms are capable of assembling entities with complex topologies. These algorithms can follow or trace the segment itself over multiple tiles and steps, even if it forms highly complex shapes. But tile-based algorithms may introduce additional artefacts. The segmentation result of the combined algorithms is heavily dependent on the performance of the SAM segmentation.

2.3.3. Prompt Selection and Accumulator Integration

As mentioned above, SAM allows queries using various prompts such as point prompts (seed points and bounding boxes) and dense prompts (masks and brushes). Multiple studies [24,30] have shown that, depending on the input data, higher segmentation quality can be achieved by using multiple prompts, such as point prompts distributed evenly over the segment region or negative point prompts, which are not considered part of the segment. Additionally, the use of rectangular prompts consisting of two anchor points often leads to adequate segmentation results.

Given that the main objective of this study is to evaluate the applicability of SAM in the automated NDT domain, we have opted to solely assess single point prompts and dense prompts as they can be easily automated.

We placed a single point prompt at the exact centre of the tile. The centre point of a tile was either chosen by a seed point or deemed highly likely to belong to the current segment, due to the iterative processing of the tiles.

For dense prompts, we utilized the SAM output stored in the accumulator, which was shifted by the relative position of the current point prompt. This requires SAM to complete the segmentation proposal at the edge of the current tile. Since our tile step size was $[1, 20]$ voxels, the overlap between the tiles and the dense prompt with the expected segmentation proposal was high, allowing SAM to only predict a relative slim border of new voxels. Figure 8 illustrates an idealized schematic of such an operation. In the case of dense prompts, we also include a corresponding point prompt at the centre of the tile as more prompts tend to increase the segmentation performance [24].

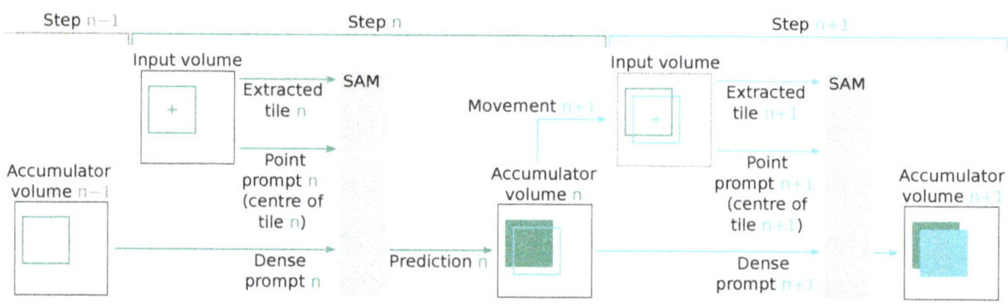

Figure 8. Schematic view of two subsequent inference steps, denoted as n (represented by ■) and $n+1$ (represented by ■), which use the modified accumulator volume from the previous step to create a dense SAM prompt. In step n, the content of the accumulator volume of the previous step $n-1$ is used to generate a dense SAM prompt $n+1$. This prompt, along with the point prompt n and the extracted input volume tile n, is used by SAM to compute prediction n. Subsequently, the accumulator volume is updated to the state n based on this prediction. In the subsequent step $n+1$, the accumulator volume n is used to determine the movement $n+1$ to the tile $n+1$. Tile $n+1$ significantly overlaps with tile n. SAM is parametrized with the extracted input volume tile $n+1$, point prompt $n+1$, and dense prompt $n+1$ to compute prediction $n+1$, which is used to update the accumulator volume $n+1$.

3. Results

3.1. Evaluation of SAM Segmentation Quality in NDT Slice Data-Sets

In an initial test of SAM's segmentation quality for CT NDT data, we applied SAM to segment individual slices from NDT volumetric data-sets. We used three pre-trained SAM models, vit_h, vit_l, and vit_b, based on Vision Transformers (ViT) arranged in descending order of size. Additionally, we tested three fine-tuned versions of the vit_b model, each adapted to the Me 163 data-set with unique target configurations. For each of the three data-sets introduced in Section 2.1, randomly selected slices were selected and segmented, which accounted for approximately 0.5% of all available validation data-sets. Each example underwent the preparation steps outlined in Section 2.2 before being processed by SAM. SAM then tried to segment the entity located at the exact centre of each slice using point prompts. Examples of typical segments can be seen in Figure 9. Notably, SAM demonstrated good segmentation performance for the marbles and corn kernels data-sets, while the segmentation quality was significantly inferior for the individual segments of the Me 163 data-set. To quantify the segmentation performance across data-sets and models, Table 2 presents the mean loss values and standard deviations for slice-wise predictions made by multiple SAM model configurations. The statistics in this table show that while the vit_b model yields the lowest loss for the corn kernels data-set, with a mean loss of 0.10, the application of vit_b with a ConstantValueBackground modification achieved the best performance on the Me 163 data-set, reducing the mean loss to 0.36.

Table 2. Mean loss value (and standard deviation) over all slice-wise predictions on the validation data-sets by multiple models for the graphs in Figure 10. Models yielding the optimum performance for each data-set are denoted in bold. Models vit_h, vit_l, and vit_b denote pre-trained SAM models that utilise Vision Transformers (ViT) as their foundation, ordered from largest to smallest. The remaining models represent fine-tuned versions of vit_b applied to the Me 163 data-set, each employing distinct target configurations.

	Marbles	**Corn**	**Me 163**
vit_h	**0.03** (0.06)	0.11 (0.10)	0.49 (0.34)
vit_l	**0.03** (0.06)	0.11 (0.10)	0.46 (0.34)
vit_b	**0.03** (0.07)	**0.10** (0.10)	0.44 (0.32)
vit_b ForegroundOnly	0.41 (0.27)	0.66 (0.24)	0.49 (0.26)
vit_b ConstantValueBackground	0.15 (0.10)	0.44 (0.16)	**0.36** (0.25)
vit_b ConnectedComponentBackground	0.51 (0.24)	0.51 (0.19)	0.57 (0.23)

Figure 10 demonstrates the segmentation dynamics of the individual models on the different data-sets. These plots represent the loss of the segmentation proposals generated by SAM for the entities at the centre of each layer of the corresponding validation data-set. The loss values are determined with respect to the reference data-set. From left to right, the loss values are sorted in ascending order, so that the nearly correctly segmented segments are on the left side of the graph, while the difficult and often incorrectly segmented segments are on the right side. The seed points of the segments were chosen in such a way that each of them corresponds to a foreground voxel, so the networks are not tasked with segmenting the background. The different colours in the plots correspond to different networks.

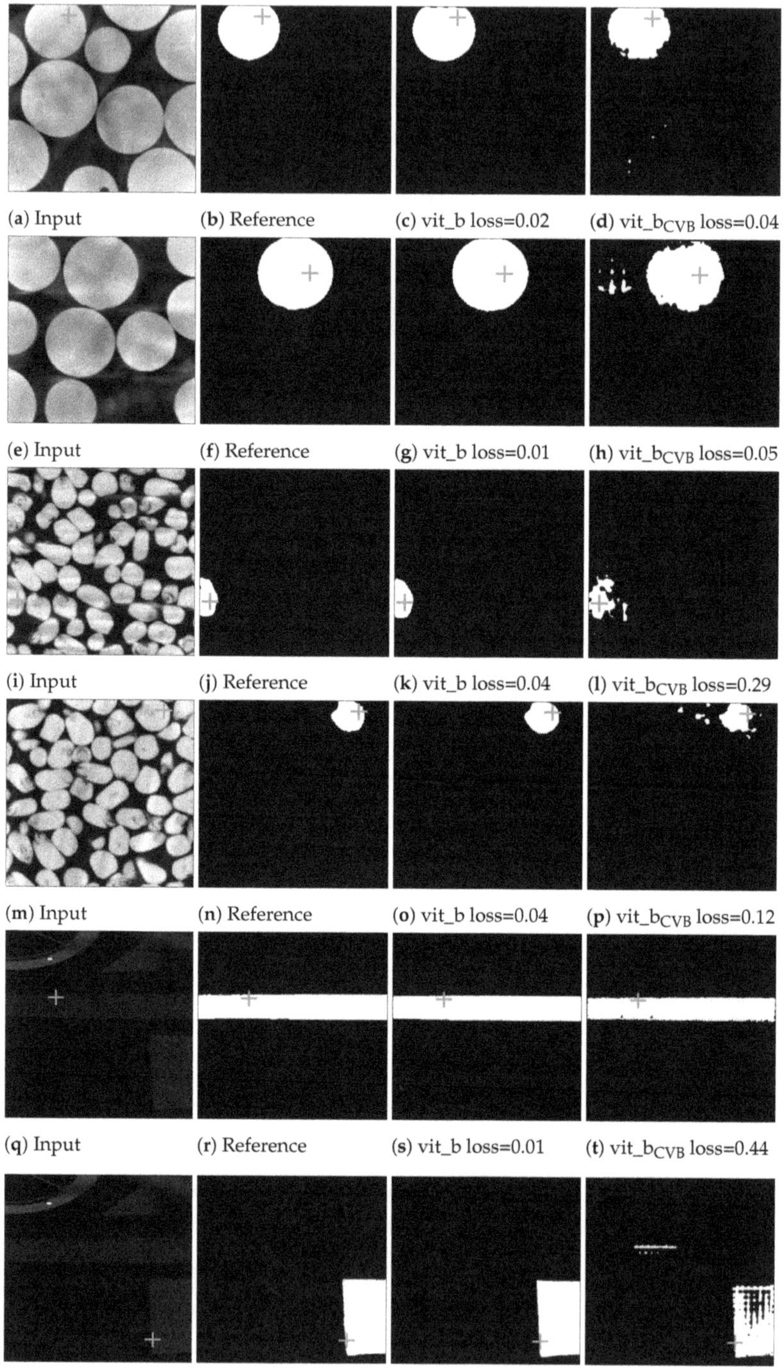

(**a**) Input (**b**) Reference (**c**) vit_b loss=0.02 (**d**) vit_b$_{CVB}$ loss=0.04

(**e**) Input (**f**) Reference (**g**) vit_b loss=0.01 (**h**) vit_b$_{CVB}$ loss=0.05

(**i**) Input (**j**) Reference (**k**) vit_b loss=0.04 (**l**) vit_b$_{CVB}$ loss=0.29

(**m**) Input (**n**) Reference (**o**) vit_b loss=0.04 (**p**) vit_b$_{CVB}$ loss=0.12

(**q**) Input (**r**) Reference (**s**) vit_b loss=0.01 (**t**) vit_b$_{CVB}$ loss=0.44

(**u**) Input (**v**) Reference (**w**) vit_b loss=0.01 (**x**) vit_b$_{CVB}$ loss=0.13

Figure 9. Segmented examples of the corn and marbles data-set. The green crosses mark the position of the currently used point prompt. The last column depicts the result of the vit_b model, which was fine-tuned on the Me 163 data-set.

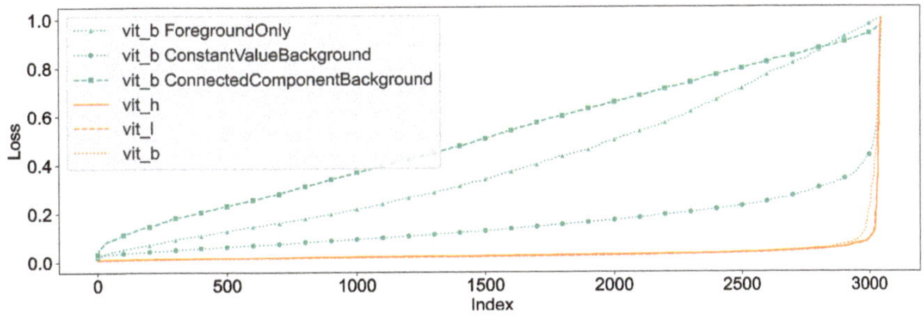

(**a**) Marbles

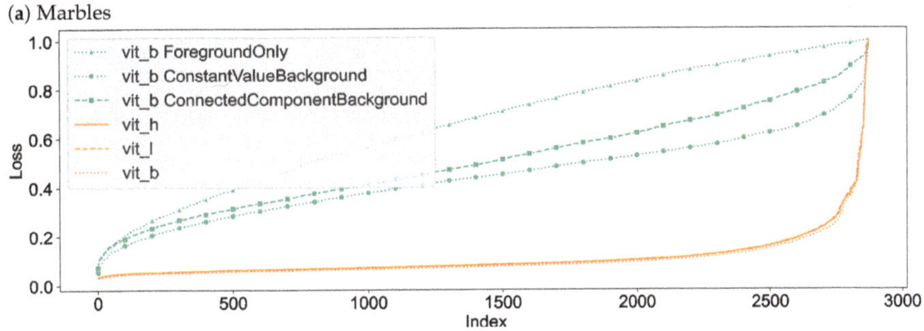

(**b**) Corn

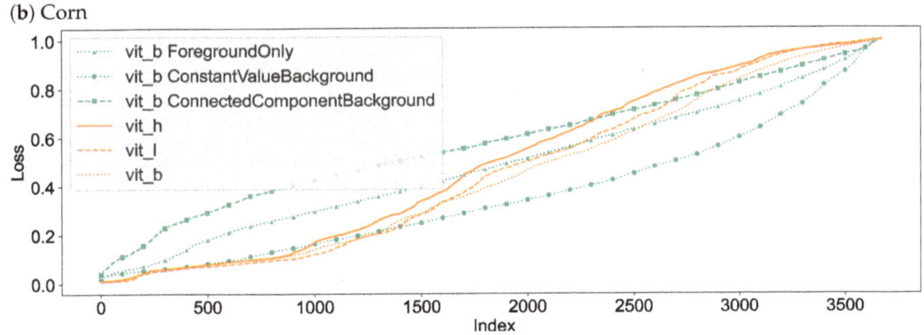

(**c**) Me 163

Figure 10. Graphs depicting the slice segmentation performance of the six evaluated SAM models on the three different testing data-sets. From left to right, the index of each segmented slice sorted by their loss value. In an ideal case, only a horizontal line close to the loss value of 0 would be visible.

It can be observed that the unchanged SAM networks perform very well in segmenting the marble and corn data-sets. The few entities which exhibit lower segmentation quality in these data-sets and are located on the right edge are often due to insufficient quality in the reference segmentation data-set, as illustrated in Figures 11 and 12. A slightly lower segmentation quality can be observed for the corn data-set, which consists of a higher count of entities that are also not as homogeneous in colour compared to the marble data-set.

Figure 10c demonstrates that the segmentation quality for the Me 163 data-set is notably lower compared to the previously mentioned data-sets. Figure 13 displays some typical error patterns in the original trained SAM images. Both under-segmentation and over-segmentation occur and segments are sometimes partially or not recognized at all.

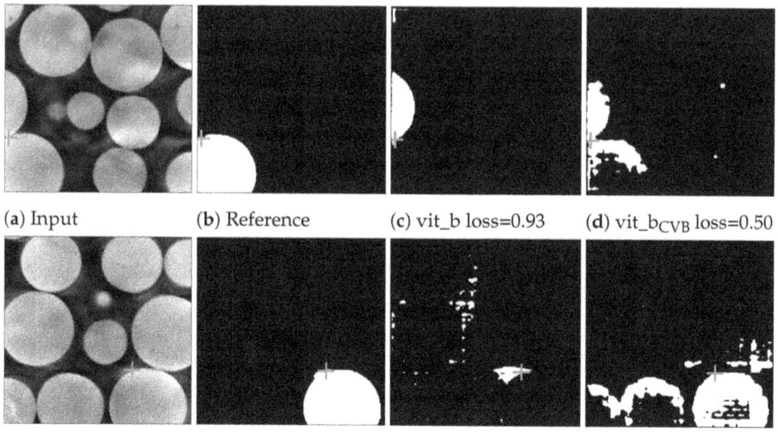

(a) Input (b) Reference (c) vit_b loss=0.93 (d) vit_b$_{CVB}$ loss=0.50

(e) Input (f) Reference (g) vit_b loss=0.50 (h) vit_b$_{CVB}$ loss=0.26

Figure 11. Error cases for the marble data-set. Here, the reference segmentation, which was generated by a connected component analysis, is erroneous. In (**b**), the point prompt (marked with a green cross) lies on the boundary of two marbles and vit_b segments the upper marble instead of the lower marble. In (**f**), the point prompt lies inside an artefact region.

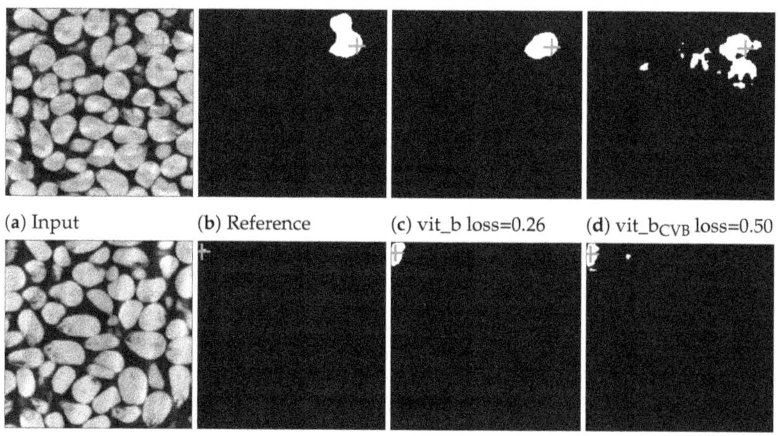

(a) Input (b) Reference (c) vit_b loss=0.26 (d) vit_b$_{CVB}$ loss=0.50

(e) Input (f) Reference (g) vit_b loss=0.90 (h) vit_b$_{CVB}$ loss=0.88

Figure 12. Error cases of the corn data-set. In the first case in (**b**), two kernels were erroneously segmented together in the reference segmentation. In contrast, in (**f**), the reference segmentation only appears erroneous as the current slice only depicts one voxel. The next slice in the input volume contains the kernel this voxel belongs to. The green crosses mark the position of the currently used point prompt.

Among the different not fine-tuned SAM models, the smallest model vit_b showed the most promising results. While it was sometimes outperformed by the other two original SAM models, vit_l and vit_h, in the well-segmented slices, it still had a higher segmentation quality in the moderately segmented slices. Therefore, we decided to use vit_b as the base model for fine-tuning and volumetric segmentation experiments.

Among the subsequently trained networks, vit_b$_{CVB}$ exhibits the highest quality in Figure 10c. It is based on vit_b and uses *ConstantValueBackground* (*CVB*) (see Section 2.2) for background examples. In simple cases, it matches the segmentation quality of non fine-tuned SAM variants. A considerable improvement in segmentation quality on the challenging entities could be achieved through training, although not to a satisfactory level.

This model was chosen as the representative of our fine-tuned model for further tests on our data.

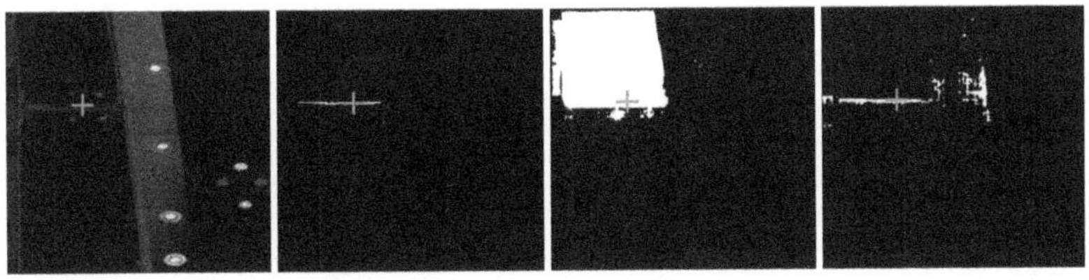

(**a**) Input (**b**) Reference (**c**) vit_b loss=0.99 (**d**) vit_b$_{CVB}$ loss=0.86

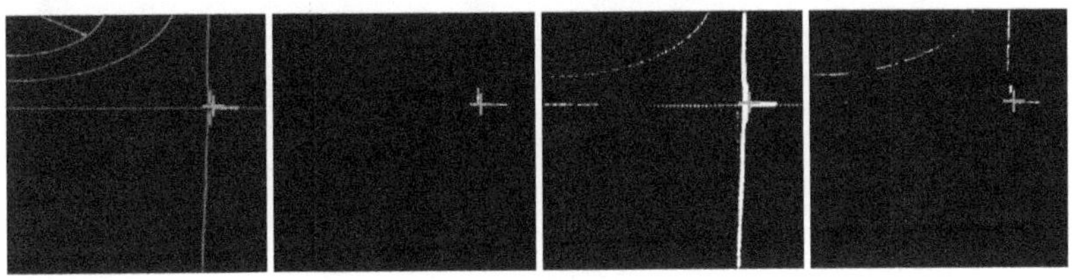

(**e**) Input (**f**) Reference (**g**) vit_b loss=0.99 (**h**) vit_b$_{CVB}$ loss=0.76

Figure 13. Poorly performing cases for SAM vit_b segmenting thin metal sheets in the Me 163 data-set as well as the better but still not optimal segmentation results achieved by the model fine-tuned on the Me 163 data-set.

3.2. Tile-Based Algorithms and Artefact Mitigation

Figures 14 and 15 showcase the segmentation results of a volumetric inference run using the proposed SAM algorithm on a small subset of the marble and corn data-sets for the two tile sizes $48 \times 48 \times 48$ voxels and $1024 \times 1024 \times 1024$ voxels. These results exhibit segmentation errors in the form of erroneous segmented edges as well as tiling artefacts, resulting in a textured appearance of the segment with noticeable gaps.

Notably, for a tile size of $48 \times 48 \times 48$ voxels, the marble example in Figure 15b demonstrates tiling artefacts. Since the volumetric inference algorithm with the small tile size cannot segment the entire marble in a single step, it must combine multiple steps, which can introduce and propagate errors. These artefacts can be cleaned up using a morphological closing operation as a post-processing step.

In contrast, segmentations using a larger tile size of $1024 \times 1024 \times 1024$ voxels exhibit fewer of these textured artefacts. However, segmentations may extend beyond the actual segment due to segmentation errors, as illustrated in Figure 14c, where thin segments protrude vertically and horizontally beyond the intended boundaries. These protrusions often occur within the initially segmented slices that include the seed point of the current segment. In the green upper right marble of the example in Figure 15c, the adjacent slices directly connected to the seed point were misclassified as not belonging to the marble, resulting in an early termination of the slice-wise segmentation process.

The inference algorithm with a tile size of $1024 \times 1024 \times 1024$ voxels can only attempt to segment the segment once as, due to its high field of view, it performs a single volumetric step per seed point. In contrast, the inference algorithm with a tile size of $48 \times 48 \times 48$ voxels iterates over the volume in multiple steps, providing the ability to compensate for weak and erroneous segmentations in subsequent steps. However, this approach tends to under-segment when a neighbouring segment has already been partially segmented in a previous step.

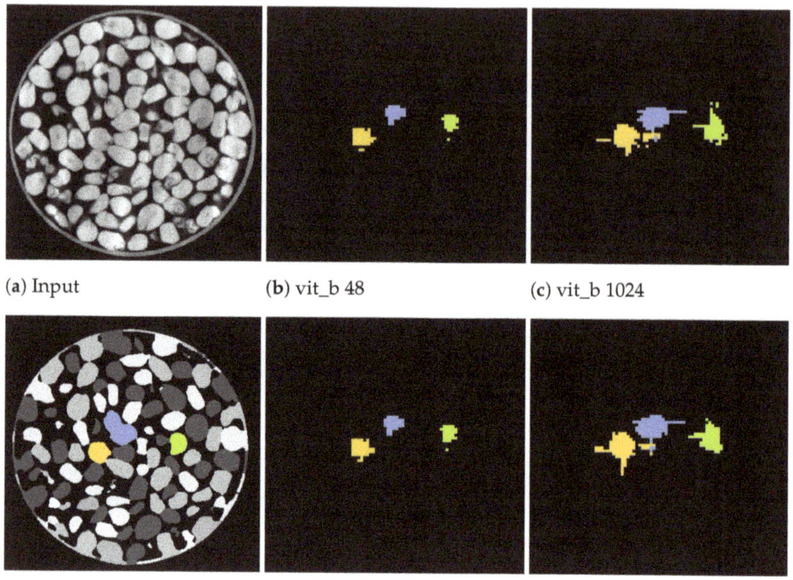

(a) Input (b) vit_b 48 (c) vit_b 1024

(d) Reference (e) vit_b 48 (postprocessed) (f) vit_b 1024 (postprocessed)

Figure 14. Slices from a volumetric inference run on three corn kernels of the corn data-set. The input volume (**a**), reference volume (**d**), and the proposed segmentations generated by the proposed algorithm using the two tile sizes: 48 × 48 × 48 voxels (**b**) and 1024 × 1024 × 1024 voxels (**c**). Additionally, the postprocessed volumes are depicted in (**e**,**f**).

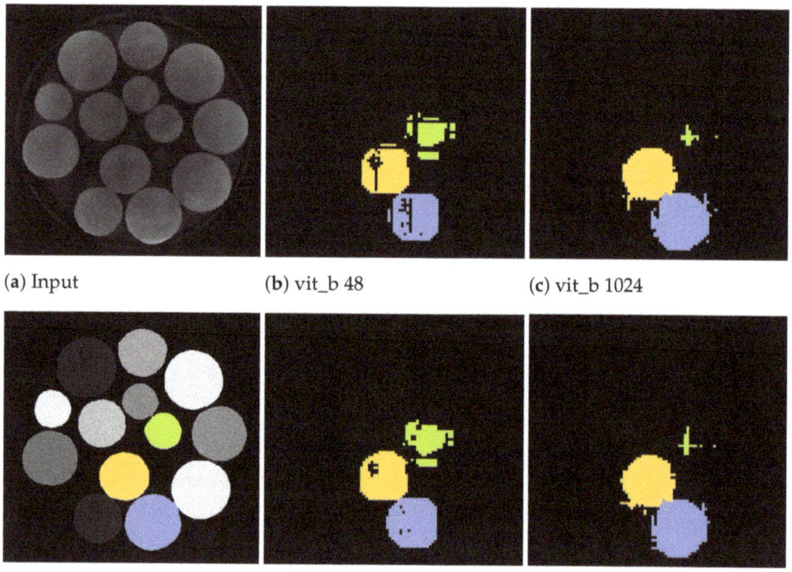

(a) Input (b) vit_b 48 (c) vit_b 1024

(d) Reference (e) vit_b 48 (postprocessed) (f) vit_b 1024 (postprocessed)

Figure 15. Slices from a volumetric inference run on three marbles of the marbles data-set. The input volume (**a**), reference volume (**d**), and the proposed segmentations generated by the proposed algorithm using the two tile sizes: 48 × 48 × 48 voxels (**b**) and 1024 × 1024 × 1024 voxels (**c**). Additionally, the postprocessed volumes are depicted in (**e**,**f**).

Due to the suboptimal quality of the segmentation, it proves problematic to compute a definitive overall numerical assessment of the complete segmentation. This difficulty arises from the ambiguity in assigning each segment unambiguously to a reference segment, a result of widespread under-segmentation or over-segmentation, which gives rise to various possible interpretations. Figure 16 shows the correlation matrices for the result of four inference runs on the Me 163 testing data-sets. Two of the inference runs were performed using the default SAM model vit_b, while the other two were performed using the fine-tuned model vit_b_{CVB}. Two of the four experiments used a tile size of $48 \times 48 \times 48$ voxels and the other two used a tile size of $1024 \times 1024 \times 1024$ voxels. Each experiment was fine-tuned on the validation data-set using [31].

The correlation matrices show the IoU of each reference segment in relation to each detected segment. The reference segments are sorted from top to bottom based on their voxel count, with the segment having the largest voxel count at the top. Similarly, the columns representing the detected segments are sorted so that the segment with the highest IoU, if compared with the largest reference segment, is on the left side. The segment with the highest IoU if compared with the second largest reference segment is then placed in the second column and so on. Each detected segment can only be linked to one reference segment once. In an ideal case, we would see a bright diagonal line from the upper left corner to the lower right corner of the matrix, indicating a perfect match between the reference and detected segments. Segments outside this diagonal indicate segmentation errors. Vertical lines indicate under-segmentation, where reference segments extend over multiple detected segments. Horizontal lines indicate over-segmentation, where reference segments are falsely split into multiple detected segments.

The individual parameters of the four inference runs can be found in Table 3. Figure 17 displays correlation matrices from Figure 16 but constrained to the detected segments with the highest IoU.

Table 3. Parameters optimized on the Me 163 validation data-set for the default vit_b and fine-tuned vit_b_{CVB} SAM model for the tile sizes of $48 \times 48 \times 48$ voxels and $1024 \times 1024 \times 1024$ voxels. (FG = foreground; – = not applicable; Options marked with * indicate volumetric SAM parameters as seen in Table 1; Options marked with × indicate FFN related parameters).

	vit_b 48	**vit_b 1024**	**vit_b_{CVB} 48**	**vit_b_{CVB} 1024**
best IoU	0.15	0.17	0.07	0.09
movement step *	1	–	1	–
seed FG count *	2	2	1	1
slice FG count *	3	1	1	1
FG threshold *	0.3	0.2	0.2	0.5
prompt type *	centre and dense	centre	centre and dense	centre and dense
SAM output channel *	index 1	max IoU	max IoU with FG	max IoU with FG
slice merge rule *	IoU to previous slice > 0.5	IoU to previous slice > 0.25	IoU to previous slice > 0.5	always
slice median *	✓	×	×	×
CCA *	✓	✓	×	×
volume median *	×	✓	×	✓
check step width ×	13	–	19	–
accumulator update ×	FG	FG	always	always
restrict movement ×	FG (128 steps)	eroded FG	eroded FG (128 steps)	FG

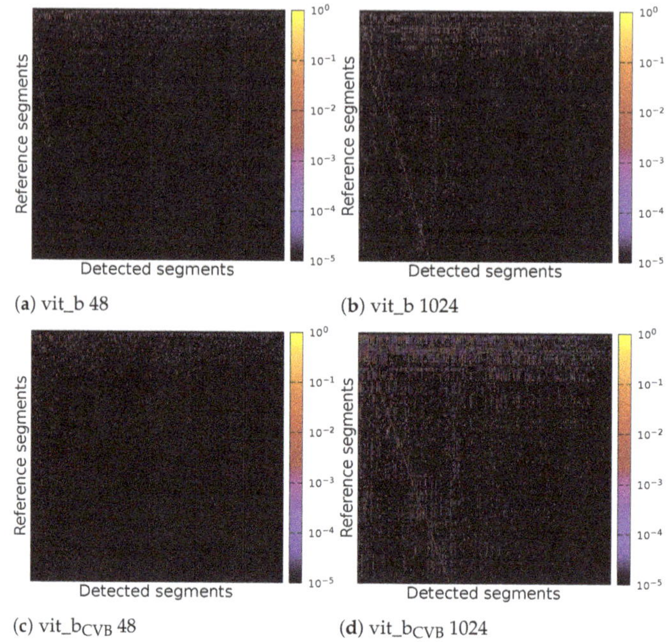

(**a**) vit_b 48 (**b**) vit_b 1024

(**c**) vit_b$_{CVB}$ 48 (**d**) vit_b$_{CVB}$ 1024

Figure 16. Correlation matrix of default and fine-tuned volumetric SAM with multiple tiles of size $48 \times 48 \times 48$ voxels or a single tile of size $1024 \times 1024 \times 1024$ voxels of the Me 163 testing data-set.

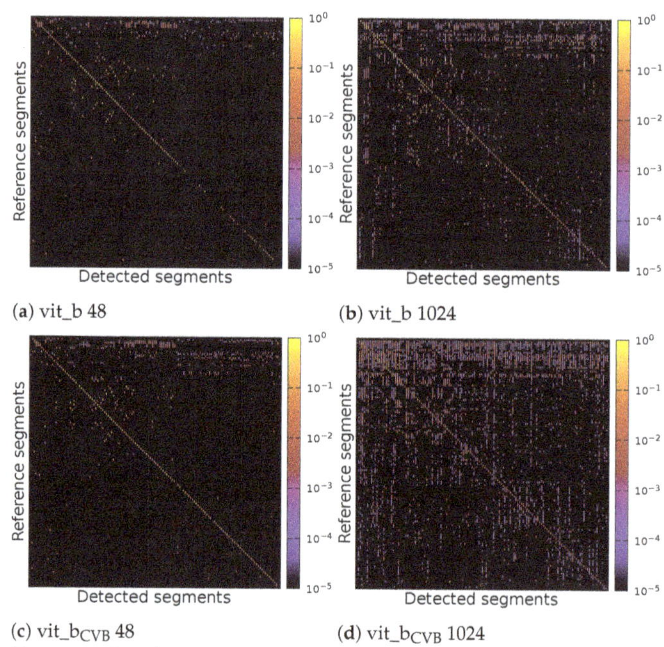

(**a**) vit_b 48 (**b**) vit_b 1024

(**c**) vit_b$_{CVB}$ 48 (**d**) vit_b$_{CVB}$ 1024

Figure 17. Correlation matrix of default and find-tuned volumetric SAM with multiple tiles of size $48 \times 48 \times 48$ voxels or a single tile of size $1024 \times 1024 \times 1024$ voxels of the Me 163 testing data-set. Detected segments have been limited to the best matches for each reference segment.

As can be seen, the vit_b$_{\mathrm{CVB}}$ models tends to generate more noise outside the main diagonal. Figure 17d especially depicts many over- and under-segmented segments. This can also be observed in the corresponding segmentation volume slice shown in Figure 18e,j. The correlation matrix of the fine-tuned vit_b$_{\mathrm{CVB}}$ model with tile size $48 \times 48 \times 48$ voxels in Figure 17c seems to perform best with respect to diagonal segments. But comparing the corresponding segmentation volume slice in Figure 18h shows that this model, tile, and parameter combination tends to miss most of the foreground segments. It seems that the default vit_b model with tile size $1024 \times 1024 \times 1024$ voxels produces the visually best results, followed by the fine-tuned vit_b$_{\mathrm{CVB}}$ model with tile size $48 \times 48 \times 48$ shown in Figure 18c.

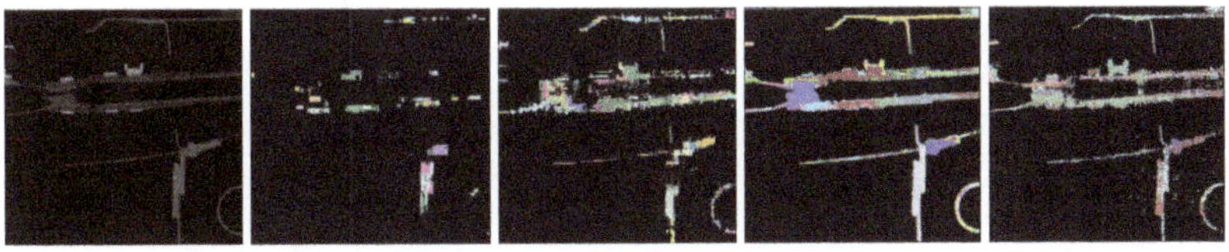

(**a**) Input (**b**) vit_b 48 (**c**) vit_b$_{\mathrm{CVB}}$ 48 (**d**) vit_b 1024 (**e**) vit_b$_{\mathrm{CVB}}$ 1024

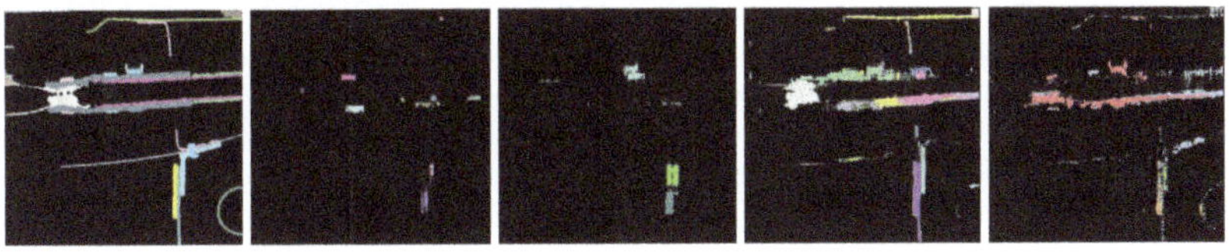

(**f**) Reference (**g**) vit_b 48 (main diagonal) (**h**) vit_b$_{\mathrm{CVB}}$ 48 (main diagonal) (**i**) vit_b 1024 (main diagonal) (**j**) vit_b$_{\mathrm{CVB}}$ 1024 (main diagonal)

Figure 18. Exemplary slices of the proposed volumetric inference output performed by default and fine-tuned SAM models on the Me 163 reference data-set (**a**,**f**). For the remaining figures, the top row (**b**–**e**) shows all segments depicted in Figure 16 while the bottom row (**g**–**j**) only shows the segments corresponding to the main diagonal in Figure 17.

Figure 19 presents multiple renderings of the seven largest reference segments in the Me 163 testing data-set, alongside their corresponding segment predictions generated by different SAM snapshots using the volumetric algorithm and fine-tuned parameters. The *true positive* voxels are coloured green, the reference segments are coloured blue, and the *false positive* voxels are coloured orange. It is evident that the volumetric segmentation of the data-sets using tiles of size $1024 \times 1024 \times 1024$ voxels yields visually more appealing segments compared to using a tile size of $48 \times 48 \times 48$ voxels.

The predicted segmentation using the tile size of $48 \times 48 \times 48$ voxels often appears *empty*, as only a small count of voxels has been segmented correctly. This is because the segmentation quality of the algorithm is too poor to generate connected tiles and so often, only a limited amount of steps (see Section 2.3.2) will be iterated for each segment. The segments are interrupted and only found in pieces. However, using a tile size of $48 \times 48 \times 48$ voxels also often leads to under-segmentation. Figure 20 exemplifies this, showcasing two orthogonal slices from the fine-tuned vit_b$_{\mathrm{CVB}}$ model's segmentation output. On the left, we present the reference segments and on the right, the corresponding predictions. Here, three adjacent segments were mistakenly connected by a single predicted segment.

But even the segmentation with a tile size of $1024 \times 1024 \times 1024$ voxels is often insufficient, as both large-scale under-segmentations and over-segmentations occur, as can be seen from the correlation matrices in Figure 17 and the cross-sectional images in Figure 18j.

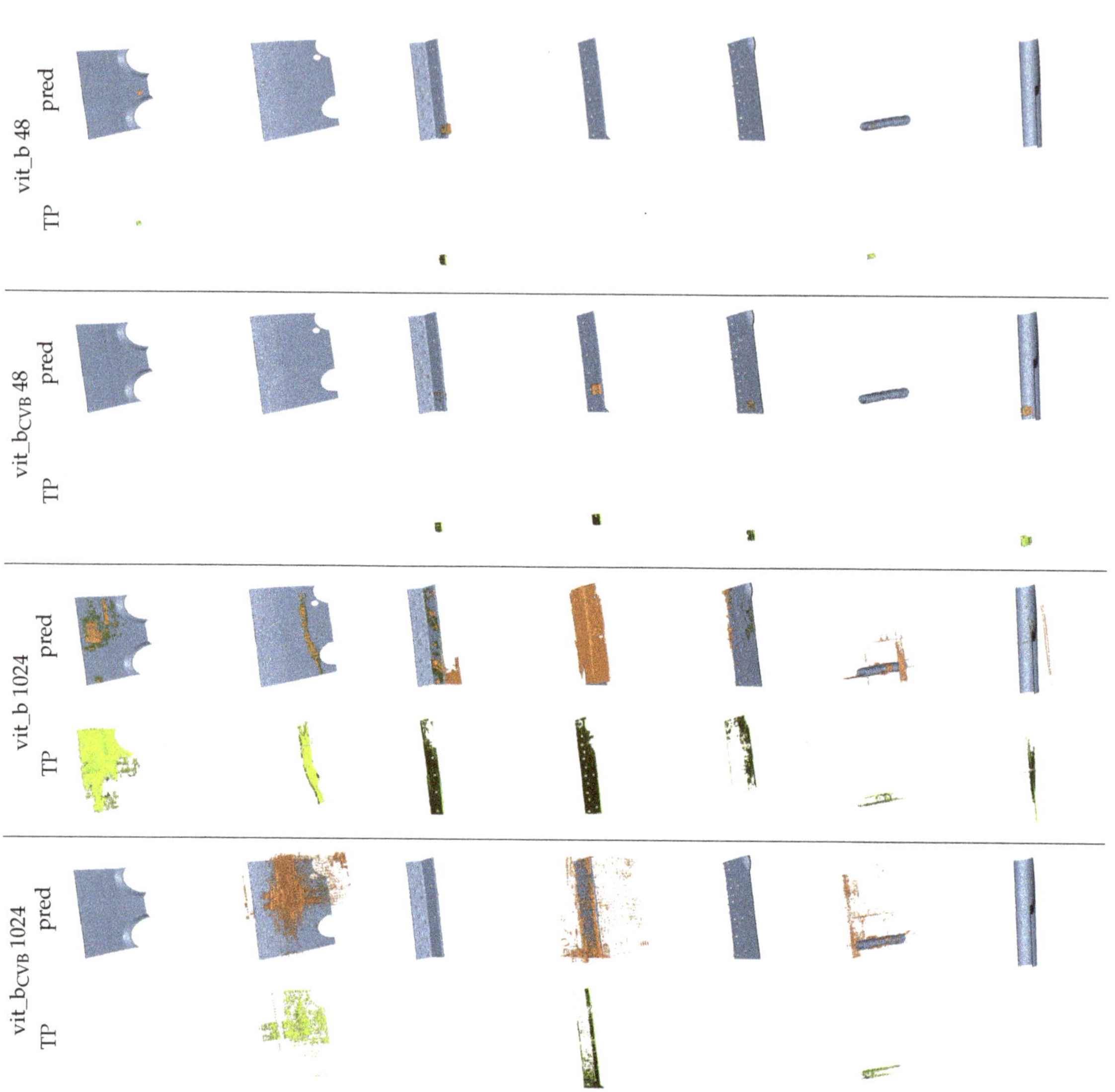

Figure 19. Renderings of the seven largest segments of the reference data-set and their corresponding predictions (pred) created with different snapshots of SAM and the volumetric algorithm. The colour coding is as follows: blue ■ reference segment, green ■ true positives (TP), and orange ■ false positives.

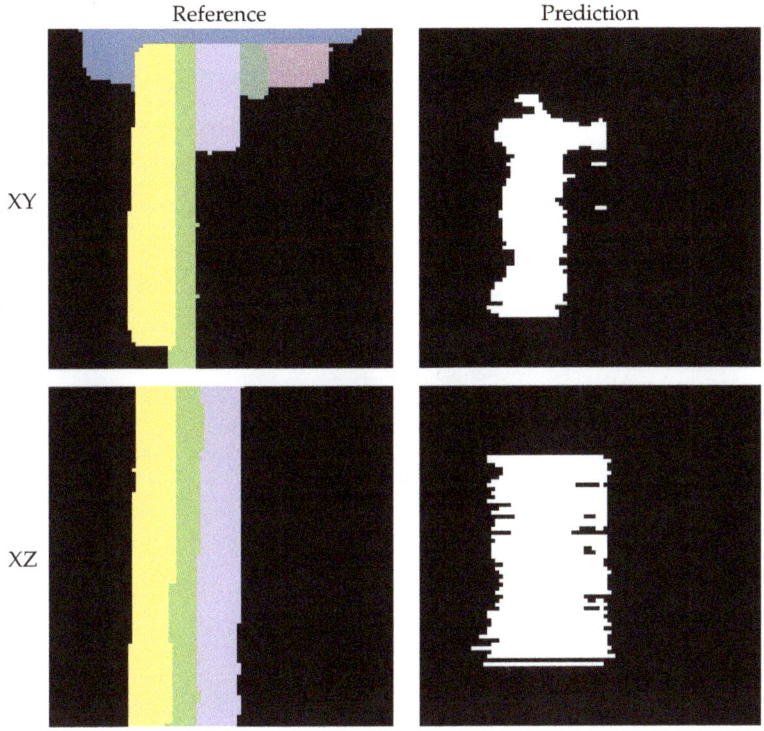

Figure 20. Slices obtained using the fine-tuned vit_b$_{CVB}$ model and tile size of 48 × 48 × 48 voxels. Due to under-segmentation, the predicted segment erroneously intersects and merges multiple reference segments.

4. Discussion

The transferability of the SAM model to instance segmentation of volumetric XXL-CT data-sets requires careful consideration. The presented results indicate that its two-dimensional image-based segmentation quality is insufficient for this specific problem domain. This limitation becomes particularly evident when dealing with the concatenation of numerous intertwined cross-sectional images in the volumetric case. The low contrast and high noise in these images pose challenges in accurately delineating individual segments. Additionally, using domain specific fine-tuning and improving slice-wise predictions did not yield substantial improvements for volumetric predictions.

One potential source of error in the presented method might be the limited computational resources allocated for both fine-tuning and subsequent hyperparameter search. A more thorough optimization process could potentially improve the results. Furthermore, the availability of labelled training data-sets of sufficient quality in this problem domain was relatively limited for training the vision transformers included in SAM. Specifically, the absence of neighbouring voxels when adding the 512 voxel wide border around the data-set for the Me 163 data-set may have possibly contributed to a decrease in segmentation quality.

Additionally, considering improved algorithms for merging the slice-wise predictions could be an initial step in the further development process. Previous studies [26–28] have demonstrated ample opportunities for the development of more sophisticated algorithms in this area. Implementing and embedding such algorithms into the processing pipeline has the potential to significantly enhance the segmentation quality.

5. Conclusions

The primary objective of this study was the exploration and possible applicability of the SAM algorithm for general image delineation to instance segmentation in XXL-CT volumetric data-sets.

In conclusion, our study highlights the potential of SAM for instance segmentation in XXL-CT volumetric data-sets, while acknowledging that there is still significant room for improvement. Furthermore, our research contributes to the following areas: (1) the evaluation of SAM on data-sets from the field of non-destructive testing based on CT image data, (2) the exploration of various methods for integrating and fusing the output from image-based SAM with volumetric data-sets, (3) the introduction of a tile-based approach for segmenting objects of arbitrary size, and (4) the utilization of dense prompts for tile combination using an accumulator. Separately and in combination, these contributions provide novel insights to the community and hence establish a foundation for further advancements in this field.

Author Contributions: Conceptualization, R.G.; Data curation, R.G.; Investigation, R.G. and S.R.; Methodology, R.G.; Resources, R.G.; Software, R.G. and S.R.; Supervision, T.W.; Validation, R.G.; Visualization, R.G.; Writing—original draft, R.G.; Writing—review and editing, S.R. and T.W. All authors have read and agreed to the published version of the manuscript.

Funding: This work was supported by the Bavarian Ministry of Economic Affairs, Regional Development and Energy through the centre for Analytics—Data—Applications (ADA-Center) within the framework of "BAYERN DIGITAL II" (20-3410-2-9-8).

Institutional Review Board Statement: Not applicable.

Informed Consent Statement: Not applicable.

Data Availability Statement: The Me 163 data-sets presented in this study are openly available in [7]. The bulk material data-sets will be made available by the authors on request.

Acknowledgments: The authors gratefully acknowledge the scientific support and HPC resources provided by the Erlangen National High-Performance Computing Center (NHR@FAU) of the Friedrich—Alexander Universität Erlangen—Nürnberg (FAU) under the NHR project b179dc. NHR funding is provided by federal and Bavarian state authorities. NHR@FAU hardware is partially funded by the German Research Foundation (DFG)—440719683. Additionally, the authors also extend their appreciation to Moritz Ottenweller for his valuable assistance during the manuscript revision process.

Conflicts of Interest: The authors declare no conflicts of interest. The funders had no role in the design of the study; in the collection, analyses, or interpretation of data; in the writing of the manuscript; or in the decision to publish the results.

References

1. Salamon, M.; Reims, N.; Böhnel, M.; Zerbe, K.; Schmitt, M.; Uhlmann, N.; Hanke, R. XXL-CT capabilities for the inspection of modern Electric Vehicles. In Proceedings of the International Symposium on Digital Industrial Radiology and Computed Tomography, Fürth, Germany, 2–4 July 2019.
2. Kolkoori, S.; Wrobel, N.; Hohendorf, S.; Redmer, B.; Ewert, U. Mobile High-energy X-ray Radiography for Nondestructive Testing of Cargo Containers. *Mater. Eval.* **2015**, *73*, 175–185.
3. Kolkoori, S.; Wrobel, N.; Hohendorf, S.; Ewert, U. High energy X-ray imaging technology for the detection of dangerous materials in air freight containers. In Proceedings of the 2015 IEEE International Symposium on Technologies for Homeland Security (HST), Waltham, MA, USA, 14–16 April 2015; pp. 1–6. [CrossRef]
4. Gruber, R.; Gerth, S.; Claußen, J.; Wörlein, N.; Uhlmann, N.; Wittenberg, T. Exploring Flood Filling Networks for Instance Segmentation of XXL-Volumetric and Bulk Material CT Data. *J. Nondestruct. Eval.* **2021**, *40*, 1. [CrossRef]
5. Gruber, R.; Reims, N.; Hempfer, A.; Gerth, S.; Wittenberg, T.; Salamon, M. *Fraunhofer EZRT XXL-CT Instance Segmentation Me163*; Zenodo: Geneva, Switzerland, 2024. [CrossRef]
6. Gruber, R.; Engster, J.C.; Michen, M.; Blum, N.; Stille, M.; Gerth, S.; Wittenberg, T. Instance Segmentation XXL-CT Challenge of a Historic Airplane. *arXiv* **2024**, arXiv:cs.CV/2402.02928.
7. Gruber, R.; Reims, N.; Hempfer, A.; Gerth, S.; Salamon, M.; Wittenberg, T. An annotated instance segmentation XXL-CT data-set from a historic airplane. *arXiv* **2022**, arXiv:cs.CV/2212.08639.

8. Kirillov, A.; Mintun, E.; Ravi, N.; Mao, H.; Rolland, C.; Gustafson, L.; Xiao, T.; Whitehead, S.; Berg, A.C.; Lo, W.Y.; et al. Segment Anything. *arXiv* **2023**, arXiv:cs.CV/2304.02643.
9. Hafiz, A.M.; Bhat, G.M. A survey on instance segmentation: State of the art. *Int. J. Multimed. Inf. Retr.* **2020**, *9*, 171–189. [CrossRef]
10. Wen, C.; Matsumoto, M.; Sawada, M.; Sawamoto, K.; Kimura, K.D. Seg2Link: An efficient and versatile solution for semi-automatic cell segmentation in 3D image stacks. *Sci. Rep.* **2023**, *13*, 7109. [CrossRef] [PubMed]
11. Zhao, T.; Olbris, D.J.; Yu, Y.; Plaza, S.M. NeuTu: Software for Collaborative, Large-Scale, Segmentation-Based Connectome Reconstruction. *Front. Neural Circuits* **2018**, *12*, 00101. [CrossRef]
12. Ohtake, Y.; Yatagawa, T.; Suzuki, H.; Kubo, S.; Suzuki, S. Thickness-Driven Sheet Metal Segmentation of CT-Scanned Body-in-White. *e-J. Nondestruct. Test.* **2023**, *28*, 27743. [CrossRef]
13. Dosovitskiy, A.; Beyer, L.; Kolesnikov, A.; Weissenborn, D.; Zhai, X.; Unterthiner, T.; Dehghani, M.; Minderer, M.; Heigold, G.; Gelly, S.; et al. An Image is Worth 16 × 16 Words: Transformers for Image Recognition at Scale. *arXiv* **2021**, arXiv:cs.CV/2010.11929.
14. Li, W.; Hsu, C.Y.; Wang, S.; Yang, Y.; Lee, H.; Liljedahl, A.; Witharana, C.; Yang, Y.; Rogers, B.M.; Arundel, S.T.; et al. Segment Anything Model Can Not Segment Anything: Assessing AI Foundation Model's Generalizability in Permafrost Mapping. *Remote. Sens.* **2024**, *16*, 797. [CrossRef]
15. Noe, S.M.; Zin, T.T.; Tin, P.; Kobyashi, I. Efficient Segment-Anything Model for Automatic Mask Region Extraction in Livestock Monitoring. In Proceedings of the 13th IEEE International Conference on Consumer Electronics—Berlin, ICCE-Berlin 2023, Berlin, Germany, 3–5 September 2023; IEEE: Piscataway, NJ, USA, 2023; pp. 167–171. [CrossRef]
16. Carraro, A.; Sozzi, M.; Marinello, F. The Segment Anything Model (SAM) for accelerating the smart farming revolution. *Smart Agric. Technol.* **2023**, *6*, 100367. [CrossRef]
17. Weinberger, P.; Schwarz, L.; Fröhler, B.; Gall, A.; Heim, A.; Yosifov, M.; Bodenhofer, U.; Kastner, J.; Senck, S. Unsupervised Segmentation of Industrial X-ray Computed Tomography Data with the Segment Anything Model. *Res. Sq.* 2024, *preprint*. [CrossRef]
18. Xu, B.; Yu, S. Improving Data Augmentation for YOLOv5 Using Enhanced Segment Anything Model. *Appl. Sci.* **2024**, *14*, 1819. [CrossRef]
19. Liu, Z. Optimizing road sign detection using the segment anything model for background pixel exclusion. *Appl. Comput. Eng.* **2024**, *31*, 150–156. [CrossRef]
20. Januszewski, M.; Kornfeld, J.; Li, P.H.; Pope, A.; Blakely, T.; Lindsey, L.; Maitin-Shepard, J.; Tyka, M.; Denk, W.; Jain, V. High-precision automated reconstruction of neurons with flood-filling networks. *Nat. Methods* **2018**, *15*, 605–610. [CrossRef] [PubMed]
21. LeCun, Y.; Boser, B.; Denker, J.S.; Henderson, D.; Howard, R.E.; Hubbard, W.; Jackel, L.D. Backpropagation Applied to Handwritten Zip Code Recognition. *Neural Comput.* **1989**, *1*, 541–551. [CrossRef]
22. LeCun, Y. Generalization and network design strategies. *Connect. Perspect.* **1989**, *19*, 143–155.
23. Otsu, N. A Threshold Selection Method from Gray-Level Histograms. *IEEE Trans. Syst. Man Cybern.* **1979**, *9*, 62–66. [CrossRef]
24. Ma, J.; He, Y.; Li, F.; Han, L.; You, C.; Wang, B. Segment Anything in Medical Images. *arXiv* **2023**, arXiv:2304.12306.
25. Loshchilov, I.; Hutter, F. Decoupled Weight Decay Regularization. *arXiv* **2017**, arXiv:1711.05101. [CrossRef].
26. Zhang, Y.; Liao, Q.; Ding, L.; Zhang, J. Bridging 2D and 3D segmentation networks for computation-efficient volumetric medical image segmentation: An empirical study of 2.5D solutions. *Comput. Med. Imaging Graph.* **2022**, *99*, 102088. [CrossRef]
27. Xia, Y.; Xie, L.; Liu, F.; Zhu, Z.; Fishman, E.K.; Yuille, A.L. Bridging the Gap between 2D and 3D Organ Segmentation with Volumetric Fusion Net. In Proceedings of the Medical Image Computing and Computer Assisted Intervention—MICCAI 2018, Granada, Spain, 16–20 September 2018; Frangi, A.F., Schnabel, J.A., Davatzikos, C., Alberola-López, C., Fichtinger, G., Eds.; Springer: Cham, Switzerland, 2018; pp. 445–453.
28. Zheng, H.; Qian, L.; Qin, Y.; Gu, Y.; Yang, J. Improving the slice interaction of 2.5D CNN for automatic pancreas segmentation. *Med. Phys.* **2020**, *47*, 5543–5554. [CrossRef]
29. Huang, Y.; Yang, X.; Liu, L.; Zhou, H.; Chang, A.; Zhou, X.; Chen, R.; Yu, J.; Chen, J.; Chen, C.; et al. Segment Anything Model for Medical Images? *arXiv* **2023**, arXiv:2304.14660.
30. Mazurowski, M.A.; Dong, H.; Gu, H.; Yang, J.; Konz, N.; Zhang, Y. Segment anything model for medical image analysis: An experimental study. *Med. Image Anal.* **2023**, *89*, 102918. [CrossRef]
31. Akiba, T.; Sano, S.; Yanase, T.; Ohta, T.; Koyama, M. Optuna: A Next-generation Hyperparameter Optimization Framework. In Proceedings of the 25th ACM SIGKDD International Conference on Knowledge Discovery and Data Mining, Anchorage, AK, USA, 4–8 August 2019.

applied sciences

Article

Dhad—A Children's Handwritten Arabic Characters Dataset for Automated Recognition

Sarab AlMuhaideb *, Najwa Altwaijry, Ahad D. AlGhamdy, Daad AlKhulaiwi, Raghad AlHassan, Haya AlOmran and Aliyah M. AlSalem

Computer Science Department, College of Computer and Information Sciences, King Saud University, Riyadh 11362, Saudi Arabia; ntwaijry@ksu.edu.sa (N.A.); 441201150@student.ksu.edu.sa (A.D.A.); 441201060@student.ksu.edu.sa (D.A.); 441201024@student.ksu.edu.sa (R.A.); 441200957@student.ksu.edu.sa (H.A.); 441203529@student.ksu.edu.sa (A.M.A.)
* Correspondence: salmuhaideb@ksu.edu.sa

Abstract: This study delves into the intricate realm of recognizing handwritten Arabic characters, specifically targeting children's script. Given the inherent complexities of the Arabic script, encompassing semi-cursive styles, distinct character forms based on position, and the inclusion of diacritical marks, the domain demands specialized attention. While prior research has largely concentrated on adult handwriting, the spotlight here is on children's handwritten Arabic characters, an area marked by its distinct challenges, such as variations in writing quality and increased distortions. To this end, we introduce a novel dataset, "Dhad", refined for enhanced quality and quantity. Our investigation employs a tri-fold experimental approach, encompassing the exploration of pre-trained deep learning models (i.e., MobileNet, ResNet50, and DenseNet121), custom-designed Convolutional Neural Network (CNN) architecture, and traditional classifiers (i.e., Support Vector Machine (SVM), Random Forest (RF), and Multilayer Perceptron (MLP)), leveraging deep visual features. The results illuminate the efficacy of fine-tuned pre-existing models, the potential of custom CNN designs, and the intricacies associated with disjointed classification paradigms. The pre-trained model MobileNet achieved the best test accuracy of 93.59% on the Dhad dataset. Additionally, as a conceptual proposal, we introduce the idea of a computer application designed specifically for children aged 7–12, aimed at improving Arabic handwriting skills. Our concluding reflections emphasize the need for nuanced dataset curation, advanced model architectures, and cohesive training strategies to navigate the multifaceted challenges of Arabic character recognition.

Keywords: deep learning; pre-trained models; child handwriting recognition; Dhad; Hijja

Citation: AlMuhaideb, S.; Altwaijry, N.; AlGhamdy, A.D.; AlKhulaiwi, D.; AlHassan, R.; AlOmran, H.; AlSalem, A.M. Dhad—A Children's Handwritten Arabic Characters Dataset for Automated Recognition. *Appl. Sci.* **2024**, *14*, 2332. https://doi.org/10.3390/app14062332

Academic Editors: Zahid Mehmood Jehangiri, Mohsin Shahzad and Uzair Khan

Received: 3 January 2024
Revised: 3 March 2024
Accepted: 8 March 2024
Published: 10 March 2024

1. Introduction

Arabic is a widely spoken language, with over 360 million people using it as their primary language [1]. In the domain of language processing and technological applications, the recognition of handwritten Arabic characters, especially in the realm of children's script, poses unique challenges [2,3]. Arabic, being a Semitic language, presents inherent complexities in its script, demanding advanced algorithms for precise recognition. These challenges emanate from factors such as the semi-cursive nature of Arabic writing, distinct character shapes based on their position in a word, and the presence of diacritical marks representing short vowels and other phonetic features. The significance of addressing the issue of Arabic handwritten recognition, particularly concerning children, arises from the increasing integration of technology in their educational and recreational activities. The ubiquitous use of smartphones and tablet devices by children, employing touchscreens and styluses for various purposes, including handwriting, underscores the need for automated recognition techniques tailored to the unique characteristics of children's handwriting [4–6].

While the existing literature has made notable strides in Arabic handwritten recognition, the focus has predominantly been on adult handwriting. Researchers have explored diverse datasets such as the Arabic Handwritten Characters Dataset (AHCD) [7] and the Database of Arabic Handwritten Characters and Ligature (DBAHCL) [8], achieving commendable accuracy rates using both conventional methods (e.g., Support Vector Machine (SVM), K-Nearest Neighbour (KNN)) and advanced techniques (e.g., Convolutional Neural Network (CNN) and Artificial Neural Network (ANN)). Notably, CNNs [9,10] have emerged as powerful tools for feature extraction, demonstrating superior performance compared to traditional machine learning approaches. In the context of children's handwriting recognition, the existing Hijja dataset [4,11] is the only resource facilitating the training of deep learning classification models. However, children's handwriting introduces additional complexities, including variations in writing quality, increased variances, and more substantial distortions. Recognizing these distinctions is imperative for developing effective applications in education, interactive learning, and other practical domains tailored to children.

Limited research has been conducted in recognizing children's written Arabic characters using the Hijja dataset. The existing literature uses conventional and deep learning approaches towards the classification of children written Arabic characters. Altwaijry and Al-Turaiki [4] introduced the unique Hijja dataset, training a CNN model with 88% accuracy, but lacked detailed investigations into existing powerful models. Alkhateeb et al. [12] implemented a custom CNN model, achieving 92.5% accuracy on the Hijja dataset. Nayef et al. [13] introduced the Optimized Leaky Rectified Linear Unit (OLReLU)-CNN model, attaining 90% accuracy on Hijja. Alwagdani and Jaha [5] explored custom CNN models, emphasizing the impact of diverse training datasets and achieving an impressive average accuracy of 92.78% on recognizing children's handwritten characters, while also proposing supplementary features for enhanced discrimination. Alheraki et al. [14] tailored a custom CNN for achieving 91% accuracy on Hijja. Recently, Bin Durayhim et al. [15] implemented a custom CNN and pre-trained VGG16 models, reporting a remarkable 99% accuracy on the Hijja dataset and introducing the Mutqin application for children's practice. These studies collectively highlight the evolving landscape of deep learning applications in recognizing children's Arabic handwriting. However, concerns about model generalization and sensitivity persist, and further exploration is warranted in this context.

This paper introduces a new dataset, "Dhad", following procedures similar to those for Hijja to ensure consistency. The Dhad dataset features improved sample quality, enhanced preprocessing to remove noisy elements, and a greater number of samples. This manuscript systematically addresses the problem by investigating the potential of existing pre-trained powerful CNN models using the transfer learning technique. Furthermore, it explores the performance of simpler CNN models trained from scratch and classification on deep visual features. In summary, the anticipated contributions of this manuscript include the following:

1. Introduction of the new "Dhad" dataset to facilitate the training of deep learning models for children's handwritten Arabic characters.
2. Investigation of the potential of pre-trained powerful CNN models for children's handwritten Arabic character classification.
3. Examination of the performance of a simple CNN model trained from scratch for children's handwritten Arabic character classification.
4. Exploration of the classification performance on CNN-extracted features using conventional machine learning models including SVM and Random Forest (RF).
5. Discussion of the practical use-case of the trained classification model, emphasizing the potential utility of children's handwritten Arabic characters recognition.

2. Background to Deep Learning Models

2.1. ResNet50

He et al. [16] introduced an innovative approach to training highly deep neural networks, proposing a novel framework based on residual learning. Instead of training networks to learn unreferenced functions, the authors suggested a reinterpretation of layers as residual learning functions by referencing the input of the layer. This concept of residual learning played a crucial role in the optimization of deep networks, enabling the attainment of enhanced accuracy with deep models. To express this mathematically, if we denote the desired mapping function as $H(x)$, in the context of residual learning, stacked non-linear layers are designed to fit another mapping function $F(x) := H(x) - x$, where x represents the input to the layer. This approach significantly contributed to the effectiveness of training deep networks by explicitly capturing the residual information between the desired and actual mappings.

2.2. *MobileNet*

Howard et al. [17] introduced a new type of CNN called MobileNets, tailored for high-performance applications on advanced hardware. The key innovation involves using depth-wise separable convolutions to efficiently build deep networks. This method introduces two global hyperparameters, allowing customization for specific problems while balancing accuracy and latency. Depth-wise separable convolution, a specialized type of convolution, breaks down the standard convolution process into two steps. First, a depth-wise convolution is applied, and then, a 1×1 point-wise convolution combines the results from the previous layer. Importantly, each layer in the network is followed by BatchNormalization and Rectified Linear Unit (ReLu) non-linearity. This separation into depth-wise and point-wise convolutions helps improve computational efficiency while maintaining the network's performance.

2.3. *DenseNet121*

Huang et al. [18] introduced DenseNet, a novel class of densely connected convolutional networks that builds upon the idea of residual connections present in traditional networks. The key innovation involves establishing connections from each layer to every other layer in the feedforward direction. This architectural choice means that each layer receives the feature maps of all preceding layers as input, resulting in a network with $L(L+1)/2$ connections, in contrast to the L connections in traditional networks with L layers. The advantages of densely connected networks include improved feature propagation, efficient feature reuse, a substantial reduction in the number of network parameters, and mitigation of the vanishing-gradient problem. Unlike the approach in residual networks, where feature maps are added before feeding into the next layer, DenseNet combines feature maps through concatenation. Mathematically, if a network comprises L layers, each with a non-linear transformation represented by a composite function F_l, the output x_l for the densely connected layer can be understood as the concatenation of feature maps from the previous layers. In practical terms, this means that each layer's output includes information from all preceding layers, promoting rich information flow and enhancing the network's ability to learn complex representations.

2.4. *Custom CNN*

A custom CNN model was developed, motivated from the literature [4,5,13–15] to investigate the performance of a simpler network trained from scratch for the children's handwritten character classification. The model is particularly tailored for grayscale images with a size of 32×32 pixels (Figure 1). The architecture is constructed as a sequential stack of layers using TensorFlow and Keras packages. The initial layer applies 64 convolutional filters of the size 3×3 with ReLU activation and the "same" padding, preserving spatial dimensions. Subsequent max pooling layers with 2×2 pool sizes and strides of two reduce the spatial dimensions. This pattern repeats with increased filter counts in deeper convolutional layers (128 and 256 filters). Dropout layers with a rate of 0.3 are strategically inserted to mitigate overfitting. Following flattening, two densely connected layers with 512 and 1024 units, respectively, deploy ReLU activation. The output layer, activated by softmax, consists of 29 units, aligning with the classification task's classes. The model is compiled using the Adam optimizer, categorical cross-entropy as the loss function, and accuracy as the performance metric.

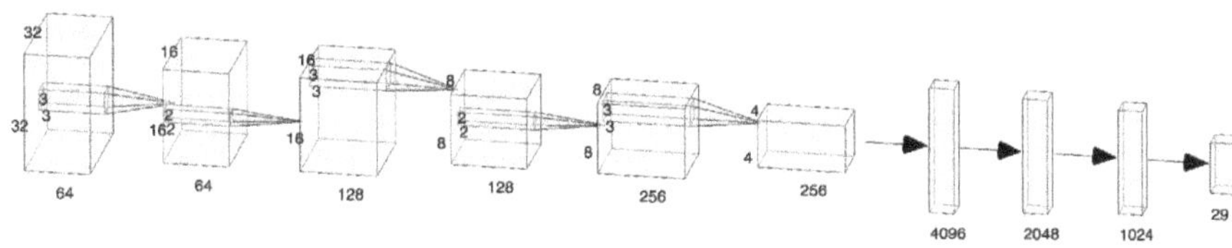

Figure 1. The architecture of our custom CNN model.

3. Related Work

This section presents a summary of the literature in the context of children's handwriting classification. First, a brief overview of how technology evolved for the children's handwriting classification is presented. In the second section, a more targeted review of more recent deep learning-based research in the context of children's written Arabic character classification is provided to demonstrate the state of the art in this domain.

3.1. Children's Handwriting Classification

The classification of children's handwriting has been an active area of research for decades, driven by the need for objective and automated assessment tools. Unlike adult handwriting, which tends towards standardization, children's writing exhibits a wide range of variations due to age, developmental stage, and individual learning styles. Early attempts at automated character recognition (OCR) for children's handwriting often relied on template matching techniques. Pioneering works employed pre-defined templates representing ideal character shapes. An input character would be compared to these templates, with the closest match assigned as the recognized character. However, this approach proved ineffective for children's handwriting due to its inherent lack of conformity [19,20]. However, there are a number of limitations identified in the template matching approaches, particularly for characters with significant variations in form.

Researchers recognized the limitations of template matching and explored alternative approaches. Utilizing statistical and structural features for character classification has been explored by researchers [21–23]. This involved analysing features like line endings, crossings, and loops within handwritten characters. While offering more flexibility than rigid templates, these methods still faced limitations. The emergence of deep learning techniques, particularly CNNs, has revolutionized the field. Unlike previous methods, CNNs excel at extracting intricate features from the data. This allows them to effectively capture the natural variations in children's handwriting, leading to more robust and accurate character-level classification. Further advancements in deep learning architectures, such as recurrent neural networks (RNNs), have shown promise in handling the sequential nature of handwriting data.

3.2. Deep Learning-Based Classification of Children's Arabic Handwriting

A summary of the latest research related to children's handwritten Arabic character classification is presented in this section to demonstrate the state of the art. This section is organized in chronological order to better understand the developments over the years.

Alkhateeb et al. [12] in 2020 implemented a custom CNN model for the classification of Arabic characters using the AHCR, AHCD, and Hijja datasets. The authors reported an accuracy of 92.5% for the Hijja dataset. Altwaijry and Al-Turaiki [4] in 2021 introduced the Hijja dataset, which stands out as a unique collection focusing exclusively on letters written by children. This dataset, comprising 47,434 characters from 591 participants aged 7–12, filled a notable gap in existing resources, particularly for understanding the nuances of children's handwriting. A CNN model was developed and trained on the Hijja dataset, which achieved a test accuracy of 88%. However, the work lacked a detailed investigation on existing powerful models and ignored the practical implications of the research.

Nayef et al. [13] in 2022 introduced an OLReLU combined with a CNN architecture and a batch normalization layer to enhance performance in scenarios with imbalanced positive and negative vectors. Four datasets, including the AHCD, self-collected data, Modified National Institute of Standards and Technology (MNIST), and AlexU Isolated Alphabet (AIA9K), were used. The proposed model was able to achieve 90% accuracy for the Hijja dataset. Alwagdani and Jaha [5] recently in 2023 investigated the problem in more detail using custom developed CNN models and hybrid approaches. The authors made use of datasets from both adults (i.e., AHCD) and children (i.e., Hijja) to explore the performance, with a particular emphasis on the impact of different training datasets. The authors further investigated the problem of classifying by deploying a conventional machine learning pipeline of extracting visual features and classifying using classical machine learning models (e.g., SVM, KNN, and RF). The findings reveal that training the model on a combination of children's and adult datasets yields the best performance, achieving an impressive average accuracy of 92.78% in recognizing children's handwritten characters. Moreover, authors extended their investigation to the classification of writers into two groups (i.e., children and adults) using the proposed CNN model. The initial results showed an average accuracy of 89.28%, indicating the presence of confusable similarities in writing styles between adults and children. To enhance discrimination performance, the study suggested supplementary features based on Histogram of Oriented Gradients (HOGs) and statistical measures, which, when combined with CNN features, result in a significantly improved accuracy of 92.29%.

Alheraki et al. [14] in 2023 implemented a custom CNN architecture tailored for recognizing children's Arabic handwritten characters. The authors made use of the AHCD and Hijja datasets to train the model and achieved accuracies of 97% and 91% for AHCD and Hijja, respectively. Additionally, the authors introduced an innovative approach using character strokes as a filter to further enhance recognition accuracy. This method, combined with CNNs, demonstrated effectiveness in improving performance. The research compared the proposed model with the pre-trained EfficientNetV0 and reported better performance for the custom model. Moreover, a multi-model approach, integrating information about the number of strokes in a character, achieved an average prediction accuracy of 96% when Hijja was merged with AHCD. Bin Durayhim et al. [15] in 2023 implemented two deep learning-based models (i.e., custom CNN and pre-trained VGG16) for children's handwriting character recognition using Hijja and AHCD. The custom CNN model was reported to outperform the VGG16 model and other models from the literature, achieving 99% accuracy on the Hijja dataset. Additionally, the paper introduced Mutqin, a prototype tablet application designed for children to practice Arabic handwriting and spelling, incorporating the best-performing CNN model. The application was evaluated through user acceptance testing, considering effectiveness, efficiency, and satisfaction, with positive results indicating good performance.

A notable progress has been made in the literature in regard to addressing the unique challenges posed by children's handwriting. Several studies have implemented various deep learning models to recognize isolated Arabic characters, particularly targeting children's writing styles. The introduced models include custom CNNs, OLReLU with CNN architectures, pre-trained models (i.e., VGG16 and EfficeintNetv0) and hybrid approaches combining CNNs with classical machine learning models (e.g., SVM, KNN, and RF). These studies have utilized datasets such as the AHCD and the specifically designed Hijja dataset, which exclusively feature letters written by children. The accuracy reported in these works range from 88% to 99%, showcasing the effectiveness of these models in handling the intricacies of children's handwriting.

However, certain limitations and gaps persist in the current research landscape. Notably, there is a lack of consistent exploration and discussion of transfer learning models, specifically ResNet50 and MobileNet, which have demonstrated success in other domains. Additionally, studies report inconsistent performance for almost similar CNN structures, indicating a potential non-reliability of simple CNN models for this specific problem. There is no justification in the literature with regard to significantly varied performances for almost same CNN architectures with minor hyperparamter variations. Furthermore, the practical applications of the proposed models are not extensively discussed across the studies, with only one work introducing a prototype tablet application named Mutqin for children's practice.

4. Dhad Dataset Formation

In this section, we describe our collected dataset, Dhad (the dataset is available at https://github.com/daadturki1/Dhad/ (accessed on 1 October 2023)), and the steps taken to preprocess and prepare the dataset for the training of deep learning models. The Dahd dataset collection process was based on the procedures described by Altwaijry and Al-Turaiki [4]. The dataset was collected from Arabic-speaking school children between 7 and 12 years old within the Riyadh region in 2019. In total, 55,587 samples for all 29 letter classes in all different forms were collected. The count of each collected letter in all its forms after discarding noisy input is reported in Table 1. In our image processing workflow, the handwritten letters without dots were identified and cropped using the `findContours()` method available in the OpenCV library [24]. Specifically, this method was utilized to locate the outer contour of each object present in the image, after which we proceeded to crop the image around the largest identified contour. Conversely, for handwritten letters containing dots, the `findNonZero()` method was employed to identify all black pixels in the image. Subsequently, the smallest possible rectangle that encompassed all black pixels was cropped to isolate the desired letter. After that, the images were resized to 32×32 pixels. To eliminate the noise within the scanned images, we used a Gaussian filter with a kernel of size 5×5 to blur specific portions of the image. Then, a high-pass filter was used to sharpen the edges. Lastly, the binarization technique was used to convert the RGB image to a binary level with a global thresholding algorithm. Figure 2 depicts sample images from the classes "mīm" and "nūn", showing the different preprocessing stages. Four data augmentation techniques were used with a range of 0.2: height and width shift, shear range, zoom range, and rescale. The dataset was then normalized, shuffled, and split into 60%, 20%, and 20% for training, validation, and testing, respectively. Overall, 30,922, 10,300, and 10,333 samples were used for the training, validation, and testing sets, respectively. Figure 3 presents the dataset collection workflow.

Table 1. The different letter forms and the total number of images for each class after data cleansing in the collected dataset.

No.	Class	Form	Count	No.	Class	Form	Count
1	·alif	ا، أ، إ، ـأ، ـإ	2869	16	ṭā·	ط، طـ، ـطـ، ـط	1925
2	bā·	ب، بـ، ـبـ، ـب	1899	17	ẓā·	ظ، ظـ، ـظـ، ـظ	1886
3	tā·	ت، تـ، ـتـ، ـت	1920	18	ʿayn	ع، عـ، ـعـ، ـع	1906
4	ṯā·	ث، ثـ، ـثـ، ـث	1734	19	ġayn	غ، غـ، ـغـ، ـغ	2012
5	ğīm	ج، جـ، ـجـ، ـج	1891	20	fā·	ف، فـ، ـفـ، ـف	2024
6	ḥā·	ح، حـ، ـحـ، ـح	1921	21	qāf	ق، قـ، ـقـ، ـق	2030
7	khā·	خ، خـ، ـخـ، ـخ	1869	22	kāf	ك، كـ، ـكـ، ـك	2019
8	dāl	د، ـد	931	23	lām	ل، لـ، ـلـ، ـل	2011
9	ďāl	ذ، ـذ	915	24	mīm	م، مـ، ـمـ، ـم	1955
10	rā·	ر، ـر	898	25	nūn	ن، نـ، ـنـ، ـن	1913
11	zāy	ز، ـز	944	26	hā·	ه، هـ، ـهـ، ـه	2180
12	sīn	س، سـ، ـسـ، ـس	1845	27	wāw	و، ـو	872
13	šīn	ش، شـ، ـشـ، ـش	1876	28	yā·	ي، يـ، ـيـ، ـي	1903
14	ṣād	ص، صـ، ـصـ، ـص	1709	29	hamzah	ء، ؤ، ئ، ـئ	1846
15	ḍād	ض، ضـ، ـضـ، ـض	1852				
						Total	51,555

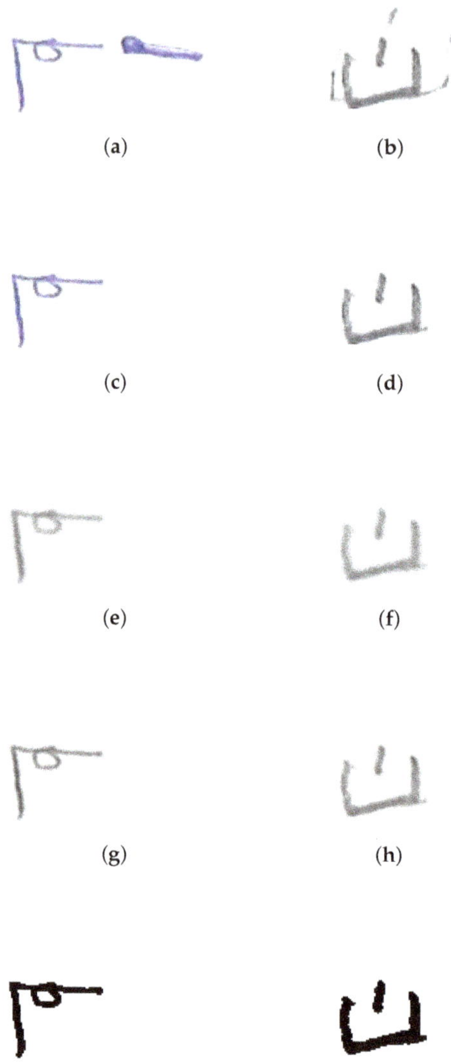

(**a**) (**b**) (**c**) (**d**) (**e**) (**f**) (**g**) (**h**) (**i**) (**j**)

Figure 2. Sample images from the letters mīm and nūn showing the consecutive data cleansing and preprocessing steps. (**a**) Letter mīm after scanning and cropping. (**b**) Letter nūn after scanning and cropping. (**c**) Letter mīm after cleansing. (**d**) Letter nūn after cleansing. (**e**) Letter mīm after applying a Gaussian filter. (**f**) Letter nūn after applying the Gaussian filter. (**g**) Letter mīm after applying a high-pass filter. (**h**) Letter nūn after applying a high-pass filter. (**i**) Letter mīm after applying binarization. (**j**) Letter nūn after binarization.

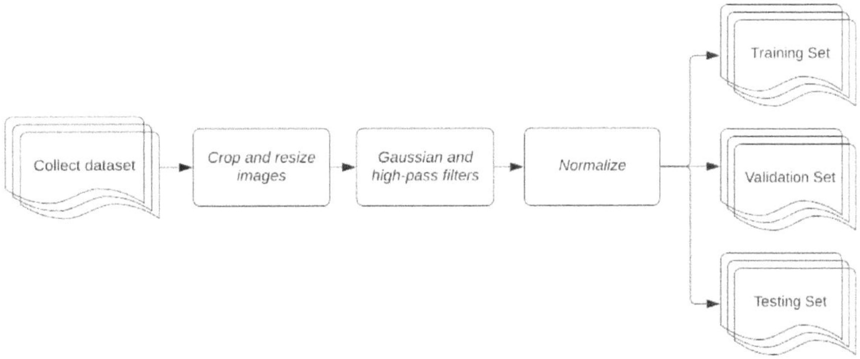

Figure 3. Dhad dataset collection and preparation workflow.

5. Experimental Design

To investigate the problem in details, in total, three experiments were performed:

- Experiment One—Pre-Trained Models: The first experiment was designed to explore the potential of existing pre-trained powerful CNN models in classifying children's written Arabic characters. The literature suggests that well-established CNN models pre-trained with large image datasets like that of ImageNet perform superior in comparison to training from scratch. In this context, the ResNet50, MobileNet, and DenseNet121 models are implemented for both the Hijja and Dhad datasets.
- Experiment Two—Custom CNN Model: The second experiment was designed to investigate the performance of a simpler custom CNN model for this problem. The development of custom CNN models has already been reported in the literature; however, varied performances are reported each time. In this experiment, we developed a simple CNN model inspired from the literature and implemented it for both the Hijja and Dhad datasets.
- Experiment Three—Classification on Deep Visual Features: The third experiment was designed to study the performance of classical classification models including SVM, RF, and MLP trained on deep visual features extracted by a deep learning CNN model (i.e., MobileNet trained over ImageNet and MobileNet trained in Experiment One).

6. Experimental Protocols and Evaluation Measures

We employed the OpenCV–4.9.0 library and the Python Imaging Library (Pillow 10.2.0) in our study, as they provide the necessary functions for data preprocessing and data augmentation and used the TensorFlow 2 [25] and Keras 3 [26] Python libraries to implement the architectures. We also used Google Colaboratory Pro [27] to speed up the training by granting access to K80, P100, T4 GPU, and 32 GB RAM. Gridsearch [28] was used for hyperparameter tuning. A dataset split of 60:20:20 was used for training, validation, and testing purposes using the conventional hold-out approach. All the pre-trained models were fine-tuned for 30 epochs, while the custom CNN model was trained from scratch for 100 epochs to ensure convergence. For all the models, the Adam optimizer was used with categorical cross-entropy loss, and a training batch size of 4 was used.

The performance of the trained models was evaluated for both the training and validation phases using standard evaluation measures. The training performance was assessed based on the training loss curves, validation loss curves, training accuracy curves, and validation accuracy curves. Further, for the best epoch model based on the validation loss, the results for all four measures were also reported in tabular format. The testing performance was assessed using test accuracy, test loss, F1 score, precision score, recall score, and J-index. In addition to these quantitative measures, confusion matrices, and Area Under Curve (AUC) curves were used to analyse the class-wise performances of trained models. For the pre-trained models, to better understand the performance, layer visualizations were also plotted.

Let TP and TN denote the numbers of true positives and true negatives, respectively, and FP and FN denote the numbers of false positives and false negatives, respectively. Accuracy (Acc) is defined as the proportion of correctly predicted examples (1). The loss ($Loss$) quantifies the degree of misclassification by determining the proportion of incorrect predictions relative to the total predictions made by the model (2). Precision (P) is the fraction of correctly classified positive examples among all positively classified examples (3). Meanwhile, recall or sensitivity measures the ratio of correctly classified positive examples to the true positive examples (4). The $F1$ score is calculated as the harmonic mean of the precision and recall; thus, it combines both precision and recall in a single value (5). The Jaccard Similarity Index (often referred to simply as $J\text{-}Index$) is a measure of how close the predicted labels are to the actual labels (6).

$$Acc = \frac{TP + TN}{TP + TN + FP + FN} \tag{1}$$

$$Loss = \frac{FP + FN}{TP + TN + FP + FN} \tag{2}$$

$$P = \frac{TP}{TP + FP} \tag{3}$$

$$R = \frac{TP}{TP + FN} \tag{4}$$

$$F1 = 2 \times \frac{P \times R}{P + R} \tag{5}$$

$$J\text{-}Index = \frac{TP}{TP + FP + FN} \tag{6}$$

7. Results

In addressing the challenges of handwritten Arabic characters among children, a series of experiments were conducted and their outcomes are presented in this section. The results encompass both numerical evaluations and graphical representations. To offer a holistic understanding of the classifier's efficacy, the performance metrics are delineated for both training and testing phases, facilitating a nuanced assessment of its capabilities across familiar and novel scenarios.

7.1. Experiment One—Pre-Trained CNN Models

In Experiment One, where ResNet50, MobileNet, and DenseNet121 underwent fine-tuning on the Dhad and Hijja datasets, several intriguing training dynamics were observed (see Figures 4 and 5). The training accuracy curves consistently exhibited a positive exponential trajectory, reflecting the models' progressive refinement and learning. Concurrently, the training loss curves showcased a negative exponential pattern, suggesting a consistent reduction in training errors—both patterns emblematic of typical training behaviour. While a nuanced performance advantage was discerned in favor of ResNet50 from the training curves, this superiority was marginal. However, the validation phase painted a slightly different picture. Although the validation curves initially mirrored the training trajectories, a noticeable degradation in performance became evident after a certain epoch. This divergence, particularly conspicuous in the validation loss curves post-epoch 7, is a clear manifestation of overfitting—a phenomenon exacerbated by the datasets' inherent simplicity. Such insights underscore the importance of leveraging validation metrics as they provide a clearer lens into the model's generalization prowess. Intriguingly, when evaluating based on validation performance, DenseNet121 emerged marginally superior, although the performance disparities among the three models remained modest, positioning them comparably in terms of efficacy on these datasets.

A nuanced perspective on the models' training outcomes becomes apparent in the comparative analysis derived from Table 2. MobileNet emerged as the frontrunner for the Dhad dataset, boasting a validation loss of 0.2278 and a commendable accuracy of 0.9396. Conversely, for the Hijja dataset, DenseNet121 showcased its prowess with metrics of 0.4359 for validation loss and an accuracy score of 0.8920.

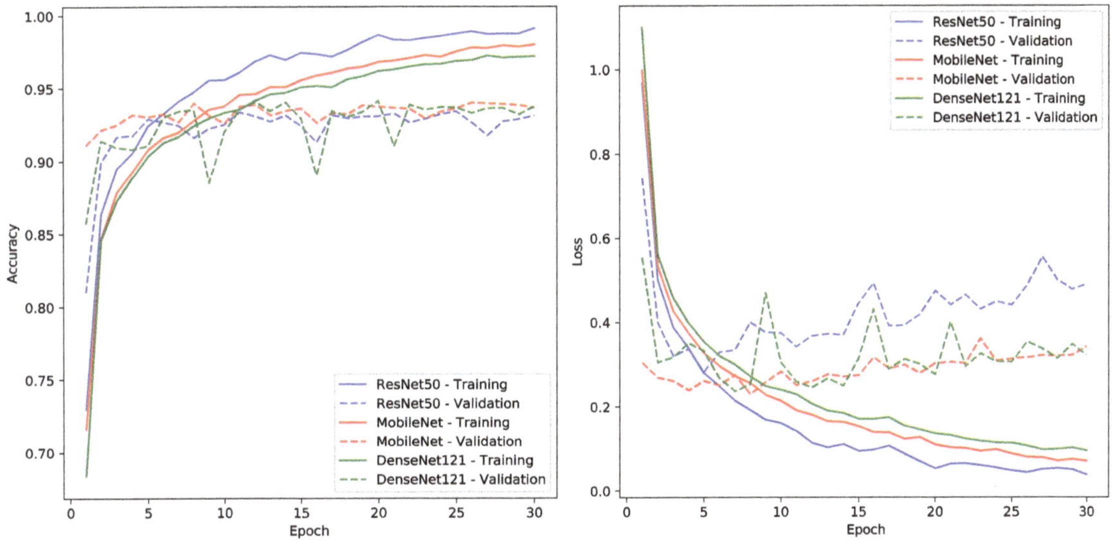

Figure 4. Training accuracy and loss curves for pre-trained models on Dhad dataset.

Delving deeper into the model performances, both MobileNet and DenseNet121 exhibited closely matched capabilities, with only marginal differences in their efficacy. In stark contrast, ResNet50's performance trajectory leaned more towards pruning, hinting at potential redundancy or inefficiencies in its architecture. This behaviour can be attributed to ResNet50's heavier design, which might have rendered it more susceptible to overfitting, especially given the datasets' inherent simplicity. In contrast, MobileNet's leaner architecture seemingly conferred upon it a more adaptive and resilient nature, enabling it to outperform its counterparts.

A comparative examination between the Dhad and Hijja datasets further illuminates this discussion. Predominantly, the pre-trained models showcased superior performance metrics on the Dhad dataset, underscoring its superior quality and efficacy in facilitating model training. Such observations align with the hypothesis positing Dhad's utilization of enhanced preprocessing methodologies, likely resulting in a cleaner, noise-attenuated dataset conducive for effective model learning. This superior data quality inherently empowered the models, enabling them to achieve heightened accuracies and reduced losses on the Dhad dataset compared to its Hijja counterpart.

The testing phase (see Table 3) further substantiated the models' capabilities, revealing outcomes that closely mirrored their validation performance. Such consistency underscores the models' adeptness at capturing generalized features, enabling them to maintain consistent performance across previously unseen datasets. Specifically, for the Dhad dataset, MobileNet continued to demonstrate its efficacy, registering a test accuracy of 0.9359, a test loss of 0.2468, and an impressive $F1$ score of 0.94. On the other hand, for the Hijja dataset, DenseNet121 emerged as the optimal performer, achieving a test accuracy of 0.8883, a test loss of 0.4919, and an $F1$-score of 0.89.

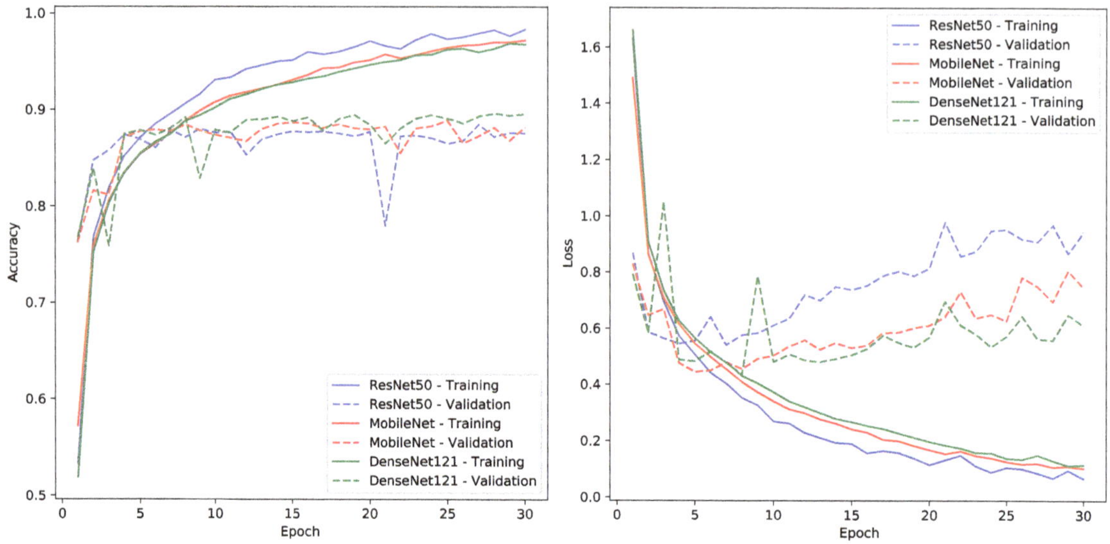

Figure 5. Training accuracy and loss curves for pre-trained models on Hijja dataset.

Table 2. Training performance of pre-trained models on Dhad and Hijja datasets.

	Training Loss	Training Accuracy	Validation Loss	Validation Accuracy
		Dhad Dataset		
MobileNet	0.0691	0.9796	0.2278	0.9396
DenseNet121	0.0941	0.9717	0.2357	0.9342
ResNet50	0.0371	0.9908	0.2810	0.9289
		Hijja Dataset		
MobileNet	0.1035	0.9722	0.4466	0.8774
DenseNet121	0.1154	0.9680	0.4359	0.8920
ResNet50	0.0674	0.9838	0.5419	0.8789

Figures 6 and 7 present the confusion matrices for the trained models on the Dhad and Hijja datasets, respectively. These matrices serve as pivotal tools for gauging the models' class-specific performances and identifying potential areas of misclassification.

Table 3. Test performance of pre-trained models on Dhad and Hijja datasets.

	Test Accuracy	Test Loss	*F*1 Score	Precision	Recall	J-Index
		Dhad Dataset				
MobileNet	0.9359	0.2468	0.94	0.94	0.94	0.88
DenseNet121	0.9306	0.2510	0.93	0.93	0.93	0.87
ResNet50	0.9228	0.3043	0.92	0.92	0.92	0.86
		Hijja Dataset				
MobileNet	0.8781	0.4677	0.88	0.88	0.88	0.78
DenseNet121	0.8883	0.4619	0.89	0.89	0.89	0.80
ResNet50	0.8705	0.6026	0.87	0.87	0.87	0.77

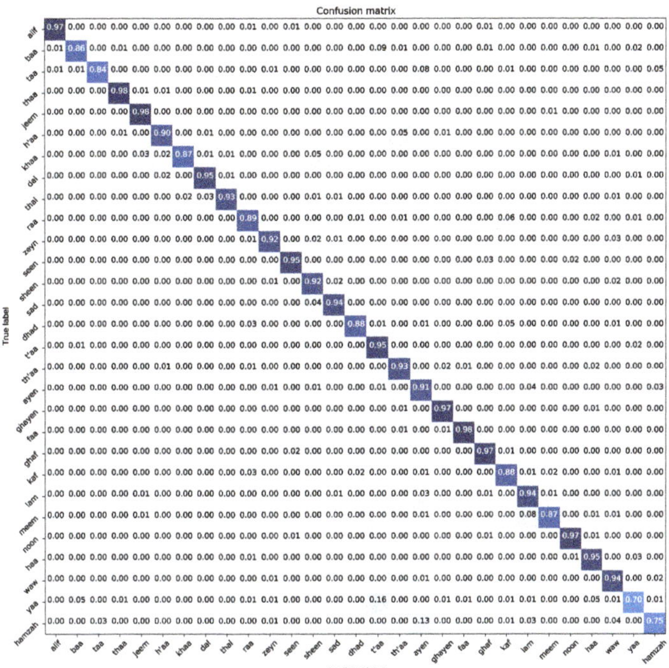

(**a**)

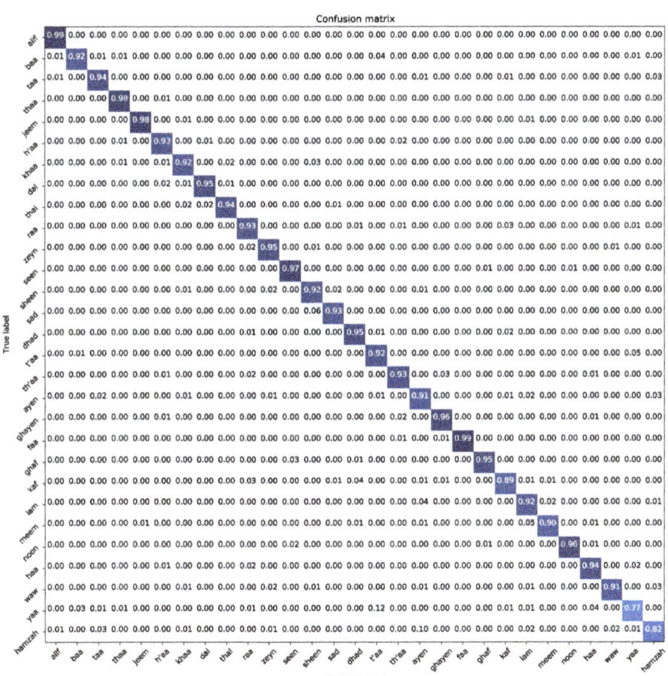

(**b**)

Figure 6. *Cont.*

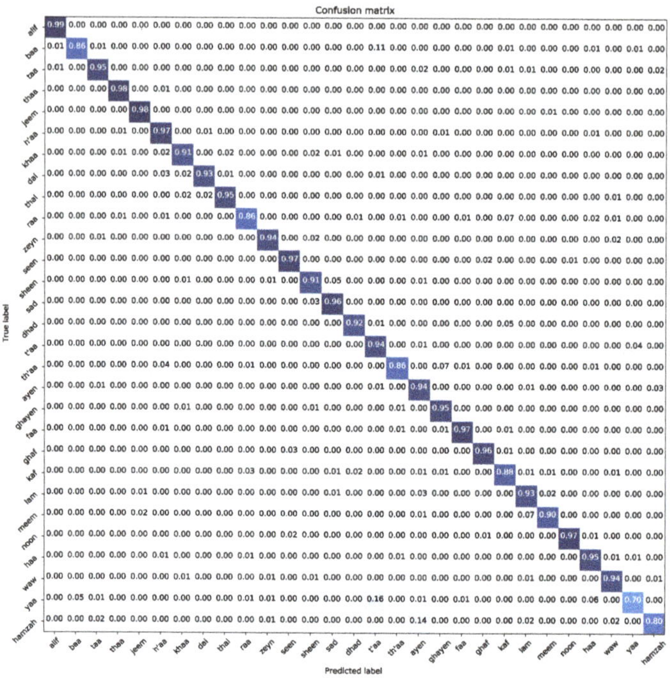

(c)

Figure 6. Confusion matrix for pre-trained models on Dhad dataset. (**a**) ResNet50, (**b**) MobileNet, and (**c**) DenseNet121.

In the context of the Dhad dataset, a detailed examination reveals MobileNet's commendable class-wise equilibrium, characterized by minimal misclassifications across various categories. Notably, there's a discernible pattern of misclassification, where 12% of the "yaa" samples are erroneously categorized as "t'aa" and 10% of the "hamzah" samples are mislabeled as "ayen". Such misclassifications likely stem from the intricate visual similarities inherent to these characters, underscoring the inherent challenges of handwritten character recognition tasks.

Turning our attention to the Hijja dataset, DenseNet121 emerges as the model with the most consistent overall performance. However, a deeper dive into the confusion matrix reveals a higher incidence of misclassifications. Two salient observations include the misclassification of 10% of "lam" samples as "ayen" and 7% of "hamzah" instances being inaccurately labeled as "waw". Such misclassifications further emphasize the intricacies and challenges posed by handwritten Arabic character recognition, necessitating continuous refinement and optimization strategies for enhanced accuracy.

Figure 8 provides an insightful glimpse into the inner workings of the trained models through layer visualizations, shedding light on their training efficacy and decision-making processes. The two visualization techniques employed, Grad-CAM and SmoothGrad, serve distinct purposes in elucidating model behaviour. While Grad-CAM accentuates the pivotal regions within images that significantly influenced predictions, SmoothGrad offers a more granular perspective by pinpointing the specific pixels most instrumental in the decision-making process.

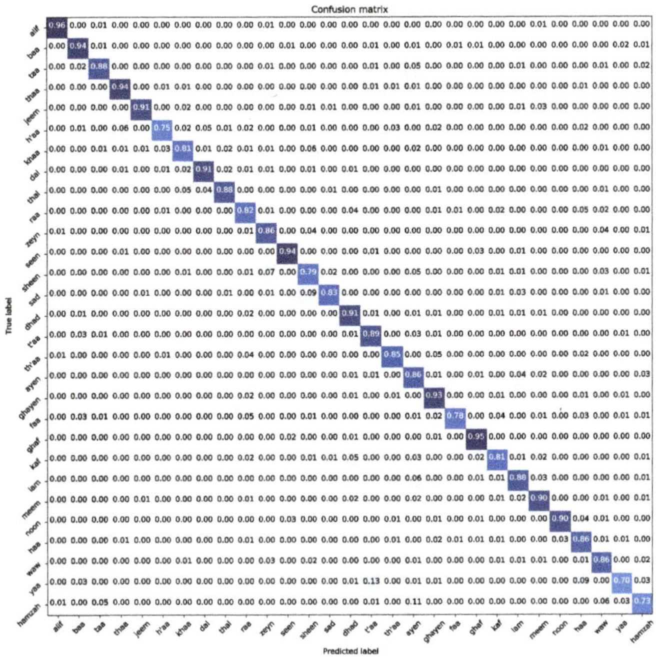

(**a**)

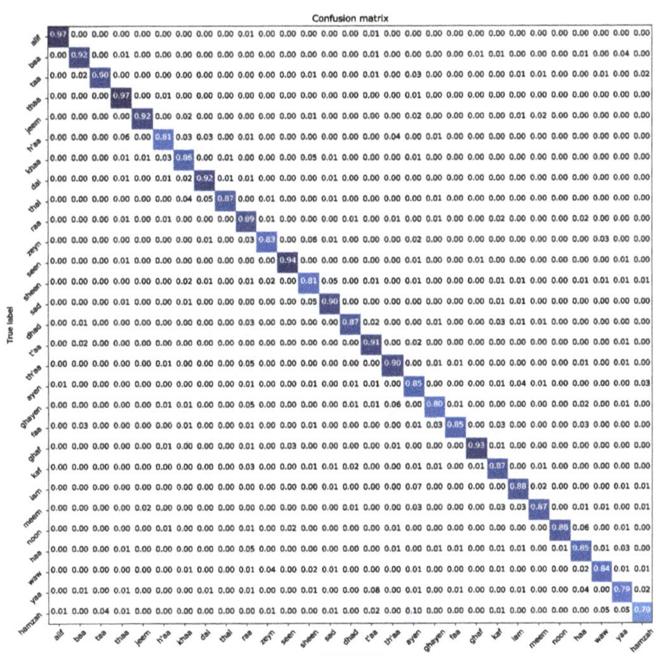

(**b**)

Figure 7. *Cont.*

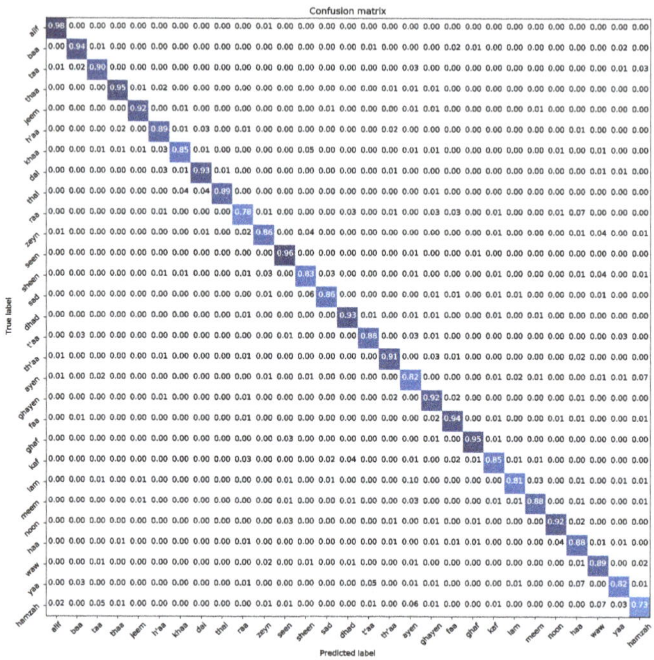

(c)

Figure 7. Confusion matrix for pre-trained models on Hijja dataset. (**a**) ResNet50, (**b**) MobileNet, (**c**) DenseNet121.

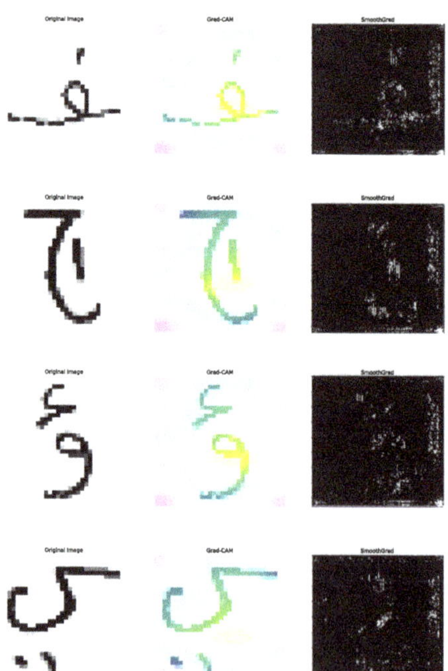

Figure 8. Layer visualizations for DenseNet121 model on Dhad dataset samples.

Upon meticulous examination of the visualizations, certain patterns and discrepancies come to the fore. Notably, for characters such as "faa" and "jeem", the models appear adept at capturing and leveraging the character-relevant pixels, indicative of robust training and feature extraction capabilities. However, a discernible shortfall becomes evident in the case of the "yaa" character. Here, the model seemingly overlooks or inadequately emphasizes crucial pixels during the prediction phase, suggesting potential areas for model refinement or additional training data augmentation to enhance accuracy and consistency.

Figures 9 and 10 present the AUC curves, offering a comprehensive overview of the discriminatory power and overall performance of the pre-trained models on the Dhad and Hijja datasets, respectively. AUC serves as a robust metric, encapsulating the model's ability to distinguish between different classes. Upon detailed examination of these curves, a pattern of closely matched performances across models emerges. Specifically, for the Dhad dataset, MobileNet slightly outperforms its counterparts, boasting an impressive AUC value of 0.9986. Conversely, on the Hijja dataset, DenseNet121 delivers a commendable performance, albeit marginally trailing behind MobileNet with an AUC of 0.9954.

From the experiments, it can be clearly observed that the models exhibited overfitting during training for both the Dhad and Hijja datasets for almost all the implemented models. Although the superior performance of the pre-trained models was recorded, it is important to take into consideration the overfitting problem. In general, this problem usually occurs when either the dataset is too small in comparison to the model complexity or the dataset is way too simple for the model. The literature suggests that dropout and data augmentation techniques can be used to overcome the overfitting problem. To further investigate this, in this experiment, we scoped the problem for only the MobileNet model on the Dhad dataset as a use case. We have tried different dropout ratios to observe the performance. Furthermore, we have also used data augmentation with the dropout. To be specific, we trained the model using the 0.2, 0.4, and 0.6 dropout values. In terms of data augmentation, we used rotation, width shift, height shift, shear, zoom, and nearest fill. Figures 11 and 12 show the trends for training and validation loss curves for both cases to understand. First, talking about the dropout variations, it can be observed from Figure 11 that a dropout percentage of 0.4 resulted in slightly better performance and a stable validation loss curve, whereas the dropout of 0.2 and 0.6 percentages degraded the performance. This suggests that not all the dropout percentages result in better performance; rather, an optimal value needs to be identified. It can be concluded that as suggested by the literature, dropout can be introduced to improve overfitting. In regard to the data augmentation and dropout variations, it can be observed from Figure 12 that the introduction of data augmentation did improve the overall training performance and resulted in emergence at lower loss values, but it did not really address the overfitting problem. However, when data augmentation was used with optimal dropout values i.e., 0.4, it resulted in a stable and improved validation loss curve. As a summary of this investigation, it can be concluded that overfitting is very common for smaller and simpler datasets. Dropout and data augmentation approaches can be used to improve overfitting to some extent; however, on a larger scale, the dataset needs to be introduced with noise and challenges to avoid this problem.

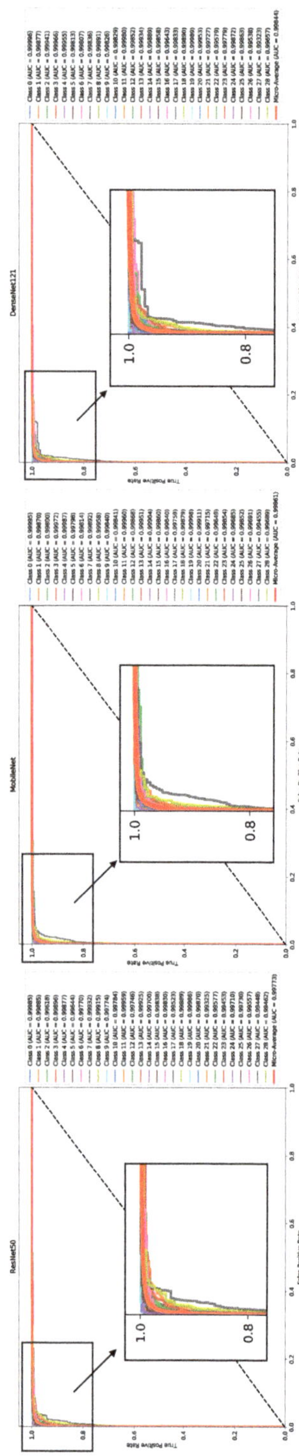

Figure 9. AUC curves for pre-trained models on Dhad dataset.

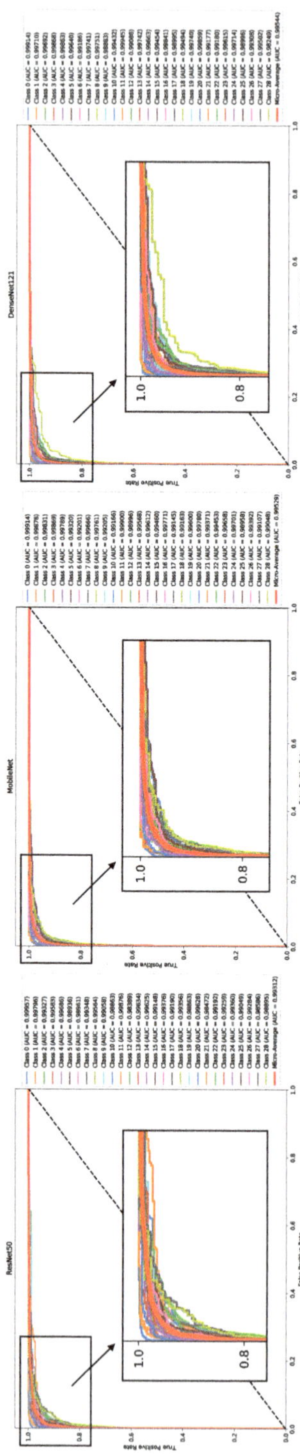

Figure 10. AUC curves for pre-trained models on Hijja dataset.

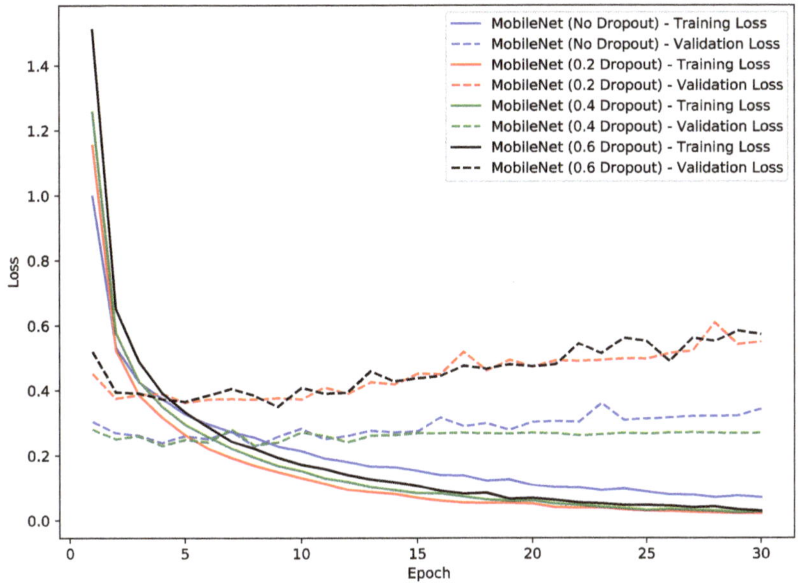

Figure 11. Loss curves for different dropout variations with MobileNet on Dhad dataset.

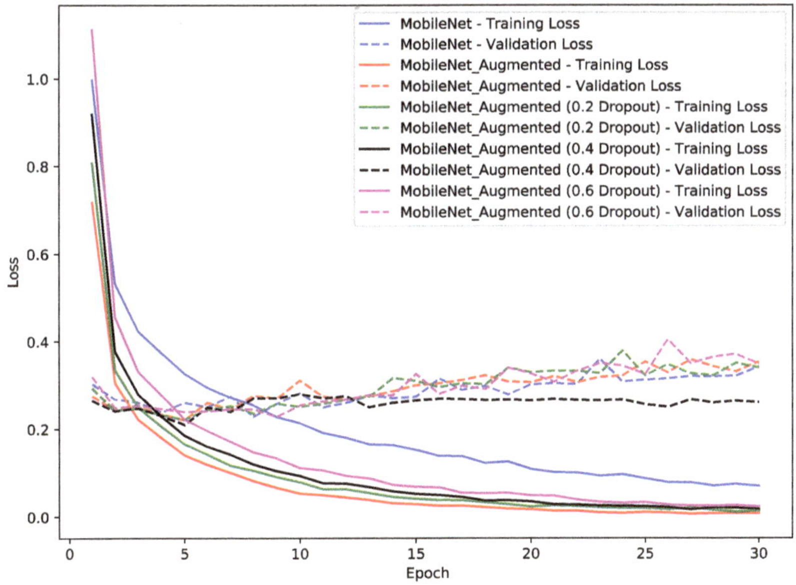

Figure 12. Loss curves for different dropout variations and data augmentation with MobileNet on Dhad dataset.

7.2. Experiment Two—Custom CNN Model

In Experiment Two, a custom CNN model, drawing inspiration from the existing literature [4,5,13–15], was meticulously crafted and subsequently trained on both the Dhad and Hijja datasets. A detailed analysis of the model's training dynamics, as depicted in Figure 13, offers invaluable insights into its performance and adaptability.

Upon scrutinizing the training accuracy plots, a discernible positive exponential trend is evident, aligning with typical training behaviour. However, an intriguing observation is the model's accelerated convergence, achieving desirable accuracy at a slightly quicker pace compared to other models. Nevertheless, the validation phase unraveled some concerns. While the validation accuracy initially mirrored the training trajectory, a conspicuous divergence emerged after the 20th epoch, signaling the onset of overfitting.

This overfitting propensity is further accentuated in the loss curves. After the 20th epoch, a palpable uptick in validation loss becomes evident, corroborating the overfitting suspicions. Such concerns are further compounded upon examining the convergence metrics; as detailed in Table 4, the model's loss values upon convergence are unexpectedly elevated for both the Dhad and Hijja datasets. Specifically, for the Dhad dataset, the model managed to attain a validation accuracy of 89% but manifested a relatively elevated validation loss of 0.3862. Conversely, the Hijja dataset witnessed a more pronounced performance disparity, with the model registering a diminished accuracy of 75% accompanied by a markedly higher validation loss of 0.8382.

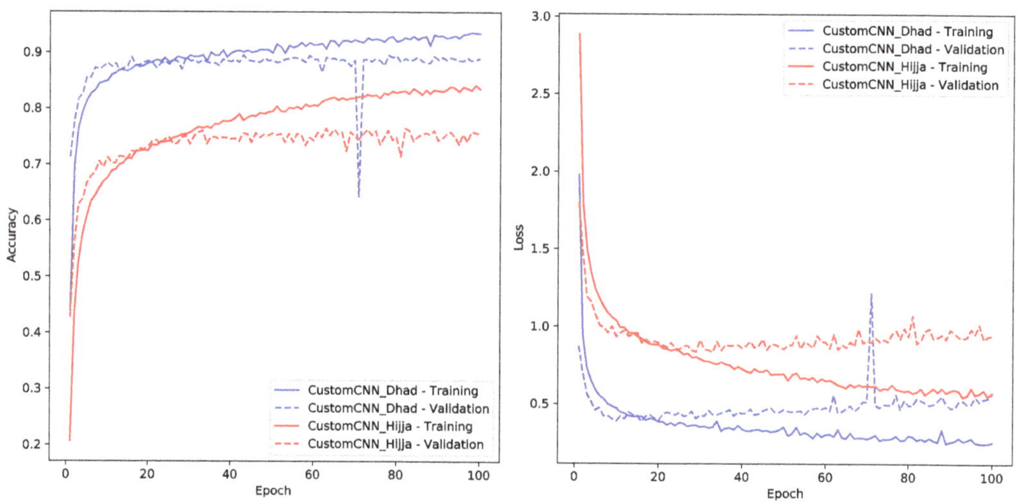

Figure 13. Training accuracy and loss curves for custom CNN model on Dhad and Hijja datasets.

Table 4. Training performance of custom CNN model on Dhad and Hijja datasets.

	Training Loss	Training Accuracy	Validation Loss	Validation Accuracy
Dhad Dataset	0.2554	0.9344	0.3862	0.8919
Hijja Dataset	0.5761	0.8354	0.8382	0.7515

Table 5 provides a comprehensive overview of the custom model's test performance metrics on both the Dhad and Hijja datasets. A cursory examination of these results reveals a coherent alignment with the model's validation trajectory, reaffirming the standard train–validate–test paradigm, where the performances across validation and test phases remain largely congruent.

For the Dhad dataset, the custom model demonstrated a commendable test accuracy of 88%, coupled with a test loss metric of 0.3988. Additionally, the model's $F1$ score stood impressively at 0.89, underscoring its proficiency in maintaining a harmonious balance between precision and recall. Conversely, when evaluated on the Hijja dataset, the model's performance exhibited a discernible decline, registering a test accuracy of 74%. The associated test loss and $F1$ score metrics further elucidate this observation, standing at 0.8693 and 0.75, respectively.

Table 5. Test performance of custom CNN model for Dhad and Hijja datasets.

	Test Accuracy	Test Loss	*F*1 Score	Precision	Recall	J-Index
Dhad Dataset	0.8854	0.3988	0.89	0.89	0.89	0.79
Hijja Dataset	0.7484	0.8693	0.75	0.75	0.75	0.6

The nuanced performance trajectories across the two datasets are further corroborated by the AUC curves, meticulously depicted in Figure 14. The Dhad dataset witnessed a marginally superior performance, with the model achieving an AUC of 0.99, indicative of its robust discriminatory prowess. In stark contrast, the Hijja dataset, although exhibiting a commendable AUC value of 0.98, revealed a more scattered performance distribution across classes, emphasizing the inherent challenges and intricacies associated with character recognition tasks on this dataset.

Figure 15 provides a comprehensive confusion matrix for both the Dhad and Hijja datasets. While the model's performance on the Dhad dataset appears balanced with minimal misclassifications, a discernible decline is evident on the Hijja dataset, characterized by widespread misclassifications across various classes. Particularly challenging are the class pairs "t'aa–yaa" and "ayen–hamza", likely due to their visual resemblance, underscoring the inherent complexities in Arabic character recognition and highlighting areas for potential model enhancement.

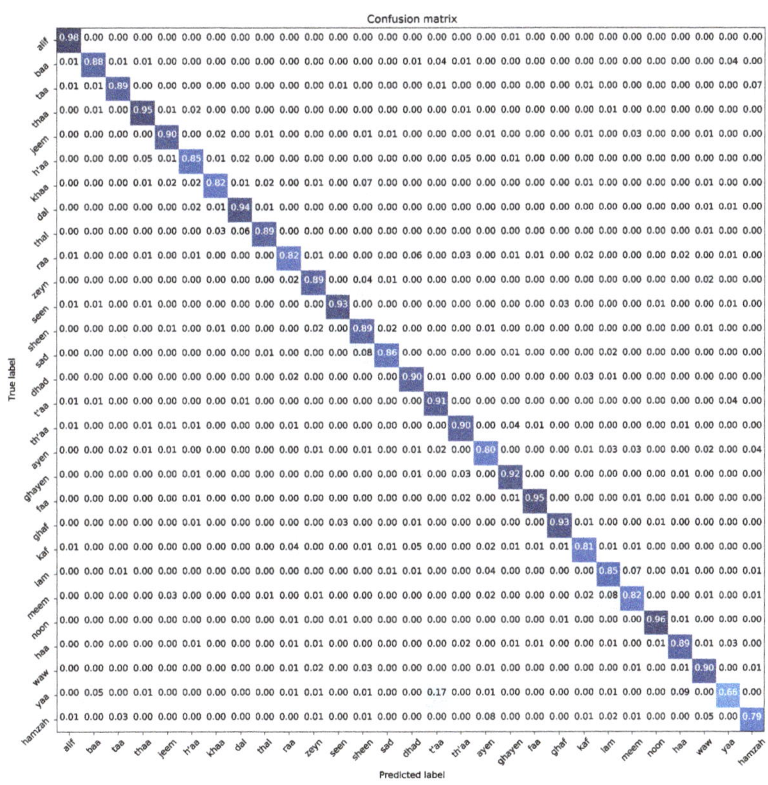

(a)

Figure 14. *Cont.*

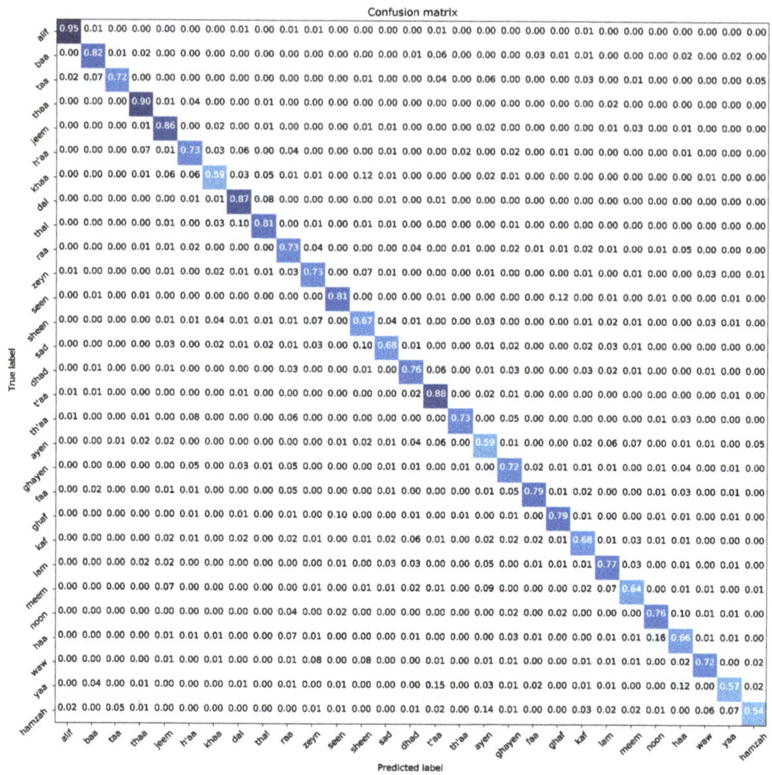

(**b**)

Figure 14. Confusion matrix for custom CNN model on Dhad and Hijja datasets. (**a**) Dhad and (**b**) Hijja.

7.3. Experiment Three—Classification of Deep Visual Features

In Experiment Three, a sophisticated two-stage approach was devised to optimize the classification process, blending the strengths of deep learning feature extraction with the precision of traditional classifiers. The foundational component of this pipeline was the MobileNet architecture, renowned for its prowess in extracting intricate features from complex datasets. By utilizing MobileNet's capabilities, the experiment aimed to transform the raw data into a more discernible and compact representation, thereby facilitating more effective subsequent classification. In this context, we have used the MobileNet model pre-trained over the ImageNet and MobileNet models trained in Experiment One.

Following the feature extraction phase, the extracted features were then subjected to three distinct conventional classifiers: SVM, RF, and MLP. SVM, a discriminative classifier, operates by finding the optimal hyperplane that best separates the data into distinct classes, making it particularly adept at handling high-dimensional feature spaces. Conversely, RF, an ensemble learning method, constructs multiple decision trees during training and outputs the class that is the mode of the classes of individual trees for classification tasks, thereby leveraging the wisdom of multiple trees to enhance accuracy and robustness. On the other hand, MLP is known for its fully connected neural architecture to extract the hidden patterns from the input feature vector.

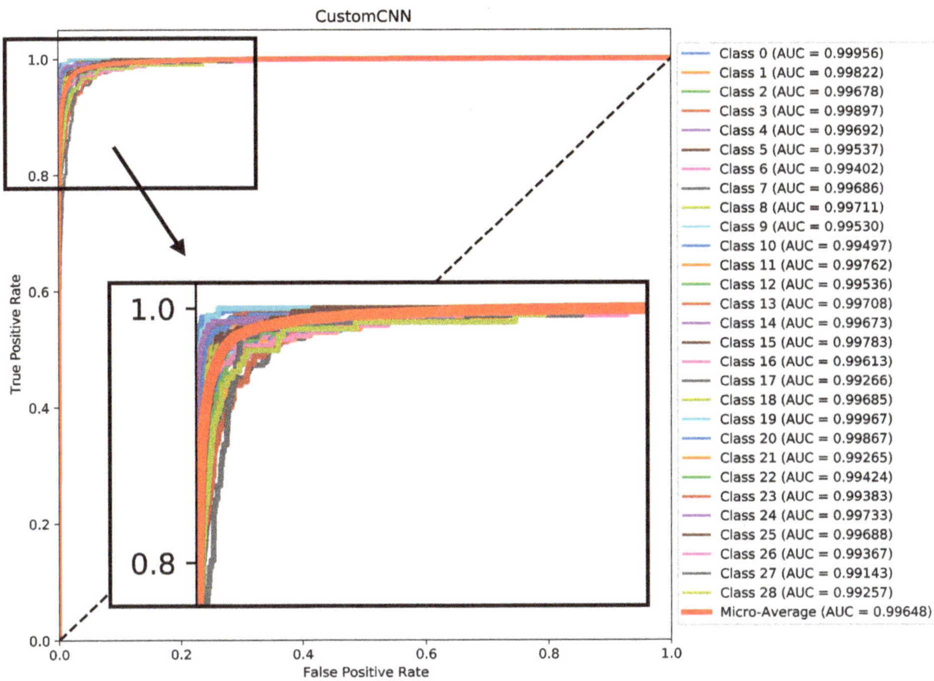

(**a**)

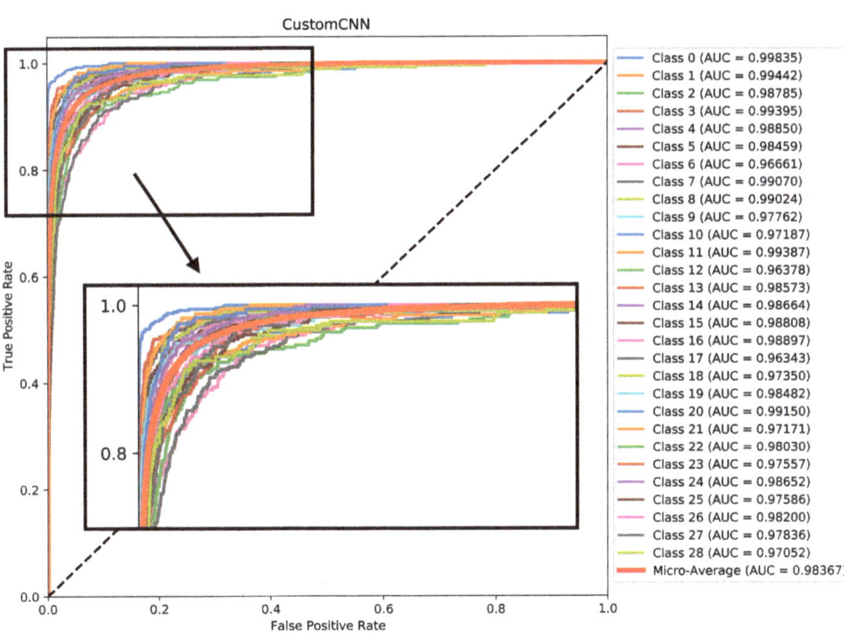

(**b**)

Figure 15. AUC curves of custom CNN model for Dhad and Hijja datasets. (**a**) Dhad and (**b**) Hijja.

Delving into the results encapsulated in Table 6, a discernible pattern emerges. For the Dhad dataset, the MobileNet + SVM ensemble manifested as the optimal configuration, demonstrating its prowess with a validation accuracy of 89%, which was corroborated by the test accuracy standing at a commendable 88%. Further fortifying its performance credentials, the ensemble yielded an $F1$ score of 0.88, underscoring its balanced precision and recall capabilities. Similarly, when transposed to the Hijja dataset, the MobileNet+SVM configuration continued its dominance, albeit with slightly diminished metrics. A validation accuracy of 73% and a corresponding test accuracy of 72% were achieved, along with an $F1$ score of 0.73, signifying a robust performance despite the dataset's inherent complexities. In context to the use of ImageNet pre-trained and Experiment One trained model, it can be observed that the ImageNet pre-trained model resulted in better performance.

Table 6. Test performance of two-stage classification on deep visual features pipeline on Dhad and Hijja datasets.

	Validation Accuracy	Test Accuracy	$F1$ Score	Precision	Recall	J-Index
		Dhad Dataset				
MobileNet (ImageNet Pre-Trained) + SVM	0.8894	0.8848	0.88	0.88	0.88	0.79
MobileNet (ImageNet Pre-Trained) + RF	0.7775	0.7803	0.78	0.78	0.78	0.64
MobileNet (ImageNet Pre-Trained) + MLP	0.0801	0.0810	0.02	0.02	0.02	0.05
MobileNet (Experiment One) + SVM	0.7118	0.7100	0.71	0.71	0.71	0.71
MobileNet (Experiment One) + RF	0.6662	0.6656	0.66	0.65	0.66	0.50
MobileNet (Experiment One) + MLP	0.0556	0.0556	0.01	0.01	0.01	0.03
		Hijja Dataset				
MobileNet (ImageNet Pre-Trained) + SVM	0.7322	0.7256	0.73	0.73	0.73	0.57
MobileNet (ImageNet Pre-Trained) + RF	0.5346	0.5270	0.53	0.52	0.53	0.36
MobileNet (ImageNet Pre-Trained) + MLP	0.0770	0.0771	0.04	0.04	0.04	0.06
MobileNet (Experiment One) + SVM	0.3937	0.3839	0.37	0.38	0.37	0.24
MobileNet (Experiment One) + RF	0.3705	0.3591	0.34	0.35	0.34	0.22
MobileNet (Experiment One) + MLP	0.0578	0.0570	0.01	0.01	0.03	0.03

Figures 16–21 provide a detailed representation of the confusion matrices derived from the SVM, RF, and MLP classifiers when employed with deep visual features from the ImageNet pre-trained and Experiment One trained MobileNet model on both the Dhad and Hijja datasets. Complementing these visual representations, the findings elucidated in the table corroborate the classifiers' performance metrics. Notably, SVM emerges as the superior performer across all the cases.

However, when contextualized within the datasets, a nuanced observation surfaces. The Dhad dataset consistently showcases enhanced performance metrics in comparison to its Hijja counterpart. This disparity in performance underscores the Dhad dataset's superior quality, likely attributed to meticulous data curation, reduced noise levels, or other preprocessing enhancements. Such insights are pivotal, as they not only validate the efficacy of the classification pipeline but also emphasize the pivotal role of dataset quality in influencing model performance and outcomes.

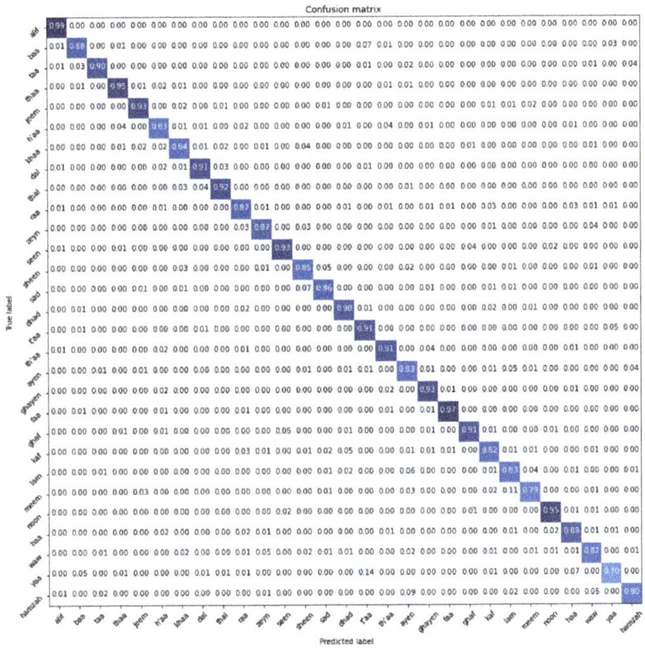

(**a**)

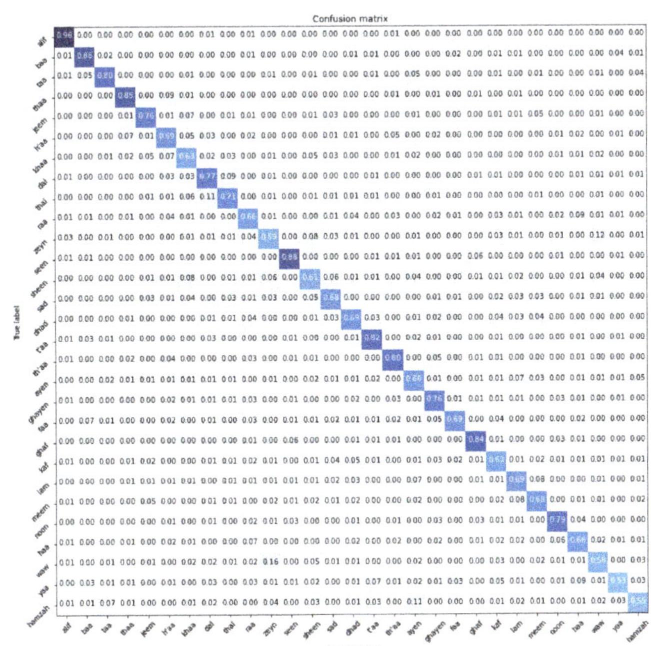

(**b**)

Figure 16. Confusion matrix of MobileNet (ImageNet pre-trained) + SVM pipeline on Dhad and Hijja datasets. (**a**) Dhad and (**b**) Hijja.

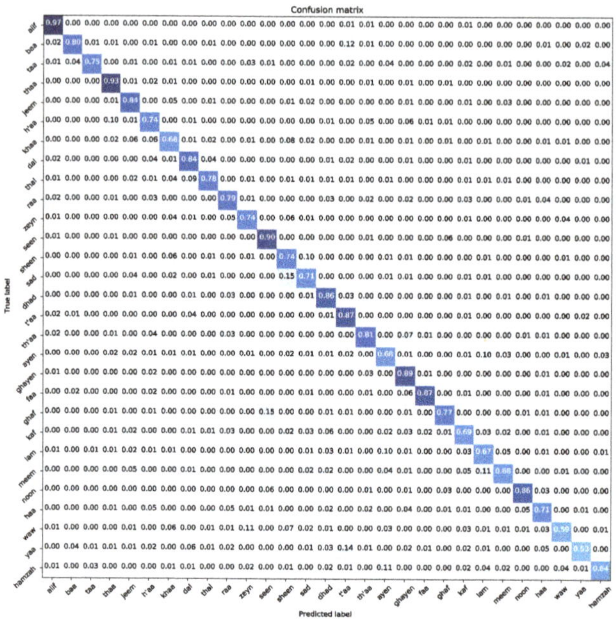

(**a**)

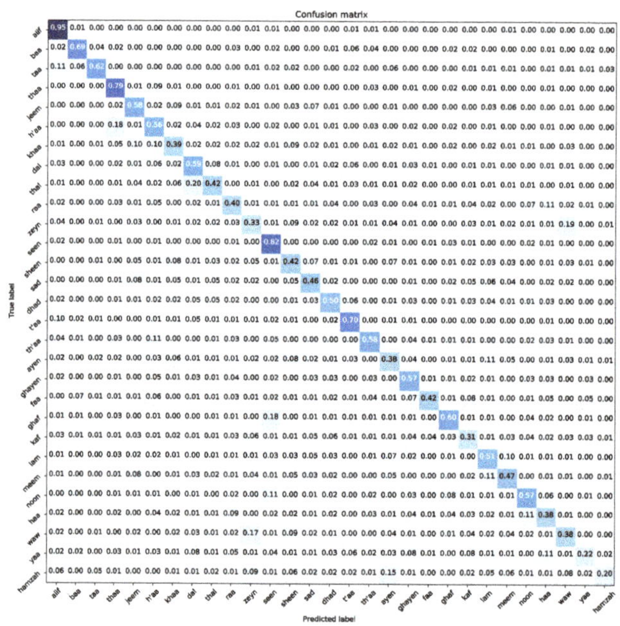

(**b**)

Figure 17. Confusion matrix of MobileNet (ImageNet pre-trained) + RF pipeline on Dhad and Hijja datasets. (**a**) Dhad and (**b**) Hijja.

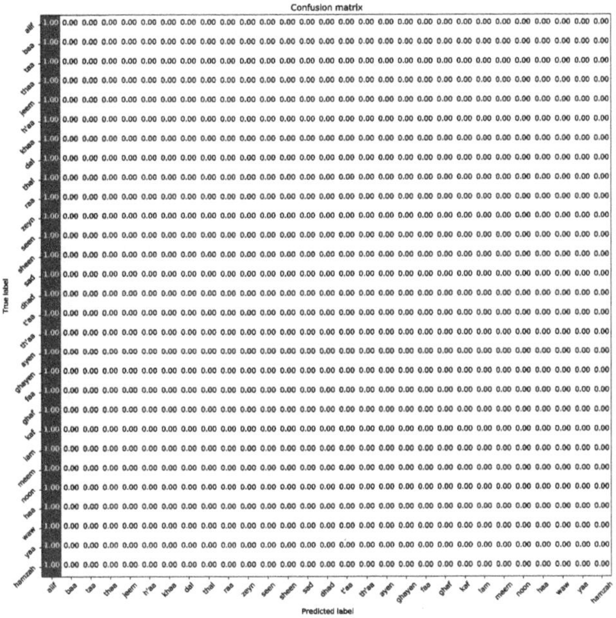

(a)

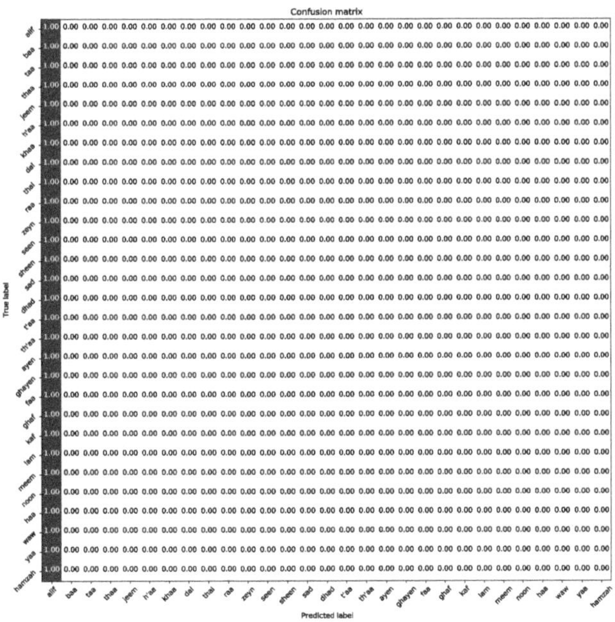

(b)

Figure 18. Confusion matrix of MobileNet (ImageNet pre-trained) + MLP pipeline on Dhad and Hijja datasets. (**a**) Dhad and (**b**) Hijja.

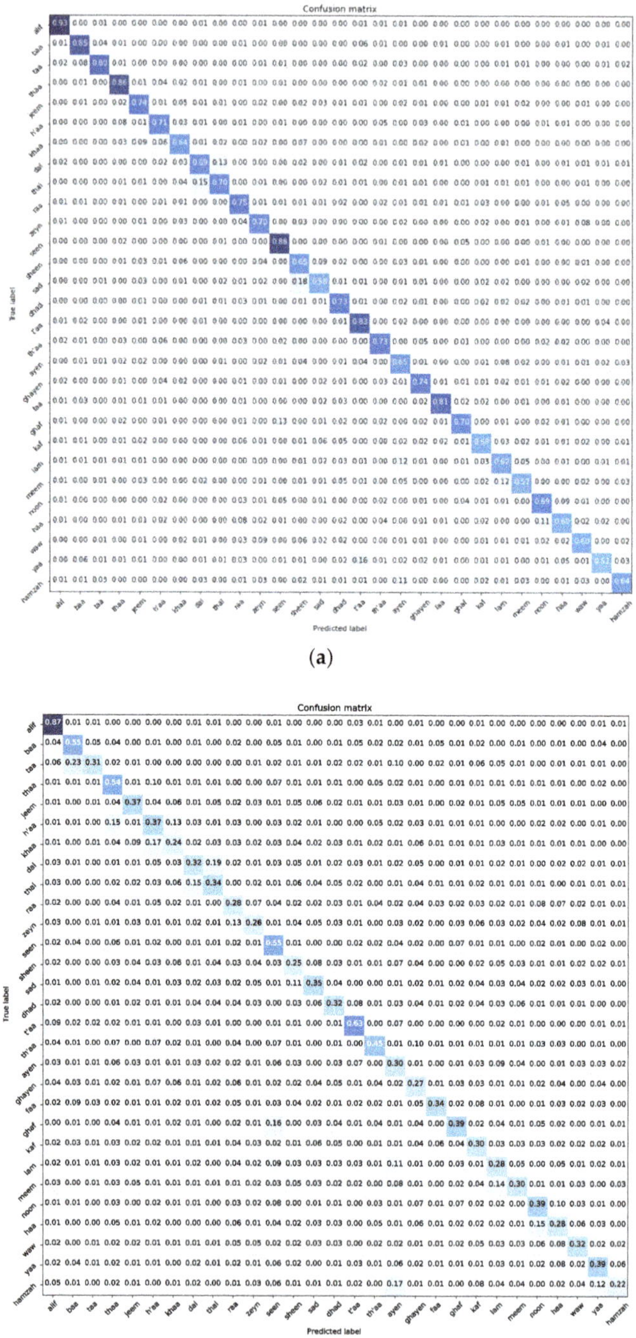

(**a**)

(**b**)

Figure 19. Confusion matrix of MobileNet (Experiment One) + SVM pipeline on Dhad and Hijja datasets. (**a**) Dhad and (**b**) Hijja.

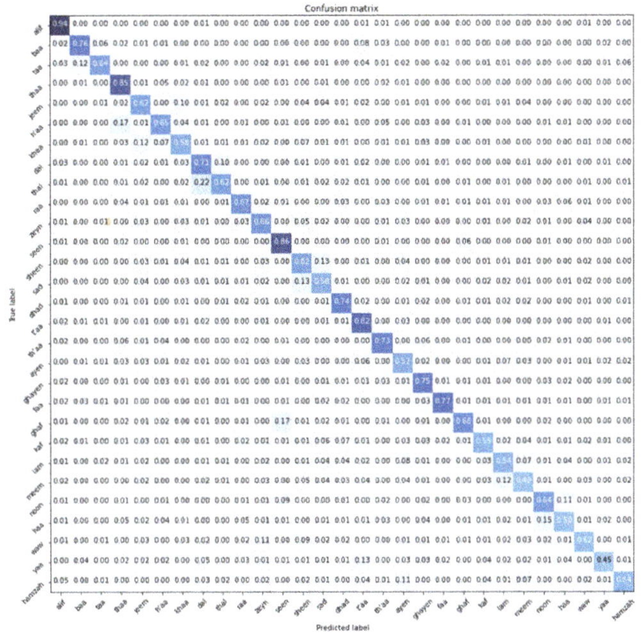

(**a**)

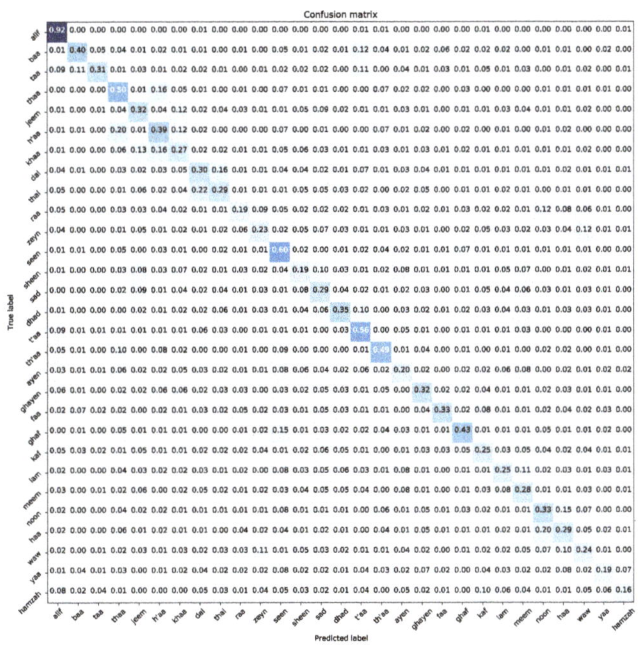

(**b**)

Figure 20. Confusion matrix of MobileNet (Experiment One) + RF pipeline on Dhad and Hijja datasets. (**a**) Dhad and (**b**) Hijja.

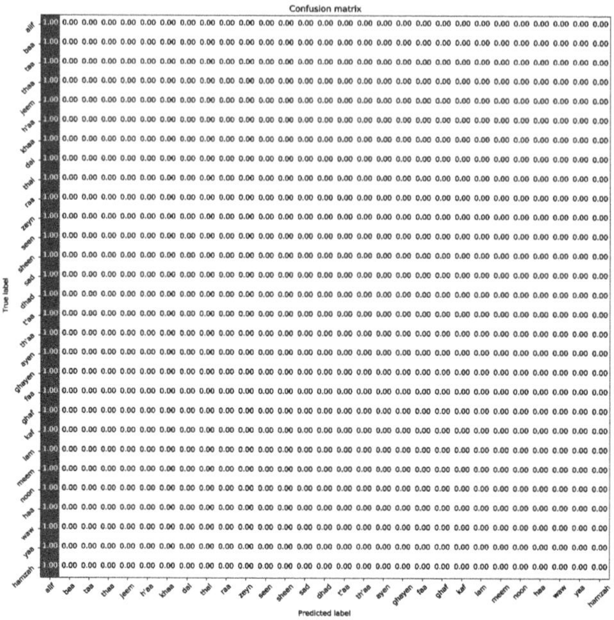

(**a**)

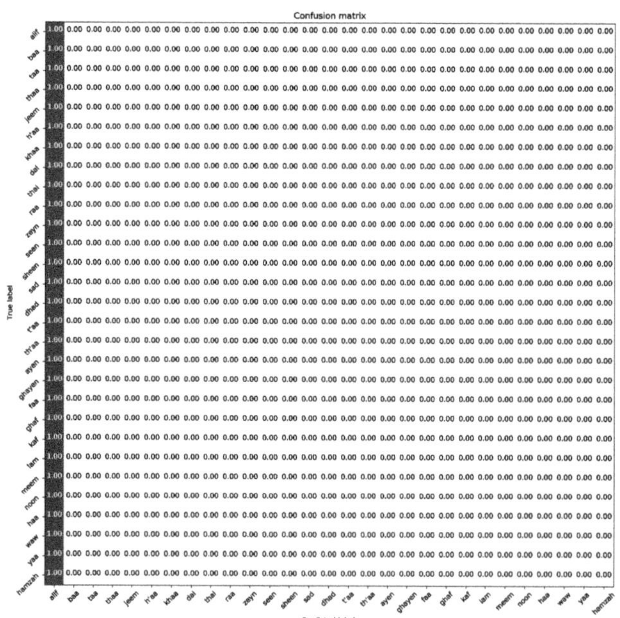

(**b**)

Figure 21. Confusion matrix of MobileNet (Experiment One) + MLP pipeline on Dhad and Hijja datasets. (**a**) Dhad and (**b**) Hijja.

8. Discussion

The exploration of handwritten Arabic character recognition, particularly among children, presents both challenges and opportunities. The results obtained from the experiments provided valuable insights into the effectiveness of deep learning models and traditional classifiers for this specific task. In this discussion, we delve deeper into the insights derived from the results, critically analysing them against the existing literature, evaluating the methodologies employed and highlighting potential avenues for future research.

- Fine-tuning of Existing Models: An important insight from the performed experiments is the unparalleled efficacy of fine-tuning existing deep learning architectures. This approach resonates with the existing literature [29–31], highlighting the potential of harnessing pre-trained models fine-tuned for application-specific tasks. The flexibility of fine-tuning, which combines using general features with adjusting to specific dataset details, highlights its essential importance. Especially in situations with limited computing power, its ability to achieve impressive results quickly becomes clearly noticeable.
- Custom CNN Model: In comparison to the established deep architectures, our simpler custom CNN achieved good results in classification. However, the performance did not exceed the fine-tuned pre-trained models. These findings align with what is commonly discussed in current research [32,33], emphasizing that simpler models can be easily affected by minor changes. This highlights the need for cautious interpretations and emphasizes the extra computing work needed when starting from scratch with new models.
- Two-stage Pipeline with Conventional Classifiers: Our exploration of a two-part process, combining deep visual feature extraction with traditional classification methods, resulted in less-than-ideal results. These results are consistent with existing research, highlighting the importance of end-to-end deep learning models trained effectively at once. The shortcomings arising from separate feature extraction and classification emphasize the need for unified model training, bringing together all elements to better achieve the main goal.
- Dataset Dynamics: At the heart of the model's performance variations lies the quality of the dataset. Our studies highlight the superiority of the Dhad dataset when compared to the Hijja one, likely due to clearer pixels and reduced interference. These insights highlight the crucial importance of careful dataset preparation, underscoring its fundamental role in shaping the best possible model results.
- Navigating Class Confounders: A recurring pattern throughout our experimental journey centres on the differentiation between certain class pairs, particularly "t'aa–yaa" and "ayen–hamzah". The blending of visual similarities among these classes leads to frequent misclassifications, highlighting the need for future efforts to develop more detailed training samples. Tackling this challenge requires a focused effort to enhance the dataset with diverse class examples, enhancing the model's ability to accurately distinguish categories.

While the experiments offer valuable insights, they are not devoid of limitations. The use of a limited number of datasets, potential biases in data curation, and the absence of real-world noise simulations may limit the external validity of the findings. Furthermore, the focus on specific architectures and classifiers suggests a comprehensive exploration of the deep learning and traditional machine learning models. Potential future research directions can be as follows:

1. Enhanced dataset curation, incorporating diverse writing styles, variations, and real-world noise simulations.
2. Comparative evaluations encompassing a broader spectrum of architectures, optimization techniques, and data augmentation strategies.
3. Exploration of ensemble methodologies, blending the strengths of multiple models to foster enhanced recognition capabilities.

9. Al-Khatta—An Early Intervention Tool for Arabic Handwriting Improvement

In envisioning a future application, a model specifically trained to classify Arabic characters holds immense potential for a highly impactful use-case "Al-Khatta" for the enhancement of Arabic handwriting skills in children aged 7 to 12. This software application can seamlessly integrate the trained model with the aim of revolutionizing handwriting improvement through innovative features. The model's real-time analysis capabilities will enable the application to deliver immediate feedback on handwritten input, fostering a dynamic and responsive learning environment. The trained model will play a pivotal role in identifying areas of difficulty within specific characters, empowering the app to generate personalized practice exercises tailored to each child's unique handwriting challenges. This forward-looking approach ensures a targeted and individualized learning experience, effectively addressing the diverse needs of young learners and fostering accelerated proficiency in Arabic handwriting.

Moreover, the application can incorporate a progress-tracking functionality, providing insightful data on a child's development across various exercises and over time. This feature will empower educators and parents with a comprehensive understanding of learning patterns, facilitating informed and targeted guidance to further support the child's progress. To maintain engagement in this envisioned future, the application can employ gamified elements and rewards, contributing to a positive reinforcement learning experience. By infusing an element of enjoyment into the learning process, the application aims to keep children motivated and enthusiastic about refining their Arabic handwriting skills.

The UI/UX development of the application can utilize HTML, CSS, and JavaScript for web-based applications or consider platform-specific frameworks such as Flutter for cross-platform mobile applications. In the realm of model development, PyTorch, TensorFlow, and Python libraries can be harnessed, with a dedicated GPU machine ensuring efficient training. For swift real-time performance in mobile deployment, models like MobileNet can be employed, while larger models like DenseNet121 may be considered for potential offline analysis.

10. Conclusions

In conclusion, our comprehensive exploration into the classification of handwritten Arabic characters among children reveals intriguing dynamics in model performance and dataset efficacy. While fine-tuned pre-existing models showcased commendable accuracy, particularly MobileNet on the Dhad dataset and DenseNet121 on the Hijja dataset, their performance trajectories underscored the challenges of overfitting, especially with datasets of inherent simplicity. The nuances observed in misclassifications, notably between visually similar characters, highlight the intricacies inherent to Arabic character recognition. A concept of computer application to facilitate the handwriting improvement in children is also discussed as a practical use-case of Arabic children's handwritten character recognition. Moving forward, addressing these challenges will demands multi-pronged approach: refining dataset quality, exploring advanced model architectures, and integrating robust training strategies to enhance generalization and accuracy.

Author Contributions: Conceptualization, N.A.; methodology, S.A. and N.A.; software, A.D.A., D.A., and R.A.; validation, H.A. and A.M.A.; data curation, D.A., R.A., H.A. and A.M.A.; writing—original draft preparation, S.A.; writing—review and editing, N.A.; visualization, A.D.A. All authors have read and agreed to the published version of the manuscript.

Funding: This work was funded by King Saud University, Riyadh, Saudi Arabia, through Researchers Supporting Project number (RSPD2024R857).

Institutional Review Board Statement: This study was approved by the "Standing Committee on Ethics of Scientific Research" at King Saud University (No. KSUKSU-HE-19-363).

Data Availability Statement: The datasets generated and/or analysed during the current study are available in the Github repository, https://github.com/daadturki1/Dhad (accessed on 1 October 2023).

Acknowledgments: The authors thank the anonymous reviewers for their constructive comments.

Conflicts of Interest: The authors declare no competing interests.

References

1. Eberhard, D.M.; Simons, G.F.; Fennig, C.D. *Ethnologue: Languages of the World*; SIL International: Dallas, TX, USA 2023.
2. Nahar, K.M.; Alsmadi, I.; Al Mamlook, R.E.; Nasayreh, A.; Gharaibeh, H.; Almuflih, A.S.; Alasim, F. Recognition of Arabic Air-Written Letters: Machine Learning, Convolutional Neural Networks, and Optical Character Recognition (OCR) Techniques. *Sensors* **2023**, *23*, 9475. [CrossRef]
3. Kasem, M.S.; Mahmoud, M.; Kang, H.S. Advancements and Challenges in Arabic Optical Character Recognition: A Comprehensive Survey. *arXiv* **2023**, arXiv:2312.11812.
4. Altwaijry, N.; Al-Turaiki, I. Arabic handwriting recognition system using convolutional neural network. *Neural Comput. Appl.* **2021**, *33*, 2249–2261. [CrossRef]
5. Alwagdani, M.S.; Jaha, E.S. Deep Learning-Based Child Handwritten Arabic Character Recognition and Handwriting Discrimination. *Sensors* **2023**, *23*, 6774. [CrossRef]
6. Zakraoui, J.; Saleh, M.; Al-Maadeed, S.; AlJa'am, J.M. A study of children emotion and their performance while handwriting Arabic characters using a haptic device. *Educ. Inf. Technol.* **2023**, *28*, 1783–1808. [CrossRef]
7. El-Sawy, A.; Loey, M.; El-Bakry, H. Arabic handwritten characters recognition using convolutional neural network. *WSEAS Trans. Comput. Res.* **2017**, *5*, 11–19.
8. Lamghari, N.; Raghay, S. Recognition of Arabic Handwritten Diacritics using the new database DBAHD. In *Journal of Physics: Conference Series*; IOP Publishing: Bristol, UK, 2021; Volume 1743, p. 012023.
9. LeCun, Y.; Bottou, L.; Bengio, Y.; Haffner, P. Gradient-based learning applied to document recognition. *Proc. IEEE* **1998**, *86*, 2278–2324. [CrossRef]
10. Fukushima, K.; Miyake, S. Neocognitron: A Self-Organizing Neural Network Model for a Mechanism of Visual Pattern Recognition. In *Competition and Cooperation in Neural Nets*; Amari, S.I., Arbib, M.A., Eds.; Springer: Berlin/Heidelberg, Germany, 1982; pp. 267–285.
11. Al-Turaiki, I.; Altwaijry, N. Hijja Dataset. 2019. Available online: https://github.com/israksu/Hijja2 (accessed on 10 January 2024).
12. Alkhateeb, J.H. An effective deep learning approach for improving off-line arabic handwritten character recognition. *Int. J. Softw. Eng. Comput. Syst.* **2020**, *6*, 53–61.
13. Nayef, B.H.; Abdullah, S.N.H.S.; Sulaiman, R.; Alyasseri, Z.A.A. Optimized leaky ReLU for handwritten Arabic character recognition using convolution neural networks. *Multimed. Tools Appl.* **2022**, *81*, 2065–2094. [CrossRef]
14. Alheraki, M.; Al-Matham, R.; Al-Khalifa, H. Handwritten Arabic Character Recognition for Children Writing Using Convolutional Neural Network and Stroke Identification. *Hum.-Centric Intell. Syst.* **2023**, *3*, 147–159. [CrossRef]
15. Bin Durayhim, A.; Al-Ajlan, A.; Al-Turaiki, I.; Altwaijry, N. Towards Accurate Children's Arabic Handwriting Recognition via Deep Learning. *Appl. Sci.* **2023**, *13*, 1692. [CrossRef]
16. He, K.; Zhang, X.; Ren, S.; Sun, J. Deep residual learning for image recognition. In Proceedings of the IEEE Conference on Computer Vision and Pattern Recognition, Las Vegas, NV, USA, 27–30 June 2016; pp. 770–778.
17. Howard, A.G.; Zhu, M.; Chen, B.; Kalenichenko, D.; Wang, W.; Weyand, T.; Andreetto, M.; Adam, H. Mobilenets: Efficient convolutional neural networks for mobile vision applications. *arXiv* **2017**, arXiv:1704.04861.
18. Huang, G.; Liu, Z.; Van Der Maaten, L.; Weinberger, K.Q. Densely connected convolutional networks. In Proceedings of the IEEE Conference on Computer Vision and Pattern Recognition, Honolulu, HI, USA, 21–26 July 2017; pp. 4700–4708.
19. Santosh, K.; Nattee, C. Template-based Nepali natural handwritten alphanumeric character recognition. *Sci. Technol. Asia* **2007** *12*, 20–30.
20. Moetesum, M.; Diaz, M.; Masroor, U.; Siddiqi, I.; Vessio, G. A survey of visual and procedural handwriting analysis for neuropsychological assessment. *Neural Comput. Appl.* **2022**, *34*, 9561–9578. [CrossRef]
21. Das, N.; Reddy, J.M.; Sarkar, R.; Basu, S.; Kundu, M.; Nasipuri, M.; Basu, D.K. A statistical–topological feature combination for recognition of handwritten numerals. *Appl. Soft Comput.* **2012**, *12*, 2486–2495. [CrossRef]
22. Sharma, A.K.; Thakkar, P.; Adhyaru, D.M.; Zaveri, T.H. Handwritten Gujarati character recognition using structural decomposition technique. *Pattern Recognit. Image Anal.* **2019**, *29*, 325–338. [CrossRef]
23. Mukherji, P.; Rege, P.P. Shape feature and fuzzy logic based offline devnagari handwritten optical character recognition. *J. Pattern Recognit. Res.* **2009**, *4*, 52–68. [CrossRef] [PubMed]
24. Itseez. Open Source Computer Vision Library. 2015. Available online: https://github.com/itseez/opencv (accessed on 15 January 2024).
25. Abadi, M.; Agarwal, A.; Barham, P. TensorFlow: Large-Scale Machine Learning on Heterogeneous Systems. 2015. Available online: http://tensorflow.org/ (accessed on 7 March 2024).
26. Chollet, F., Keras. 2015. Available online: https://keras.io (accessed on 15 February 2024).
27. Bisong, E., Google Colaboratory. In *Building Machine Learning and Deep Learning Models on Google Cloud Platform: A Comprehensive Guide for Beginners*; Apress: Berkeley, CA, USA, 2019; pp. 59–64. [CrossRef]

28. LaValle, S.M.; Branicky, M.S.; Lindemann, S.R. On the relationship between classical grid search and probabilistic roadmaps. *Int. J. Robot. Res.* **2004**, *23*, 673–692. [CrossRef]
29. Iqbal, U.; Barthelemy, J.; Li, W.; Perez, P. Automating visual blockage classification of culverts with deep learning. *Appl. Sci.* **2021**, *11*, 7561. [CrossRef]
30. Iqbal, U.; Barthelemy, J.; Perez, P.; Davies, T. Edge-computing video analytics solution for automated plastic-bag contamination detection: A case from remondis. *Sensors* **2022**, *22*, 7821. [CrossRef]
31. Barthélemy, J.; Verstaevel, N.; Forehead, H.; Perez, P. Edge-computing video analytics for real-time traffic monitoring in a smart city. *Sensors* **2019**, *19*, 2048. [CrossRef] [PubMed]
32. Riaz, M.Z.B.; Iqbal, U.; Yang, S.Q.; Sivakumar, M.; Enever, K.; Khalil, U.; Ji, R.; Miguntanna, N.S. SedimentNet—A 1D-CNN machine learning model for prediction of hydrodynamic forces in rapidly varied flows. *Neural Comput. Appl.* **2023**, *35*, 9145–9166. [CrossRef]
33. Iqbal, U.; Barthelemy, J.; Perez, P. Prediction of hydraulic blockage at culverts from a single image using deep learning. *Neural Comput. Appl.* **2022**, *34*, 21101–21117. [CrossRef]

applied sciences

MDPI

Article

Comprehensive Analysis of Mammography Images Using Multi-Branch Attention Convolutional Neural Network

Ebtihal Al-Mansour, Muhammad Hussain *, Hatim A. Aboalsamh and Saad A. Al-Ahmadi

Department of Computer Science, CCIS, King Saud University, Riyadh 11451, Saudi Arabia; hatim@ksu.edu.sa (H.A.A.); salahmadi@ksu.edu.sa (S.A.A.-A.)
* Correspondence: mhussain@ksu.edu.sa

Abstract: Breast cancer profoundly affects women's lives; its early diagnosis and treatment increase patient survival chances. Mammography is a common screening method for breast cancer, and many methods have been proposed for automatic diagnosis. However, most of them focus on single-label classification and do not provide a comprehensive analysis concerning density, abnormality, and severity levels. We propose a method based on the multi-label classification of two-view mammography images to comprehensively diagnose a patient's condition. It leverages the correlation between density type, lesion type, and states of lesions, which radiologists usually perform. It simultaneously classifies mammograms into the corresponding density, abnormality type, and severity level. It takes two-view mammograms (with craniocaudal and mediolateral oblique views) as input, analyzes them using ConvNeXt and the channel attention mechanism, and integrates the information from the two views. Finally, the fused information is passed to task-specific multi-branches, which learn task-specific representations and predict the relevant state. The system was trained, validated, and tested using two public domain benchmark datasets, INBreast and the Curated Breast Imaging Subset of DDSM (CBIS-DDSM), and achieved state-of-the-art results. The proposed computer-aided diagnosis (CAD) system provides a holistic observation of a patient's condition. It gives the radiologists a comprehensive analysis of the mammograms to prepare a full report of the patient's condition, thereby increasing the diagnostic precision.

Keywords: breast cancer; mammography; deep learning; multi-label classification; convolutional neural network (CNN)

Citation: Al-Mansour, E.; Hussain, M.; Aboalsamh, H.A.; Al-Ahmadi, S.A. Comprehensive Analysis of Mammography Images Using Multi-Branch Attention Convolutional Neural Network. *Appl. Sci.* **2023**, *13*, 12995. https://doi.org/10.3390/app132412995

Academic Editors: Zahid Mehmood Jehangiri, Mohsin Shahzad and Uzair Khan

Received: 20 October 2023
Revised: 26 November 2023
Accepted: 30 November 2023
Published: 5 December 2023

1. Introduction

Breast cancer is a malignant transformation and proliferation of breast cells [1]. According to the American Cancer Society [2], it is ranked as the second most prevalent cancer among women in the United States. The survival of patients depends on whether it is diagnosed at an early stage [3]. Screening programs help to identify breast cancer at early stages to facilitate early treatment, which increases patients' survival rates. In contrast, delayed diagnoses allow the disease to spread, and the cancer can grow to a stage where treatment is no longer possible. Mammography is a breast-imaging technique that is usually used to detect abnormal tissues in the breast, thereby aiding the early diagnosis of abnormalities found in a patient.

Mammography screening involves many breast views, the most common being the craniocaudal (CC) and mediolateral oblique (MLO) views. Radiologists usually use these two views to observe breast tissues from different angles and detect abnormal tissues. A mammogram aids in identifying the breast density type according to the Breast Imaging Reporting and Data System (BI-RADS), the kind of abnormality (e.g., masses and calcifications), and the level of severity of the abnormality (benign or malignant) [4]. Several approaches for automated breast cancer diagnosis are available [5]. However, the existing studies mainly focus on a single view. The research presented in [6] demonstrated that two views are more helpful in improving a diagnosis; however, this method requires special treatment to avoid redundant data existing in both views. Therefore, using a fusion technique helps to handle this case.

In addition, most studies model the diagnosis problem as a single-label classification, such as detecting the breast density type [7–10], identifying the masses as benign or malignant [11–14], diagnosing microcalcifications as benign or malignant [15,16], or classifying both masses and calcifications as benign or malignant [17,18]. Single-label classification ignores the interdependencies between different conditions; for instance, a breast with high density is more likely to be malignant than a breast with low density. In addition, single-label classification requires developing multiple methods, each focusing on one aspect of the problem. These issues can be overcome by using a method based on multi-label classification for identifying breast cancer in an initial phase. A method that formulates the diagnosis as a multi-label classification problem can help to diagnose a patient's condition comprehensively. It leads to better diagnosis results by considering additional aspects, such as the correlation between the density type, lesion type, and states of the lesions, which radiologists usually perform. It can assist radiologists in their decision making by providing a comprehensive report of the patient's condition, increasing the precision of their diagnosis.

Given the above discussion, we formulate a diagnosis as a multi-label classification using two views. Inspired by the success of advanced CNN and transformer models [19–25], we designed the proposed method using four modules: a feature extraction module, an attention module, a fusion module, and a multi-label classification module. First, the feature extraction module employs the state-of-the-art CNN and transformer models, such as the Swin transformer and ConvNeXt, and extracts features from two views, CC and MLO. The attention module concentrates on relevant features and suppresses irrelevant features. The information extracted from the two views is fused using the fusion module. Finally, the multi-label classification module takes the fused features as input and simultaneously predicts the density type, abnormality type, and severity levels.

The key contributions of this research paper are as follows:

- We propose a method based on the multi-label classification of two-view mammography images—with CC and MLO views—that diagnoses the patient's condition comprehensively.
- We employ channel attention for selectively emphasizing the most informative channels of the input feature maps while suppressing the less informative ones.
- We propose a multi-branch deep architecture, which takes the features from two views as input and performs multi-label classification.
- We thoroughly evaluated the proposed method on two public-domain benchmark datasets, INBreast and CBIS-DDSM.

This paper is organized as follows: Section 2 discusses the 'Related Work' on breast cancer diagnosis. Section 3 details the 'Proposed Method.' The 'Evaluation Method' is described in Section 4, while Section 5 covers the 'Experiments and Results.' Section 6 includes the 'Discussion,' followed by 'Limitations and Future Work' in Section 7. This paper concludes in Section 8 with the 'Conclusions,' and the 'Nomenclature' used throughout the paper is provided at the end.

2. Related Work

Significant research has contributed to developing and improving advanced CAD systems, especially for detecting and diagnosing breast cancer using mammography. Within this particular context, many studies have focused on solving related problems such as mass classification as non-cancerous (benign) or cancerous (malignant), the classification of microcalcifications as non-cancerous (benign) or cancerous (malignant), the classification of both microcalcifications and masses as non-cancerous (benign) or cancerous (malignant); classification as masses and microcalcifications; the classification of mammograms based on breast density; and the multi-label classification of mammograms. This section provides an in-depth review of the recent state-of-the-art (SOTA) research and methods that address these problems.

2.1. Mass Classification as Benign or Malignant

A whole mammogram image or the regions of interest (ROIs) are classified as benign or malignant, and there are many methods for this purpose.

Some recent methods classify mammogram images containing masses as benign or malignant. Chen et al. [11] developed a method using two mammography views, MLO and CC, for both breasts, extracting spatial and frequency domain features. They utilized particle swarm optimization (PSO) and support vector machine (SVM) for feature selection and classification, achieving an AUC-ROC of 0.79 for two-view and 0.75 for four-view images. Das et al. [13] implemented adaptive contrast enhancement in mammogram images, followed by segmentation and artificial neural network classification, resulting in a high accuracy of 97.2%. Sun et al. [12] introduced a multi-dilated CNN that integrates multiple views and optimizes classification accuracy by modifying the cross-entropy cost function, achieving accuracies of 82.02% and 63.06%, respectively. Nagarajan et al. [14] employed bi-dimensional empirical mode decomposition and GLCM for feature extraction, leading to AUC-ROC values of 0.9 and 0.96. Ayana et al. [26] presented a novel model employing a transformer for feature extraction combined with transfer learning, tackling the issue of imbalanced data and achieving near-perfect classification accuracy. Yu, Xiang et al. [27] developed VGG19-DF with a dRVFL classifier, showing an average AUC of 0.93 and an accuracy of 81.71%.

2.2. Microcalcification Classification as Either Benign or Malignant

In order to classify microcalcifications, the research has considered full mammogram images or segmented ROIs. Some recent methods classify mammogram images containing macrocalcifications as benign or malignant.

George et al. [15] proposed a multi-scale connected chain graph method for classifying microcalcifications, achieving up to 90% accuracy. Mabrouk et al. [16] enhanced mammogram images using various mechanisms and integrated feature extractions followed by ANN, KNN, and SVM classification, resulting in an accuracy of 0.96. Gerbasi et al. [28] introduced DeepMiCa, a U-Net-based network for the segmentation and classification of microcalcifications, achieving an AUC of 95%. Sarvestani et al. [29] enhanced extracted ROIs using a fuzzy system and Gabor filtering, achieving a 93% accuracy rate in classifying microcalcifications.

2.3. Mass and Microcalcification Classification as Benign or Malignant

In the task of classifying masses and microcalcifications as benign or malignant, two approaches can be employed: either classifying the ROIs corresponding to segmented masses and microcalcifications or performing classification on the entire mammogram image to determine its benign or malignant nature.

Li et al. [17] enhanced the DenseNet architecture for classifying mammogram images, achieving a 94.55% accuracy rate. Mohanty et al. [18] proposed a method using block-based discrete-wavelet packet transform and principal component analysis, enhanced with a kernel extreme learning machine classifier, achieving accuracy rates above 99%. Jabeen et al. [30] developed an automated framework for breast cancer classification from mammogram images, employing a novel image enhancement technique and the EfficientNet-b0 model fine-tuned via deep transfer learning. The framework, which included advanced feature extraction and optimization using the Equilibrium-Jaya controlled Regula Falsi algorithm, was tested on the CBIS-DDSM and INBreast datasets, achieving notable accuracies of 95.4% and 99.7%, respectively. Chakravarthy et al. [31] combined deep learning with metaheuristic techniques to classify mammography images, achieving up to 97.36% accuracy. Azour et al. [32] utilized ensemble learning techniques with a combination of multiple deep-learning models, achieving an accuracy of approximately 82.4%.

2.4. Multi-Label Classification of Mammograms

Few studies have addressed the multi-label classification of mammograms, with studies only [33] investigating this issue in recent years.

Chougrad et al. [33] introduced a CAD system for the multi-label classification of mammogram images. They employed VGG16-CNN with fine-tuning techniques and a label powerset classifier, demonstrating promising results across various datasets.

2.5. Analysis

The above studies indicate significant research addressing breast cancer detection in mammogram images from various perspectives and formulating different problems, such as classifying masses as non-cancerous or cancerous, microcalcifications as non-cancerous or cancerous, and masses and microcalcifications together as benign or malignant. These works achieved a favorable performance for the above-mentioned problems.

Only a few studies have focused on solving the problem of the multi-label classification of mammogram images [33], which simultaneously identifies the risk/density grade, abnormality type (e.g., mass or microcalcification), and state of the lesion (benign or malignant). The research presented in [19] adopted VGGNet and used transfer learning to fine-tune the model using ROIs before using the model as a feature extractor and multi-label classifier. Although this method provides favorable results and uses new techniques such as deep learning and transfer learning, it entails some limitations, such as using a simple CNN architecture and single-view ROIs as input. According to the study presented in [6], using multiple views can enhance prediction performance compared with using a single view. In addition, integrating residual learning into a CNN helps to overcome many challenges, such as vanishing gradients, overfitting, and complex correlations between labels [21].

Table 1 summarizes the existing research in the field. Jafari et al. [34] presented a CNN-based breast cancer detection method for mammography images, extracting features from various CNN models and selecting key features. Tested on the RSNA, MIAS, and DDSM datasets, it achieved the highest accuracy with an NN classifier: 92% for RSNA, 94.5% for MIAS, and 96% for DDSM.

Table 1. Comprehensive summary of the existing works.

Paper	Method	Dataset	Performance
	Mass classification as benign or malignant		
Chen et al. [11], 2019	Fifty-nine features like shape, density, FFT, and DCT for feature extraction; PSO for feature selection; SVM as classifier	FFDM	Sen. = 81 Spe. = 77
Das et al. [13], 2020	Power-law transformation + shift invariant extrema characterization + ANN	MIAS, DDSM	ACC = 97.2 Sen. = 98.4
Nagarajan et al. [14], 2019	GLCM and GLRM from MBEMD and SVM/LDA	MGM	ACC = 90 AUC = 0.92
Ayana et al. [26], (2023)	Transformer	DDSM	AUC = 1 ± 0
Yu, Xiang, et al. [27], (2023)	CNN	DDSM	AUC = 0.93 ACC = 81.71
Sun et al. [12], 2019	CNN with dilated CONV layers	MIAS DDSM	ACC = 63.06 ACC = 82.02

Table 1. *Cont.*

Paper	Method	Dataset	Performance
Microcalcification classification as benign or malignant			
George et al. [15]	Topology, graph connectivity, multi-scale morphology, and KNN	DDSM	ACC = 86.47 AUC = 0.899
Mabrouk et al. [16]	HS, WT, ME, HE, Otsu, Shape, GLCM, invariant moment features, ANN, KNN, and SVM	MIAS	ACC = 96 Sen. = 98
Gerbasi et al. [28], 2023	UNet for segmentation + ResNet18	DDSM	AUC = 0.95
Sarvestani et al. [29], 2023	Fuzzy system + Gabor filtering for image enhancement + ANN for classification	DDSM	ACC = 93 Sen. = 95
Mass and microcalcification classification as benign or malignant			
Li et al. [17], 2019	DenseNet with the Inception structure.	FFDM	94.55
Mohanty et al. [18]	BDWPT + PCA + WC-SSA-KELM	MIAS	ACC = 99.28 Sen. = 99.44 AUC = 0.994
Chakravarthy et al. [31], 2023	Resnet18 + wKNN + PSO + DFOA + CSOA	MIAS	ACC = 84.35
F. Azour et al. [32], 2023	VGG, Resnet, Inception-v3, DensNet, MobileNet, and EfficientNet		ACC = 82.4
Jabeen et al. [30], 2023	Image enhancement, EfficientNet-b0, feature optimization and selection, and ML classifiers	CBIS-DDSM and INBreast	ACC = 95.4% and 99.7%
Jafari et al. [34], 2023	Feature extraction from various CNNs, feature selection using mutual information, and classification with NN, kNN, RF, and SVM	RSNA, MIAS, and DDSM	Acc = 92%, 94.5%, and 96%
Multi-label classification of mammograms			
Chougrad et al. [33], 2020	VGG16	DDSM, BCDR INBreast, and MIAS	Exact match: 0.822, 0.802 0.827, and 0.782

3. Proposed Method

We address the problem of simultaneously identifying the breast density type (according to BI-RADS), abnormality type (mass or calcification), and severity level/pathology (benign or malignant) from mammogram images. This is a multi-label classification problem. First, we formally define and formulate the problem and then present the details of the proposed method.

3.1. Problem Formulation

To screen a patient for breast cancer detection, two commonly used views of a patient's mammogram are the MLO and CC views. The problem is identifying the breast density, severity level/pathology, and findings from the two views. Based on the BI-RADS (Breast Imaging Reporting and Data System) guidelines, there are four density levels, BI-RADS I, BI-RADS II, BI-RADS III, and BI-RADS IV, which are used to classify breast density, where BI-RADS I represents the category with the lowest density, while BI-RADS IV corresponds to the category with the highest density. Additionally, there are two main abnormality types, masses, and calcifications, which are important to identify. Finally, the severity level/pathology means whether the case for the abnormality type is benign or malignant.

We represent an ROI as $x \in \mathbb{R}^{m\times n}$, where m and n represent the resolution of the ROI. x_{MLO} and x_{CC} stand for the ROIs extracted from the MLO view and the CC view, respectively. There are eight categories: BI-RADS I (1), BI-RADS II (2), BI-RADS III (3), BI-RADS IV (4), mass (5), calcification (6), benign (7), and malignant (8). The first four categories correspond to density types, the next two categories represent abnormality types, and the last two categories stand for severity levels (pathology). In view of this, the label for a pair of ROIs $(x_{MLO}, x_{CC}) \in \mathbb{R}^{m\times n} \times \mathbb{R}^{m\times n}$ corresponding to a patient is $l = \left(l_d, l_f, l_p\right)$, where $l_d \in Y_d = \{0,1\}^4$, $l_f \in Y_f = \{0,1\}^2$, and $l_p \in Y_p = \{0,1\}^2$; Y_d, Y_f, and Y_p are the label spaces of the density type, abnormality type/findings, and severity level/pathology, respectively, in one-hot encoding; and 0 means absent, and 1 means present. For example, if $l = \left(l_d, l_f, l_p\right)$, where $l_d = [0\,1\,0\,0]$, $l_f = [1\,0]$, and $l_p = [1\,0]$, then the density level is 2, the finding is a mass, and the case is benign. It is a multi-label classification problem. We need to design a mapping $\varphi : \mathbb{R}^{m\times n} \times \mathbb{R}^{m\times n} \to Y_d \times Y_f \times Y_p$, such that $\varphi(x_{MLO}, x_{CC}) = \left(l_d, l_f, l_p\right)$.

We employ deep learning techniques to design the mapping φ that extracts discriminative features from the input ROIs and associates them with three labels in an end-to-end manner. In the following subsections, we give the details of the deep-learning-based method for modeling φ.

3.2. Dataset Description

Our model utilizes two benchmark mammography datasets: the Curated Breast Imaging Subset of DDSM (CBIS-DDSM) [35] and INBreast [36]. Both datasets are publicly available and extensively annotated. Both cover various breast densities, abnormalities, and pathologies in different imaging views (MLO and CC). These datasets include a variety of breast densities (BI-RADS I-IV), abnormalities, and pathologies across the MLO and CC imaging views. They provide detailed annotations of ROIs and clinical findings such as masses, calcifications, and architectural distortions and categorize lesions as benign or malignant. For our research, we focused on cases where ROIs are present in both views, suitable for fusion methods while excluding cases with ROIs in only one view to ensure the relevance and comprehensiveness of our data.

3.3. Two-View-Based Deep Model

This section outlines the model architecture for simultaneously classifying mammography images into multiple labels, density, severity level, and abnormality type, using a channel-attention-based multi-task learning framework. It employs two branches, CC and MLO, with the Swin and ConvNeXT models as feature extractors.

The model consists of a feature extraction module and a fusion module. The feature extraction module, operating in the CC and MLO views, uses the Swin and ConvNeXT backbones to extract features crucial for all classification tasks, aiding in multiple-label prediction. A channel attention mechanism refines the focus on relevant features and reduces the less informative ones, ensuring selective emphasis on essential channels.

Subsequently, the fusion module merges features from both views, enhancing the multi-label classification by incorporating diverse information. Integrated features pass through three fully connected layers, each dedicated to a specific classification task (density, severity level, or abnormality type). These layers function as classifiers, producing prediction labels for the input mammographic views, thus enabling simultaneous multi-label classification.

This architecture integrates channel attention, feature extraction, and fusion mechanisms, facilitating the learning of both shared and task-specific features, thereby improving efficiency in the multi-label classification of mammography images. This section provides an overview of the model's structure, with the subsequent sections detailing the data preprocessing, evaluation metrics, and deep learning model architecture. Figure 1 visually represents this model, highlighting its components and their interplay.

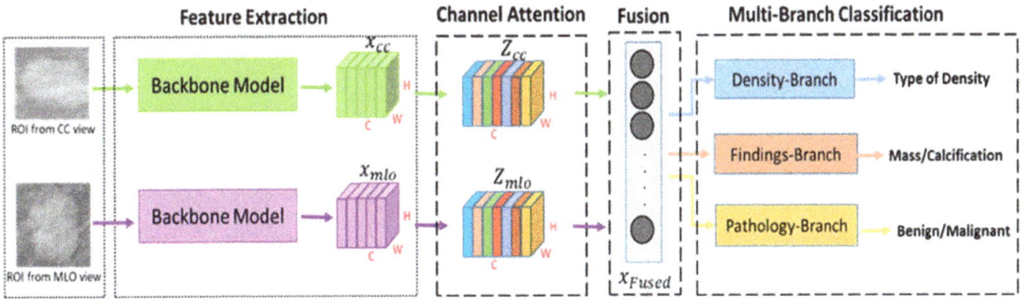

Figure 1. The overall architecture of the proposed system.

3.3.1. Details of the Model Architecture

This subsection delves into the design and functionality of our multi-label classification model for mammography images, building on the initial overview. We examine the feature extraction and fusion modules in detail, highlighting their roles in processing CC and MLO views for adequate classification. The model integrates channel attention and multi-task learning, focusing on pertinent features and learning shared and task-specific characteristics, enhancing breast cancer diagnosis via precise mammography classification.

3.3.2. Preprocessing

To prepare mammography images for model training, we first implement preprocessing, including resizing the ROIs to 224 × 224 pixels, aligning them with the input requirements of our backbone architectures like Swin and ConvNeXT. The normalization of ROI pixel values, based on ImageNet dataset standards, ensures consistency in feature representation.

3.3.3. Backbone Model

Our proposed architecture uses two advanced models, ConvNeXt and the Swin transformer, for feature extraction in mammography image classification. These models were chosen for their superior performance in various computer vision tasks, marking them as leading solutions in CNN and transformer models.

The Swin transformer excels in capturing both local and global image features due to its hierarchical structure, shifted windows, and feature attention mechanism, leading to highly discriminative and informative representations. This feature makes it well-suited for feature extraction in multi-label classification.

The ConvNeXt model is known for its modularity, efficiency, and scalability, with a deep architecture that addresses gradient issues, providing stable and effective feature representations. We utilize pre-trained models on a 21 K image dataset. Their extracted features, refined through a channel attention mechanism, are combined to integrate features from both views, enhancing task-specific feature extraction for classification.

The depth of the Swin transformer and ConvNeXt models allows for the capture of distinct features crucial for each classification task, significantly boosting the model's performance in the multi-label classification of mammography images. Our final model uses ConvNeXt-L as a feature extractor, omitting its last fully connected layer and retaining the feature map from the penultimate ConvNeXt block (dimension 7,71536). The architecture, including omitted layers, is illustrated in Figure 2, with critical features indicated for clarity.

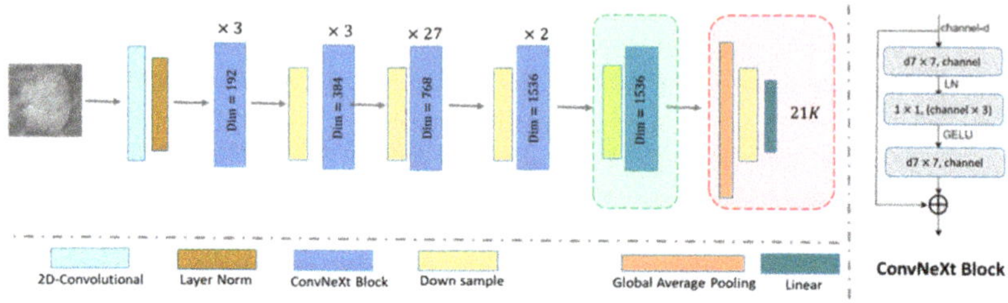

Figure 2. This diagram illustrates ConvNeXt-L, our backbone model, highlighting customized layers (red boundaries and shading) and key features in the penultimate layer (green box) for our mammography image classification task.

3.3.4. Chanel Attention Block

Not all feature channels of a feature map are equally important for the current task. Some channels may contain highly informative features that are directly relevant to the task, while others may contain less informative or redundant features that can potentially distract the network from focusing on the relevant information.

To address this, we incorporate a squeeze-and-excitation block after the last feature map generated by the backbone model. By doing so, we enable the model to adaptively weigh the importance of each channel, giving more attention to the informative channels and suppressing the less relevant ones. This dynamic attention mechanism assists the network in making better-informed decisions during the classification process. Figure 3 shows the channel attention block used.

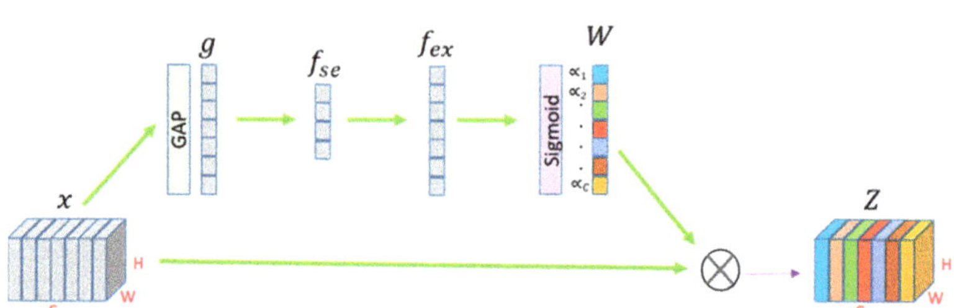

Figure 3. Channel attention block.

The channel attention block takes the input feature map $x \in \mathbb{R}^{H \times W \times C}$, where C is the number of channels, H is the height, and W is the width. First, it applies a squeeze operation:

$$g = GlobalAveragePool(x) \tag{1}$$

$$f_{se} = ReLU\ (W_1.g + b_1) \tag{2}$$

where $g \in \mathbb{R}^C$, $f_{se} \in \mathbb{R}^{C/c'}$, W_1 and b_1 are the weights and biases of the FC layer, which squeezes g to f_{se} to incorporate interdependencies. Then, it applies the excitation operation:

$$w = \sigma(W_2 \,.\, f_{se} + b_2) \tag{3}$$

where W_2 and b_2 are the weights and biases of the FC layer that adaptively recalibrate f_{se}, and $w = [\propto_1, \propto_2, \ldots, \propto_C]^T$, where $\propto_c$ signifies the channel-wise excitation factor for channel c. Once the channel-wise excitation factors w are computed, they are used to attend to the corresponding channels:

$$Z = w \bigodot x = [z_1, z_2, \ldots, z_C] \tag{4}$$

3.3.5. Fusion Layer

Fusing features from both CC and MLO views enhances mammography classification in multi-label tasks. CC views offer a lateral perspective of breast tissue, while MLO views provide an oblique angle, capturing additional tissue. Merging these views gives classifiers a more comprehensive understanding of breast tissue, thus improving model performance.

For feature fusion, methods like concatenation, average-wise, and element-wise operations were considered. Following an ablation study, we chose average-wise operations for fusing CC and MLO view features. This involves computing the global average pooling for channel attention feature maps from both the CC and MLO branches, each resulting in a 1D vector. We then average these vectors to form the final fused feature representation. This approach aggregates relevant information from both views, balancing their differences for a robust feature representation for classification.

The dimension of the channel attention feature map from the backbone model's last layer is $H \times W \times C$, with H *and* $W = 7$ *and* $C = 1536$. In this context, we denote the number of channels as D ($D = 1536$) to differentiate from previous sections.

After applying GAP across H × W, we obtain the vector with dimension 1x1536. The output of the fusion model is given below:

$$x_{Fused} = g(g_{GAP}(z_{cc}), g_{GAP}(z_{mlo})) \tag{5}$$

where $x_{Fused} \epsilon R^D, z_{cc}, z_{mlo} \epsilon R^{W \times H \times D}, g_{GAP}(z_{cc}) = \propto = [\propto_1, \propto_2, .., \propto_d]^T, g_{GAP}(z_{MLO}) = \beta = [\beta_1, \beta_2, .., \beta_d]^T$ *and* $g(\propto, \beta) = \left[\frac{\propto_1+\beta_1}{2}, \ldots, \frac{\propto_d+\beta_d}{2}\right]$.

The function is g_{GAP}, and g represents the global average pooling and point-wise average operation.

The diagram in Figure 4 shows the details of the fusion layer.

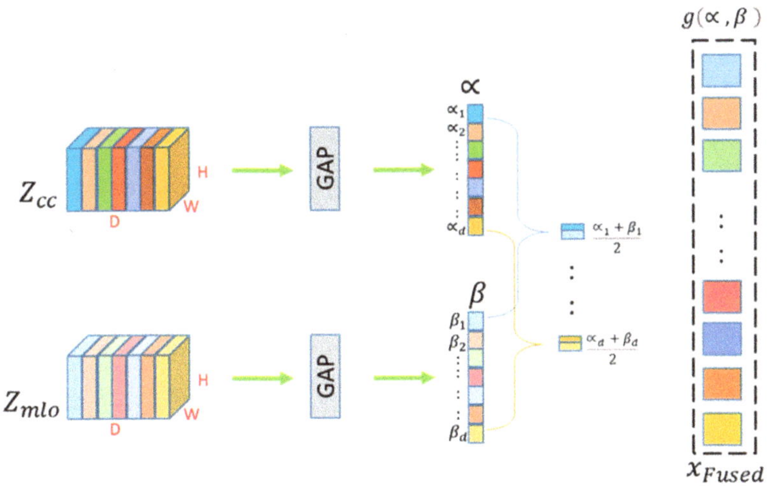

Figure 4. In the fusion block, GAP is calculated separately for the CC and MLO feature maps produced by attention block, followed by average CC and MLO feature correspondence to the same position.

3.3.6. Multi-Branch Classification

Our system classifies input ROIs into three distinct groups: density, severity level, and abnormality type. To enhance feature representation and task-specific learning, we integrated three branches of fully connected layers at the end of the fusion model.

Each branch in the multi-branch architecture focuses on learning features specific to a particular task, allowing the model to extract more effective task-specific features. In the post-fusion layer, we apply three parallel FC layers dedicated to density, severity level, and abnormality type, enabling distinct learning for each task.

The architecture includes a series of FC layers leading to a classification layer, with configurations determined via an ablation study. We use FC 256, FC 128, and dense layers with ReLU activation, batch normalization, and then the classification layer. This setup helps in learning task-specific features for efficient classification.

The final classification layer makes predictions for the three groups. Density classification uses four output neurons for BI-RADS classes, while abnormality type and severity level classifications use two output neurons each for classifying into mass or calcification and benign or malignant, respectively.

We employ the softmax activation function in each group's output layer for these tasks. This function transforms the network's output into class-specific probabilities, aiding in accurate class determination. The choice of softmax is guided by our multi-label classification needs and the nature of the tasks. Figure 5 illustrates the multi-branch classification block.

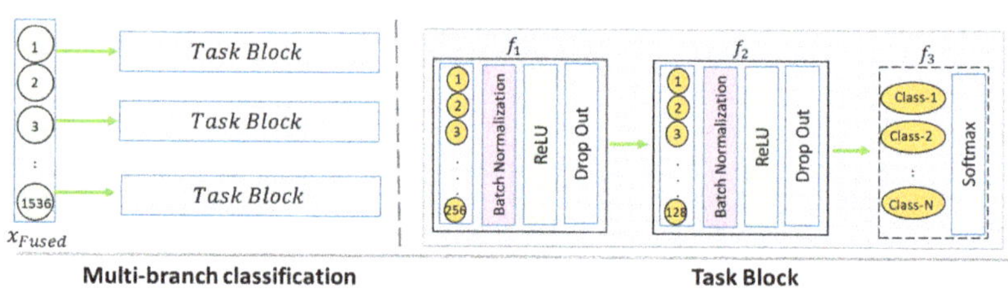

Figure 5. Multi-branch classification block. It takes a fused feature and passes it to three parallel task blocks. It presents a multi-branch neural network architecture with a unified task block structure replicated across three tasks: density, pathology, and severity level categories. The left side enumerates sequential task blocks, indicative of a deep and comprehensive processing approach. On the right is a breakdown of the task block. The output layer employs a softmax function, tailored with four classes for the density task and two classes for both pathology and severity levels.

In this block, we employ a series of mathematical equations to map the input data to the output, with the overarching goal of achieving multi-label classification. Specifically, we utilize these equations to model and predict probability distributions for three distinct branches: density (denoted as d), severity level/pathology (denoted as p), and abnormality type/findings (denoted as n). The equations for each branch are presented as follows.

For a given branch,

$$b\ (b \in \{d, p, n\}) :\ P^b = Softmax\left(f_3{}^b\ o\ f_2{}^b\ o\ f_1{}^b\ \left(x_{Fused}\right)\right) \tag{6}$$

In this equation, P^b represents the probability distribution specific to branch b, and $\{f_3{}^b, f_2{}^b, f_1{}^b\}$ denote the functions modeled with fully connected layers (FC1, FC2, and FC3) tailored to each branch. The output is a probability vector p^b, and the predicted class label l^b is determined as max $\max_{1 \le i \le k^b} p_i{}^b$. , where k^b is the number of classes associated with branch b.

Each function $f_i{}^b$ follows a consistent structure, comprising a fully connected layer, batch normalization, ReLU activation, and dropout. The number of neurons and layer-specific parameters may vary across branches. Additionally, in each branch, we implement a two-stage projection on the fusion features, initially reducing their dimension from D to D_1, and then further compressing them to D_2, which are variables determined according to the specific branch's requirements.

In each branch, we employ a two-stage projection on the fusion features, initially reducing their dimension from 1536 to 256, and then compressing them to a 128-dimensional space. These projections are followed by a classification layer specific to each task, which has varying numbers of classes (ranging from 2 to 4 depending on the task). This strategy yields the following benefits:

1. Dimensionality reduction: Reducing fusion features from 1536 to 256 and, further, to 128 dimensions decreases the data complexity, mitigating computational overhead and overfitting risks while retaining essential information.
2. Targeted feature learning: This helps the model learn crucial task-specific features by mapping them onto a lower-dimensional space, enhancing class discrimination.
3. Task-specific classification: Post-projection, a classification layer for each task, accommodating 2 to 4 classes, transforms features into class probabilities for precise task-specific classification.

The entire process is implemented using functions f_1 , f_2, and f_3, encompassing the operations described. This strategy integrates dimensionality reduction, task-driven feature learning, and specialized classification to optimize model performance across different tasks.

4. Evaluation Method

This section describes the evaluation methods for our proposed model, including datasets, challenges, the evaluation protocol and metrics, and model training.

4.1. Model Training

Our model, trained simultaneously across all branches, uses weighted cross-entropy loss for each branch, with an average calculated for the final loss. Key considerations include handling unbalanced multi-label data and utilizing three output branches for specific tasks.

We employed an RMSprop optimizer with dual learning rates for pre-trained (1×10^{-4}) and new layers (1×10^{-3}). The training involved 400 epochs, a batch size of 128, learning rate reduction on a plateau, dropout (factor 0.2), and weight decay (1×10^{-6}) for regularization. The stopping strategy had a patience of 40 epochs.

Pre-trained weights from the ImageNet-21K dataset were used for transfer learning. Data augmentation included random rotations, width and height shifts, horizontal flips, and zoom, coupled with normalization.

4.2. Evaluation Protocol and Metrics

The datasets were split into 80:20 for training and testing, with a 10% validation set, using 5-fold cross-validation [37]. The evaluation metrics included mean average precision, F1-score, Hamming loss, coverage, ranking loss, and exact match [38,39].

The system was implemented using TensorFlow, Keras, and PyTorch in Anaconda Navigator (2022) on an Intel(R) Core(TM) i9-9900K CPU with a GPU with 32 GB memory and 64.0 GB RAM.

5. Experiments and Results

This section outlines the experiments to evaluate our multi-label classification model using the CBIS-DDSM and INBreast datasets. We tested the model with various SOTA backbone models, assessing its performance using metrics like F1-score, mean average precision, and exact match. The focus was on classifying breast cancer and abnormalities in terms of density, severity level, and abnormality type. Additionally, an ablation study examined different fusion methods and configurations for a multi-branch block, culminating in a comprehensive assessment of our proposed fusion model.

5.1. Ablation Study

We conducted this study to analyze which fusion technique is more suitable and which configuration is the best for a multi-branch block.

5.1.1. Which Fusion Technique Is Suitable?

We conducted this study to analyze and evaluate the model's performance when using different common fusion techniques. The goal was to determine which fusion method was more suitable for our model. We examined three fusion methods: concatenation, element-wise addition, and averaging. Table 2 shows the results of each technique.

Table 2. Performance results for CBIS-DDSM with different fusion methods.

Fusion Method	F1%	mAP%	EM%
Concatenation	94.29% ± 0.011	90.54% ± 0.016	86.69% ± 0.016
Multiply	93.97% ± 0.01	89.8% ± 0.013	85.6 % ± 0.023
Average	94.7% ± 0.01	91.3% ± 0.017	86.9% ± 0.019

The table compares three fusion techniques: concatenation, element-wise, and average-wise, using metrics like F1-score, mAP, and EM. Average-wise outperformed others in all metrics, indicating its effectiveness in integrating features from two views. Consequently, we selected average-wise as our preferred fusion method.

5.1.2. How Many Hidden Layers in Multi-Branch Block

We conducted a study on how many deep layers we needed for our model. The table below shows the model's performance for a multi-branch block in a fusion model with different numbers of hidden layers. In this study, we examined up to five layers. Table 3 shows that the highest F1-score, mAP, and EM were achieved with two fully connected layers, with scores of 94.72% and 91.33 for F1-score, mAP, and EM, respectively.

Table 3. The performance of a multi-branch block fusion model with different configurations.

# Layers	# of Neurons in Each Layer	F1%	mAP%	EM%
1	256	94.44	90.69	86.69
2	256 and 128	94.72	91.33	86.86
3	512, 256, and 128	94.27	90.5	86.45
4	1024, 512, 256, and 128	94.24	90.4	86.53
5	2048, 1024, 512, and 256	94.48	90.9	86.69

This study recommends using two hidden layers to provide a good balance and trade-off between the model's performance and complexity, as it achieved the best performance among all evaluated metrics.

5.1.3. Which Backbone Model Is Better?

The question was which of the backbone models used in the experiments performs the best among selected SOTA pre-trained CNN and transformer models? We used the CBIS-DDSM dataset to test those two models. As shown in Table 4, ConvNeXt was selected as a backbone model for our proposed model as it obtained the best result. This decision was made because ConvNeXt outperformed the Swin transformer model in terms of F1-score, RL, and Cov.

Table 4. The performance of different backbone models on the DDSM dataset.

Model	F1%	HL	mAP%	RL	Cov	EM%
ConvNeXt	0.910	0.07	0.85	0.12	4.11	0.78
Swin transformer	0.90	0.07	0.85	0.13	4.12	0.78

5.1.4. The Effect of Fusion

Fusing features from both CC and MLO views enhances mammography classification, as they capture different breast tissue aspects. CC views offer a lateral perspective, while MLO views include additional tissue through an oblique angle. This fusion provides a more comprehensive tissue analysis, potentially improving model performance. Our experiments using SOTA backbone models compared the efficacy of single- and dual-view approaches for multi-label classification. The results in Table 5 show that dual-view fusion surpasses single-view classifiers in performance on the CBIS-DDSM dataset.

Table 5. The influence of the overall performance when comparing the single view and two views of the CBIS-DDSM dataset.

View	Model	F1%	mAP%	EM%
Single view	Swin	90%	85%	78%
	ConvNeXt	91.0%	85%	78%
Dual view	Swin	94.18%	90.15%	86.28%
	ConvNeXt	94.54%	90.88%	87.27%
	ConvNeXt with attention	94.72%	91.33%	86.86%

5.1.5. The Effect of the Attention Module

In deep learning models, not all channels in the input image contribute equally to classification, as some may contain irrelevant or noisy data. To address this, we used a channel attention mechanism in our backbone model and fusion process. This experiment demonstrated that channel attention positively impacts performance. Table 6 shows that integrating this mechanism into the ConvNeXt fusion model improves most evaluation metrics. The F1-score increased from 94.54% to 94.72%, and the HL decreased from 0.0364 to 0.0355, while mAP and RL showed slight improvements.

Table 6. This table shows the impact of incorporating the attention mechanism into the fusion model.

Model	Chanel Attention	F1%	HL	mAP%	RL	Cov	EM%
ConvNeXt	No	94.54%	0.0364	90.88%	0.071	3.636	87.27%
ConvNeXt	Yes	94.72%	0.0355	91.33%	0.070	3.657	86.86%

5.1.6. The Analysis of Features

In this section, we assess the performance of the proposed model on the test data by analyzing the distribution of the features learned by each classification branch. Figure 6 shows how the features are discriminative for each label. This indicates that the proposed model can extract the discriminative features with less overlapping.

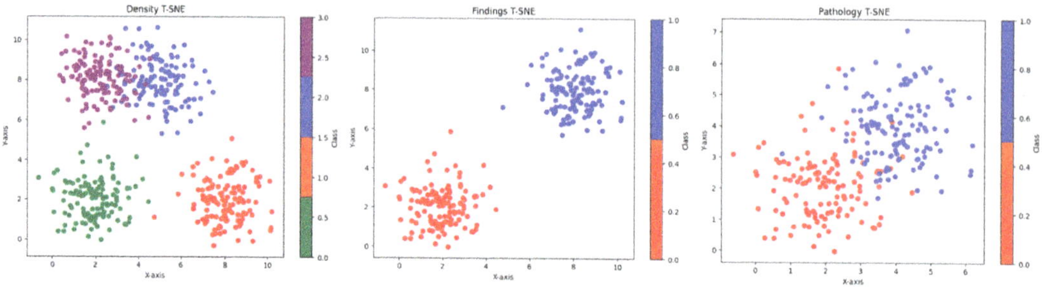

(a) (b) (c)

Figure 6. Visualization of the distribution of features (**a**), (**b**), and (**c**) indicate the distribution of the features among density, abnormality type (findings), and severity level (pathology), respectively.

6. Discussion

We employed the two-views (CC and MLO) technique to construct and evaluate a proposed multi-label classification of breast cancer into eight labels corresponding to three groups, simultaneously, i.e., density (I, II, III, or IV), abnormality type/findings (mass or calcification), and severity level/pathology (benign or malignant). The CBIS-DDSM benchmark dataset was used to decide which technique is more suitable for this task; for example, it was used to evaluate the effect of different configurations, backbone models, fusion methods, and so on. The SOTA backbone model was used as the core component of the model to enable the automatic extraction of the features without human intervention.

By relying on both views, the system can detect and identify the abnormalities that might not exist and are visible in a single view. Furthermore, certain irregularities might only be seen in one view and not the other; accordingly, using both views raises the chance of identifying the abnormalities and decreases the possibility of missing and neglecting any potential concerns and important information.

We examined the SOTA CNN and transformer models, including ConvNeXt and the Swin transformer, and concluded that ConvNeXt was the best model for our proposed CAD system.

After evaluating the performance of each model, we found that ConvNeXt was the most appropriate option for our specific tasks. ConvNeXt outperformed other architectures due to its use of depth-wise separable convolutions. These convolutions required fewer learnable parameters than traditional convolutional layers, making the network more effective and reducing overfitting. In addition, it can be adaptive to the data by dynamically building networks with a varying number of layers, depending on the complexity of the data. This makes the architecture more adaptable and flexible to several tasks and allows it to extract important features more efficiently.

Moreover, ConvNeXt utilizes multi-scale processing. This allows for extracting the features from multiple levels of abstraction. This helps the model to extract and capture extra related information and enhance the performance of the introduced system.

Three different fusion techniques were examined. The average-wise method was selected based on the results as it gave the best performance among the methods.

As the proposed system classifies the input ROIs into three non-overlapped groups, integrating three branches into the end of the backbone model assists in enhancing the performance as it helps improve the feature representation and task-specific learning. Each branch of the multi-branch architecture focuses on learning features specific to a particular task, leading to more effective feature representation and task-specific learning. This helps the model capture more task-related features, yielding a better performance on that task, leading to improved performance.

On the other hand, incorporating a channel attention mechanism for each task in the classification layer of a deep neural network has a slightly positive impact on the model's performance. Channel attention leads the model to focus on the significant relevant channels and suppress noninformative ones.

Incorporating channel attention into ConvNeXt enables the selective enhancement of the most informative channels in the feature maps. It leads to a more effective representation of features and, eventually, results in enhanced model performance.

6.1. Performance Comparison with the SOTA Methods

As shown in the related work section, few studies have addressed the multi-label classification of mammograms, with only [33] investigating this issue in recent years.

The multi-label classification of mammogram images proposed by Chougrad et al. [33] simultaneously classifies a mammogram into its abnormality type/findings (mass/microcalcifications), severity level/pathology (benign/malignant), and density class (I–IV). They used ROIs as input for the deep learning module, transfer learning to initiate the VGG16-CNN weights with a fine-tuning technique, and a label powerset classifier for classification. The proposed algorithm considers the correlation between labels. They evaluated their method using the CBIS-DDSM [35], BCDR [40], INBreast [36], and MIAS [41] datasets with multiple metrics.

On the other hand, the introduced system outperformed the SOTA methods as we used the fusion method and SOTA backbone model in addition to the multi-branch and channel attention techniques. It achieved a higher performance in all metrics on both the CBIS-DDSM and INBreast datasets. It is observed that fusing two view inputs improves the overall performance across all performance metrics when compared with a single view. This is because several irregularities and features might be visualized better in one view than in the other, and combining the features from two views leads to a comprehensive assessment of the breast situation. In addition, utilizing the SOTA ConvNeXt performs better than using other backbone models.

Table 7 gives the performance metrics for the proposed and existing methods on the INBreast and CBIS-DDSM datasets.

Table 7. The comparison of the proposed method with SOTA method on CBIS-DDSM and INBreast datasets.

Reference	Method	F1	HL	mAP	RL	Cov	EM
			CBIS-DDSM				
Chougrad et al. [33]	VGG16	93.5% ± 0.019	0.047 ± 0.022	89.5% ± 0.017	0.087 ± 0.025	3.895 ± 0.320	82.2% ± 0.041
Proposed method	ConvNeXt	94.7% ± 0.01	0.036 ± 0.007	91.3% ± 0.017	0.07 ± 0.012	3.66 ± 0.095	86.9% ± 0.019
			INBreast				
Chougrad et al. [33]	VGG16	94.2% ± 0.102	0.042 ± 0.092	88.7% ± 0.140	0.082 ± 0.125	3.723 ± 0.147	82.7% ± 0.092
Proposed method	ConvNeXt	95.1% ± 0.016	0.032 ± 0.013	92.8% ± 0.025	0.065 ± 0.021	3.55 ± 0.174	88.9% ± 0.035

For the CBIS-DDSM dataset, Chougrad et al. [33], who applied the VGG16 architecture, attained an F1-score of 0.935, a Hamming loss (HL) of 0.047, a mean average precision (mAP) of 0.895, a ranking loss (RL) of 0.087, a coverage (Cov) of 3.895, and an exact match (EM) of 0.822. The proposed method, which utilized the ConvNeXt architecture with a fusion technique, outperformed the existing method, achieving a higher F1-score of 0.947, a lower HL of 0.036, a higher mAP of 0.913, a lower RL of 0.07, a lower Cov of 3.66, and a higher EM of 0.869.

For the INBreast dataset, Chougrad et al.'s method [33] achieved the following performance metrics when testing their proposed method using the INBreast dataset: an F1-score of 0.935, a hamming loss of 0.047, a mean average precision of 0.895, a ranking loss of 0.087, a coverage of 3.895, and an exact match of 0.822. The proposed method outperformed their method, achieving a higher F1-score of 95.13, a lower HL of 0.032, a higher mAP of 0.928, a lower RL of 0.06565, a lower Cov of 3.557143, and a higher EM of 0.88857.

These results indicate the superiority of the proposed method compared with the existing methods for both the CBIS-DDSM and INBreast datasets.

On the CBIS-DDSM dataset, our proposed method achieved improvement for all matrices. For the F1-score, the proposed method improved by 1.33% compared with Chougrad et al.'s method [33]. For HL, the proposed method significantly improved by 23.4% compared with Chougrad et al.'s method [33]. The proposed method showed a 4.60% improvement in mAP compared with Chougrad et al.'s [33] method. For RL, the proposed method significantly improved by 16.09% compared with Chougrad et al.'s method [33]. Regarding Cov, the proposed method showed a slight enhancement of 0.60% compared with Chougrad et al.'s [33] method. The proposed method improved by 5.06% in EM compared with Chougrad et al.'s method [33].

On the other hand, regarding the INBreast dataset, the proposed system outperformed the existing method in the INBreast dataset, with an improvement of 0.99% in the F1-score, 23.1% in Hamming loss, 4.64% in mAP, 20% in ranking loss, 4.45% in the coverage score, and 7.42% in exact match.

Overall, the proposed method achieved notable improvements compared with the previous methods. The results in Table 7 indicate that the proposed method using the ConvNeXt architecture with two views performs better than the method using both the VGG16 and Efficientnetb3 architectures among all the performance metrics. The higher F1-score, EM, and mAP, and the lower HL, Cov, and RL show that the proposed method achieves better accuracy and precision in predicting the presence of abnormalities in mammograms and classifying breast density.

In addition, the ROC curves shown in Figure 7, for each category—density, abnormality type (case), and severity level (pathology)—demonstrate the model's classification effectiveness. For 'Density,' the model's ability to differentiate between multiple density classes yielded an AUC score of 0.91, demonstrating its discriminative power. The 'Abnormality type (case)' category exhibited near-perfect classification with an AUC of 0.99, indicative of the model's exceptional accuracy in case determination. Similarly, the 'severity level (pathology)' category showed an AUC of 0.96, reflecting the model's high proficiency in identifying pathological features. These AUC values, significantly exceeding the 0.5 threshold of random chance, underscore the model's potential to provide reliable and accurate diagnostic assistance in mammographic analysis.

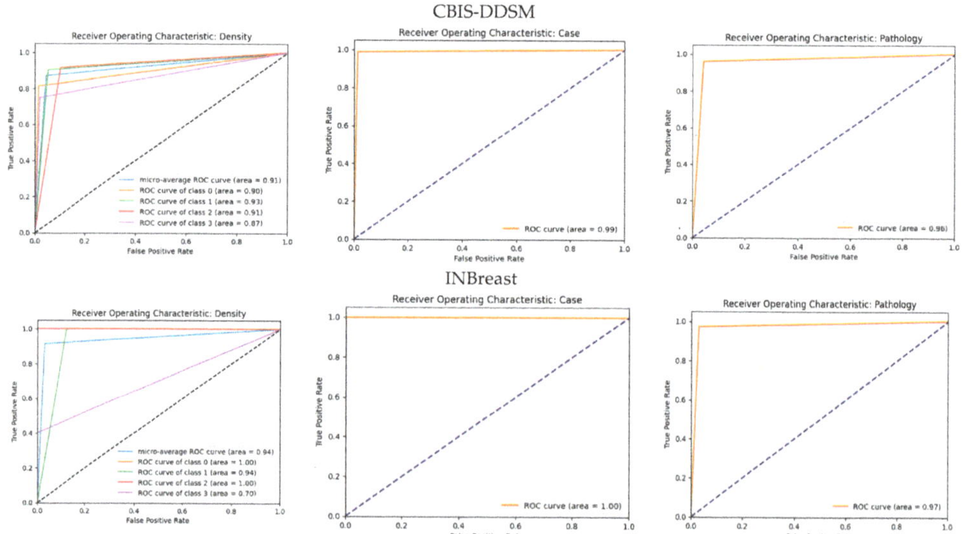

Figure 7. ROC curves for each case.

6.2. Comparative Analysis with Recent Deep Networks

Our research advances the multi-label classification of mammograms by integrating state-of-the-art techniques. It contrasts with previous studies like that of Chougrad et al. [33], which primarily relied on the simpler CNN model with the VGG16 architecture. Our model employs the ConvNeXt architecture, utilizing depth-wise separable convolutions. This design requires fewer learnable parameters, reducing the risk of overfitting and proving more adaptability in dynamically building networks based on data complexity, a clear advantage over the more rigid structure of VGG16.

Unlike traditional CNN models like VGG16, our approach incorporates advanced techniques like residual learning and transformer mechanisms. It also incorporates a channel attention mechanism based on squeeze-and-excitation, focusing the model on the most significant features and suppressing the less important ones in the input feature maps. A fusion method is utilized to integrate features from both CC and MLO views, providing a holistic analysis and a step forward from traditional single-view analyses, leading to a more complete representation of mammograms.

The dual-view technique (with CC and MLO views) significantly enhances our model's ability to detect abnormalities that might be visible in one view but not the other. It is a notable improvement over single-view analysis. Our exploration of various fusion methods revealed the average-wise method as the most effective. Incorporating a multi-branch architecture and channel attention techniques leads to more effective feature representation and task-specific learning. These innovative approaches contrast with traditional single-branch architectures, enhancing our model's accuracy and precision in predicting abnormalities.

6.3. Quantitative Analysis of Proposed Model

In this subsection, we provide the qualitative analysis of our model's performance, as illustrated in Figure 8. This analysis complements our quantitative findings and gives insight into the practical application of our method.

Figure 8 is structured into two columns, each representing a progressive increase in breast tissue density from left to right. Column 1 corresponds to the lowest density (BI-RADS I), while column 2 represents the highest density (BI-RADS IV). Within each column, rows 1 and 2 illustrate the differences between benign and malignant calcifications, and rows 3 and 4 distinguish between benign and malignant masses. This arrangement demonstrates the algorithm's capability to analyze and accurately differentiate breast abnormalities across tissue densities, showcasing its robustness and precision in lower- and higher-density scenarios.

Additionally, our model's inference efficiency is noteworthy, processing each image in an average of 286 milliseconds on advanced hardware. This speed is crucial for rapid and accurate breast tissue analysis in clinical settings. The combination of our model's diagnostic precision demonstrated in the annotated images and its quick processing time solidifies its potential as an effective tool in medical image analysis.

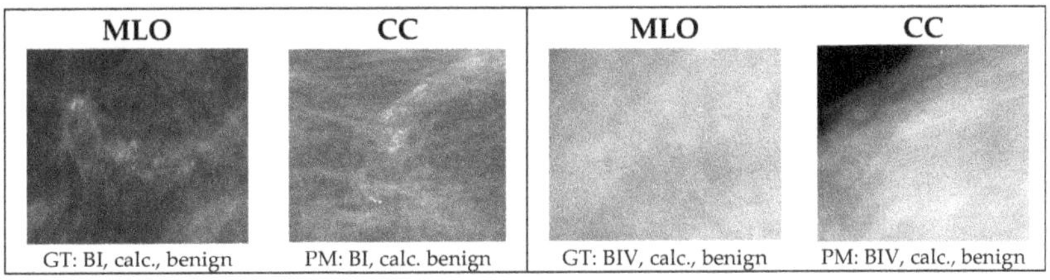

Figure 8. *Cont.*

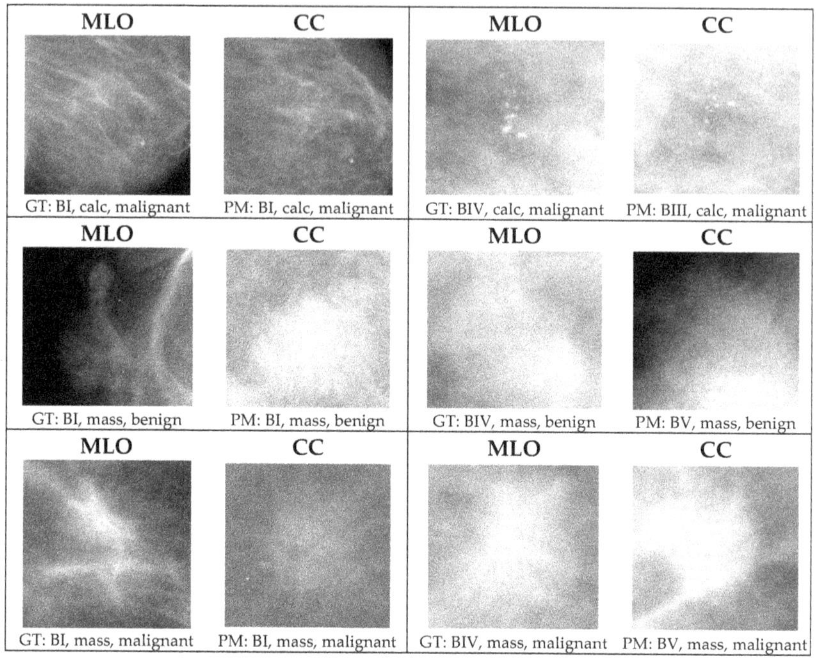

Figure 8. The ground truth (GT) labels and the labels predicted with the proposed model (PM) of each pair from two views.

7. Limitations and Future Work

The current work concentrates only on masses and calcifications in mammograms. It does not predict other abnormalities like asymmetries and architectural distortions. Furthermore, it is not able to reveal the regions in mammograms that play key roles in decision making, i.e., the interpretability of a decision.

In future studies, we aim to expand the scope of the algorithm by including abnormalities like asymmetries and architectural distortions and explore various fusion techniques for integrating CC and MLO views and different channel attention mechanisms, including SKNet, to enhance its performance in a multi-label classification system.

Additionally, we plan to implement spatial attention mechanisms for revealing the regions in mammogram images that play crucial roles in decision making and improving accuracy by utilizing complete contextual information. Future work will also investigate multi-label classification algorithms for more effective breast cancer diagnosis and risk assessment, particularly those centered on problem transformation and adaptation techniques.

The benchmark datasets, which we used to develop and evaluate the proposed system, are annotated according to the fourth edition of BI-RADS. However, the fifth edition of BI-RADS is available now, and there is a need to annotate the datasets according to this edition and evaluate the system's performance.

8. Conclusions

This research paper presented an innovative deep-learning-based model precisely designed to utilize the power of dual mammogram views: the craniocaudal (CC) and mediolateral oblique (MLO) views. This model's main objective is to diagnose comprehensively by simultaneously classifying mammograms based on their density, severity level/pathology, and abnormality type/findings. To achieve this, our model incorporates the state-of-the-art ConvNeXt as its backbone model. The design of this model is based on techniques like residual learning and transformer mechanisms, setting a solid foundation for advanced deep learning techniques. We utilized a channel attention mechanism based on squeeze-and-excitation to improve the ability of the model to concentrate on the most significant features and suppress the less important ones in the input feature maps. We employed an average-element-wise fusion method to consolidate the features' importance from both the CC and MLO views. This fusion method operates as a new layer within the model, seamlessly integrating the information extracted from the two views. This collaborative data merging ensures that no essential details are neglected, and the model acquires a holistic understanding of the mammogram image. Recognizing the diverse nature of the classification tasks, we introduced a multi-task/multi-branch architecture. This architecture tailors the feature-learning process to the unique requirements of each task: density, abnormality type, and lesion severity level. These tasks each have their distinct path within the architecture, facilitating more effective feature representation and task-specific learning. This enables our model to provide more accurate diagnoses for specific medical aspects. Employing multi-label learning helps enhance the model's ability to learn task-specific features and improves the model's performance. The proposed method was evaluated using benchmark datasets, the CBIS-DDSM and INBreast datasets, and outperformed SOTA. The proposed model is limited to masses and microcalcifications, and its extension to include other abnormalities will be the subject of future work.

Author Contributions: Conceptualization, E.A.-M. and M.H.; data curation, E.A.-M.; formal analysis, E.A.-M. and S.A.A.-A.; funding acquisition M.H.; methodology, E.A.-M. and M.H.; project administration, M.H. and H.A.A.; resources, M.H. and H.A.A.; software, E.A.-M.; supervision, M.H. and H.A.A.; validation, E.A.-M.; visualization, E.A.-M.; writing—original draft, E.A.-M.; writing—review and editing, M.H. and S.A.A.-A. All authors have read and agreed to the published version of the manuscript.

Funding: The authors extend their appreciation to the Deputyship for Research and Innovation of the Ministry of Education in Saudi Arabia for funding this research work under project no. IFKSUOR3–482-2.

Institutional Review Board Statement: Not applicable.

Informed Consent Statement: Not applicable.

Data Availability Statement: Public-domain datasets were used for the experiments. The CBIS-DDSM dataset is available at: https://wiki.cancerimagingarchive.net/display/Public/CBIS-DDSM (accessed on 29 November 2023). The InBreast dataset is available at: https://biokeanos.com/source/INBreast, https://www.kaggle.com/datasets/martholi/inbreast. (accessed on 29 November 2023).

Conflicts of Interest: The authors declare no conflict of interest.

Nomenclature

The following is a list of symbols and terms used in this manuscript along with their definitions for reference.

x	Input feature map
g	GlobalAveragePool function
W_1 , b_1	Weights and biases of the first fully connected layer
f_{se}	Squeezed feature map after ReLU activation
W_2, b_2	Weights and biases of the second fully connected layer
f_{ex}	Excited feature map
σ	Sigmoid activation function
W	Channel-wise excitation factors
Z	Feature map with applied attention

z_i	i^{th} channel of Z
$\propto_i$	Channel-wise excitation factor for channel i
x^a	Weighted feature map
z_{cc}, z_{mlo}	Feature maps for CC and MLO views.
g_{GAP}	GlobalAveragePool operation
x_{Fused}	Fused feature map
b	Label for breast cancer classification (density d, pathology p, and normal n)
p^b	Probability of class b after applying softmax
D	Number of channels
$\propto, \beta$	Vectors representing global average pooled features for CC and MLO views, respectively
N	Total number of instances
y , $\hat{y}_n$	Ground truth label and predicted label for the nth instance
$F1_{Score}$	Harmonic mean of precision and recall
MAP	Mean average precision
$r(x_n, y)$	Rank of label y for n^{th} instance
HL	Hamming loss
Δ	Symmetric difference operator
RL	Ranking loss
$\overline{y_n}$	Complement of the set y_n
$Coverage$	Average count of labels to examine for all reference labels
EM	Exact match

References

1. National Cancer Institute (United States). Available online: https://www.cancer.gov/about-cancer/understanding/what-is-cancer (accessed on 9 February 2015).
2. American Cancer Society. Available online: https://www.cancer.org/healthy/find-cancer-early/womens-health/cancer-facts-for-women.html (accessed on 1 August 2019).
3. American Cancer Society. Available online: https://www.cancer.org/cancer/breast-cancer/understanding-a-breast-cancer-diagnosis/breast-cancer-survival-rates.html (accessed on 8 August 2019).
4. Allison, K.H.; Abraham, L.A.; Weaver, D.L.; Tosteson, A.N.A.; Nelson, H.D.; Onega, T.; Geller, B.M.; Kerlikowske, K.; Carney, P.A.; Ichikawa, L.E.; et al. Trends in breast tissue sampling and pathology diagnoses among women undergoing mammography in the US: A report from the breast cancer surveillance consortium. *Cancer* **2015**, *121*, 1369–1378. [CrossRef] [PubMed]
5. Ramadan, S.Z. Methods Used in Computer-Aided Diagnosis for Breast Cancer Detection Using Mammograms: A Review. *J. Healthc. Eng.* **2020**, *2020*, 9162464. [CrossRef] [PubMed]
6. Abdelhafiz, D.; Yang, C.; Ammar, R.; Nabavi, S. Deep convolutional neural networks for mammography: Advances, challenges and applications. *BMC Bioinform.* **2019**, *20*, 281. [CrossRef] [PubMed]
7. Ahn, C.K.; Heo, C.; Jin, H.; Kim, J.H. A novel deep learning-based approach to high accuracy breast density estimation in digital mammography. In *Medical Imaging 2017: Computer-Aided Diagnosis*; SPIE: Bellingham, WA, USA, 2017; Volume 10134, pp. 691–697.
8. Ionescu, G.V.; Fergie, M.; Berks, M.; Harkness, E.F.; Hulleman, J.; Brentnall, A.R.; Cuzick, J.; Evans, D.G.; Astley, S.M. Prediction of reader estimates of mammographic density using convolutional neural networks. *J. Med. Imag.* **2019**, *6*, 031405. [CrossRef] [PubMed]
9. Wu, N.; Geras, K.J.; Shen, Y.; Su, J.; Kim, S.G.; Kim, E.; Wolfson, S.; Moy, L.; Cho, K. Breast density classification with deep convolutional neural networks. In Proceedings of the 2018 IEEE International Conference on Acoustics, Speech and Signal Processing (ICASSP), Calgary, AB, Canada, 15–20 April 2018; pp. 6682–6686.
10. Mohamed, A.A.; Berg, W.A.; Peng, H.; Luo, Y.; Jankowitz, R.C.; Wu, S. A deep learning method for classifying mammographic breast density categories. *Med. Phys.* **2018**, *45*, 314–321. [CrossRef] [PubMed]
11. Chen, X.; Zargari, A.; Hollingsworth, A.B.; Liu, H.; Zheng, B.; Qiu, Y. Applying a new quantitative image analysis scheme based on global mammographic features to assist diagnosis of breast cancer. *Comput. Methods Programs Biomed.* **2019**, *179*, 104995. [CrossRef]
12. Sun, L.; Wang, J.; Hu, Z.; Xu, Y.; Cui, Z. Multi-View Convolutional Neural Networks for Mammographic Image Classification. *IEEE Access* **2019**, *7*, 126273–126282. [CrossRef]
13. Das, P.; Das, A. Shift invariant extrema based feature analysis scheme to discriminate the spiculation nature of mammograms. *ISA Trans.* **2020**, *103*, 156–165. [CrossRef]
14. Nagarajan, V.; Britto, E.C.; Veeraputhiran, S.M. Feature extraction based on empirical mode decomposition for automatic mass classification of mammogram images. *Med. Nov. Technol. Devices* **2019**, *1*, 100004. [CrossRef]
15. George, M.; Chen, Z.; Zwiggelaar, R. Multiscale connected chain topological modelling for microcalcification classification. *Comput. Biol. Med.* **2019**, *114*, 103422. [CrossRef]

16. Mabrouk, M.S.; Afify, H.M.; Marzouk, S.Y. Fully automated computer-aided diagnosis system for micro calcifications cancer based on improved mammographic image techniques. *Ain Shams Eng. J.* **2019**, *10*, 517–527. [CrossRef]
17. Li, H.; Zhuang, S.; Li, D.-A.; Zhao, J.; Ma, Y. Benign and malignant classification of mammogram images based on deep learning. *Biomed. Signal Process. Control* **2019**, *51*, 347–354. [CrossRef]
18. Mohanty, F.; Rup, S.; Dash, B.; Majhi, B.; Swamy, M.N.S. An improved scheme for digital mammogram classification using weighted chaotic salp swarm algorithm-based kernel extreme learning machine. *Appl. Soft Comput.* **2020**, *91*, 106266. [CrossRef]
19. Simonyan, K.; Zisserman, A. Very deep convolutional networks for large-scale image recognition. *arXiv* **2014**, arXiv:1409.1556.
20. Li, X.; Wang, W.; Hu, X.; Yang, J. Selective kernel networks. In Proceedings of the IEEE Conference on Computer Vision and Pattern Recognition, Long Beach, CA, USA, 15–20 June 2019; pp. 510–519.
21. Chen, Y.; Li, J.; Xiao, H.; Jin, X.; Yan, S.; Feng, J. Dual path networks. In *NIPS*; Curran Associates Inc.: Red Hook, NY, USA, 2017.
22. Huang, G.; Liu, Z.; Van Der Maaten, L.; Weinberger, K.Q. Densely connected convolutional networks. In Proceedings of the IEEE Conference on Computer Vision and Pattern Recognition, Honolulu, HI, USA, 21–26 July 2017.
23. He, K.; Zhang, X.; Ren, S.; Sun, J. Deep residual learning for image recognition. In Proceedings of the IEEE Conference on Computer Vision and Pattern Recognition, Las Vegas, NV, USA, 26 June–1 July 2016.
24. Xie, S.; Girshick, R.; Dollár, P.; Tu, Z.; He, K. Aggregated residual transformations for deep neural networks. In Proceedings of the IEEE Conference on Computer Vision and Pattern Recognition, Honolulu, HI, USA, 21–26 July 2017.
25. Hu, J.; Shen, L.; Sun, G. Squeeze-and-excitation networks. *arXiv* **2017**, arXiv:1709.01507.
26. Ayana, G.; Dese, K.; Dereje, Y.; Kebede, Y.; Barki, H.; Amdissa, D.; Husen, N.; Mulugeta, F.; Habtamu, B.; Choe, S.-W. Vision-Transformer-Based Transfer Learning for Mammogram Classification. *Diagnostics* **2023**, *13*, 178. [CrossRef] [PubMed]
27. Yu, X.; Ren, Z.; Guttery, D.S.; Zhang, Y.-D. DF-dRVFL: A novel deep feature based classifier for breast mass classification. *Multimed. Tools Appl.* **2023**, 1–30. [CrossRef]
28. Gerbasi, A.; Clementi, G.; Corsi, F.; Albasini, S.; Malovini, A.; Quaglini, S.; Bellazzi, R. DeepMiCa: Automatic segmentation and classification of breast MIcroCAlcifications from mammograms. *Comput. Methods Programs Biomed.* **2023**, *235*, 107483. [CrossRef]
29. Sarvestani, Z.M.; Jamali, J.; Taghizadeh, M.; Dindarloo, M.H.F. A novel machine learning approach on texture analysis for automatic breast microcalcification diagnosis classification of mammogram images. *J. Cancer Res. Clin. Oncol.* **2023**, *149*, 6151–6170. [CrossRef]
30. Jabeen, K.; Khan, M.A.; Balili, J.; Alhaisoni, M.; Almujally, N.A.; Alrashidi, H.; Tariq, U.; Cha, J.-H. BC2NetRF: Breast cancer classification from mammogram images using enhanced deep learning features and equilibrium-jaya controlled regula falsi-based features selection. *Diagnostics* **2023**, *13*, 1238. [CrossRef]
31. Chakravarthy, S.S.; Bharanidharan, N.; Rajaguru, H. Deep Learning-Based Metaheuristic Weighted K-Nearest Neighbor Algorithm for the Severity Classification of Breast Cancer. *IRBM* **2023**, *44*, 100749. [CrossRef]
32. Azour, F.; Boukerche, A. An Efficient Transfer and Ensemble Learning Based Computer Aided Breast Abnormality Diagnosis System. *IEEE Access* **2022**, *11*, 21199–21209. [CrossRef]
33. Chougrad, H.; Zouaki, H.; Alheyane, O. Multi-label transfer learning for the early diagnosis of breast cancer. *Neurocomputing* **2020**, *392*, 168–180. [CrossRef]
34. Jafari, Z.; Karami, E. Breast Cancer Detection in Mammography Images: A CNN-Based Approach with Feature Selection. *Information* **2023**, *14*, 410. [CrossRef]
35. Lee, R.S.; Gimenez, F.; Hoogi, A.; Miyake, K.K.; Gorovoy, M.; Rubin, D.L. A curated mammography data set for use in computer-aided detection and diagnosis re- search. *Sci. Data* **2017**, *4*, 170177. [CrossRef]
36. Moreira, I.C.; Amaral, I.; Domingues, I.; Cardoso, A.; Cardoso, M.J.; Cardoso, J.S. INbreast: Toward a full-field digital mammographic database. *Acad. Radiol.* **2012**, *19*, 236–248. [CrossRef] [PubMed]
37. Solla Sara, A.; Levin, E.; Fleisher, M. Accelerated Learning in Layered Neural Networks. *Complex Syst.* **1988**, *2*, 3.
38. Multi-Label Classification by Exploiting Local Positive and Negative Pairwise Label Correlation—ScienceDirect, (n.d.). Available online: https://www.sciencedirect.com/science/article/pii/S0925231217301571 (accessed on 11 May 2018).
39. Multi-Label Learning Based on Label-Specific Features and Local Pairwise Label Correlation—ScienceDirect, (n.d.). Available online: https://www.sciencedirect.com/science/article/pii/S0925231217313462 (accessed on 11 May 2018).
40. Lopez, M.G.; Posada, N.; Moura, D.C.; Pollán, R.R.; Valiente, J.M.F.; Ortega, C.S.; Solar, M.; Diaz-Herrero, G.; Ramos, I.M.A.P.; Loureiro, J.; et al. BCDR: A breast cancer digital repository. In Proceedings of the 15th International Conference on Experimental Mechanics, Porto, Portugal, 22–27 July 2012.
41. Suckling, J.; Parker, J.; Dance, D. The Mammographic Image Analysis Society Digital Mammogram Database. In *Exerpta Medica*; International Congress Series; Excerpta Medica Foundation: Amsterdam, Netherlands, 1994; Volume 1069, pp. 375–378.

applied sciences

MDPI

Article

Multi-Pedestrian Tracking Based on KC-YOLO Detection and Identity Validity Discrimination Module

Jingwen Li [1,2], Wei Wu [1], Dan Zhang [3], Dayong Fan [3], Jianwu Jiang [1,2,*], Yanling Lu [1,2], Ertao Gao [1,2] and Tao Yue [1,2]

1 College of Geomatics and Geoformation, Guilin University of Technology, Guilin 541004, China
2 Ecological Spatiotemporal Big Data Perception Service Laboratory, Guilin 541004, China
3 Guilin Agricultural Science Research Center, Guilin 541004, China
* Correspondence: fengbuxi@glut.edu.cn

Abstract: Multiple-object tracking (MOT) is a fundamental task in computer vision and is widely applied across various domains. However, its algorithms remain somewhat immature in practical applications. To address the challenges presented by complex scenarios featuring instances of missed detections, false alarms, and frequent target switching leading to tracking failures, we propose an approach to multi-object tracking utilizing KC-YOLO detection and an identity validity discrimination module. We have constructed the KC-YOLO detection model as the detector for the tracking task, optimized the selection of detection frames, and implemented adaptive feature refinement to effectively address issues such as incomplete pedestrian features caused by occlusion. Furthermore, we have introduced an identity validity discrimination module in the data association component of the tracker. This module leverages the occlusion ratio coefficient, denoted by "k", to assess the validity of pedestrian identities in low-scoring detection frames following cascade matching. This approach not only enhances pedestrian tracking accuracy but also ensures the integrity of pedestrian identities. In experiments on the MOT16, MOT17, and MOT20 datasets, MOTA reached 75.9%, 78.5%, and 70.1%, and IDF1 reached 74.8%, 77.8%, and 72.4%. The experimental results demonstrate the superiority of the methodology. This research outcome has potential applications in security monitoring, including public safety and fire prevention, for tracking critical targets.

Keywords: KC-YOLO; object detection; identity validity discriminator; multi-pedestrian tracking

Citation: Li, J.; Wu, W.; Zhang, D.; Fan, D.; Jiang, J.; Lu, Y.; Gao, E.; Yue, T. Multi-Pedestrian Tracking Based on KC-YOLO Detection and Identity Validity Discrimination Module. *Appl. Sci.* **2023**, *13*, 12228. https://doi.org/10.3390/app132212228

Academic Editors: Zahid Mehmood Jehangiri, Mohsin Shahzad and Uzair Khan

Received: 13 October 2023
Revised: 7 November 2023
Accepted: 8 November 2023
Published: 10 November 2023

1. Introduction

Multi-pedestrian tracking (MPT) serves as a foundational task within the realm of computer vision and finds applications in numerous computer vision domains [1]. MPT involves estimating the trajectories of multiple objects of interest within video sequences, holding pivotal significance in video analytics systems for domains like surveillance security [2], automated driving, intelligent transportation [3], behavioral recognition [4], human–computer interaction, and intelligent agriculture [5,6]. While extensive research has been conducted in this field, a definitive method that can consistently perform exceptionally well in addressing the challenges posed by complex scenes with frequent occlusions in surveillance videos remains elusive [7]. The current focus for enhancing the accuracy of multi-pedestrian tracking primarily involves optimizing pedestrian detector performance, refining the extraction of representative pedestrian features, and improving data association matching algorithms [8].

For the optimization of pedestrian detector performance, Zhang [9], in his study, introduced a small-target pedestrian inspection model incorporating residual networks and feature pyramids, which dispenses with unnecessary, redundant computations in the model and solves the gradient problem in a neural network by using residual blocks with a discarded layer instead of the standard residual block, thus significantly improving the accuracy and anti-jamming ability of small-target pedestrian detection. Liu [10] introduced an enhanced detection-and-tracking framework with a semantic matching strategy based on deep learning. Integrating scene-aware affinity detection, this framework proves to be highly effective in alleviating challenges related to occlusion and similar appearances. Zhang [11] introduced an innovative approach, FairMOT, which combines CenterNet and directly embeds the Re-ID module, whose training process utilizes the cross-entropy loss function, aiding in obtaining more accurate target features. This amalgamation achieves higher precision in capturing target features, all while considering the trade-off between speed and accuracy in the multi-target tracking model. Zhang et al. [12] proposed a multi-pedestrian tracking algorithm using the Tracking-by-Detection framework. It addresses the diversity of human postures, appearance similarities, and occlusion in real-time road traffic scenes. The algorithm effectively leverages both pedestrian depth appearance features and motion features to establish correlations among the tracking targets, thus realizing the multi-objective target tracking of pedestrians. Zhou et al. [13] proposed an improved MOT approach for occlusion scenarios, combining attention mechanisms and occlusion sensing as a solution. Jia [14] designed and developed a network for learning separate representations for processing occlusion re-identification guided by semantic preference object queries in a converter without strict character image alignment or any additional supervision. To better eliminate occlusion interference, they devised a Contrast Feature Learning approach to better separate hidden features from recognition features. Bewley et al. [15] proposed the Simple Online Real-Time Tracking (SORT) method, which fuses positional and motion information in a similarity matrix for target ID association and achieved good results in short-range matching. Bewley et al. [16] proposed DeepSORT based on SORT, which adds an offline pedestrian re-identification network and achieves better results in long-distance matching by merging appearance and motion information. Zhang [17] proposed the BYTE data association method, which introduces low-confidence detection frames into data association matching and utilizes these low-confidence similarities between the detection frames and the tracking trajectories to mine out heavily occluded targets, thus maintaining the continuity of the tracking trajectories.

While significant progress has been made in enhancing detector performance, extracting more representative features, and improving data association and matching algorithms, most tracking tasks still face common challenges in complex scenarios, such as occlusion, omissions, and distractions [18]. As a result, the robustness of existing methods is in need of improvement [19].

Based on the above problems, in order to solve complex surveillance video scenes with multiple targets tracked simultaneously, we propose a method for the simultaneous tracking of multiple pedestrians based on KC-YOLO detection and an identity validity discrimination module (IVDM). We have made improvements in both the detector and the tracker. The Convolution Block Attention Module (CBAM) [20] is introduced into the detector, utilizing attention weights to allow for a more focused and refined representation of the target features, which improves the ability of the detector to capture the effective feature information of the target, which has been decisive in improving the overall precision and accuracy of the test procedure. In the tracking process, to address the issue of tracking failure due to the short-term occlusion of the target, this method constructs an IVDM after cascade matching. The target occlusion coefficient k is calculated to discriminate whether the target identity in the low-scoring detection frame after target detection and cascade matching is valid or not and to decide whether to update its appearance features so as not to generate redundant identity data, thereby ensuring the purity of the tracked pedestrian's identity and improving the overall performance of the tracking task.

To summarize, this paper's primary contributions can be outlined as follows:

- An efficient, robust, and practical multi-pedestrian tracking method based on KC-YOLO deep detection and identity validity discrimination is proposed. This method provides an effective solution for multi-pedestrian tracking tasks in complex surveillance videos. Experimental results demonstrate its high utility, making it suitable for the long-term tracking of critical targets in various scenarios, such as public safety and firefighting.

- An improved pedestrian object detector based on YOLOv5, tailored for complex environments, has been designed. This detector employs the K-means++ clustering method to select optimal detection frames and introduces the CBAM for adaptive feature refinement. The KC-YOLO network is introduced for extracting target depth features.
- A pedestrian identity validity model has been developed. To address challenges such as targets reappearing after occlusion and rapid identity switches, this model assesses the identity validity of newly generated targets. Different processing strategies are applied to targets with identity validity, enhancing the tracking accuracy while ensuring the purity of pedestrian target identities.

This paper is structured such that Section 2 introduces the summary of the work related to the proposed method in this study. The multi-Pedestrian Tracking Method Based on IVDM is discussed in Section 3. The experimental data and analyses the experimental results are highlighted in Section 4, and in Section 5 we summarize this study. Lastly, we discuss this study and provide an outlook for future research in Section 6.

2. Related Work

2.1. Target Detection Methods

Target detection serves as the foundational component in the domain of multi-target tracking. The role of the detector is to furnish the tracker with the positional information of objects within the image, typically yielding the detection frame of the object. Presently, target detection algorithms achieving high accuracy are frequently implemented on the bedrock of Deep Convolutional Neural Networks (CNNs) [21]. Unlike traditional methods, deep-learning-based object recognition utilizes CNNs to autonomously capture recognizable object features. This automatic extraction process allows the model to learn complex patterns and representations from the input data, thereby improving the recognition accuracy and efficiency. In addition, hierarchical learning using CNNs allows the model to recognize features at different levels of abstraction, resulting in more justifiable and precise feature extraction. It has diverged based on detection principles, segregating into two types of methodologies [7].

Two-stage target detection involves generating candidate regions and subjecting them to a two-fold classification process. Region proposals utilizing a CNN (R-CNN) [22] input fixed-size images into a neural network to facilitate training and object feature extraction. While it attains higher detection precision compared to traditional object-detection methods, it does suffer from computational intensity and tardiness in object detection. Extending from the R-CNN algorithm, Fast R-CNN and Faster R-CNN emerge. Although these methods enhance detection accuracy compared to traditional approaches, the bifurcation between candidate region generation and classification engenders sluggish algorithmic operation, hampering real-time target detection realization. Efforts to enhance real-time capabilities still grapple with the challenge of duplicated computation. Additionally, R-CNN is hamstrung by its fixed input image size. To mitigate the pre-input image-scaling computational burden, the Spatial Pyramid Pool Network (SPPNet) was conceived, albeit only partially reducing superfluous computations. Among the R-CNN family, Faster R-CNN currently stands out with the swiftest and closest-to-real-time detection performance. This efficiency is pivotal for applications demanding rapid and accurate object detection, yet it remains encumbered in meeting the demands of intricate target detection scenarios.

One-stage detection algorithms eschew the region proposal phase and promptly produce class probabilities and the positional coordinates of objects. Representative algorithms include the YOLO family [23], the Single-Shot Multi-Frame Detector (SSD), and RetinaNet. YOLO's framework implements the detection process by allowing the model to directly predict the bounding box and class probability of each cell, which distinguishes this method from two-stage object-detection models. By doing so, YOLO achieves a more efficient and faster differentiation and correlation process, making it particularly suitable for real-time applications such as video analysis and object tracking. Notwithstanding its strengths, YOLO demonstrates suboptimal detection accuracy for smaller objects. SSD capitalizes on feature maps of varying dimensions for object detection, rectifying YOLO's shortcomings in smaller-object detection. The contemporary YOLO family of algorithms collectively refines the detection accuracy without compromising on high detection speed.

However, in intricate traffic environments, existing object-detection algorithms still cannot simultaneously ensure real-time performance and capture as many feature points as possible.

2.2. Attention Mechanisms

The attention mechanism is a mechanism that mimics human attentional processes, and it is widely used in deep learning [24]. This mechanism enables the rapid extraction of key information from the environment and allows the observer to scrutinize the details of the object. After the attention mechanism, different regions will have their own weights so that the system can focus on the important information. This mechanism was initially introduced in the sphere of computer vision and is now utilized across various domains, such as natural language processing, speech recognition, and recommendation systems. The versatility of attention mechanisms lies in their ability to enhance model performance by focusing on relevant information while reducing the computational burden associated with processing unnecessary or redundant data. Therefore, they constitute a crucial component of advanced machine-learning models in diverse domains [25]. Based on their different scopes of action, attention mechanisms have been classified into three categories.

The spatial attention mechanism originates from the rationale that certain regions within input images are extraneous to recognition or segmentation tasks. The mechanism processes only regions pertinent to the task, preserving task-relevant regions while suppressing extraneous ones. An exemplary embodiment, the Spatial Transformation Network (STN) by Google DeepMind, learns preprocessing operations from input data that align with the specific task [13].

In the detection task, the input images pass through both the spatial and channel dimensions, one after the other. The network provides a significantly more comprehensive understanding of the underlying information based on the inter-channel dependencies [19]. The prominent channel attention model, SENet, compresses the input feature map spatially while preserving its channel dimension. SENet devises channel weights, adapts them during training, and then utilizes them to amplify crucial channel information while dampening insignificant channel data. Consequently, the network's feature extraction efficiency is notably enhanced [6].

Hybrid attention mechanisms amalgamate spatial and channel methods. However, certain models inadequately address the inherent interplay between features, rendering them unable to concurrently process both spatial and channel features. In this domain, representative models include the CBAM and dual-attention networks.

2.3. Multi-Objective Tracking Methods

Multi-objective tracking (MOT) can be classified into a detection-based tracking framework [26] and a joint detection-and-tracking-based framework, depending on the method. The detection-based tracking framework is a common approach to tracking multiple targets; it relies on target detection as the first step in locating and identifying targets in each frame and crops the objects according to the enclosing frame to obtain all of the targets in the image. Then, it is transformed into a target association problem between neighboring frames, and a similarity matrix is constructed based on IOU, appearance, etc., and solved by methods such as the Hungarian algorithm. As the performance of target detection has improved by leaps and bounds, the field of MOT has revolved around detection-based tracking frameworks for quite some time [27]. Representative methods are SORT and DeepSORT. SORT is an algorithm for tracking objects in a video sequence in real time, which is similar to many modern tracking methods [23]. Consider the case where two targets are occluded. The trajectories of the matched targets cannot be matched for detection, and the targets temporarily disappear. When a target that disappeared briefly reappears later, the target will regain its ID number to stop changing. To enhance the SORT algorithm, the researchers added cascade matching and state estimation to it. In recent years, several joint detection tracking approaches [11] have been introduced to jointly enhance detection and a few other components. The joint tracker provides equivalent performance with minimal computational cost. However, any inconsistencies or inaccuracies in any of the components can propagate errors, which can degrade the overall tracking performance, due to the fact that there are too many components, causing this type of method to not perform very well. Therefore, the detection-based tracking framework remains the most suitable multi-target tracking method in terms of tracking accuracy.

2.4. Current Issues in Multi-Pedestrian Tracking

In today's day and age, multi-pedestrian tracking still presents many challenges. For example, the following are three of the more common challenges:

- Robustness: In complex scenarios characterized by rapidly changing lighting conditions, frequent occlusions, and dynamic blurring, the robustness of multi-pedestrian tracking algorithms tends to be compromised. To tackle this, we constructed the KC-YOLO detection model as the detector in our research. This model optimizes the selection of detection frames and implements adaptive feature refinement, thereby enhancing the robustness and accuracy of the detection algorithm.
- Long-term tracking: Tracking targets in long temporal sequences presents several challenges, as it requires addressing cross-frame target re-identification and scene updates. To tackle this, we employ cascade matching for target re-identification, which effectively reduces instances of target loss caused by occlusion and scene updates [28].
- Algorithm efficiency: In real-world applications, multi-pedestrian tracking algorithms often need to process a large volume of data in real time. Overcoming these challenges is crucial for improving multi-pedestrian tracking methods and ensuring their effectiveness in diverse and complex practical environments. In our research work, we introduced an identity validity discrimination module into the tracking algorithm. This module is designed to assess and remove erroneous data resulting from incomplete or unclear features, reducing the unnecessary data-processing workload.

3. Multi-Pedestrian Tracking Method Based on IVDM

In the context of multi-pedestrian tracking, detection and tracking tasks are both independent and closely related to each other [29]. We adopt the KC-YOLO detection model to detect pedestrians in complex traffic environments, where the apparent information may be incomplete and unclear. Then, we introduce the IVDM as part of the improved approach to realizing the tracking of multiple pedestrian targets. The integrated detector–tracker structure is shown in Figure 1.

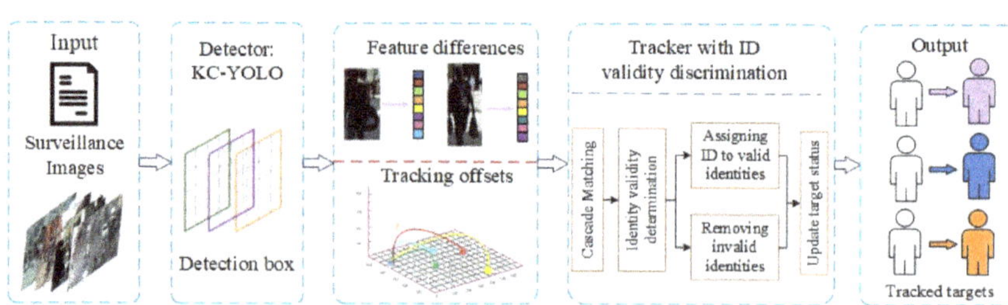

Figure 1. Integrated detector–tracker structure.

We utilize the KC-YOLO network as our detector to extract the adaptive deep features of the targets. We combine this with trajectory matching based on Kalman filtering predictions. For tracking, we employ DeepSORT, which incorporates pedestrian identity validity discrimination. This combination allows us to perform accurate and efficient multi-pedestrian tracking.

3.1. Construction of KC-YOLO Detection Model

We propose a model called KC-YOLO, which is applied to complex scenes in surveillance video and uses the YOLOv5 [30] detection model as the base algorithm.

The core steps of the KC-YOLO model are as follows:

- Determine the optimal anchor frame that is compatible with the input pedestrian image;
- Extract the deep features of the pedestrian image through the KC-YOLO network. Use the attention mechanism to highlight its salient information and achieve adaptive feature refinement.

We introduce the CBAM into the backbone and neck parts of the detection network for the following reasons: the backbone part is the key part for extracting pedestrian features, while the neck part fuses the features and sends them to the head for prediction, and the introduction of the CBAM here can improve the feature extraction ability of the network more effectively. The structure of the improved KC-YOLO network is shown in Figure 2.

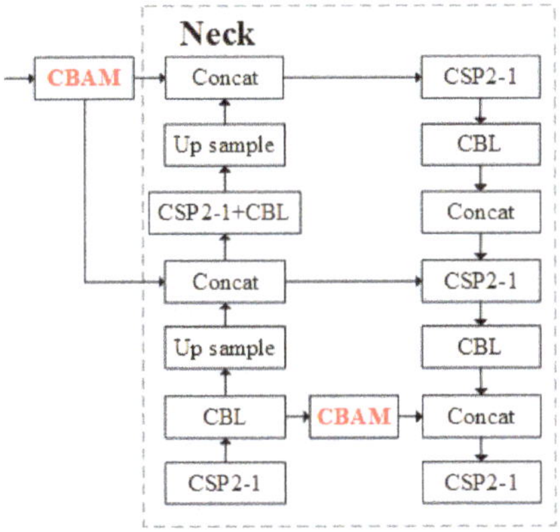

Figure 2. KC-YOLO network model structure (Concat is primarily responsible for combining addition and residual convolution operations; CBL is a convolutional block; Up sample means that usampling operations are performed; CSP2_1 divides the input feature map into two parts).

Concat is primarily responsible for combining addition and residual convolution operations. Through feature fusion, it allows the detection network to simultaneously utilize the extracted shallow and deep features. The main purpose of the upsample structure is to perform upsampling operations; CBL is a convolutional block. Within CSP2_1, the input feature map is divided into two parts. One part is processed through a subnetwork, while the other part undergoes further processing directly. These two sets of feature maps are then concatenated and used as input for the next layer. By combining the features processed by the subnetwork with those processed directly, a series of convolution operations are performed. This approach effectively integrates low-level detail features with high-level abstract features, thereby improving the feature extraction efficiency.

3.1.1. Optimal Pedestrian Detection Frame Determination

In the context of pedestrian detection, YOLOv5 defaults to using k-means clustering to generate anchor frames. However, before performing k-means clustering, it is crucial to initialize k cluster centers, as the convergence can be significantly affected by uninitialized cluster centers. To address this issue, we employ the k-means++ clustering method [31]. Here is how it works:

- Initially, a random sample point is selected from the dataset as the first initial cluster center.
- Then, the shortest distance between each sample point and the currently existing cluster centers is calculated.
- Finally, each sample point is chosen as the next cluster center with a probability proportional to the shortest distance. The sample point with the highest probability is selected as the next cluster center.

This approach provides a more reliable initialization method, improving the stability and convergence of the clustering process, which, in turn, optimizes the selection of detection frames. The formula for the calculation is as follows:

$$P(\mathrm{x}) = \frac{D(X_i)^1}{\sum\limits_{i=1}^{n} D(X_i)^1} \tag{1}$$

where C_i represents the first initial cluster center; *D(X)* denotes the shortest distance between each sample point and the currently existing cluster centers; and *P(X)* represents the probability of each sample point being selected as the next cluster center.

3.1.2. Deep Feature Extraction

Deep features extracted by convolutional neural networks can provide an effective description of the high-level semantic information of an image, and the CBAM is an attention mechanism module used to enhance the performance of convolutional neural networks with significant results. In order to improve the feature extraction capability of the detection network [32], we introduce the CBAM [20] into the detection model.

The CBAM depicted in Figure 3 comprises both the Channel Attention Module (CAM) and the Spatial Attention Module (SAM). The CAM is employed to enhance the weights of important features while reducing the weights of irrelevant features. It begins by subjecting the input feature map to max pooling and average pooling along the channel dimension. The output results are then fused through an MLP network, and subsequently, weight coefficients are obtained by applying the Sigmoid activation function.

$$M_c(F) = \sigma(MLP(AvgPool(F)) + MLP(MaxPool(F))) = \sigma\left(W_1\left(W_0\left(F^c_{avg}\right)\right) + W_1(W_0(F^c_{\max}))\right) \tag{2}$$

where σ is the Sigmoid activation function; $w_0 \in R^{C_r \times C}$ and $w_1 \in R^{C \times C_r}$ are the weights, and r is the contraction rate; *MLP* stands for a neural network; $M_c(F)$ is obtained by performing element-wise summation and applying the Sigmoid activation operation on the shared fully connected layer; and F^c_{avg} and $F^c_{\max}$ are the two features obtained by pooling the extracted features.

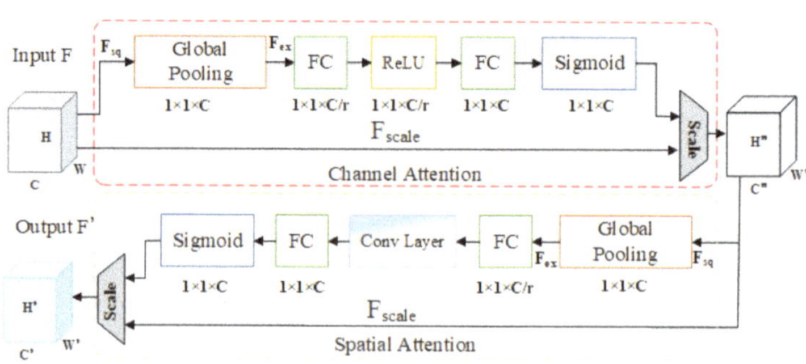

Figure 3. The structure of CBAM (Global Pooling is the global maximum pooling layer; FC is Rectified Linear Unit; Sigmoid is an S-type activation function).

The SAM focuses on the intrinsic relationships within the spatial dimensions of the input feature map. It takes the output from the CAM and performs max pooling and average pooling along the channel direction. The results obtained are then processed through a convolutional layer with a kernel size of 7 × 7. Finally, the SAM's feature map is obtained by applying the Sigmoid activation function. The calculation is performed with the following equation:

$$M_s(F) = \sigma\left(f^{7\times7}([AvgPool(F); MaxPool(F)])\right) = \sigma\left(f^{7\times7}\left(\left[F^c_{avg}; F^c_{\max}\right]\right)\right) \quad (3)$$

where $f^{7\times7}$ denotes the convolution kernel size; $M_s(F)$ is obtained by the logistic activation function.

3.2. Multi-Pedestrian Tracking Methods Based on IVDM

Pedestrian tracking not only provides trajectory information but also provides valuable information for behavioral analysis. However, in crowded scenes, a large number of targets may be occluded, resulting in missing and blurred features, which seriously affects the function of detection-based tracking methods [16]. When the video surveillance fields of view do not overlap and the pedestrians are heavily occluded, the "1-n" pedestrian identity phenomenon results. Existing tracking algorithms still lack a flexible approach to dealing with heavily occluded targets and thus perform poorly in complex scenarios where heavy occlusion occurs frequently [33]. To address the above situation, based on the improved detection method in the previous section, we introduce pedestrian identity validity judgment into the pedestrian tracking process, which performs "occlusion perception-occlusion ratio k calculation-pedestrian identity validity discrimination" on unmatched targets between different frames (Figure 4).

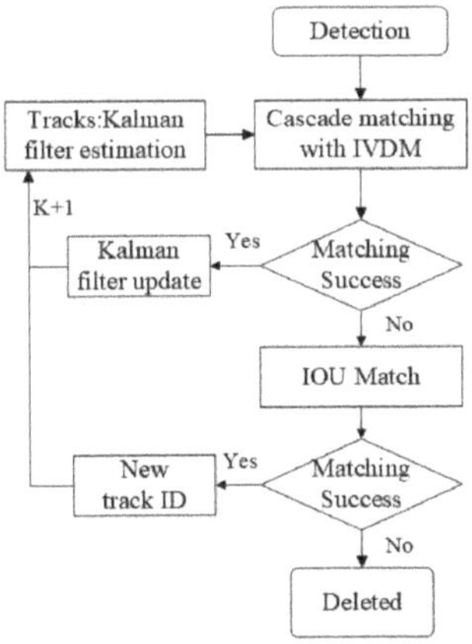

Figure 4. Multi-pedestrian tracking method based on IVDM.

The module performs a series of calculation and discrimination operations on unmatched targets between different frames, such as "occlusion perception-calculation of occlusion ratio k-pedestrian identity validity discrimination", which determines the degree of occlusion of pedestrians detected by the surveillance video based on the magnitude of the coefficient k of the proportion of occlusion of pedestrians in the frame and categorizes the occluded pedestrians into valid ID_Y and invalid ID_N through k. In essence, the above process is used to discern whether or not the identity of the detected pedestrian has validity. The most successful associations in pedestrian tracking often occur in the cascade matching section. Therefore, we have incorporated the IVDM into the cascade matching process, as depicted in Figure 5.

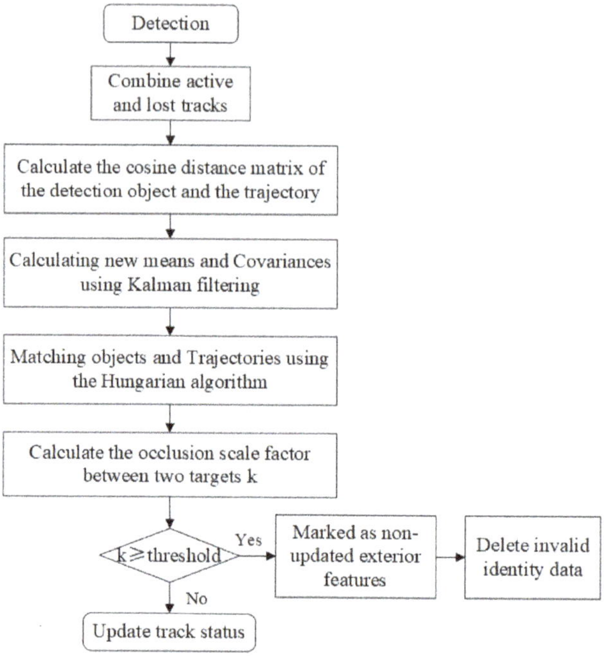

Figure 5. Cascade matching with IVDM.

3.2.1. Occlusion-Aware Detection

For occlusion-aware detection, traditional Intersection Over Union (IOU) cross-ratio algorithms calculate the overlap ratio and filter targets that satisfy the requirements by setting a threshold [23].

Figure 6a,b show that the IOU algorithm is effective in discriminating pedestrians when they have similar body size ratios. However, real applications mostly involve complex scenes, and the size of the pedestrian detection frame produces a very large error due to the different distances of the camera from the ground. The IOU algorithm has very little utility in this case (Figure 6c), which is why it cannot be used as a calculation standard to show the occlusion of pedestrians and small targets in real applications [13]. Therefore, we propose the identity validity discriminant coefficient k, which calculates the ratio of the extent of the occluded portion of an occluded pedestrian to its detection frame and can more accurately discriminate the degree of the pedestrian's occlusion.

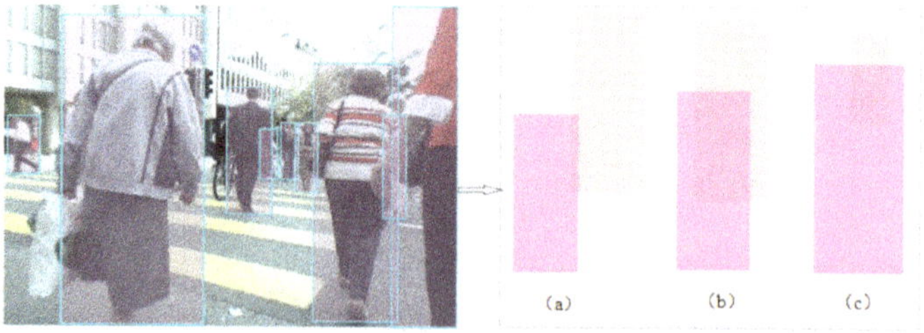

Figure 6. Target occlusion diagram: (**a**) similarly proportioned pedestrian non-influential screening; (**b**) similarly proportioned pedestrian-impacted screening; (**c**) disparately proportioned pedestrian-impacted screening.

3.2.2. Determination of the Shading Scale Factor k

In order to express the derivation of the occlusion ratio coefficient more intuitively, we define the coordinates of the detection frame. As shown in Figure 7, (x_{a_1}, y_{a_1}) denotes the coordinates of the upper-left corner of the blocked pedestrian detection frame; (x_{a_2}, y_{a_2}) denotes the coordinates of the upper-right corner of the blocked pedestrian detection frame; (x_{b_1}, y_{b_1}) denotes the coordinates of the upper-left corner of the blocked pedestrian detection frame; (x_{b_2}, y_{b_2}) denotes the coordinates of the upper-right corner of the blocked pedestrian detection frame; (x_1, y_1) is the upper-left corner of the blocking section; and (x_2, y_2) is the upper-right corner of the blocking section, calculated with the following equation:

$$\begin{cases} x_1 = \max(x_{a_1}, x_{b_1}), y_1 = \max(y_{a_1}, y_{b_1}) \\ x_2 = \max(x_{a_2}, x_{b_2}), y_2 = \max(y_{a_2}, y_{b_2}) \\ S = (x_{a_2} - x_{a_1} + 1.0).(y_{a_2} - y_{a_1} + 1.0) \\ S_0 = \max(x_2 - x_1 + 1.0).\max(y_2 - y_1 + 1.0) \end{cases} \tag{4}$$

S denotes the range of the occluded target frame; S_0 denotes the range of the occluded region.

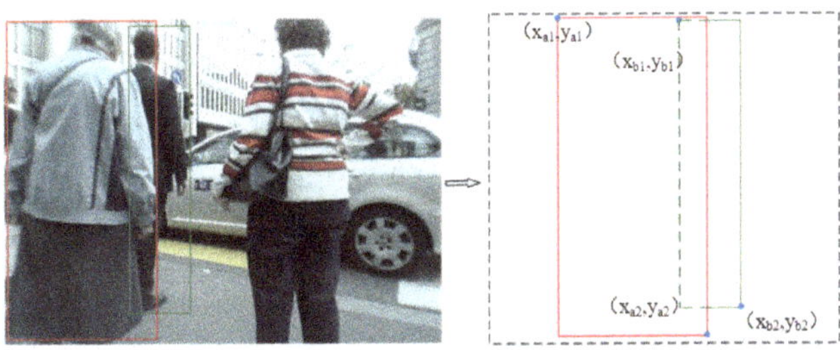

Figure 7. Occlusion frame coordinate plot.

Based on the obtained data for each attribute of the pedestrian detection frame, the unmatched target occlusion ratio coefficient k after cascade matching is derived from the ratio of the occluded area range. S_0 and the occluded target frame range S are calculated with the following equation:

$$k = \frac{S_0}{S} \tag{5}$$

3.2.3. Identity Validity Determination Module

When two pedestrians form an occlusion, in general, if the center of mass of one target is detected within the detection frame coordinates of the other target, the identity of the pedestrian is determined to be invalid due to the effect of the occlusion, and then the identity validity = 0; otherwise, identity validity = 1. However, when occlusion is generated, the above method will not be able to accurately determine the degree of occlusion of the pedestrian if the center of mass of the pedestrian is not within the coordinates of the other pedestrian detection frames, and then the specific degree of occlusion of the pedestrian needs to be calculated. The degree of occlusion of the pedestrian is determined if the target occlusion ratio coefficient k is greater than a threshold value, and then identity validity = 1; otherwise, identity validity = 0.

In this study, the original model was tested on consecutive frames from the MOT16, MOT17, and MOT20 datasets. Based on the experimental responses in Figure 8, when $k > 0.535$, the proportion of tracking failures due to occlusion increases significantly.

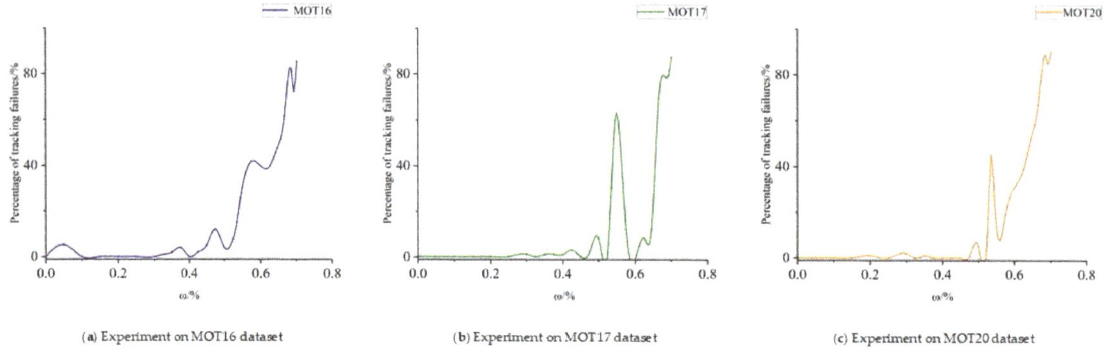

Figure 8. Percentage of tracking failures for different levels of occlusion: (**a**) Experiment on MOT16 dataset; (**b**) Experiment on MOT17 dataset; (**c**) Experiment on MOT20 dataset.

The identity validity is binarized by the occlusion ratio coefficient k as the identity validity score of the corresponding pedestrian, where 1 indicates that the target identity is invalid and 0 indicates that the pedestrian identity is valid, and the relationship of the identity validity score calculation is calculated with the following equation:

$$\mathrm{e}_i = \begin{cases} 1, \ k \geq \omega \\ 0, \ else \end{cases} \tag{6}$$

4. Experiments and Analyses

Here, we statistically summarize the results of the experiments and analyze them in depth, leading to well-reasoned conclusions.

4.1. Experimental Environment

In this study, we used Pytorch [34] for code writing, and we conducted the experiments on a server configured with Intel$^{(R)}$ Xeon$^{(R)}$ CPU E5-2680 V4 @ 2.40GHz (Intel, made in Malaysia) and NVIDIA GeForce RTX 3090 GPUs (Msi, made in China).

4.2. Experimental Dataset and Evaluation Index

4.2.1. Experimental Data

We opted for the MOT series datasets, CrowdHuman dataset, and MIX datasets, commonly utilized in pedestrian tracking tasks, to conduct our experiments. This choice enhances the credibility of our proposed method's effectiveness. Below is an introduction to the three datasets:

- MOT series datasets: These are datasets on the Open Data Lab platform and are mainly targeted at pedestrian tracking tasks in dense scenes.
- CrowdHuman dataset [35]: It is for pedestrian detection. Unlike other mainstream human detection datasets, the pedestrian targets in the CrowdHuman dataset are much denser, more crowded, and even have serious overlaps. According to the data provided in the citation, the CrowdHuman dataset has an average of 22.64 figures per image, which is far more than other human detection datasets.
- MIX datasets: They are diverse and comprehensive, covering different types of pedestrian detection and tracking scenarios. This comprehensiveness allows researchers to test the robustness and effectiveness of algorithms in a variety of real-world situations. Using these datasets, researchers can conduct multimodal data studies, explore commonalities and differences between different datasets, and lay the foundation for improving multi-target tracking and pedestrian detection algorithms.

4.2.2. Evaluation Metrics

- Pedestrian Detection Evaluation Metrics: These are quantitative measures utilized to assess the performance and accuracy of algorithms and models designed to detect pedestrians in images or videos. They offer in-depth insights into a system's ability to recognize pedestrians within a given dataset. Common evaluation metrics for pedestrian detection include precision, recall, and *mAP*. Precision is the ratio of true positives to the total number of predicted positives, where true positives are the instances where the prediction is correct. Recall calculates the ratio of instances where the prediction is correct to the total number of actual positives. *mAP* is the sum of the average precision values for all classes divided by the number of classes. In other words, it represents the average of the average precisions for all classes in the dataset.

  The mathematical expressions for the above evaluation indicators are as follows:

$$Precision = \frac{TP}{TP + FP} \tag{7}$$

$$\text{Rec}all = \frac{TP}{TP + FN} \tag{8}$$

$$mAP = \frac{1}{R}\int_0^1 P(R)dR \tag{9}$$

where *TP* represents the true positives, *FP* represents the false positives, and *FN* represents the false negatives.

- Pedestrian Tracking Evaluation Metrics: In order to test our proposed multi-target pedestrian tracking method, we use five criteria as evaluation metrics: the multi-objective tracking accuracy (*MOTA*) [36], which is commonly expressed as a percentage, ranging from 0% to 100%, where a higher score indicates the superior performance of the tracking algorithm; the ratio of the average of the number of correctly recognized ground-truth detections to the number of computed detections (IDF_1) [37]; the Majority of Tracked (*MT*); Major Lost Targets (*ML*); and Identity Switches (*IDS*).

  The mathematical expressions are as follows:

$$MOTA = 1 - \frac{\sum_t (FN_t + FP_t + IDS_{w_t})}{\sum_t GT_t} \times 100\% \tag{10}$$

$$IDF_1 = \frac{2IDTP}{2IDTP + IDFP + IDFN} \times 100\% \tag{11}$$

where *t* represents the index of each frame of the video, and *GT* is the number of real labeled targets in the image. IDS_w denotes the total number of *ID* switches occurring in the tracked target in frame *t*.

4.3. Analysis of Experimental Results

4.3.1. Comparison Experiments

We conducted the following comparative experiments on different multi-pedestrian tracking algorithms using the MOT16, MOT17, and MOT20 datasets. Table 1 shows some common algorithms for the target tracking task and the experimental results of our introduced algorithm.

Table 1. Experimental results on the MOT16 dataset.

Method	MOTA/%	IDF1/%	MT/%	ML/%	IDs
SORT [15]	59.8	53.8	25.4	22.7	1423
JDE [38]	64.4	55.8	35.4	20.0	1544
CNNMTT [39]	65.2	62.2	32.4	21.3	946
CTrackV1 [40]	67.6	57.2	32.9	23.1	5529
FairMOT [11]	73.7	72.4	44.7	15.9	1074
DeepSORT [16]	74.8	73.6	45.2	15.4	1022
Our Method	75.9	74.8	42.5	18.3	816

Table 1 shows that our multi-pedestrian tracking method has an absolute advantage over many methods. In terms of evaluation metrics, *MOTA* is improved by 1.1%, and *IDF1* is improved by 1.2%, with enhanced robustness compared to the original tracking method. It is worth noting that the IDs of our method are significantly lower than those of DeepSORT; presumably, improved models may improve the predictive power of the tracking method compared to the original model, making the tracking results more accurate and significantly improving the problem of ID hopping. The smaller number of *IDs* makes the tracking results of the model more practical in real applications. After a series of comparisons, it leads to the conclusion that our tracking algorithm has significant advantages in all aspects of performance.

In Table 2, we can see that our multi-pedestrian tracking method significantly improves the experimental metrics on the MOT17 dataset, with improvements of 2.1% and 4.4% for *MOTA* and *IDF1*. We believe that the proposed IDVM is better at presenting false and erroneous identity data and thus can be applied to complex surveillance video scenarios with frequent occlusions.

Table 2. Experimental results on the MOT17 dataset.

Method	MOTA/%	IDF1/%	MT/%	ML/%	IDs
SST [41]	52.4	49.5	21.4	30.7	8431
TubeTK [42]	63.0	68.6	31.2	24.2	4137
CenterTrack [33]	67.8	64.7	34.6	24.6	2583
FairMOT [11]	73.1	72.7	41.1	19.0	2964
TransMOT [38]	75.1	74.6	40.8	22.6	2340
ByteTack [17]	77.4	76.1	39.9	20.2	2236
DeepSORT [16]	76.4	73.4	39.1	21.0	1898
Our Method	78.5	77.8	38.6	19.9	1586

In Table 3, we compare the original DeepSORT method with the multi-pedestrian tracking method that we propose based on DeepSORT on the MOT20 dataset for comparative tests. The *MOTA* and *IDF1* indices of our proposed method are improved by 3.3% and 4.5%, respectively. Therefore, the improved method significantly improves the power to extract the apparent features of the pedestrian target, which leads to more accurate feature extraction and makes the overall performance of this tracking method significantly better. In addition to this, the other two metrics are also significantly improved, which illustrates that the robustness of the tracker to act on the same target during the pedestrian tracking process has been improved. Meanwhile, the pedestrian IVDM not only reduces the appearance feature contamination problem but also improves the tracking robustness for whether or not to update the appearance features after discrimination.

Table 3. Experimental results on the MOT20 dataset.

Method	MOTA/%	IDF1/%	MT/%	ML/%	IDs
DeepSORT [16]	66.8	67.9	68.7	8.4	2269
Our Method	70.1	72.4	69.2	8.7	1689

4.3.2. Pedestrian Detection Algorithm Ablation Experiments

In order to validate our proposed improvement strategy for YOLOv5, ablation experiments were carried out on datasets such as CrowdHuman to judge the effect of each enhancement point. The results of effectiveness experiments for each component of YOLOv5 are as follows.

As shown in Table 4, the KC-YOLO model's accuracy value is improved by 6%, and AP is improved by 4%. The improved model greatly improves the ability to extract intra-pedestrian detection deformations and appearance features, thus capturing more accurate features.

Table 4. YOLOv5 ablation experiment.

k-Means++	CBAM	Precision	Recall	AP
×	×	0.85	0.78	0.85
×	√	0.89	0.83	0.88
√	×	0.88	0.84	0.85
√	√	0.91	0.85	0.89

In Figure 9, YOLOv5+K-means+++CBAM is the KC-YOLO model proposed in this study. At the beginning of training, the values of *AP* and accuracy reach more than 0.8, which is mainly due to the pre-trained model when training YOLOv5. After using the K-means++ clustering method, both *AP* and accuracy are inevitably improved compared to YOLOv5, while the *Loss* value has a small decrease and gradually converges, which indicates that our improvements to the model are positively oriented and the effects are evident. After embedding the CBAM, with increasing epochs, *AP* and accuracy increase significantly, and the *Loss* becomes smaller and converges gradually, which indicates that the improved model is more desirable.

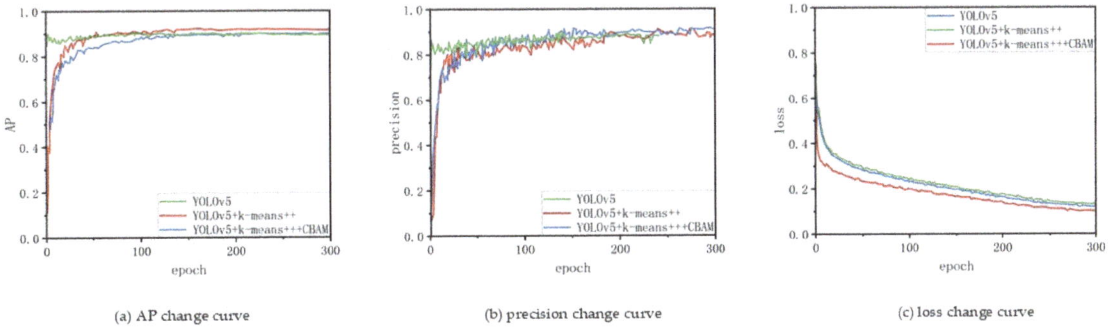

Figure 9. Pedestrian ablation test results: (**a**) AP change curve; (**b**) precision change curve; (**c**) loss change curve.

4.3.3. Pedestrian Tracking Algorithm Ablation Experiments

To check whether the IVDM in our proposed multi-pedestrian tracking method is a positive improvement, we conducted ablation experiments on the IVDM.

The results of ablation experiments with the IVDM are shown in Table 5. They clearly show that the addition of the IVDM improves the *MOTA* of the multi-pedestrian tracking method by 3.0% and the *IDF1* by 2.8%, and the *IDs* are also significantly reduced. The role of the IVDM is mainly to eliminate false and erroneous identity data due to occlusion and to maintain the purity of the original tracking target while reducing the generation of redundant data. After the introduction of the IVDM, the overall performance of the task is improved.

Table 5. Experimental ablation study of IVDM on the MOT20 dataset.

Method	MOTA/%	IDF1/%	MT/%	ML/%	IDs
Without IVDM	67.1	69.6	67.6	9.1	2638
With IVDM	70.1	72.4	69.2	8.7	1689

To illustrate more intuitively the advantages of the improvements made to the tracking method in this study, we present visual comparisons of the tracking experiments conducted on the MOT16 training dataset. Figure 10 compares the demonstrations of the multi-pedestrian tracking method with and without the IVDM.

(a)

(b)

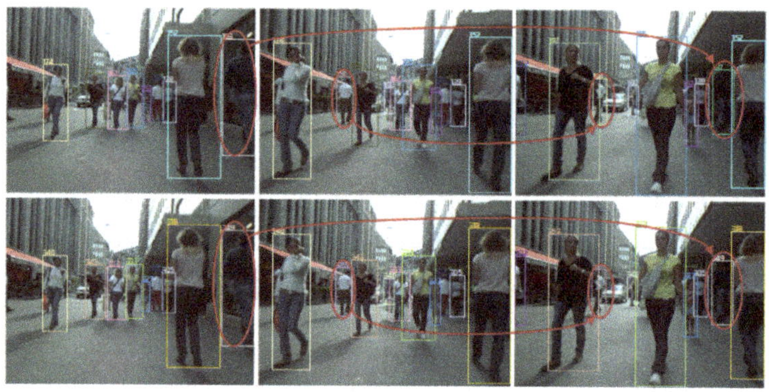

Figure 10. Visualization of multi-pedestrian tracking results: (**a**) without IVDM; (**b**) with IVDM.

As shown in Figure 10, when using the multi-pedestrian tracking without adding the IVDM, the target pedestrians with ID number 154 and ID number 183 changed their ID numbers due to brief occlusion, which also means that the tracking failed and generated redundant and incorrect ID numbers at the same time. However, the problem of tracking failures due to transient occlusion is well solved after adding the IVDM.

5. Discussion

This study introduces a method for the multi-object tracking of pedestrians across multiple cameras in complex scenes, and it exhibits a higher tracking accuracy compared to existing methods in practical applications. However, like any research, our work has certain limitations that need to be considered. One major limitation is the potential influence of environmental factors on the accuracy of our model. For instance, the spacing between cameras could impact the accuracy of our tracking algorithm. Additionally, further research on the algorithm using different datasets can enhance its robustness and generalizability.

Furthermore, in order to enable rapid and accurate tracking of critical targets in applications such as public safety and fire protection systems, our next step will involve considering the design of a more lightweight model to reduce storage and computational requirements. These studies will contribute to expanding the applicability of our approach and assist in the development of more efficient and powerful pedestrian tracking algorithms.

6. Conclusions

In this research, we have developed a multi-pedestrian tracking method based on deep detection and identity validity assessment, specifically designed for complex surveillance video scenarios where issues like target occlusion are frequent.

We have constructed the KC-YOLO network as the detector, which employs the k-means++ clustering method to select the optimal target detection frames. Additionally, we have integrated a convolutional attention mechanism into the target detection algorithm, utilizing attention weights for adaptive feature refinement. This effectively suppresses secondary features to highlight crucial target characteristics, enhancing the robustness of target detection in complex scenes, where target features may become less distinct due to occlusion. The robustness of the detector has been verified through experiments.

In the target tracker, we have introduced the IVDM, which performs occlusion-aware processing on pedestrian targets after feature extraction by the detector. In cases where target identities are compromised due to occlusion-induced errors, we use the occlusion coefficient "k" to assess the validity of the identity. Based on the output of this module, we determine whether pedestrian targets possess valid identities, influencing the decision to update the appearance features of the current dynamic target.

Here are the experimental results on the MOT16 dataset: MOTA is 75.9%, and IDF1 is 74.8%. Compared to SORT, there is a 20.1% increase in MOTA and a 21.0% increase in IDF1. In comparison to CNNMTT, MOTA has improved by 10.7%, and IDF1 has seen a 12.6% improvement. When contrasted with the prototype DeepSORT method, MOTA has increased by 1.1%, and IDF1 has increased by 1.2%. The most noteworthy aspect is the substantial reduction in IDS, maintaining a high level of tracking continuity. For the MOT17 dataset, MOTA is 78.5%, and IDF1 is 77.8%. For the MOT20 dataset, the results show a MOTA of 70.1% and an IDF1 of 72.4%. When contrasted with the prototype DeepSORT method, the MOTA and IDF1 indices of our proposed method are improved by 3.3% and 4.5%. These experiments confirm that our research outperforms several advanced MOT algorithms across nearly all metrics. This study provides a stable and efficient approach to multi-pedestrian tracking in complex scenarios, significantly reducing the number of ID switches to ensure the continuity of tracking trajectories. This approach is particularly well suited for public safety and fire protection departments, enabling the continuous tracking of critical targets in crowded scenes with severe occlusion.

Author Contributions: Conceptualization, W.W. and J.L.; methodology and validation, J.J., D.F. and D.Z.; formal analysis, Y.L., E.G. and T.Y.; investigation, W.W. and D.Z.; writing—original draft preparation, J.L. and Y.L.; writing—review and editing, T.Y. and E.G. All authors have read and agreed to the published version of the manuscript.

Funding: This research was funded by the National Natural Science Foundation of China, grant number 41961063; Guilin Technology Application and Promotion Project, 2022, grant 20220138-2; and Guilin Key R&D Project, 2022, grant 20220109.

Institutional Review Board Statement: Not applicable.

Informed Consent Statement: Not applicable.

Data Availability Statement: Publicly available datasets were analyzed in this study. This data can be found here: https://github.com/xuanwang-91/Framework-for-Pedestrian-Detection-Tracking-and-Re-identifcation.git, accessed on 10 February 2023.

Conflicts of Interest: The authors declare no conflict of interest.

Abbreviations

MPT	Multi-pedestrian tracking
SORT	Simple Online Real-Time Tracking
CBAM	Convolution Block Attention Module
CNNs	Deep Convolutional Neural Networks
R-CNN	Region proposals utilizing CNN
SPPNet	Spatial Pyramid Pool Network
SSD	Single-Shot Multi-Frame Detector
STN	Spatial Transformation Network
MOT	Multi-objective tracking
IOU	Intersection Over Union
MOTA	Multi-objective tracking accuracy
ML	Major Lost Targets
MT	Majority of Tracked

IDS	Identity Switches
IVDM	Identity validity discrimination module

References

1. Xiao, C.; Luo, Z. Improving multiple pedestrian tracking in crowded scenes with hierarchical association. *Entropy* **2023**, *25*, 380. [CrossRef] [PubMed]
2. Pouyan, S.; Charmi, M.; Azarpeyvand, A.; Hassanpoor, H. Propounding first artificial intelligence approach for predicting robbery behavior potential in an indoor security camera. *IEEE Access* **2023**, *11*, 60471–60489. [CrossRef]
3. Zhang, Q. Multi-object trajectory extraction based on YOLOv3-DeepSort for pedestrian-vehicle interaction behavior analysis at non-signalized intersections. *Multimed. Tools Appl.* **2023**, *82*, 15223–15245. [CrossRef]
4. Geng, P.; Xie, H.; Shi, H.; Chen, R.; Tong, Y. Pedestrian Fall Event Detection in Complex Scenes Based on Attention-Guided Neural Network. *Math. Probl. Eng.* **2022**, *2022*, 4110246. [CrossRef]
5. Lin, Y.; Hu, W.; Zheng, Z.; Xiong, J. Citrus Identification and Counting Algorithm Based on Improved YOLOv5s and DeepSort. *Agronomy* **2023**, *13*, 1674. [CrossRef]
6. Osman, Y.; Dennis, R.; Elgazzar, K. Yield Estimation and Visualization Solution for Precision Agriculture. *Sensors* **2021**, *21*, 6657. [CrossRef]
7. Yang, J.; Ge, H.; Yang, J.; Tong, Y.; Su, S. Online pedestrian multiple-object tracking with prediction refinement and track classification. *Neural Process. Lett.* **2022**, *54*, 4893–4919. [CrossRef]
8. Li, X.; Hu, W.; Shen, C.; Zhang, Z.; Dick, A.; Hengel, A.V.D. A survey of appearance models in visual object tracking. *ACM Trans. Intell. Syst. Technol.* **2013**, *4*, 1–48. [CrossRef]
9. Meinhardt, T.; Kirillov, A.; Leal-Taixe, L.; Feichtenhofer, C. Trackformer: Multi-object tracking with transformers. In Proceedings of the IEEE/CVF Conference on Computer Vision and Pattern Recognition, New Orleans, LA, USA, 18–24 June 2022; pp. 8844–8854. [CrossRef]
10. Liu, C.J.; Lin, T.N. DET: Depth-enhanced tracker to mitigate severe occlusion and homogeneous appearance problems for indoor multiple-object tracking. *IEEE Access* **2022**, *10*, 8287–8304. [CrossRef]
11. Zhang, Y.; Wang, C.; Wang, X.; Zeng, W.; Liu, W. Fairmot: On the fairness of detection and re-identification in multiple object tracking. *Int. J. Comput. Vis.* **2021**, *129*, 3069–3087. [CrossRef]
12. Yi, Z.; Shen, Y.; Zhao, Q. Multi-Person tracking algorithm based on data association. *Optik* **2019**, *194*, 163124. [CrossRef]
13. Zhou, X.; Chan, S.; Qiu, C.; Jiang, X.; Tang, T. Multi-Target Tracking Based on a Combined Attention Mechanism and Occlusion Sensing in a Behavior-Analysis System. *Sensors* **2023**, *23*, 2956. [CrossRef] [PubMed]
14. Jia, M.; Cheng, X.; Lu, S.; Zhang, J. Learning disentangled representation implicitly via transformer for occluded person re-identification. *IEEE Trans. Multimed.* **2022**, *25*, 1294–1305. [CrossRef]
15. Bewley, A.; Ge, Z.; Ott, L.; Ramos, F.; Upcroft, B. Simple online and realtime tracking. In Proceedings of the 2016 IEEE International Conference on Image Processing (ICIP), Phoenix, AZ, USA, 25–28 September 2016; pp. 3464–3468. [CrossRef]
16. Wojke, N.; Bewley, A.; Paulus, D. Simple online and realtime tracking with a deep association metric. In Proceedings of the 2017 IEEE International Conference on Image Processing (ICIP), Beijing, China, 17–20 September 2017; pp. 3645–3649. [CrossRef]
17. Zhang, Y.; Sun, P.; Jiang, Y.; Yu, D.; Weng, F.; Yuan, Z.; Luo, P.; Liu, W.; Wang, X. Bytetrack: Multi-object tracking by associating every detection box. In Proceedings of the European Conference on Computer Vision, Tel Aviv, Israel, 23–27 October 2022; Springer Nature: Cham, Switzerland, 2022; pp. 1–21.
18. Ning, C.; Menglu, L.; Hao, Y.; Xueping, S.; Yunhong, L. Survey of pedestrian detection with occlusion. *Complex Intell. Syst.* **2021**, *7*, 577–587. [CrossRef]
19. Wang, Z.; Li, Z.; Leng, J.; Li, M.; Bai, L. Multiple pedestrian tracking with graph attention map on urban road scene. *IEEE Trans. Intell. Transp. Syst.* **2022**, *24*, 8567–8579. [CrossRef]
20. Woo, S.; Park, J.; Lee, J.Y.; Kweon, I.S. Cbam: Convolutional block attention module. In Proceedings of the European Conference on Computer Vision (ECCV), Munich, Germany, 8–14 September 2018. [CrossRef]
21. Hu, X.; Xu, X.; Xiao, Y.; Chen, H.; He, S.; Qin, J.; Heng, P.A. SINet: A scale-insensitive convolutional neural network for fast vehicle detection. *IEEE Trans. Intell. Transp. Syst.* **2018**, *20*, 1010–1019. [CrossRef]
22. Long, J.; Shelhamer, E.; Darrell, T. Fully convolutional networks for semantic segmentation. In Proceedings of the IEEE Conference on Computer Vision and Pattern Recognition, Boston, MA, USA, 7–12 June 2015; pp. 3431–3440. [CrossRef]
23. Yan, J.; Du, S.; Wang, Y. Multi-Pedestrian Tracking in Crowded Scenes by Modeling Movement Behavior and Optimizing Kalman Filter. *IEEE Access* **2022**, *10*, 118512–118521. [CrossRef]
24. Fu, J.; Liu, J.; Tian, H.; Li, Y.; Bao, Y.; Fang, Z.; Lu, H. Dual attention network for scene segmentation. In Proceedings of the IEEE/CVF Conference on Computer Vision and Pattern Recognition, Long Beach, CA, USA, 15–20 June 2019; pp. 3146–3154. [CrossRef]
25. Zhou, Q.; Zhong, B.; Zhang, Y.; Li, J.; Fu, Y. Deep alignment network based multi-person tracking with occlusion and motion reasoning. *IEEE Trans. Multimed.* **2018**, *21*, 1183–1194. [CrossRef]
26. Du, Y.; Zhao, Z.; Song, Y.; Zhao, Y.; Su, F.; Gong, T.; Meng, H. Strongsort: Make deepsort great again. *IEEE Trans. Multimedia* **2023**, 1–14. [CrossRef]

27. Bochkovskiy, A.; Wang, C.Y.; Liao, H.Y.M. Yolov4: Optimal speed and accuracy of object detection. *arXiv* **2020**, arXiv:2004.10934. [CrossRef]
28. Li, H.; Liu, Y.; Wang, C.; Zhang, S.; Cui, X. Tracking algorithm of multiple pedestrians based on particle filters in video sequences. *Comput. Intell. Neurosci.* **2016**, *2016*, 8163878. [CrossRef] [PubMed]
29. Mykhaylo, A. People-tracking-by-detection and people-detection-by-tracking. In Proceedings of the 2008 IEEE Conference on Computer Vision and Pattern Recognition (CVPR'08), Anchorage, AK, USA, 23–28 June 2008. [CrossRef]
30. Zhu, X.; Lyu, S.; Wang, X.; Zhao, Q. TPH-YOLOv5: Improved YOLOv5 based on transformer prediction head for object detection on drone-captured scenarios. In Proceedings of the IEEE/CVF International Conference on Computer Vision, Montreal, BC, Canada, 11–17 October 2021; pp. 2778–2788. [CrossRef]
31. Hämäläinen, J.; Kärkkäinen, T.; Rossi, T. Improving scalable K-means++. *Algorithms* **2020**, *14*, 6. [CrossRef]
32. Li, Z.H.; Chen, J.; Bi, J. Multiple object tracking with appearance feature prediction and similarity fusion. *IEEE Access* **2023**, *11*, 52492–52500. [CrossRef]
33. Chen, K.; Song, X.; Zhai, X.; Zhang, B.; Hou, B.; Wang, Y. An integrated deep learning framework for occluded pedestrian tracking. *IEEE Access* **2019**, *7*, 26060–26072. [CrossRef]
34. Paszke, A.; Gross, S.; Massa, F.; Lerer, A.; Bradbury, J.; Chanan, G.; Killeen, T.; Lin, Z.; Gimelshein, N.; Antiga, L. Pytorch: An imperative style, high-performance deep learning library. *Adv. Neural Inf. Process. Syst.* **2019**, 32. [CrossRef]
35. Shao, S.; Zhao, Z.; Li, B.; Xiao, T.; Yu, G.; Zhang, X.; Sun, J. Crowdhuman: A benchmark for detecting human in a crowd. *arXiv* **2018**, arXiv:1805.00123. [CrossRef]
36. Bernardin, K.; Stiefelhagen, R. Evaluating multiple object tracking performance: The clear mot metrics. *EURASIP J. Image Video Process.* **2008**, *2008*, 246309. [CrossRef]
37. Hua, G.; Jégou, H. (Eds.) *Computer Vision–ECCV 2016 Workshops: Amsterdam, The Netherlands, October 8–10 and 15–16 2016, Proceedings, Part II*; Springer: Berlin/Heidelberg, Germany, 2016. [CrossRef]
38. Wang, Z.; Zheng, L.; Liu, Y.; Wang, S. Towards Real-Time Multi-Object Tracking. In Proceedings of the European Conference on Computer Vision, Glasgow, UK, 23–28 August 2020; Springer International Publishing: Cham, Switzerland, 2020; pp. 107–122. [CrossRef]
39. Mahmoudi, N.; Ahadi, S.M.; Rahmati, M. Multi-target tracking using CNN-based features: CNNMTT. *Multimed. Tools Appl.* **2019**, *78*, 7077–7096. [CrossRef]
40. Peng, J.; Wang, C.; Wan, F.; Wu, Y.; Wang, Y.; Tai, Y.; Wang, C.; Li, J.; Huang, F.; Fu, Y. Chained-tracker: Chaining paired attentive regression results for end-to-end joint multiple-object detection and tracking. In Proceedings of the Computer Vision–ECCV 2020: 16th European Conference, Glasgow, UK, 23–28 August 2020; Proceedings, Part IV 16. Springer International Publishing: Cham, Switzerland, 2020; pp. 145–161. [CrossRef]
41. Sun, S.; Akhtar, N.; Song, H.; Mian, A.S.; Shah, M. Deep affinity network for multiple object tracking. *IEEE Trans. Pattern Anal. Mach. Intell.* **2019**, *43*, 104–119. [CrossRef]
42. Pang, B.; Li, Y.; Zhang, Y.; Li, M.; Lu, C. Tubetk: Adopting tubes to track multi-object in a one-step training model. In Proceedings of the IEEE/CVF Conference on Computer Vision and Pattern Recognition, Seattle, WA, USA, 13–19 June 2020; pp. 6308–6318. [CrossRef]

applied sciences

MDPI

Article

Bidirectional-Feature-Learning-Based Adversarial Domain Adaptation with Generative Network

Chansu Han [1], Hyunseung Choo [2,*] and Jongpil Jeong [3,*]

[1] Department of Electrical and Computer Engineering, Sungkyunkwan University, Suwon-si 16419, Republic of Korea; b2451184@naver.com
[2] Department of AI System Engineering, Sungkyunkwan University, Suwon-si 16419, Republic of Korea
[3] Department of Smart Factory Convergence, Sungkyunkwan University, Suwon-si 16419, Republic of Korea
* Correspondence: choo@skku.edu (H.C.); jpjeong@skku.edu (J.J.); Tel.: +82-10-8540-6171 or +82-31-290-7145 (H.C.); +82-10-9700-6284 or +82-31-299-4267 (J.J.)

Abstract: Studying domain adaptation is a recent research trend. Generally, many generative models that researchers have studied perform well on training data from a specific domain. However, their ability to be generalized to other domains might be limited. Therefore, a growing body of research has utilized domain adaptation techniques to address the problem of generative models being vulnerable to input from other domains. In this paper, we focused on generative models and representation learning. Generative models have received a lot of attention for their ability to generate various types of data such as images, music, and text. In particular, studies utilizing generative adversarial neural networks (GANs) and autoencoder structures have received a lot of attention. In this paper, we solved the domain adaptation problem by reconstructing real image data using an autoencoder structure. In particular, reconstructed image data, considered a type of noisy image data, are used as input data. How to reconstruct data by extracting features and selectively transforming them in order to reduce differences in characteristics between domains entails representative learning. Considering these research trends, this paper proposed a novel methodology combining bidirectional feature learning and generative networks to innovatively approach the domain adaptation problem. It could improve the adaptation ability by accurately simulating the real data distribution. The experimental results show that the proposed model outperforms the traditional DANN and ADDA. This demonstrates that combining bidirectional feature learning and generative networks is an effective solution in the field of domain adaptation. These results break new ground in the field of domain adaptation. They are expected to provide great inspiration for future research and applications. Finally, through various experiments and evaluations, we verify that the proposed approach outperforms the existing works. We conducted experiments for representative generative models and domain adaptation techniques and found that the proposed approach was effective in improving data and domain robustness. We hope to contribute to the development of domain-adaptive models that are robust to the domain.

Keywords: adversarial domain adaptation; bidirectional feature learning process; generative network; adversarial learning

Citation: Han, C.; Choo, H.; Jeong, J. Bidirectional-Feature-Learning-Based Adversarial Domain Adaptation with Generative Network. *Appl. Sci.* **2023**, *13*, 11825. https://doi.org/10.3390/app132111825

Academic Editors: Zahid Mehmood Jehangiri, Mohsin Shahzad and Uzair Khan

Received: 5 October 2023
Revised: 26 October 2023
Accepted: 27 October 2023
Published: 29 October 2023

1. Introduction

The field of domain adaptation is currently witnessing rapid progress in innovative research to address the problem of data mobility across different domains. In particular, recent advances in machine learning and deep learning techniques have attracted attention on how to overcome distributional differences between domains and improve the generalization performance of models [1–6].

In particular, a growing body of research has focused on the relationship between domain adaptation and generative models. This research seeks to understand why gaps between adversarial domains occur and how to counteract them in order to make models more robust. There has been a large body of research showing that approaches using generative models are useful for improving a model's generalization performance across domains [7]. Generative models mimic real-world data distributions to generate data in the target domain for adaptation, especially using generative adversarial neural networks (GANs) or autoencoder structures [7–13].

This paper utilizes an autoencoder structure in the generation model to reconstruct and utilize real image data. At this time, reconstructed image data can be treated as a kind of noisy image data caused by reconstruction error. They are used as input data. In addition, representation learning is a field that studies how to reconstruct data through feature extraction and selective feature transformation to address differences in features across domains. In view of these recent trends and research developments, this paper is expected to play an important role in presenting a new methodology that utilizes cross-domain bidirectional feature learning and generative networks to solve domain adaptation problems and overcome the limitations of existing studies.

That is how I came across the model shown in panel (a) in Figure 1. This model is described in an interesting paper that allowed the authors to start studying domain adaptation. It is called DANN [1] for short, and it is a great paper to look at alongside its successor, the ADDA model [7], shown in panel (b) of Figure 1. Both models are adversarial models, but there is a difference. While DANN behaves adversarially when computing loss, the ADDA model initially learns to associate a discriminator with two domains to determine domain labels. If the domains had completely different characteristics, the loss would be extremely high at the beginning of training. However, the more the model is trained, the better the model will perform, i.e., it will be a better design. With this in mind, we wanted to build a model that is adversarial and has a bit more complexity.

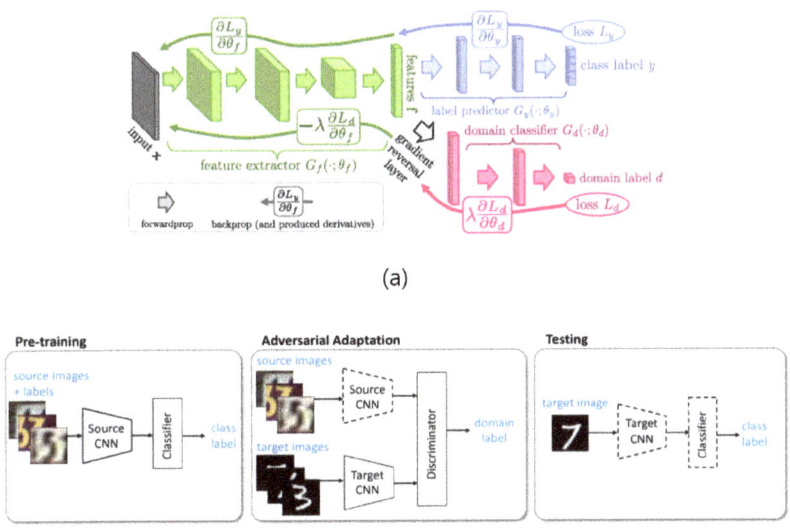

(a)

(b)

Figure 1. The different structures of DANN (**a**) and ADDA (**b**).

Table 1 summarizes what DANN and ADDA have in common and their differences from the model we propose in this paper. The reason why this table is important is because everything we will introduce in this paper is in this table. It starts with the question of whether there is a generative network or not. Then, there is adversarial learning, bidirectional feature learning, using a pretrained network (this is also related to training time, which works a little differently in this work and will be discussed later), sharing weights, using a hyperparameter, etc.

Table 1. Comparison of recent methods (o denotes method is used in model; x denotes it is not used).

Method	DANN	ADAA	BiFLP-AdvDA
Generative network	x	x	o
Adversarial learning	o	o	o
Bidirectional feature learning	x	x	o
Pretrained network	x	o (CNN)	o (autoencoder)
Weight sharing	x	x	o
Hyperparameter	λ	λ	λ, m (margin)

In the modern world, the rapid growth of data has led to the production of many different kinds of data in many different fields. However, these data often originate from different domains, each with unique characteristics and statistical distributions. This makes it difficult for machine learning models to generalize to new domains, which hinders the effective application of models in the real world. In response, domain adaptation research has gained increasing importance. In particular, recent research trends have explored innovative methods to address the problem of cross-domain portability. These approaches are mainly centered on generative models and representation learning [14,15]. Generative models focus on overcoming distributional differences while generating data in the target domain. Studies utilizing generative adversarial neural networks (GANs) or autoencoder structures have received much attention. Representation learning also focuses on performing adaptation in a way that reduces differences in characteristics between domains through feature extraction and transformation. However, most of the existing research has focused on unidirectional feature transformations. In the learning processes of previous studies that learn by transforming features themselves, limitations and problems arise when trying to adapt the target domain to a model trained on the source domain. In contrast, this paper explores the domain adaptation problem from a new angle by combining a bidirectional feature learning process and generative network and proposes an innovative methodology to perform adaptation while mutually preserving features between two domains. It is expected to more accurately consider the distribution of real data, overcome the limitations of existing methods in effectively solving the domain adaptation problem, deepen our understanding of the relationship between generative models and domain adaptation and features, and contribute to the development of more robust and stable models.

The starting point of this paper is the need for domain adaptation and the current state of research. The problem of the cross-domain mobility of data collected from various fields limits the performance of machine learning models. As a solution, existing research has mainly focused on unidirectional feature transformations. However, starting from the idea that the relationship between domains could be bi-directional, we tried to introduce a bi-directional feature learning process. We found that the bi-directional feature learning method, which is the training method of our proposed model, has been used in several studies. However, the most important concepts in the field of domain adaptation are learning in the direction of minimizing empirical risk [16], feature conversion using image-to-image translation [10,17], transfer learning [18–21], reducing the gap between domains with generalization and robustness of the model and generative networks [12], adversarial learning with discriminators, adversarial learning without discriminators [22], pseudo-labeling for domain adaptation in the absence of discriminative classifiers [23], and so on, all of which seem to be similar to domain adaptation. However, they have slightly different contributions. Thus, there is still a lot of potential for further development. To advance this idea, it is necessary to strengthen the role of generative networks. Generative models need to be improved to simulate more realistic data distributions. Although such research studies have been ongoing, they have not been able to move beyond a one-way learning process. To this end, we propose a model combining generative networks and bidirectional feature learning processes to perform both functions together [24]. Furthermore, bidirectional feature learning can serve as a potential way to enforce interdependence between two domains. In this way, we devised a novel network architecture that could achieve good adaptation while preserving features between the two domains. From the perspective of generative models, new ideas can be generated on how to reconfigure different data to adapt to new domains. This will allow the model to acquire generalization capabilities that allow it to adapt in more diverse situations or in different environments. Through this process of idea generation, the model proposed in this paper can be further enriched and innovated.

This paper proposes a novel approach in the field of domain adaptation by combining bidirectional feature learning and generative networks. The main idea is to enhance cross-domain adaptation through bidirectional feature learning and to improve the adaptation ability by more accurately simulating actual data distribution through generative networks.

Based on the evaluation results, the proposed model demonstrated significantly higher accuracy than existing models such as DANN and ADDA. Through experiments, it was found that the proposed model exhibited outstanding performance in domain adaptation. These results strongly indicate that combining bidirectional feature learning and generation networks is an effective method for domain adaptation. As a major contribution and result, this paper not only presents an innovative solution in the domain adaptation field, but also provides a model that is superior to existing models such as DANN and ADDA. Thus, this research breaks new ground in domain adaptation and provides significant inspiration for future research and applications. Furthermore, the experimental results demonstrated that the proposed approach could minimize performance degradation of the generative model while aiding in the generation of desired outputs. This indicates the effectiveness of this paper in improving the stability and reliability of the generative model simultaneously. These evaluation results are of great significance in presenting the validity and practicality of the proposed approach in the paper. They are expected to contribute to research aimed at enhancing the security and robustness of generative models.

The primary contribution of this paper is the development of a novel approach for domain adaptation in generative models and representation learning. By combining bidirectional feature learning and generative networks, we significantly improve the adaptation ability, accurately simulating the real data distribution. The experimental results showcase the superiority of our approach over traditional methods, marking a substantial advancement in the field of domain adaptation and enhancing the data and domain robustness in generative models.

This article is structured into five sections. The first section introduces the background, the problem statement, the derivation process, and a brief summary of the idea. The second section presents a summary of previous research and relevant techniques. The third section provides an overall description of the proposed model. The fourth section explains and presents the results of experiments conducted to evaluate the model's performance.

2. Related Work

This section describes the background of the previous research that leads to the proposal of this paper.

2.1. Adversarial Domain Adaptation

Domain adaptation is an important topic to address the problem of reducing the generalization ability of a model due to differences in the data distribution between different domains. Domain adaptation is a type of transfer learning, which can be viewed as a transfer of knowledge from one domain to another when there is a gap between two domains with different characteristics, and closing the gap to make them similar. However, there is a limitation, in that a model trained on one domain usually does not perform as well as expected when tested on data from a domain other than the one it was trained on due to differences in the trained domain. Therefore, researchers have been thinking about how to make a model perform well in a domain that is completely different from the domain it was trained on. This was how domain adaptation was approached.

When there are two completely different datasets or two datasets with some similarity, if the similarity between the two domains is not large, it is expressed as how much domain shift has occurred when quantified. Therefore, in the field of domain adaptation, a lot of research studies have been conducted to study and solve this domain shift as much as possible. Also, when the domain shift is reduced as much as possible and the performance is good even in different domains, we say that the generalization is good. In other words, domain adaptation is the process of generalizing a model as much as possible.

Recent advances in machine learning and deep learning techniques have attracted attention on how to overcome distributional differences between domains and improve the generalization performance of models. In the field of domain adaptation, two main approaches have been used to solve the problem.

First, models that transform features between domains in a way that minimizes distributional differences between domains have been widely studied. This approach attempts to achieve adaptation by transforming data from the target domain into a distribution similar to the source domain. There are various approaches to this method. First of all, many models such as BDA [3], JAN [19], ADDA, and so on can calculate the MMD, which is the difference in probability distribution. They can also learn by reducing the distribution difference between domains as much as possible [24]. If the probability distribution difference between domains is previously calculated and learned, there are also models that can transform features themselves or directly use features to reduce the gap between domains. There are also models that can use a generative model to generate the target data and then let domains adapt and check performances of discriminators by calculating how similar they are.

Second, instead of addressing distributional differences, a common approach is to train the model to adapt to the target domain. This approach allows the model to overcome differences between domains while still being able to recognize and utilize characteristics of the target domain. To this end, research has focused on training models using labeled information from the target domain, typically using classifiers or regression models.

In these domain adaptation studies, various methods were used to reduce the gap between two domains or make them similar. Among them, the method of learning adversarially using the class label of the domain and the relationship with the domain label is mainly used [1,2,5,7,11,22]. Therefore, this method is usually referred to as adversarial domain adaptation or domain adversarial adaptation.

2.2. Bidirectional Feature Learning

Bidirectional feature learning aims to transform features between the source and target domains in both directions while preserving information. To achieve this, it is mainly studied by combining a generative network and a feature transformation process, which is a key strategy used to enhance the adaptive ability while mutually preserving characteristics between domains.

Bidirectional feature learning provides better adaptive capabilities than traditional unidirectional feature transformation. In other words, when the model performs a conversion from one domain to another, the key point is to achieve conversion while preserving features between the two domains [25,26]. In particular, recent attempts have been made to develop more powerful adaptive models by combining bidirectional feature learning with self-supervised learning, meta-learning, and so on. This can be seen as an effort to overcome the limitations of existing methods and provide more practical domain adaptation solutions.

Such bidirectional feature learning can act as a potential way to strengthen the interdependence between two domains. In this way, we have devised a new network architecture that can preserve features between the two domains and still achieve good adaptation. This methodology uses an approach that deepens our understanding of the relationship between domains and trains the model to more effectively translate characteristics between the two domains while minimizing the loss of information.

However, unlike other studies, bidirectional feature learning works a little differently in this paper. Although the fields of application are different, even a comparison with one of the existing studies, bidirectional LSTM [27], shows that the LSTM is doubly connected to learn effectively. However, in this paper, extracted features are not intertwined, but simply used to calculate two loss functions. Thus, it is expressed that it is based on the bidirectional feature learning process. In this respect, the role of the bidirectional feature learning process, which is used slightly differently in this paper, can be seen through a comparison of evaluation indicators.

2.3. Generative Network

Generative models play an important role in this work. They are models that can simulate or generate distributions of real-world data. In particular, with recent advances in deep learning, generative models have attracted attention for overcoming distributional differences between domains and generating or transforming data in new domains. In the past, various generative networks such as Variational Auto-Encoders (VAEs) [8] and Generative Adversarial Networks (GANs) [10,28] have been introduced in related research. VAEs are used to learn the distribution of data in a latent space to generate different variations, while GANs use a competitive network of generators and discriminators to produce data mimicking the actual data distribution.

Autoencoder has been used in many fields for a while because it has a wide range of applications, such as utilizing convolutional layers for its application and having deep hidden layers by stacking multiple layers. In this paper, we used convolutional autoencoder (CAE), an autoencoder that utilizes convolutional layers [4]. In addition, GANs generate fake data that is almost indistinguishable from real data through competitive learning between generators and discriminators, which has the great advantage of reducing distribution differences between domains, and converting while maintaining features of the data. As a result, GANs have become a key tool in the field of domain adaptation. Recent research trends are moving towards combining GANs with VAE or CAE to develop more robust generative models. In addition, various variants that take into account characteristics of the data and the relationship between domains are proposed, which are utilized to better simulate distributional differences between domains and perform data transformations.

This paper emphasizes the role and importance of generative networks and proposes a new methodology to apply to the domain adaptation problem. In this paper, we combined autoencoder and GAN to develop a generative model considering the distribution of real data. We then applied it to the domain adaptation problem to overcome the distribution difference between the domains. In this way, realistic and stable adaptation results were obtained. The effectiveness of the proposed method was demonstrated through comparison with related studies. As a result, the accuracy of the generative network on the dataset was 98.5$\sim$99.23%, which showed a very high classification accuracy. An error of 0.77$\sim$1.5% was assumed to be some noise. The model was designed and tested to be robust against noise. This shows an innovative solution in the field of domain adaptation. It is expected to provide an important direction for strengthening the stability and generalization ability of models not only in domains, but also in the face of noise.

3. BiFLP-AdvDA

3.1. Model Overview

This paper is a model for performing adversarial domain adaptation. The architecture of the proposed model is shown in Figure 2. First, for preprocessing the existing dataset and constructing the model, we pretrained a generative model with a convolutional autoencoder structure on the input data X_s which is from source domain. And the other input data is X_t which is from target domain. We then loaded and used the pretrained model. The inputs X_s and X_t are passed through the pretrained generative network to reconstruct the input image data and generate a theoretically identical image. However, in practice, they are not identical. For example, if you apply the generative network that is customized to the MNIST dataset to the SVHN dataset, which is an RGB image dataset, the same image cannot be generated. Thus, results may vary depending on the source domain. However, results may change when adjusting parameters. In this paper, the classification accuracy of the generative network was 98.5∼99.23%. Images were generated by the well-trained generative model. The image data X_s, X_t of the generated source domain and target domain were taken as input data and features of those images were extracted with the feature extractor. This prepared us to perform the adversarial domain adaptation (AdvDA) problem. Weights were shared for consistency in experiments [29].

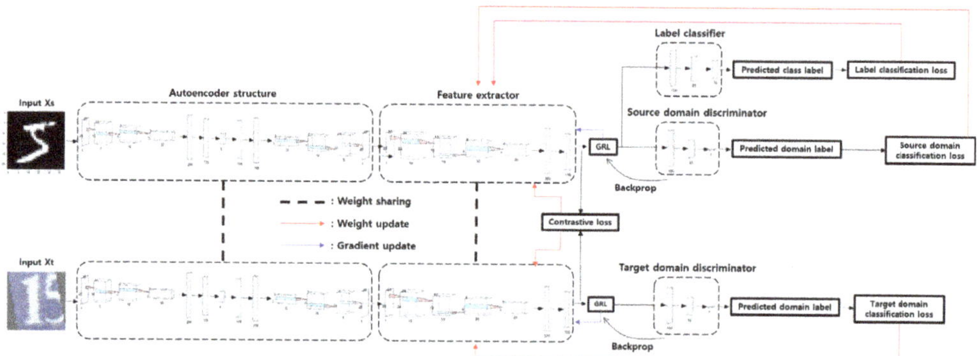

Figure 2. The architecture of our proposed model BiFLP-AdvDA.

For the input layer, we used a convolutional autoencoder structure, and used three convolutional layers for the encoder and decoder parts, and encoded by increasing the number of channels from 1 channel to 8 channels, 16 channels, and 32 channels, and reconstructed by returning to 32 channels, 16 channels, 8 channels, and 1 channel. The amount of error in this process is considered as noise for the robustness of the model and data in this paper. After that, the rest of the learning process proceeds through the feature extractor.

It trains a label classifier with features extracted from the source domain to classify the class label of the source data. In the process, it calculates the label loss, or label classification loss, which shows how badly it classifies the label. Label classification loss uses cross-entropy loss as a general classification loss function. Therefore, this paper learns to maximize the performance of the label classifier by minimizing the label classification loss and learns a domain discriminator to determine the source domain or target domain using features. As this paper is a domain adaptation problem, it is a problem for the target domain to learn well without label information for the model learned for the source domain. Therefore, it can be said that this part is adversarial, in that the domain discriminator learns to distinguish the source domain and the target domain as well as possible, while learning to extract domain-invariant features that fool the domain discriminator into not distinguishing as well as possible with features extracted with the feature extractor. Therefore, if the label classifier has a classification loss, the domain discriminator uses the same general classification loss as the label classifier as a domain classification loss. A single device is required to classify the label of the source domain well despite a difference in distribution between the source domain and the target domain and to prevent the two domains (i.e., the source domain and the target domain) from being distinguished.

The device is the gradient reversal layer (GRL) [2] proposed in the DANN paper. This GRL acts as an intermediary between the feature extractor and the domain discriminator. When learning, it still passes linearly when forward. However, when performing back-propagation, it reverses the direction of the gradient by reversing the sign of the gradient. The inverted gradient is inserted between the domain classifier and the feature extractor. This plays a role in training the feature extractor to minimize the domain information while minimizing the domain classification loss. This allows the feature extractor to extract features with minimal domain information. The hyperparameter alpha, which controls how much domain information to include, is not a trainable parameter. In this paper, it takes the form of bidirectional feature learning, in addition to mutually conservative learning, in the direction of minimizing similarity difference between features while preserving characteristics between domains by including very little domain information.

In addition to learning methods limited to a single domain, the model in this paper induces smooth learning for a series of domains with completely different characteristics such as their distribution. In addition to reducing the gap between domains, the data itself is subjected to a slight noise effect using a pretrained generative network. By reconstructing the input image data, we propose a method to improve the robustness of both the domains and data.

3.2. Loss Functions

This paper aims to reduce the gap between domains and improve the robustness of the model using a generative model with an autoencoder structure and an adversarial domain adaptation (BiFLP-AdvDA) approach based on bidirectional feature learning to extract features that can finally be learned regardless of the domain. This study also aims to perform domain adaptation to minimize the gap between domains so that characteristics of domains become similar. For these purposes, four loss functions and a total of five loss functions are used until total loss using all of them. Formulas and brief descriptions of the loss functions are presented below.

Label classification loss was used to perform a classification task using data from the source domain and to minimize the difference between the actual and predicted labels. The following is the formula for label classification loss:

$$L_{\mathrm{cls}}(y, \hat{y}_s) = -\sum_i y_i \cdot log(\hat{y}_s, i) \tag{1}$$

where y denotes actual classified class labels of the source domain, which can be binary or multi-class depending on the problem; $\hat{y}_s$ denotes predicted class labels of the source domain containing probabilities for each class; i denotes an index variable representing classes, varying depending on the number of possible classes in a classification task; y_i denotes the value corresponding to class i in the actual label vector, the value of $\hat{y}_s$; and i in the log function corresponds to class i in the predicted probability vector, representing the likelihood of belonging to that class.

Domain classification loss was used to train the domain discriminator to misclassify the domain as much as possible. This helps the feature extractor to extract features that the domain discriminator will use to prevent the domain discriminator from classifying as well as possible. As a loss function, unlike label classification loss, we used binary cross-entropy loss (BCE loss) because the task was to classify into one of two domains. The following is the formula for domain classification loss:

For the source domain,

$$L_{\mathrm{src}}(D_s, \hat{D}_s) = -\sum_i (D_i \cdot log(\hat{D}_s, i) + (1 - D_i) \cdot log(1 - \hat{D}_s, i)) \tag{2}$$

For the target domain,

$$L_{\text{tar}}(D_t, \hat{D}_t) = -\sum_i (D_i \cdot log(\hat{D}_t, i) + (1 - D_i) \cdot log(1 - \hat{D}_t, i)) \tag{3}$$

where L_{src} indicates the source domain classification loss function; D_i denotes the actual classified source domain labels obtained using the domain discriminator; $\hat{D}_s$ denotes the predicted source domain labels; i denotes the index variable for the elements in one batch of source data; L_{tar} denotes the target domain classification loss function; D_t indicates a binary value of 0 or 1 and indicates whether the domain is the source or target; $\hat{D}_t$ denotes the predicted domain labels of the target domain; and i denotes the index variable for the elements in one batch of the target data.

Adversarial loss was used to train the domain discriminator to not discriminate between features extracted from the feature extractor. In this way, it learns to reduce the gap between domains. Here is the formula for adversarial loss:

For the source domain,

$$L_{\text{adv\_src}}(L_{\text{cls}}, L_{\text{src}}) = L_{\text{cls}} + \alpha * L_{\text{src}} \tag{4}$$

For the target domain,

$$L_{\text{adv\_tar}}(L_{\text{cls}}, L_{\text{tar}}) = L_{\text{cls}} + \alpha * L_{\text{tar}} \tag{5}$$

where L_{cls} denotes the classification loss for the source domain; L_{src} denotes the source domain classification loss; alpha denotes a hyperparameter representing the weight or importance given to the source loss in the combination; L_{cls} denotes the classification loss for the source domain, as already mentioned; and L_{tar} denotes the target domain's classification loss.

The adversarial loss function, represented by Equations (4) and (5), was calculated by utilizing Equations (1) and (2) or Equation (3) as appropriate. First of all, Equation (1) is a task to classify the label of the source domain as best as possible, as mentioned earlier, and the corresponding loss function is learned to be maximized; while Equations (2) and (3) are the main tasks to learn to distinguish whether the input data are from the source domain or the target domain as much as possible. Corresponding loss functions are learned to be minimized. In this respect, it is similar to the existing minmax loss. Thus, it is called adversarial loss. In addition, the alpha value of GRL, borrowed from the model proposed for the domain adaptation task, which is the subject of this paper, is multiplied by the domain classification loss to adjust how much to use characteristics of the domain. Contrastive loss is a loss function to measure similarity between two features [30,31], which is defined as the Euclidean distance between two features with the following formula:

$$D_{st}(f(X_{\text{src}}), f(X_{\text{tar}})) = \sqrt{\sum_{i,j=1}^{n} ((f(X^i_{\text{src}}) - f(X^j_{\text{tar}}))^2)} \tag{6}$$

$$L_{\text{contrastive}}(y, D_{st}) = y * D^2_{st} + (1 - y) * max(m - D_{st}, 0)^2 \tag{7}$$

where f indicates the feature extraction function mapping the input data to feature embeddings; X_{src} and X_{tar} denotes the original source data X_s and the original target data X_t reconstructed by the pretrained autoencoder; D_{st} indicates the Euclidean distance, obtained by measuring the distance of features between the source data X_{src} and the target data X_{tar} using the feature embeddings generated by the function; f and n denote the size of one batch of the source and target domains and the number of data points within that batch; i and j denote the index variables of the source and target data; X^i_{src} denotes the ith data point of one batch of X_{src}; X^j_{tar} denotes the jth data point of one batch of X_{tar}; the means of D are different for each loss function. One of the D's, in Equations (2) and (3), indicates the domain discriminator and the other, in Equations (6) and (7), indicates the Euclidean distance; y denotes a binary label that takes the value 0 or 1, 1 if the two features are similar and 0 if they are different, for example, in traditional research, the loss function is computed in such a way that if a pair of data enters as input a pair of data of the same class, the value of y is 1, while if a pair of data enters as input a pair of data of a different class, the value of y is 0; and m denotes the margin value. This hyperparameter m is not learnable. However, it can be tuned to an appropriate value through experimentation.

The total loss function is calculated by summing the loss functions of all components. After all, minimizing it is the goal of generalizing the model across domains and guiding it to extract similar features. Here is the formula for the total loss function.

$$L_{\text{total}} = \cdot L_{\text{adv\_src}} + \cdot L_{\text{adv\_tar}} + \cdot L_{\text{contrastive}} \quad (8)$$

In the learning process of the model proposed in this paper, the loss function is finally composed of adversarial loss and contrastive loss, which are arbitrarily configured for the model as in Equation (8). By learning the loss function for the source domain and the target domain adversarially, the feature extractor extracts domain-invariant features that work well in both domains, i.e., features with similar characteristics to both domains. The contrastive loss is calculated according to the Euclidean distance between the two extracted features so that the similarity of the data points in the vector space increases. Eventually, the boundary between the two domains is blurred.

4. Experimental Results

In this paper, we examine the effectiveness of a domain adaptation method based on bidirectional feature learning with generative networks by comparing it with various domain adaptation methods. This allows us to clearly identify its superiority and strengths. In this section, we first describe the model implementation and experimental environment, followed by experimental results and quantitative and qualitative evaluations, including the datasets used in the experiments. We also provide various evaluation metrics to prove that it performs well compared to existing studies. We use MNIST, USPS, SVHN, and EMNIST as datasets. MNIST is a validated dataset. However, the SVHN dataset is a sparse dataset with varying results depending on the preprocessing. In some cases, it does not learn at all. Thus, we need to pay attention to the preprocessing for datasets other than MNIST. More details on these experiments will be discussed in the following sections.

4.1. Experiment Configurations

To implement the model proposed in this paper, the author built the following experimental environment, which is introduced in Table 2.

Table 2. Experimental configuration of experiments for adversarial domain adaptation.

	Experimental Setup	
Operating System	Linux-5.15.109+-x86_64-with-glibc2.35	Windows 10
GPU	V100 (Google Colab)	RTX 3070 Ti (Personal Desktop Computer)
CPU	Intel(R) Xeon(R) CPU @ 2.00 GHz	Intel(R) Core(TM) i7-10700F CPU @ 2.90 GHz
RAM	52 GB	32 GB
Language	Python 3.10	
Framework	Pytorch 2.0.1 + cu118	
Library (necessary)	NumPy, matplotlib, torchvision, pandas, sklearn	

In this paper, two experimental environments were used. The hardware environments, such as CPU, GPU, and operating system, were different, but the software environments for learning were mostly the same. First of all, we used Python version 3.10 in common and implemented and experimented with Pytorch-based models. The version of Pytorch does not matter much. We used 2.01+cu118, the most recent updated version at the time of the experiment. Other libraries included NumPy for data preprocessing and image processing and matplotlib, sklearn, and torchvision for the visualization of results.

4.2. Datasets

The Mixed National Institute of Standards and Technology (MNIST) is a dataset of handwritten digit images widely used in machine learning and computer vision. As shown in Figure 3, the dataset consists of images of handwritten digits from 0 to 9. Each image is a monochrome image with a size of 28 × 28 pixels. Each image is a 28 × 28 pixel grayscale image represented by a pixel value between 0 and 255. The entire dataset consists of a total of 70,000 images, of which 60,000 belong to the training set and the remaining 10,000 belong to the test set. This dataset is used as a representative benchmark for number recognition problems. It is widely used to compare and evaluate the performance of machine learning algorithms. Since it has already been widely used, it performs well for classification tasks. Therefore, we thought it would perform well for domain adaptation. The number of data was too large for training.

The United States Postal Service (USPS) dataset is a dataset of handwritten digit images collected from the United States Postal Service. This dataset is primarily used for recognizing handwritten zip codes. As shown in Figure 4, the data contains digit images similar in shape and organization to the aforementioned MNIST, consisting of black and white images with a size of 16 × 16. The entire dataset contains a total of 9298 images, of which 7291 belong to the training set and the remaining 2007 belong to the test set. This dataset is used in applications such as address recognition and mail sorting. It can also be utilized for domain adaptation tasks.

The Street View House Numbers (SVHN) dataset is a dataset of house number images taken in a street environment, as shown in Figure 5. This dataset is a collection of images containing numerical numbers of houses. It addresses the problem of number recognition in a realistic environment. The entire dataset consists of 604,388 images, of which 73,257 belong to the training set, 26,032 belong to the validation set, and the remaining 26,032 belong to the test set. The images were taken in a variety of street conditions, including complex backgrounds and lighting variations. The SVHN dataset is used to evaluate performance under realistic conditions. It is one of the most important datasets in the field of digit recognition.

Figure 3. Samples of MNIST dataset.

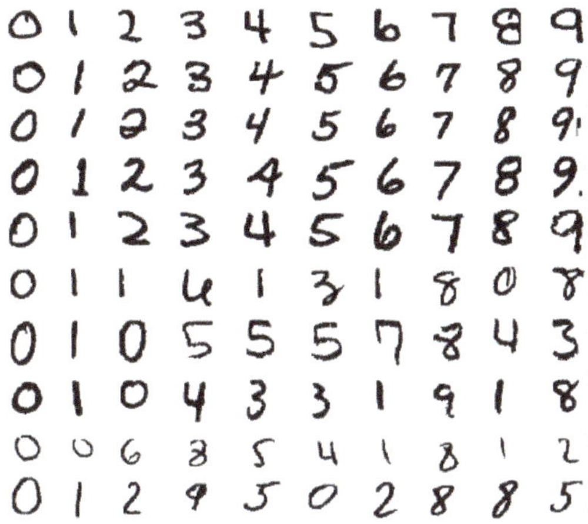

Figure 4. Samples of USPS dataset.

Figure 5. Samples of SVHN dataset.

This dataset is a good dataset for number recognition. Thus, it was also selected in this study. However, since the dataset itself is taken from real photos, even though it is labeled, the performance of the dataset may vary depending on the preprocessing.

The EMNIST (extended MNIST) dataset is a dataset of handwritten alphabet letters and numeric images. This dataset has a similar organization to MNIST, with a total of 814,255 images. Of these, 814,255 images are in the training set. The EMNIST dataset is used to recognize handwritten alphabet letters and numbers. It is commonly used to solve the problem of handwriting recognition. It is one of the most important datasets that can be applied to the problem of recognizing various alphabetic characters and numbers.

As can be seen in Figure 6, it is an extension of the MNIST dataset. Thus, the number of data is very large compared to MNIST. The reason is that it consists of grayscale images that are easy to classify, such as numeric data from 0 to 9, letters from a to z, and so on. Therefore, it was thought that it would be best to run a domain adaptation experiment with MNIST.

Figure 6. Samples of EMNIST dataset.

4.3. Performance Metrics

In machine learning and pattern recognition tasks, performance evaluation metrics play a key role in quantifying and comparing the performance of models. In this section, we will take a closer look at the main performance metrics used to evaluate the performance of classification models: precision, recall, and F1 score [32].

First, precision is a metric that indicates the percentage of samples that a model predicts as true that are actually true. More specifically, precision indicates how many results the model predicts as positive classes that are actually positive. This metric is represented by the following formula:

$$\text{Precision} = \frac{\text{True Positives}}{\text{True Positives} + \text{False Positives}} \tag{9}$$

where true positives refers to the number of samples that the model predicts are true and are actually true, while false positives refers to the number of samples that the model predicts are false but are actually true. Precision is a measure of how reliable a model's positive predictions are, which is important for reducing unnecessary misdiagnoses.

Recall is a metric that shows the percentage of samples that the model predicts as true out of those that are actually true. In other words, recall shows how many of the true positive samples in a positive class the model correctly detects. Recall is calculated with the following formula:

$$\text{Recall} = \frac{\text{True Positives}}{\text{True Positives} + \text{False Positives}} \tag{10}$$

Recall indicates how well the model detects true positives. It is particularly important for ensuring that positive cases are not missed.

The F1 score is a metric calculated as the harmonic mean of the precision and recall, which balances the accuracy and precision of the model. It is a useful metric for evaluating a model's performance from different angles. The F1 score is calculated using the following formula:

$$\text{F1 score} = 2 * \frac{\text{Precision} * \text{Recall}}{\text{Precision} + \text{Recall}} \tag{11}$$

The F1 score represents a balance between precision and recall, as shown in Equation (11). It evaluates a model's performance by considering both the accuracy of its positive class predictions and the detection rate of actual positive samples. This provides a more comprehensive view of a model's performance than considering accuracy alone. This concludes our detailed discussion of performance evaluation metrics. These metrics can be used to evaluate the performance of a model. They can help us to select and refine the right model for a particular problem.

4.4. Results

In this section, we will describe experiments conducted using a bidirectional-feature-learning-based domain adaptation method with a generative network. We will also provide the overall experimental environment and experimental results. As a quantitative evaluation, we have tried to prove the validity of this paper by comparing the proposed model with the DANN and ADDA models that already exist. To prove how well the domain adaptation is achieved, in addition to the comparison between models, the proposed model was trained only in the source domain and the test accuracy was measured and compared. This shows that it is an effective way to realize better model adaptation. By using the loss function, which was not included in the models implemented in the previous studies mentioned above, the interaction between the generative network and the feature conversion process could reduce the difference in the distribution of data, increase the similarity of the extracted features, and strengthen the interdependence between the two domains. This approach was successfully implemented through numerical comparison of several evaluation indicators. As a qualitative evaluation, the comparison of data points in the source and target domains before and after learning provided a visualization of how well the domain adaptation was achieved. This suggests that bidirectional feature learning can effectively reduce distributional differences between datasets, significantly reducing the distance between unadapted and adapted data.

Figure 7 shows a visualization of data points without adaptation and with adaptation. It could be seen from actual experiments that the two-way feature-learning-based domain adaptation method with a generative network outperformed the traditional one-way feature conversion method. Compared to traditional domain adaptation techniques, the method proposed in this paper showed higher accuracy and stability in quantitative evaluation. Each row shows the results of experiments conducted on different datasets divided into source and target domains, with red indicating the target domain and blue indicating the source domain. From the top, each row was organized in the following order: MNIST → USPS, SVHN → MNIST, EMNIST → MNIST. In addition, each column could be divided into cases without domain adaptation and with domain adaptation in each experimental environment. From the left, column 1 represents the case without domain adaptation and column 2 represents the case with domain adaptation. It can be seen from Figure 7 that the model proposed in this study performed well in domain adaptation.

The domain adaptation results in this paper are shown in Table 3. We can see that the proposed method outperformed other methods by a large margin, except for the SVHN to MNIST experimental results. This figure shows the test accuracy of the experiments, which is a value between 0 and 1. The maximum value of 1 indicates 100% accuracy. Therefore, the results of this model showed a high accuracy, of more than 90%, except for the one case mentioned above. Even the comparison of figures with existing studies and the results of learning and testing only the source domain showed a lot of differences. This indicates that the two-way approach of feature transformation and feature learning through generative networks can improve the generalization ability of the model by more effectively reducing distributional differences between domains.

Furthermore, Tables 4–6 show that the ADDA paper is part of a follow-up study published after the DANN paper. It does not change the fact that it performs better than DANN to some extent. However, the moment when the DANN model has a higher accuracy value than the ADDA model is the EMNIST to MNIST experiment shown in Table 3. On the other hand, the proposed model has a value of more than 0.98 for evaluation metrics, with the highest value close to 0.998, which means that the performance is good. We focused on finding the optimal value for the number of epochs through multiple experiments and set the number of epochs to 150. As a result, the main model took 1 h and 14 min, the ADDA model took 42 min by pretraining a CNN on the source domain to reduce the time, and the DANN model took 55 min.

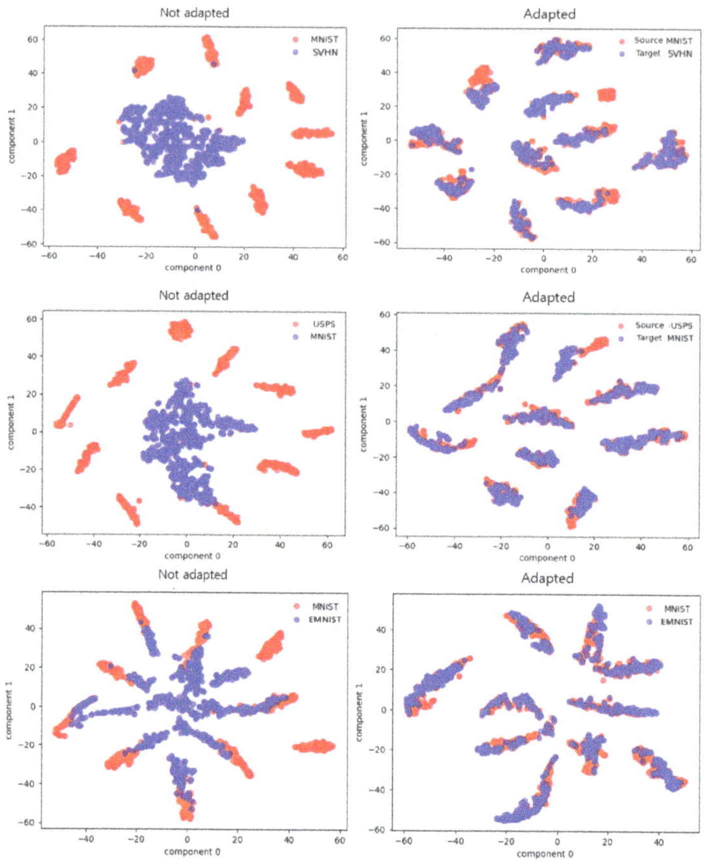

Figure 7. Visualization of data points without domain adaptation and with domain adaptation.

Table 3. Test accuracy of adversarial domain adaptation between other models for target to source domain.

Method	MNIST → USPS	SVHN → MNIST	EMNIST → MNIST
Source only	0.601	0.489	0.752
DANN	0.771	0.739	0.984
ADDA	0.894	0.760	0.955
BiFLP-AdvDA	**0.961**	**0.859**	**0.990**

Table 4. Adversarial domain adaptation for MNIST (target domain) to USPS (source domain) dataset.

	MNIST → USPS		
Method	Precision	Recall	F1 Score
Source only	0.714	0.709	0.712
DANN	0.901	0.895	0.899
ADDA	0.920	0.905	0.911
BiFLP-AdvDA	**0.989**	**0.989**	**0.989**

Table 5. Adversarial domain adaptation for SVHN (target domain) to MNIST (source domain) dataset.

SVHN → MNIST			
Method	**Precision**	**Recall**	**F1 Score**
Source only	0.531	0.528	0.531
DANN	0.676	0.673	0.665
ADDA	0.726	0.711	0.703
BiFLP-AdvDA	**0.988**	**0.988**	**0.988**

Table 6. Adversarial domain adaptation for EMNIST (target domain) to MNIST (source domain) dataset.

EMNIST → MNIST			
Method	**Precision**	**Recall**	**F1 Score**
Source only	0.756	0.621	0.630
DANN	0.821	0.832	0.831
ADDA	0.895	0.894	0.894
BiFLP-AdvDA	**0.998**	**0.996**	**0.996**

The qualitative results also show the superiority of bidirectional feature learning in a visual quality evaluation. Figure 7 shows a visualization of the experimental results when only the source domain was trained and tested, on the left, and when the entire model was trained and tested, on the right. Comparing the two, it can be seen that the proposed model achieved excellent results in domain adaptation. This can be attributed to the fact that the generated network reflected the actual distribution of the data well. The effect of the bidirectional feature transformation also strengthened the interdependence between the two domains.

Taken together, these experimental results show that the use of generative models and bidirectional feature learning can significantly improve the performance compared to various existing domain adaptation studies. It is an effective way to achieve better model adaptation capabilities. This suggests that using interaction between the generative network and the feature transformation process is a successful approach to reduce distributional differences in the data and enhance interdependence between the two domains.

4.5. Discussion

In this section, we present the effectiveness of domain adaptation based on the proposed bidirectional feature learning method with generative networks through experimental results. We also describe our main contributions through comparison with existing domain adaptation techniques. The quantitative evaluation showed that the bidirectional feature learning method had higher accuracy and stability than the one-way feature conversion method. Domain adaptation results between various datasets confirmed that the bidirectional feature learning with generative network outperformed previous studies such as DANN and ADDA. Although it might not be possible to conclude that the robustness of the model was improved based on the experimental results alone, considering the flow of minimizing domain information, increasing similarity between features, and reducing gaps from one-way learning to two-way learning, it is clear that bidirectional feature learning with a generative network in the field of domain adaptation is effective because the accuracy and various evaluation metrics recorded higher values than for previous domain adaptation studies.

The visual quality evaluation visually confirmed that the bidirectional feature learning method produced good results in domain adaptation. This was determined to be a result of the generative network more accurately reflecting the actual distribution of the data. The effect of the bidirectional feature transformation strengthened the interdependence between the two domains. Taken together, this study clearly demonstrates that the combination of the bidirectional feature learning method and generative network shows higher performance than various other domain adaptation techniques. It is an effective method for reducing distribution difference between datasets and improving the adaptive ability of the model.

The limitations of this study are discussed in this section. We believe that there are four main limitations of this paper. First, this paper proposed a method to solve the domain adaptation problem by reconstructing real image data, but when the variability in the domain is high, the quality and generalization ability of the reconstructed image data may be limited. In particular, it is necessary to discuss how to overcome the limitations for the adaptation problem between complex and diverse domains. Second, the proposed method relies on the amount of training data. In the absence of a sufficient amount of labeled data, the generalization ability of the model may suffer. This is a common problem in real-world scenarios, and we need to discuss how to develop effective models even with small amounts of data. Third, although the reconstructed image data were considered as noisy image data, the performance of the model may fluctuate depending on the degree and shape of the noise. This may cause the model to be sensitive to noise, and ways to improve this instability are needed. Fourth, the proposed method can take a long time to learn and infer, and there may be issues with scalability on large datasets. There is a need to develop more efficient learning and inference methods to increase their practical applicability.

5. Conclusions and Outlook

This study proposes a domain adaptation based on a bidirectional feature learning method with a generative network. The main idea that it introduces outperforms the traditional one-way feature transformation method in the domain adaptation problem. Bidirectional feature learning can improve the efficiency of domain adaptation through an approach that reduces distributional differences in the data and strengthens the interdependence between the two domains through the interaction of the generative network and the feature transformation process. This idea emphasizes the common features between datasets. The role of the generative network is to learn the true distribution of the data more accurately, which improves the performance of the adaptation model and increases the robustness of the model to noise.

As a future direction of this research, we would like to develop domain adaptation techniques with higher performance and generalization ability. To this end, we plan to explore adaptation across more diverse datasets and domains to maximize the generalization ability of the model and study its applicability in various domains in the real world. We also want to develop more robust and reliable domain adaptation techniques by conducting research for optimizing network architecture and learning strategies with the effective combination of generative networks and feature transformations. Finally, we plan to explore the feasibility of using bidirectional feature learning in applications other than domain adaptation so that it can be applied to a wider variety of problems.

Author Contributions: Conceptualization, C.H.; methodology, C.H.; software, C.H.; validation, C.H.; formal analysis, C.H.; investigation, C.H.; resources, C.H.; data curation, C.H.; writing—original draft preparation, C.H.; writing—review and editing, C.H.; visualization, C.H.; supervision, H.C. and J.J.; project administration, H.C. and J.J. All authors have read and agreed to the published version of the manuscript.

Funding: This research was supported by the SungKyunKwan University and the BK21 FOUR (Graduate School Innovation). It was also funded by the Ministry of Education (MOE, Korea) and National Research Foundation of Korea (NRF).

Data Availability Statement: Four public datasets of the MNIST database of handwritten digits (https://www.kaggle.com/datasets/hojjatk/mnist-dataset), accessed on 1 March 2023), SVHN dataset (http://ufldl.stanford.edu/housenumbers/, accessed on 1 March 2023), and two Pytorch datasets of EMNIST Dataset (https://pytorch.org/vision/main/generated/torchvision.datasets.EMNIST.html, accessed on 1 March 2023), and USPS dataset (https://pytorch.org/vision/main/generated/torchvision.datasets.USPS.html, accessed on 1 March 2023).

Acknowledgments: This research was supported by the SungKyunKwan University and the BK21 FOUR (Graduate School Innovation) and funded by the Ministry of Education (MOE, Korea) and National Research Foundation of Korea (NRF). Moreover, this research was supported by the MSIT (Ministry of Science and ICT), Korea, under the ICT Creative Consilience Program (IITP-2023-2020-0-01821) supervised by the IITP (Institute for Information & communications Technology Planning & Evaluation).

Conflicts of Interest: The authors have no conflicts of interest relevant to this study to disclose.

References

1. Ganin, Y.; Ustinova, E.; Ajakan, H.; Germain, P.; Larochelle, H.; Laviolette, F.; Marchand, M.; Lempitsky, V. Domain-adversarial training of neural networks. *J. Mach. Learn. Res.* **2016**, *17*, 2030–2096.
2. Ganin, Y.; Lempitsky, V. Unsupervised domain adaptation by backpropagation. In Proceedings of the 32nd International Conference on International Conference on Machine Learning-ICML'15, Lille, France, 6–11 July 2015; Volume 37, pp. 1180–1189.
3. Wang, J.; Chen, Y.; Hao, S.; Feng, W.; Shen, Z. Balanced Distribution Adaptation for Transfer Learning. In Proceedings of the 2017 IEEE International Conference on Data Mining (ICDM), New Orleans, LA, USA, 18–21 November 2017; pp. 1129–1134.
4. Tzeng, E.; Hoffman, J.; Zhang, N.; Saenko, K.; Darrell, T. Deep Domain Confusion: Maximizing for Domain Invariance. *arXiv* **2014**, arXiv:1412.3474.
5. Ma, S.; Gao, S.-H.; Gao, Y. Baochang Zhang End-to-End Label-Constraint Adaptation for Adversarial Domain Alignment. In Proceedings of the IEEE/CVF Conference on Computer Vision and Pattern Recognition (CVPR), Nashville, TN, USA, 20–25 June 2021.
6. Li, S.; Li, T.; Zhang, H.; Zhang, R.; Zhang, H. Graph-Based Domain Adaptation with Attention-Guided Generative Adversarial Networks. In Proceedings of the IEEE/CVF Conference on Computer Vision and Pattern Recognition (CVPR), Nashville, TN, USA, 20–25 June 2021.
7. Tzeng, E.; Hoffman, J.; Saenko, K.; Darrell, T. Adversarial Discriminative Domain Adaptation. In Proceedings of the 2017 IEEE Conference on Computer Vision and Pattern Recognition (CVPR), Honolulu, HI, USA, 21–26 July 2017; pp. 2962–2971.
8. Kingma, D.P.; Welling, M. Auto-Encoding Variational Bayes. In Proceedings of the 2nd International Conference on Learning Representations, ICLR 2014, Banff, AB, Canada, 14–16 April 2014.
9. Maggipinto, M.; Masiero, C.; Beghi, A.; Susto, G.A. A Convolutional Autoencoder Approach for Feature Extraction in Virtual Metrology. *Procedia Manuf.* **2018**, *17*, 126–133. [CrossRef]
10. Zhu, J.Y.; Park, T.; Isola, P.; Efros, A.A. Unpaired Image-to-Image Translation using Cycle-Consistent Adversarial Networks. In Proceedings of the IEEE International Conference on Computer Vision (ICCV), Venice, Italy, 22–29 October 2017.
11. Bousmalis, K.; Silberman, N.; Dohan, D.; Erhan, D.; Krishnan, D. Unsupervised Pixel-Level Domain Adaptation with Generative Adversarial Networks. In Proceedings of the 2017 IEEE Conference on Computer Vision and Pattern Recognition (CVPR), Honolulu, HI, USA, 21–26 July 2017; pp. 95–104.
12. Volpi, R.; Morerio, P.; Savarese, S.; Murino, V. Adversarial Feature Augmentation for Unsupervised Domain Adaptation. In Proceedings of the 2018 IEEE/CVF Conference on Computer Vision and Pattern Recognition (CVPR), Salt Lake City, UT, USA, 18–23 June 2018; pp. 5495–5504.
13. Hou, J.; Ding, X.; Deng, J.D.; Cranefield, S. Unsupervised Domain Adaptation using Deep Networks with Cross-Grafted Stacks. In Proceedings of the 2019 IEEE/CVF International Conference on Computer Vision Workshop (ICCVW), Seoul, Republic of Korea, 28–29 October 2019; pp. 3257–3264.
14. Bousmalis, K.; Trigeorgis, G.; Silberman, N.; Krishnan, D.; Erhan, D. Domain separation networks. In Proceedings of the 30th International Conference on Neural Information Processing Systems (NIPS'16), Barcelona Spain, 5–10 December 2016; pp. 343–351.
15. Kim, T.; Jeong, M.; Kim, S.; Choi, S.; Kim, C. Diversify and Match: A Domain Adaptive Representation Learning Paradigm for Object Detection. In Proceedings of the 2019 IEEE/CVF Conference on Computer Vision and Pattern Recognition (CVPR), Long Beach, CA, USA, 15–20 June 2019; pp. 12448–12457.
16. Huang, L.; Zhang, C.; Zhang, H. Self-adaptive training: Beyond empirical risk minimization. In Proceedings of the 34th International Conference on Neural Information Processing Systems (NeurIPS'20), Virtual, 6–12 December 2020; Article 1624.
17. Murez, Z.; Kolouri, S.; Kriegman, D.; Ramamoorthi, R.; Kim, K. Image to Image Translation for Domain Adaptation. In Proceedings of the 2018 IEEE/CVF Conference on Computer Vision and Pattern Recognition (CVPR), Salt Lake City, UT, USA, 18–23 June 2018; pp. 4500–4509.
18. Long, M.; Cao, Y.; Cao, Z.; Wang, J.; Jordan, M.I. Transferable Representation Learning with Deep Adaptation Networks. *IEEE Trans. Pattern Anal. Mach. Intell.* **2019**, *41*, 3071–3085. [CrossRef] [PubMed]
19. Long, M.; Zhu, H.; Wang, J.; Jordan, M.I. Deep transfer learning with joint adaptation networks. In Proceedings of the 34th International Conference on Machine Learning (ICML'17), Sydney, Australia, 6–11 August 2017; Volume 70, pp. 2208–2217.

20. Tzeng, E.; Hoffman, J.; Darrell, T.; Saenko, K. Simultaneous Deep Transfer Across Domains and Tasks. In Proceedings of the 2015 IEEE International Conference on Computer Vision (ICCV), Santiago, Chile, 7–13 December 2015; pp. 4068–4076.
21. Liu, M.; Li, J.; Li, G.; Pan, P. Cross Domain Recommendation via Bi-directional Transfer Graph Collaborative Filtering Networks. In Proceedings of the 29th ACM International Conference on Information & Knowledge Management (CIKM '20), Online, 19–23 October 2020; pp. 885–894.
22. Donahue, J.; Krähenbühl, P.; Darrell, T. Adversarial Feature Learning. *arXiv* **2016**, arXiv:1605.09782.
23. Saito, K.; Ushiku, Y.; Harada, T. Asymmetric Tri-training for Unsupervised Domain Adaptation. In Proceedings of the 34th International Conference on Machine Learning (ICML'17), Sydney, Australia, 6–11 August 2017; Volume 70, pp. 2988–2997.
24. Baktashmotlagh, M.; Harandi, M.T.; Lovell, B.C.; Salzmann, M. Unsupervised Domain Adaptation by Domain Invariant Projection. In Proceedings of the 2013 IEEE International Conference on Computer Vision (ICCV), Sydney, Australia, 1–8 December 2013; pp. 769–776.
25. Li, Y.; Yuan, L.; Vasconcelos, N. Bidirectional Learning for Domain Adaptation of Semantic Segmentation. In Proceedings of the 2019 IEEE/CVF Conference on Computer Vision and Pattern Recognition (CVPR), Long Beach, CA, USA, 15–20 June 2019; pp. 6929–6938.
26. Cai, Y.; Yang, Y.; Zheng, Q.; Shen, Z.; Shang, Y.; Yin, J.; Shi, Z. BiFDANet: Unsupervised Bidirectional Domain Adaptation for Semantic Segmentation of Remote Sensing Images. *Remote Sens.* **2022**, *14*, 190. [CrossRef]
27. Schuster, M.; Paliwal, K.K. Bidirectional Recurrent Neural Networks. *IEEE Trans. Signal Process.* **1997**, *45*, 2673–2681. [CrossRef]
28. Goodfellow, I.; Pouget-Abadie, J.; Mirza, M.; Xu, B.; Warde-Farley, D.; Ozair, S.; Courville, A.; Bengio, Y. Generative Adversarial Nets. In Proceedings of the Advances in Neural Information Processing Systems (NIPS), Montreal, QC, Canada, 8–13 December 2014; pp. 2672–2680.
29. Rozantsev, A.; Salzmann, M.; Fua, P. Beyond Sharing Weights for Deep Domain Adaptation. *IEEE Trans. Pattern Anal. Mach. Intell.* **2019**, *41*, 801–814. [CrossRef] [PubMed]
30. Grill, J.B.; Strub, F.; Altché, F.; Tallec, C.; Richemond, P.H.; Buchatskaya, E.; Doersch, C.; Pires, B.A.; Guo, Z.D.; Azar, M.G.; et al. Bootstrap Your Own Latent a New Approach to Self-Supervised Learning. In Proceedings of the 34th International Conference on Neural Information Processing Systems (NeurIPS'20), Online, 6–12 December 2020; Article 1786.
31. Motiian, S.; Piccirilli, M.; Adjeroh, D.A.; Doretto, G. Unified Deep Supervised Domain Adaptation and Generalization. In Proceedings of the 2017 IEEE International Conference on Computer Vision (ICCV), Venice, Italy, 22–29 October 2017.
32. Van Rijsbergen, C.J. *Information Retrieval*, 2nd ed.; Butterworth: London, UK; Boston, MA, USA, 1979.

applied sciences

MDPI

Article

Automatic Fruits Freshness Classification Using CNN and Transfer Learning

Umer Amin, Muhammad Imran Shahzad, Aamir Shahzad, Mohsin Shahzad, Uzair Khan and Zahid Mahmood *

Department of Electrical and Computer Engineering, COMSATS University Islamabad, Abbottabad Campus, Abbottabad 22060, Pakistan; shahumar203@gmail.com (U.A.); imranshahzad@cuiatd.edu.pk (M.I.S.); ashahzad@cuiatd.edu.pk (A.S.); mohsinshahzad@cuiatd.edu.pk (M.S.); uzairkhan@cuiatd.edu.pk (U.K.)
* Correspondence: zahid0987@cuiatd.edu.pk

Abstract: Fruit Freshness categorization is crucial in the agriculture industry. A system, which precisely assess the fruits' freshness, is required to save labor costs related to tossing out rotten fruits during the manufacturing stage. A subset of modern machine learning techniques, which are known as Deep Convolution Neural Networks (DCNN), have been used to classify images with success. There have recently been many changed CNN designs that gradually added more layers to achieve better classification accuracy. This study proposes an efficient and accurate fruit freshness classification method. The proposed method has several interconnected steps. After the fruits data is gathered, data is preprocessed using color uniforming, image resizing, augmentation, and image labelling. Later, the AlexNet model is loaded in which we use eight layers, including five convolutional layers and three fully connected layers. Meanwhile, the transfer learning and fine tuning of the CNN is performed. In the final stage, the softmax classifier is used for final classification. Detailed simulations are performed on three publicly available datasets. Our proposed model achieved highly favorable results on all three datasets in which 98.2%, 99.8%, and 99.3%, accuracy is achieved on aforesaid datasets, respectively. In addition, our developed method is also computationally efficient and consumes 8 ms on average to yield the final classification result.

Keywords: classification; deep learning; object detection

Citation: Amin, U.; Shahzad, M.I.; Shahzad, A.; Shahzad, M.; Khan, U.; Mahmood, Z. Automatic Fruits Freshness Classification Using CNN and Transfer Learning. *Appl. Sci.* **2023**, *13*, 8087. https://doi.org/10.3390/app13148087

Academic Editor: Seokwon Yeom

Received: 24 June 2023
Revised: 8 July 2023
Accepted: 10 July 2023
Published: 11 July 2023

1. Introduction

With the rapid developments in computer vision and machine learning methods, several algorithms have emerged that facilitate automatic object detection and recognition of various objects [1]. These methods are also benefiting the fruit processing industries, where classification and grading of fruits freshness are crucial for the manufactures to produce high-quality products such as fruit juices or tin packs. In an open environment, fruits are sensitive to numerous viruses and fungi that worry the agricultural industry and thus result in economic pressure. Physical ordering of fruits to categorize its quality either fresh or rotten is a laborious procedure. Thus, an automated assessment of fruit quality is an active research topic, which is experiencing growing interest all over the world. Recently, several works have appeared in literature that use Deep Neural Networks (DNNs) and Convolutional Neural Networks (CNNs) to classify the fruit freshness. Fruit freshness classification methodology is primarily inspired by pattern recognition and object classification that ultimately produces features set in which fruits are categorized through extensive training and learning. Multi-fruit categorization has extensive practical applications such as multi-fruit identification tools can be utilized in self-service fruit purchasing in supermarkets. It can be handy to eliminate human selection mistakes in production lines and hence increase efficiency. Nowadays, in agriculture, multi-fruit classification can assist the breeding of various fruit species. Due to the extensive developments in deep learning architectures, computer vision-based approaches are thought to be the most intelligent and cost-effective solutions.

Studies on object classification use various approaches, for example, Support Vector Machines (SVMs), linear discriminant analysis, or k-nearest neighbors (k-NN) to improve accuracy or speed [2]. Fruit freshness is a major factor in determining the quality of fruits as it can affect their shelf life and overall nutritious value. Figure 1 shows an example of conditions of fruit from fresh to rotten stage of apples and bananas, respectively. Therefore, to determine the proper price, the customer must be able to identify the variety of the fruits that are intended to be purchased. Fruit freshness classification is also crucial for consumers, as it can help them make knowledgeable decisions about the worth of the fruit item they are purchasing. It is also important for producers and retailers as it can help them to accomplish inventory and ensure that their products are meeting quality standards.

Figure 1. Top-row: Fresh fruits and **Bottom-row**: rotten fruits.

Recently, several machine learning-based methods for fruit freshness classification have appeared in the literature [3]. Few of them use CNNs, Deep CNNs, and Faster-RCNN to accomplish the fruits classification task [4]. The CNN models are also implemented in a number of tasks, for instance, object detection, classification, and recognition [3]. Over the past few years, various CNN architectures have been developed, and they have demonstrated excellent performance for image classification. Transfer learning is one of the techniques that uses previously developed architectures on various problems to produce more accurate results. Deep learning and machine learning techniques integrated with transfer learning could also be used for image classification. This study investigates the possibility of transfer learning with regard to CNN models for the quality assessment of fruits instead of using CNN architectures from starch. In real-life, classification of fruits is normally carried out by people that we believe is ineffective for fruit farmers and fruit sellers. Therefore, the development of an accurate classification method is desired, which will significantly reduce human efforts and costs. A robust fruit freshness classification method will also reduce the industry's production time in the agriculture domain by correctly identifying fruit defects. Therefore, this study proposes a novel and automatic fruit freshness classification method using fine tuning and transfer learning of the AlexNet. The effectiveness of the proposed method is validated on three publicly available fruit datasets. Our main contributions to this manuscript are listed below.

- We develop an automatic fruits classification method that accurately classifies whether the fruits are fresh or rotten. Our developed method is based on transfer learning, which uses classical convolutional architectures such as AlexNet. The introduction of transfer learning with the AlexNet yields higher accuracy than few of the recently published works with much lower computational complexity.

- We propose an intelligent system that reliably recognizes fruits, for instance, apples, bananas, and oranges, which are later categorized as either fresh or rotten classes. Automatic and timely identification of fresh and rotten fruits will enable agriculturalists to produce large quantities of various fruits and thus put on great value to a country's economy.
- We report experiments on three well known and publicly available datasets in our simulations. Our findings are encouraging as we obtain over 99% fruit freshness classification accuracy. We are hopeful that our developed method will also be helpful to customers in supermarkets to identify fresh fruit.

Rest of this manuscript is organized as follows. In Section 2, we briefly review the recently developed fruits freshness classification methods. In Section 3, we describe our developed method. While simulation results and comparisons are listed in Section 4 followed by Section 5 that concludes our findings and also hints towards the possible future work.

2. Related Work

This section briefly discusses recent methods that aim to classify various fruits using machine learning and image processing-based methods.

In [5], a deep learning-based method to classify fruits and vegetables is developed, which is primarily based on the YOLOv4 model. This method initially recognizes the object type in an image and then classifies the object either as fresh or rotten. This model also improves the backbone of the YOLOv4 version using the Mish activation function, which results in rapid detection of objects. In [6], researchers analyze and proposed a novel design of computer vision-based method using deep learning with the Convolutional Neural Network (CNN) model to detect several fruit freshness level. The specially designed CNN model is later evaluated and extensively tested with public datasets of fruits fresh and rotten for classification. This is a nice effort and nicely handles the fruits' freshness level instantly.

In [7], published work focuses on classifying rotten and good apples. For the task of apple classification, initially texture features of apples are extracted. For instance, discrete wavelet feature, histogram of oriented gradients, and law's texture energy along with the gray level co-occurrence. Later, various classifiers are applied, for instance the SVM, the k-NN, and Linear Discriminant. Researchers' conclude that the SVM classifier yields 98.9% accuracy, which is better than few of the compared classifiers. In [8], eight deep learning models namely AlexNet, Google Net, ResNet18, ResNet50, ResNet101, VGG16, VGG19, and NasNetMobile are fine-tuned to assess the quality of fruits and vegetables. The performance of deep learning models is based on the training and validation accuracy. The model's outcome shows that the VGG19 model reached the highest validation accuracy over the original samples and the ResNet18 model achieved the highest validation accuracy based on the augmented data samples. In [9], the authors investigate the maturity status of Papaya fruit by using machine learning. To classify the fruits, the LBP, the HOG, Gray Level co-occurrence Matrix (GLCM), SVM, K-Nearest neighbor (KNN), and Naive Bayes methods are applied and compared. Seven pre-trained models are fine-tuned on the given dataset of Papaya to evaluate the performance of the robust system. The K-Nearest neighbors (KNN) with the HOG features results high accuracy with much less training rate. In [9], authors apply deep learning model on the banana different dataset. In this work, bananas' freshness was analyzed by transfer learning and established the relationship between freshness and storage dates. Banana feature extraction were extracted by Google Net. The reported classification accuracy of this model is 98.92%, which is at par with normal human detection. In [10], authors use k-means clustering along with colors, textures, and shape features to classify the apple freshness by investigating its disease. This work also uses multiclass SVM during classification stage.

In [11], a system for classifying fruits and vegetables in supermarkets is implemented, which combines backdrop removal with a split-and-merge strategy to find fruits and vegetables in pictures. This model also employs color, shape, and texture as key identifying characteristics. The feature space was condensed using the PCA. Several kernel functions, such as the MWM-SVM, the MWV-SVM, and the DAG-SVM were used during the algorithm development. The maximum accuracy of 88.2%, was attained by MWV-SVMs utilizing Gaussian radial basis kernels. In [12], authors published a method for identifying fruit flaws in retail. Cameras are positioned on the borders of a conveyor to capture orange data samples. They applied color as a feature in the RGB images and also produced color histograms. The Fisher-LDA is employed to decrease features size and to reduce noise. Next, the orange problematic zones are found using the SVM. The trial results showed that their proposed technique had a 96.7% recall rate. The automotive, commercial, and agricultural industries, as well as other worldwide businesses, have all made substantial use of it for object identification and picture categorization. Various image processing and deep learning methods are extensively used to extract and alter supervised or unsupervised features from several layers of non-linear data with the aim to classify objects to understand its patterns [13]. In [14], the developed method employed background subtraction modeling to handle diverse samples. They use a range of recent methods, which include decision trees, the k-NN, the LDA, and the SVM. Simulations showed that SVM performed better than a few techniques.

In [15], CNN is used to detect various fruits. This work is performed on a relatively small dataset, and it produced an excellent performance by yielding 98.92% detection accuracy. In [16], researchers compared performances of multi-task learning, domain adaptation, and sample selection bias. They also carried out a detailed review of the method that are used to detect and classify various objects. In [17], a deep learning-based technique is used for the freshness classification of Hog Palm fruit. This work uses four CNN-based models, which were fine-tuned on imageNet Dataset. The Dataset was augmented and used for training and hyper parameter tuning for the purpose of grid search and k-fold cross validation the results were compared in terms of different parameters listed therein. In [18], the proposed method uses VGG16 and the CNN to extract various robust fruits features. In this work, SVM, decision trees, and logistic regression models were also compared. The authors concluded that the SVM the achieved highest 99% classification accuracy than the compared methods. In [19], fruit classification was achieved by using CNN and Softmax, which yielded 97.14% accurate classification.

The aforementioned is a brief review of the recent methods that handle fruit freshness detection and classification problems. To achieve accurate results and to facilitate the humans each of these works performed experiments on standard and publicly available datasets. While a few researchers, such as [20], gathered their own dataset, which contains sixteen different species of fruits. The methods briefly described above are a nice addition to the research domain to tackle the fruits freshness problem. We believe that our work is latest addition in this domain, which aims to achieve high fruit freshness classification of different fruits. Our study indicates that machine learning algorithms are helpful to determine the freshness of perishable items, such as fruits as well vegetables. In addition, deep learning models focus to extract features. Ultimately, as we will see in the results section that testing the data on unseen data is a good indicator of the performance of the developed method. Below we detail our developed method.

3. Methodology

Our developed method has various interconnected modules as shown in Figure 2. Below, we describe the various modules that are used in our developed algorithm.

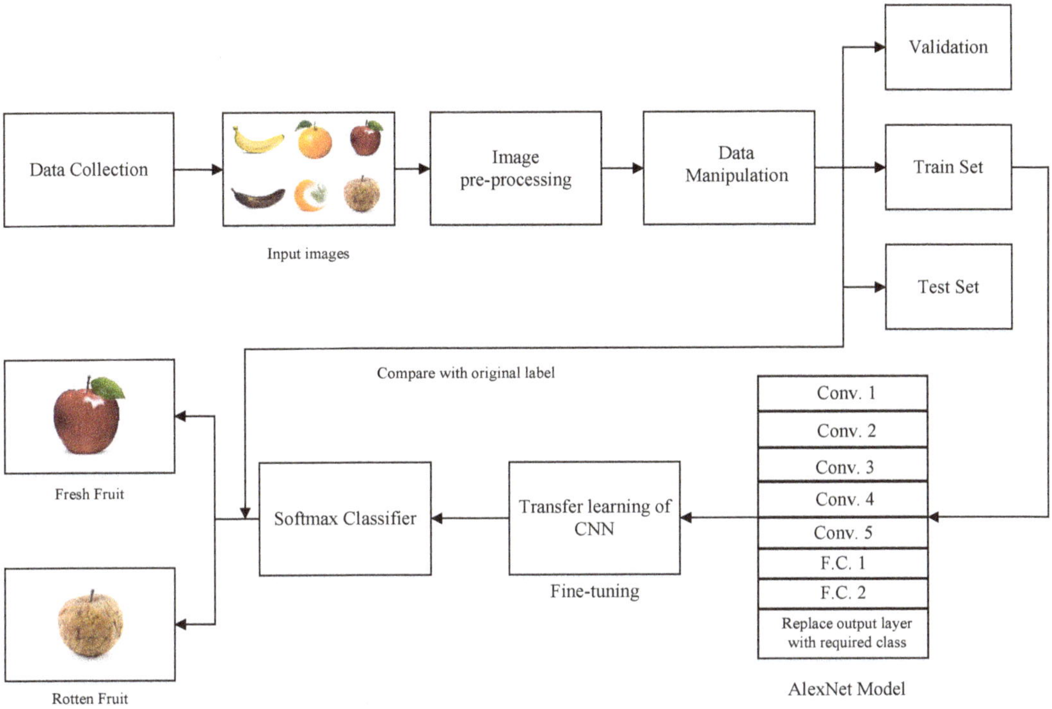

Figure 2. Flow of our developed method.

3.1. Data Collection

To develop our fruits freshness classification method, we acquire different fruits images data from Kaggle (www.kaggle.com, accessed on 5 June 2023). The Kaggle is a publicly available dataset and contains different classes of various fruits such as apples, bananas, or oranges. The fruits image dataset provided by Kaggle contains three kinds of fresh and rotten images. Moreover, the fruit dataset contains images in separate files, such as fresh apples, fresh bananas, fresh, oranges, rotten apples, rotten bananas, and rotten oranges. A few of such sample fruits images are shown in Figure 2. These images are now pre-processed by the next module.

3.2. Pre-Processing

Data pre-processing is conducted earlier than data manipulation to fit the data for Convolutional Neural Network (CNN) and various filters are employed therein. In our method, we performed the pre-processing in following manner.

Color Uniformity: To maintain uniformity, we process all images in the RGB domain. This essentially creates uniformity in each channel of the image.

Image Resizing: Since, the original dataset contains colored images of different fruits in several formats and sizes. Therefore, we resize images one by one and label it and store in separate directory. Hence, we resize images to $227 \times 227 \times 3$.

Image Augmentation: Augmentation is conducted by flipping all images to x-axis and randomly rotating images. In our work, augmentation is conducted in parallel where each image was rotated at 90°, 180°, and 270°. Hence, each image created three new images, which made the size of dataset four times its original size shown in Figure 3.

Image Labeling: Finally, the converted dataset is labelled according to each class they belong. Training and testing on the test set are conducted concurrently with validation. Meanwhile, the string labels were also changed into numeric format, which later helps the employed models to accurately predict the true labels. The pre-processed data is now further manipulated as listed below.

Figure 3. Image Augmentation by flipping.

3.3. Data Manipulation

During this phase, each image in the dataset was remodeled into a single size and scale. We observe that this strategy maintained significant data uniformity. As shown in Figure 4, we split the pre-processed datasets into three parts, which are validation, train set, and test set. In our work, 80% of the data is used for training, 10% for validation, and 10% is used in the test phase. We also state that in each class 100 images were taken randomly. Out of these 85 images of the single object contains the plane background. While 15 images of multiple objects contain complex background.

Figure 4. Few sample images of fresh and rotten fruits and vegetables.

3.4. AlexNet Architecture

This is an eight layers weighted model in which the first five are convolutional layers, while the remaining three are fully connected layers [21]. In AlexNet architecture, first layer processes the input image resolution of 150 × 150 × 3 and applies 96 convolutional filters 11 × 11 resolution. The output of first layer is processed as the input of the second layer and 256 convolutional filters of 5 × 5 resolution. Moreover, third and fourth layers apply 384 convolutional filters with a resolution of 3 × 3. While the 5th convolutional layer applies 256 kernels of 3 × 3 resolution. In AlexNet, all five layers apply maximum pooling of 2 × 2 resolution through batch normalization. The selection of an appropriate activation function encourages us to improve the accuracy of our method. Hence, while choosing the activation function, we make sure that the gradient function converges quickly and also at the infinity of the activation function is not 0. In our work, the ReLU activation function and its derivate are shown in Equations (1) and (2), respectively.

$$f(x) = max(0, x), \tag{1}$$

$$f'(x) = \begin{cases} 1, & if\ x > 0 \\ 0, & otherwise \end{cases}. \tag{2}$$

Equation (2) indicates that the derivative of the ReLU function is continually equal to 1 during the positive half-axis of x, and is constantly 0 during the negative half-axis of x. The ReLU function is used in the first five convolutional layers. In addition, the output from the first five layers is passed to three fully connected layers in which the first two fully connected layers contain 4096 units, and the last fully connected layer comprises 1000 units. Final output layer applies the Softmax activation function and consists of 6 units. After the AlexNet processes the data, in next step, transfer learning is performed as described below.

3.5. Transfer Learning of the CNN

The CNNs are networks that filter inputs for relevant information using constitutional layers. Constitutional filters along with the CNN layers are used to find neurons outputs that are linked to specific local input areas. It aids in the extraction of spatial and temporal visual characteristics. The CNN correctly extracts features from the input image of the given Dataset. There are three key components in the CNN, which are a convolution layer that learns features max pooling, which reduces the dimensionality, and finally, a fully connected layer that classifies the input image.

Transfer learning or knowledge transfer is a technique in which we use pre-trained network as a starting point to solve specified classification problems. During the transfer learning phase, we replaced a few upper layers of a fixed model base and added new layers. While a final layer of the output is replaced with the required classes and for fine tuning some of the parameters are changed such as epochs, size, and learning rate to achieve better performance. The parameters used for the experiments were set as:

No. of epochs = 10, Batch size = 32, and Learning rate = 10^{-5}.

3.6. SoftMax Classifier

Softmax or multinomial logistic regression has a unique advantage to deal the N-dimensional vectors. The Softmax is widely used in diverse fields including deep learning for various objects classification [22]. The Softmax classifier determines the probability of extracted vectors for classification. For the same data set, it gives the sum of the probability equal to 1 for all vectors as indicated by Equation (3).

$$f_j(z) = \frac{e^{zj}}{\sum_k e^{zk}}, \tag{3}$$

where z is the input vector obtained from fine-tuned network in previous stage. These results are mapped to the probability domain from the exponential domain. We select Softmax loss function for fruits classification. Our study indicates that this Softmax function has good performance and converges quickly. Mathematically, the Softmax loss function is modelled as shown in Equation (4).

$$Loss_{Softmax} = \frac{1}{N}\sum_i -log\frac{e^{zyi}}{\sum_j e^{zj}}, \tag{4}$$

where N represents the output value of last fully connected layer of the correct class (yi). Moreover, zj is the output value of the last fully connected layer of the jth class.

Equations (3) and (4) indicate that the $Loss_{Softmax}$ function uses the data from correct labels and maximizes the possibility of data. Meanwhile, it also ignores the information from the prevailing incorrect labels. Algorithm 1 shows the pseudo cod of our developed method. Algorithm 1 indicates that the fruits images, which are obtained from the Kaggle dataset, are pre-processed as described in the above section. In Algorithm 1, from lines (2) to (8) the gathered data is processed, which is later fed to the deep learning models. The data processed in initial stage is now processed by the AlexNet model as indicated in lines (9) to (15). During the AlexNet operation, all the eight layers are used. For first five convolution layers, the ReLU activation function is used. The other layers are utilized as described in above section. From lines (15) to (20), transfer learning of the CNN is performed. Meanwhile, the CNN is fine tuned in which epochs, batch sizes, and learning rate is set as shown in lines (18) and (19) of Algorithm 1.

Algorithm 1: Pseudocode of fruits freshness classification method

```
1.  Input: Obtain colored fruits images                  ▶ Data Collection from Kaggle
2.    do:
3.      Process collected data obtained in step (1).     ▶ Initial stage.
4.        Perform pre-processing operations as listed in Section 3.2.
5.          Process uni-channel images and make uniform image resizing.
6.            Perform image augmentation and image labeling.
7.              Perform data manipulation.               ▶ Isolate into validation, train, and
    test set.
8.    end
9.      begin
10.       Load AlexNet model
11.         Activate 8-layers including 5 conv layers and 3 fully connected layers.
12.           Use ReLU for first 5 conv layers.
13.             Process and utilize 8-layers as depicted in Section 3.4.
14.     end
15.       begin
16.         Transfer learning of the CNN
17.           Apply fine tuning and set the parameters as:
18.             Epochs = 10 and Batch size = 32,
19.               Set the learn rate = 10^-5
20.       end
21.   Apply softmax classifier:
22.     Use Equations (1) and (2) for classification tasks.
23. Output: Final classification result:
24.         Fresh fruit or Rotten fruit
```

In the final stages, the softmax classifier is used to predict the final status of the fruit. It will be shown in next section that our developed method is robust and accurately classifies the fruit condition instantly. Moreover, the steps shown in Algorithm 1 are simple and easy for readers to follow. In section below, we detail our findings along with useful images. In addition, we also discuss our observations during algorithm development stages.

4. Simulation Results

This section lists the experimental setup, used datasets brief description, fruits classification results, discussion, and observations in detail.

4.1. System Specifications and Experimental Setup

Our developed algorithm was executed on an Intel® NY USA Core I-i53550 machine, which has a CPU@3.30 GHz along with the facility of a NVIDIA GTX1080 graphics card. The aforesaid machine has 16 GB of RAM, which is sufficient to investigate the fruits freshness classification results that are yielded by our developed algorithm. Moreover,

Table 1 lists the parameters and experimental setup that is used throughout the simulations during the transfer learning of the CNN. As shown in Table 1, during our all simulations, the SGDM optimizer is used along with a learning rate of 10^{-5}. The L2 regularization was used in our method. Moreover, the validation frequency was set to 50 along with epochs and batch size were set to 10 and 32, respectively. The data was shuffled after the completion of every epoch. To achieve the classification output in a reasonable time, the pace of the momentum was set to 0.9. Training and test image resolution is 227 $\times$ 227 pixels.

Table 1. Experimental setup during simulations.

Parameter	Simulation Environment
Training/test image resolution	227 $\times$ 227 pixels
Optimizer	SGDM
Learning Rate	0.0001
Validation frequency	50
Epochs	10
Batch Size	32
L2 Regularization	0.0001
Gradient Threshold Method	L2norm
Gradient Threshold	Inf
Validation Patience	Inf
Shuffle	Every-epoch
Momentum	0.9

In sections below, we describe the details of the datasets that we used during our algorithm execution along with detailed qualitative and quantitative analysis. For each of these sections, a detailed discussion is also carried out along with our findings and recommendations.

4.2. Datasets Description

To simulate and validate our developed fruits freshness classification method, we choose three publicly available datasets. Below, we briefly present the details of each of the dataset.

Dataset 1: This data set contains 12,000 diverse images of fresh and rotten categories of fruits and vegetables [5]. Specifically, this dataset contains ten different classes. Prominent categories of fruits gathered in this dataset are apple, banana, orange, mango, strawberry, potato, carrot, tomato, cucumber, and bell peper. Each of the fresh categories in this dataset contain at least 600 images, while the rotten category contains minimum of 500 images. A few of the sample images from this dataset are shown in Figure 4.

Dataset 2: This data set contains total of 13,346 images of fresh and rotten fruits [19]. The foremost categories of this dataset are total of 6 classes, out of which are 3 classes for each fresh and rotten for apple, orange, and banana. Table 2 depicts the details of this dataset. From Table 2, it is clear that for fresh fruits category of apple, orange, and banana, this dataset contains at least 1400 images. While the rotten categories of apple and orange contain over 2200 images. Rotten banana in this case has the least gathering of 1595 bananas.

Dataset 3: This data set contains total of 3200 images in a duration of two weeks in March 2022 [20]. This dataset has been organized into 16 major classes, which are fresh and rotten grapes, fresh and rotten guavas, fresh and rotten jujubes, fresh and rotten pomegranates, fresh and rotten strawberry, fresh and rotten apples, fresh and rotten bananas, and fresh and rotten oranges. Few sample images of this dataset are shown in Figure 5. Developers of this dataset also provided the augmented images of these classes, which result in a total of 12,335 images.

Table 2. Statistics in Dataset 2.

Classes	Training	Testing
Fresh Apple	1693	394
Fresh Orange	1581	381
Fresh Banana	1466	388
Rotten Apple	2342	478
Rotten Orange	2224	436
Rotten Banana	1595	368
Total	10,901	2445

Figure 5. Sample images of fresh and rotten fruits from Dataset 3.

4.3. Evaluation Parameters

In our method, we use two well-known parameters, which are theaccuracy and confusion matrix as briefly described below.

Confusion Matrix: it is a popular parameter that assesses the effectiveness of any classification model. Normally, a confusion matrix is a square matrix, which indicates the predicted classes against the actual classes. The rows in a confusion matrix denote true class labels, while columns indicate predicted class labels. In our work, we use the confusion matrix for each of the three datasets to analyze the performance of our developed fruits freshness classification method. A confusion matrix provides us the flexibility to compute several classification performance matrix, such as Accuracy, as described below.

Accuracy: it is a well-known parameter and is widely used in classification and recognition related tasks. In our work, we believe that Accuracy is a good indicator to evaluate our developed classification method as it is extensively used to measure the pixels, which are correctly classified by any model. Mathematical accuracy is formulated as shown by Equation (5)

$$\text{Accuracy} = \frac{\text{True Positives} + \text{True Negitives}}{\text{True Positives} + \text{True Negitives} + \text{False Positives} + \text{False Negitives}}, \quad (5)$$

where True Positives and True Negatives belong to the true fruits positive and negative classes, respectively. While False Positives are fruits that are incorrectly classified as positives and False Negatives are fruits, which are incorrectly classified as negatives.

4.4. Fruits Classification Analysis

In this section, we discuss in detail the performance of our developed method through confusion matrixes for each of the datasets described earlier. For Dataset 1, Figure 6 shows our developed classifier's performance through a confusion matrix. The diagonal in the confusion matrix indicates true positives or actual values and shows the classification accuracy, which is achieved by our developed method. As discussed earlier that Dataset 1 contains 20 classes in which each of the fresh and rotten classes contains 10 items, respectively. For each of the fresh categories of apple, banana, and mango, 100% classification accuracy is obtained. Whereas for the fresh categories of orange and strawberry, our proposed method yields 99% accuracy.

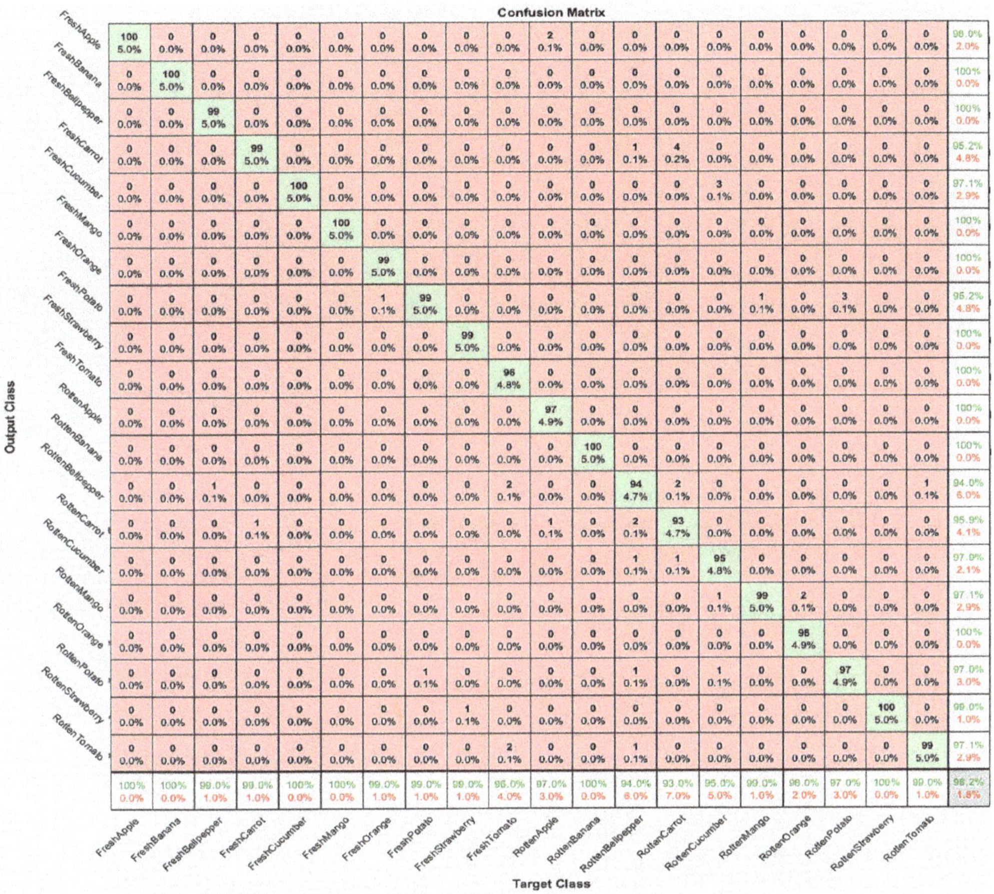

Figure 6. Confusion matrix for Dataset 1.

For the rotten categories of banana and strawberry, our classification method yields 100% accuracy. While for the aforesaid categories of apple, mango, and orange, our developed method yields 97%, 99%, and 98% classification accuracy, respectively. We note that once the training of the CNN is completed almost 98.2% classification accuracy is obtained immediately. Figure 6 also shows the accuracy of fresh and rotten vegetables, such as bell pepper, carrot, cucumber, potato, and tomato, respectively. Since our method is confined to fruit categories only, therefore, we believe that this provided accuracy comparison is useful for the research community.

Figure 7 shows the classification accuracy obtained by our developed method on Dataset 2. Since this dataset contains total of three fruits with six classes. As shown in the diagonal of Figure 7 that for each of the fresh apple, banana, and orange our developed method obtains 99.2%, 100%, and 99.7% accuracy, respectively. Similarly, for the rotten categories of these fruits, our proposed algorithm obtains 99.8%, 100%, and 99.7% classification accuracy for apple, banana, and orange, respectively. We believe that these findings are useful and encouraging. In the future, many food industries might also benefit from our findings.

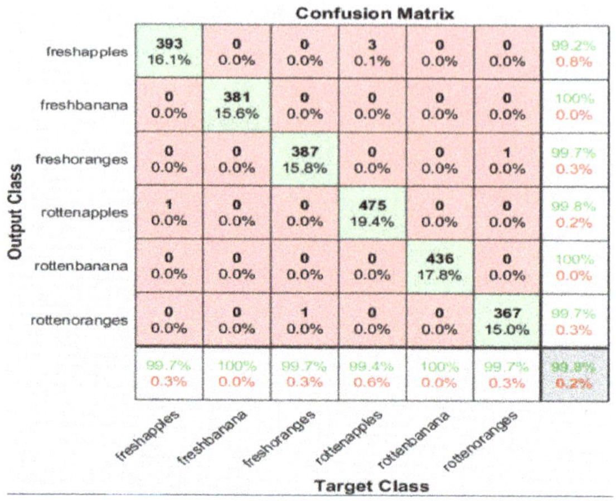

Figure 7. Confusion matrix for Dataset-2.

In our series of experiments, Figure 8 reports our experiments on Dataset 3, which was briefly described earlier. This is a huge dataset and contains several classes. As can be seen in Figure 8, for fresh categories of apple, banana, guava, jujube, orange, pomegranate, and strawberry, our developed method yields 100% classification accuracy. Moreover, for fresh category of grape, our developed algorithm yields 96% classification accuracy. Similarly, for the rotten categories of the aforementioned fruits, 100% classification accuracy is obtained for apple, banana, grape, orange, and strawberry. We are optimistic that our findings are encouraging and useful.

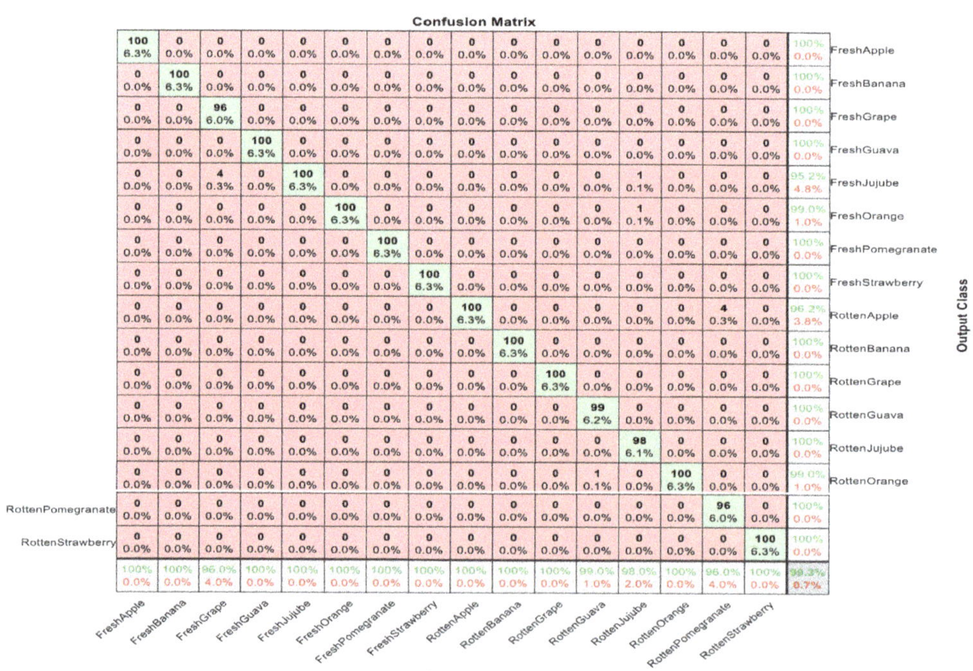

Figure 8. Confusion matrix for Dataset-3.

4.5. Comparison

In this section, we present the comparison of our developed method with several other methods on same datasets. We compare our work with three recently reported fruit freshness classification methods [5,19,20,23–25]. For fair comparison, we use the same training strategy as reported by above-described methods. Table 3 shows the mean accuracy comparison. From Table 3, a few important observations are listed below.

- On Dataset 1, the YOLO based method [5] reports 97% fruit freshness classification accuracy. On this dataset, our proposed method yields 98.2% classification accuracy.
- On Dataset 2, our developed method yields 99.8% classification accuracy and beats all the compared methods. In this dataset, the ResNet-50 based method [24] also reported an encouraging classification accuracy of 98.89%. Moreover, dataset [20] reported detailed classification results by implementing several architectures with the mean outcome of 88.77% fruit freshness classification.
- In [25], Dataset 3 is introduced. To the best of our knowledge, none of the work reports accuracy on this dataset. On this dataset, our developed method yields 99.3% fruit freshness classification. On the whole, our developed method achieves a mean accuracy of 99.1% on all three fruits datasets.
- We believe that our findings are encouraging and will be useful for various fruit packing industries. At advanced level, our method can also be used to know the fruits freshness level when they are growing with the tree.

Table 3. Mean accuracy (%) comparison.

Method		Datasets		
		Dataset 1	Dataset 2	Dataset 3
[5]	Improved YOLO	97%	–	–
[19]	Trained CNN	–	97.14%	–
[20]	MobileNetV2	_	88.62%	–
	ResNet50	_	73.26%	–
	VGG16	_	96.10%	–
	InceptionV3	_	97.10%	–
Mean Accuracy [20]		**88.77%**		
[23]	Compare different feature extraction techniques classification through SVM	_	97.61%	_
[24]	Freshness classification using RESNET50	–	98.89%	_
[25]	CNN + ResNet50	_	_	Pioneered to introduce this dataset.
Proposed Method		**98.2%**	**99.8%**	**99.3%**

4.6. Discussion

The points discussed above give good insight about the fruit freshness classification performance of our developed method. However, the discussion below sheds more light on the performance of our developed algorithm.

- Our study indicates that different networks achieve diverse accuracy outcomes on different algorithms. The MobileNetV2 trained by [20] on Dataset 2 achieves 97.14% classification accuracy. Whereas VGG16 and InceptionV3 achieve 96.10% and 97.10% classification accuracy, respectively. On Dataset 2, the ResNet50 achieved the lowest classification accuracy of 73.26%.

- Our method in general performs well on all the three datasets and achieves at least 98% classification accuracy as reported on Dataset 2 and also shown in Table 3. We believe that for Dataset 2, our proposed method has almost solved the fruit freshness classification problem by achieving the 99.8% accuracy.
- In general, all the compared methods perform well to handle the fruits freshness classification challenge and achieve at least 88% accuracy. One of the reasons for our developed method's superior performance is employment of data augmentation in pre-processing stage, which significantly mitigates overfitting on small datasets. Similarly, while using the pre-trained CNN architecture model and replacing the last layer with required targeted class also are reasons for our model's superior classification performance.
- Our method achieves high accuracy and outperforms several recently published works [5,19,20,23–25] due to intelligent selection of the AlexNet architecture. Our study indicates that the training time of AlexNet architecture is five times faster as compared to others deeper architecture speed. Moreover, the AlexNet is computationally efficient and does not require high performance workstation [26]. Similarly, in presence of other functions, for instance, *tanh, logistic, arctan, or Sigmoid* as activation functions, the AlexNet uses the ReLU activation function that drastically reduces likelihood of vanishing gradient problem.
- The layers in the AlexNet architecture contains more filters and each convolution layer is followed by a pooling layer. Such characteristics motivated us to utilize the AlexNet to address the fruits classification problem. Similarly, rotation and augmentation procedure as described in Section 3 increases the fruits images that ultimately resulted in good training of the AlexNet architecture, which later yields high accuracy.
- Recently published works, for instance, refs. [26–31] could also be investigated to develop a more robust fruits freshness classification method. Similarly, few of the [32,33] could also be investigated and optimized to develop a more robust and accurate algorithm that can detect and classify large species of several fruits.

4.7. Limitations

Although our proposed method achieves promising results in the aforementioned datasets. However, for the research community below, we briefly discuss the limitations faced by our developed algorithm.

- Since our method uses the AlexNet architecture, during the preprocessing stage our method consumes a bit more time.
- Since the AlexNet model was pre-trained on ImageNet dataset, which consists of 1000 object categories. Therefore, the model's performance and features are biased towards the visual patterns and objects that are present in the dataset. If the targeted class is significantly different from ImageNet dataset, then the pretrained feature may not generalize well, which ultimately leads to the reduced performance.
- The robustness of transfer learning depends on the similarity between the pre-training dataset and the target dataset. The AlexNet was trained on a large-scale dataset with millions of labeled images. If the target dataset is small or significantly different in terms of image content, style, or domain, then the pre-trained features may not capture the relevant patterns, which will also result in reduced classification performance.
- We observe that fruit classification remains a difficult task for a machine learning algorithm due to several reasons. For example, fruits shape, colors, and texture similarity among various fruits species. Moreover, high variations in a single fruit class that is dependent on fruit maturity phase, and the actual condition when some fruit is presented such as fruits placed inside plastic bags, sliced, or unpicked from farm. In such scenarios, our developed method might struggle to accurately classify the fruits freshness.

4.8. Computational Complexity

In Figure 9, we show the computational complexity of our developed method. We work in three different databases in fresh and rotten fruit datasets. Training took almost 67 h on all three datasets. For Dataset 1, training took 31 h on 32,667 images and tests images are 1000. While for Dataset 2, almost 17 h were consumed on 10,901 images and tests images are 2445. Finally, for Dataset 3, our method consumed 19 h with 12,335 images and tests images are 1600. As shown in Figure 9 in Dataset 1, test image consumes almost 8 ms to yield the final classification output. Similarly, for Dataset 2, slightly over 13 ms are consumed to obtain the final output. Our method requires almost 10 ms to yield the final classification result on Dataset 3. As shown in the last tower in Figure 9, on the average, our method requires almost 8.8 ms to yield the final classification result.

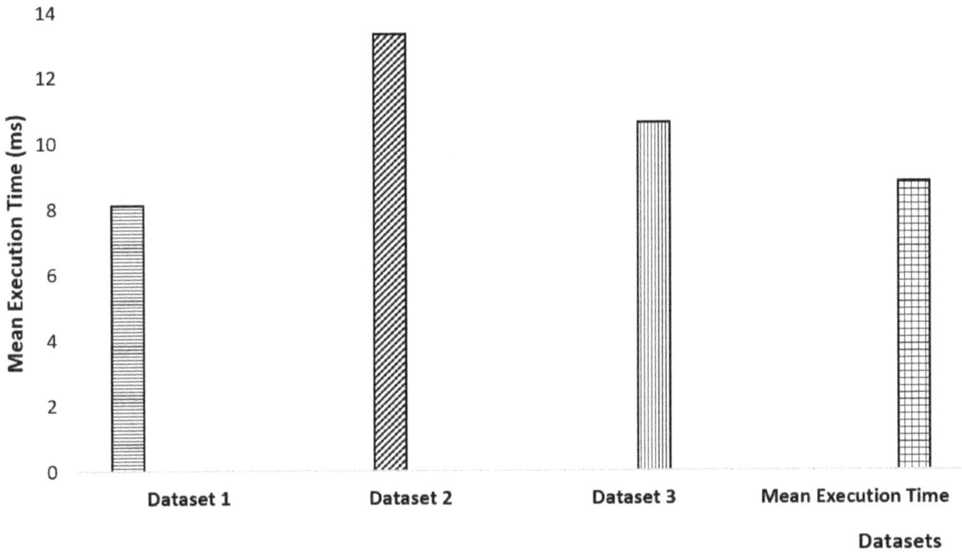

Figure 9. Computational complexity of our proposed model.

5. Conclusions

This paper focused on the use of a deep convolutional neural networks model to propose a fully automated fruit freshness classification method. To check the quality standard of fruit, the consumer first manually checks the freshness of the fruit. We used transfer learning of CNN model AlexNet to develop a robust to assess the quality of fruits. We changed some hyper parameters while fine-tuning and obtained an enhanced performance of our algorithm. We also varied other parameters, such as learning rate and batch size. We achieved higher accuracy with our fine-tuned CNN model through transfer learning produce. Our proposed model achieved an average accuracy of 99% on three publicly available fruits datasets.

In the future, we aim to increase the variety of fruits so that the farmers will easily judge fresh and rotten fruit. This will essentially help them to purchase better quality fruits from the market. We also intend to develop a user-friendly mobile application that will display the classification results of more fruits and vegetables. Moreover, we also aim to generalize the evaluation of our developed method on more classes such as extra vegetable species. Furthermore, we also aim to investigate the effects of different parameters, for instance, the activation function, pooling function optimization, and a loss function. Finally, to handle the execution of complex machine learning and deep learning-based methods, our method can also be deployed into a cloud-based framework.

Author Contributions: Conceptualization, U.A. and Z.M.; methodology, U.A.; software, U.A., M.I.S., A.S., M.S., U.K. and Z.M.; validation, U.A.; formal analysis, U.A., M.I.S., A.S., M.S., U.K. and Z.M.; investigation, U.A. and Z.M.; resources, U.A., M.I.S., A.S., M.S., U.K. and Z.M.; data curation, U.A.; writing—original draft preparation, U.A., M.I.S., A.S., M.S., U.K. and Z.M.; writing—review and editing, U.A., M.I.S., A.S., M.S., U.K. and Z.M.; visualization, U.A. and Z.M.; supervision, Z.M.; project administration, Z.M. All authors have read and agreed to the published version of the manuscript.

Funding: This research received no external funding.

Institutional Review Board Statement: Not applicable.

Informed Consent Statement: Not applicable.

Data Availability Statement: Not applicable.

Conflicts of Interest: The authors declare no conflict of interest.

References

1. Kazi, A.; Panda, S.P. Determining the freshness of fruits in the food industry by image classification using transfer learning. *Multimedia Tools Appl.* **2022**, *81*, 7611–7624. [CrossRef]
2. Mahmood, Z.; Muhammad, N.; Bibi, N.; Ali, T. A review on state-of-the-art face recognition approaches. *Fractals Complex Geom. Patterns Scaling Nat. Soc.* **2017**, *25*, 1750025. [CrossRef]
3. Fu, Y.; Nguyen, M.; Yan, W.Q. Grading Methods for Fruit Freshness Based on Deep Learning. *SN Comput. Sci.* **2022**, *3*, 264. [CrossRef]
4. Wan, S.; Goudos, S. Faster R-CNN for multi-class fruit detection using a robotic vision system. *Comput. Networks* **2019**, *168*, 107036. [CrossRef]
5. Mukhiddinov, M.; Muminov, A.; Cho, J. Improved Classification Approach for Fruits and Vegetables Freshness Based on Deep Learning. *Sensors* **2022**, *22*, 8192. [CrossRef]
6. Valentino, F.; Cenggoro, T.W.; Pardamean, B. A design of deep learning experimentation for fruit freshness detection. *IOP Conf. Ser. Earth Environ. Sci.* **2021**, *794*, 012110. [CrossRef]
7. Singh, S.; Singh, N.P. Machine learning-based classification of good and rotten apple. In *Recent Trends in Communication, Computing, and Electronics: Select Proceedings of IC3E 2018*; Springer: Singapore, 2019; pp. 377–386.
8. Turaev, S.; Abd Almisreb, A.; Saleh, M.A. Application of transfer learning for fruits and vegetable quality assessment. In Proceedings of the 2020 14th International Conference on Innovations in Information Technology (IIT), Virtual Conference, 17–18 November 2020.
9. Ni, J.; Gao, J.; Deng, L.; Han, Z. Monitoring the Change Process of Banana Freshness by GoogLeNet. *IEEE Access* **2020**, *8*, 228369–228376. [CrossRef]
10. Dubey, S.R.; Jalal, A.S. Apple disease classification using color, texture and shape features from images. *Signal Image Video Process.* **2016**, *10*, 819–826. [CrossRef]
11. Zhang, Y.; Wu, L. Classification of fruits using computer vision and a multiclass support vector machine. *Sensors* **2012**, *12*, 12489–12505. [CrossRef]
12. Wang, L.; Li, A.; Tian, X. Detection of fruit skin defects using machine vision system. In Proceedings of the 2013 Sixth International Conference on Business Intelligence and Financial Engineering, Hangzhou, China, 14–16 November 2013; pp. 44–48.
13. Dubey Ram, S.; Jalal, A.S. Application of image processing in fruit and vegetable analysis: A review. *J. Intell. Syst.* **2015**, *24*, 405–424. [CrossRef]
14. Rocha, A.; Hauagge, D.C.; Wainer, J.; Goldenstein, S. Automatic fruit and vegetable classification from images. *Comput. Electron. Agric.* **2010**, *70*, 96–104. [CrossRef]
15. Akçay, S.; Kundegorski, M.E.; Devereux, M.; Breckon, T.P. Transfer learning using convolutional neural networks for object classification within X-ray baggage security imagery. In Proceedings of the 2016 IEEE International Conference on Image Processing (ICIP), Phoenix, AZ, USA, 25–28 September 2016.
16. Pan, S.J.; Yang, Q. A survey on transfer learning. *IEEE Trans. Knowl. Data Eng.* **2009**, *22*, 1345–1359. [CrossRef]
17. Arunachalaeshwaran, V.R.; Mahdi, H.F.; Choudhury, T.; Sarkar, T.; Bhuyan, B.P. Freshness classification of hog plum fruit using deep learning. In Proceedings of the 2022 International Congress on Human-Computer Interaction, Optimization and Robotic Applications (HORA), Ankara, Turkey, 9–11 June 2022.
18. Mehta, D.; Choudhury, T.; Sehgal, S.; Sarkar, T. Fruit Quality Analysis using modern Computer Vision Methodologies. In Proceedings of the 2021 IEEE Madras Section Conference (MASCON), Chennai, India, 27–28 August 2021.
19. Kumar, T.B.; Prashar, D.; Vaidya, G.; Kumar, V.; Kumar, S.D.; Sammy, F. A Novel Model to Detect and Classify Fresh and Damaged Fruits to Reduce Food Waste Using a Deep Learning Technique. *J. Food Qual.* **2022**, *2022*, 4661108. [CrossRef]

20. Nerella, J.T.; Nippulapalli, V.K.; Nancharla, S.; Vellanki, L.P.; Suhasini, P.S. Performance Comparison of Deep Learning Techniques for Classification of Fruits as Fresh and Rotten. In Proceedings of the 2023 International Conference on Recent Advances in Electrical, Electronics, Ubiquitous Communication, and Computational Intelligence (RAEEUCCI), Chennai, India, 19–21 April 2023; pp. 1–6.
21. Sultana, N.; Jahan, M.; Uddin, M.S. An extensive dataset for successful recognition of fresh and rotten fruits. *Data Brief* **2022**, *44*, 108552. [CrossRef] [PubMed]
22. Enciso-Aragón, C.J.; Pachón-Suescún, C.G.; Jimenez-Moreno, R. Quality control system by means of CNN and fuzzy systems. *Int. J. Appl. Eng. Res.* **2018**, *13*, 12846–12853.
23. Zhu, Q.; Zu, X. A Softmax-Free Loss Function Based on Predefined Optimal-Distribution of Latent Features for Deep Learning Classifier. *IEEE Trans. Circuits Syst. Video Technol.* **2022**, *33*, 1386–1397. [CrossRef]
24. Karakaya, D.; Ulucan, O.; Turkan, M. A comparative analysis on fruit freshness classification. In Proceedings of the 2019 Innovations in Intelligent Systems and Applications Conference (ASYU), Izmir, Turkey, 31 October–2 November 2019; pp. 1–4.
25. Foong, C.C.; Meng, G.K.; Tze, L.L. Convolutional neural network based rotten fruit detection using resnet50. In Proceedings of the 2021 IEEE 12th Control and System Graduate Research Colloquium (ICSGRC), Shah Alam, Malaysia, 7 August 2021; pp. 75–80.
26. Hosny, K.M.; Kassem, M.A.; Foaud, M.M. Classification of skin lesions using transfer learning and augmentation with Alex-net. *PLoS ONE* **2019**, *14*, e0217293. [CrossRef]
27. Kang, J.; Gwak, J. Ensemble of multi-task deep convolutional neural networks using transfer learning for fruit freshness classification. *Multimed. Tools Appl.* **2022**, *81*, 22355–22377. [CrossRef]
28. Jawad, K.; Mahto, R.; Das, A.; Ahmed, S.U.; Aziz, R.M.; Kumar, P. Novel Cuckoo Search-Based Metaheuristic Approach for Deep Learning Prediction of Depression. *Appl. Sci.* **2023**, *13*, 5322. [CrossRef]
29. Aziz, R.M.; Joshi, A.A.; Kumar, K.; Gaani, A.H. Hybrid Feature Selection Techniques Utilizing Soft Computing Methods for Cancer Data. In *Computational and Analytic Methods in Biological Sciences*; River Publishers: Aalborg, Denmark, 2023; pp. 23–39.
30. Yaqoob, A.; Aziz, R.M.; Verma, N.K.; Lalwani, P.; Makrariya, A.; Kumar, P. A review on nature-inspired algorithms for cancer disease prediction and classification. *Mathematics* **2023**, *11*, 1081. [CrossRef]
31. Aziz, R.M. Cuckoo search-based optimization for cancer classification: A new hybrid approach. *J. Comput. Biol.* **2022**, *29*, 565–584. [CrossRef] [PubMed]
32. Sultan, F.; Khan, K.; Shah, Y.A.; Shahzad, M.; Khan, U.; Mahmood, Z. Towards Automatic License Plate Recognition in Challenging Conditions. *Appl. Sci.* **2023**, *13*, 3956. [CrossRef]
33. Farid, A.; Hussain, F.; Khan, K.; Shahzad, M.; Khan, U.; Mahmood, Z. A Fast and Accurate Real-Time Vehicle Detection Method Using Deep Learning for Unconstrained Environments. *Appl. Sci.* **2023**, *13*, 3059. [CrossRef]

applied sciences

MDPI

Article

PhotoMatch: An Open-Source Tool for Multi-View and Multi-Modal Feature-Based Image Matching

Esteban Ruiz de Oña [1], Inés Barbero-García [1], Diego González-Aguilera [1,*], Fabio Remondino [2], Pablo Rodríguez-Gonzálvez [3] and David Hernández-López [4]

[1] Cartographic and Terrain Engineering Department, Higher Polytechnic School of Ávila, University of Salamanca, 05003 Ávila, Spain
[2] 3D Optical Metrology (3DOM) Unit, Bruno Kessler Foundation (FBK), 38121 Trento, Italy
[3] Department of Mining Technology, Topography and Structures, Universidad de León, 24071 Ponferrada, Spain
[4] Institute for Regional Development (IDR), University of Castilla-La Mancha, 13001 Albacete, Spain
* Correspondence: daguilera@usal.es

Abstract: The accurate and reliable extraction and matching of distinctive features (keypoints) in multi-view and multi-modal datasets is still an open research topic in the photogrammetric and computer vision communities. However, one of the main milestones is selecting which method is a suitable choice for specific applications. This encourages us to develop an educational tool that encloses different hand-crafted and learning-based feature-extraction methods. This article presents PhotoMatch, a didactical, open-source tool for multi-view and multi-modal feature-based image matching. The software includes a wide range of state-of-the-art methodologies for preprocessing, feature extraction and matching, including deep learning detectors and descriptors. It also provides tools for a detailed assessment and comparison of the different approaches, allowing the user to select the best combination of methods for each specific multi-view and multi-modal dataset. The first version of the tool was awarded by the ISPRS (ISPRS Scientific Initiatives, 2019). A set of thirteen case studies, including six multi-view and six multi-modal image datasets, is processed by following different methodologies, and the results provided by the software are analysed to show the capabilities of the tool. The PhotoMatch Installer and the source code are freely available.

Keywords: photogrammetry; computer vision; artificial intelligence; feature-based matching; feature extraction methods; hand-crafted methods; learning-based methods

Citation: Ruiz de Oña, E.; Barbero-García, I.; González-Aguilera, D.; Remondino, F.; Rodríguez-Gonzálvez, P.; Hernández-López, D. PhotoMatch: An Open-Source Tool for Multi-View and Multi-Modal Feature-Based Image Matching. *Appl. Sci.* **2023**, *13*, 5467. https://doi.org/10.3390/app13095467

Academic Editor: Zahid Mehmood Jehangiri

Received: 3 March 2023
Revised: 24 April 2023
Accepted: 26 April 2023
Published: 27 April 2023

1. Introduction

Feature-based image matching is a process that provides a correspondence between two or more images connecting basically local image features. The development of automatic and accurate image-matching processes has been a traditional problem in the field of photogrammetry and computer vision [1]. At present, modern camera orientation techniques such as Structure from Motion (SfM) or Visual Simultaneous Localization and Mapping (VSLAM) also rely on the extraction of accurate and reliable homologous points between images. Particularly, these correspondence points between images are normally used within the image orientation and self-calibration process, exploiting globally inherent geometric constraints in an optimization scheme known as bundle adjustment. Image matching can be used for object recognition and tracking, including some specifically hand-crafted features [2,3] and, more recently, deep learning approaches [4–7].

The spread of smartphones with powerful cameras, as well as the development of automatic tools for the creation of 3D models from a set of images, has led to the democratization and popularization of photogrammetry and computer vision. At first, photogrammetry was applied only by experts with good knowledge and expertise and using very specialized equipment. At present, techniques such as SfM, together with multi-view stereo (MVS), allow for the creation of 3D models by end-users without specific knowledge [8–12]. The creation of 3D models from images acquired by non-experts also presents a challenge for image matching, since the basic rules and protocols for imagery acquisition are often not fulfilled [13]. These amateur users will often acquire images with low overlap, at different scales and perspectives, or even with large differences in lighting or other radiometric conditions.

Although modern matching techniques cope with images with radiometric and geometric variations, the image matching process is especially challenging in the case of multi-modal images. Multi-modal image matching is performed between images coming from different sensors or different acquisition techniques and those with significant and nonlinear radiometric distortions. The differences can be due to the use of different sensors (e.g., multispectral, thermal, depth cameras), differences in data types (e.g., drawings vs. photography, vector vs. raster), or different illumination conditions (e.g., day/night images). Multi-modal matching is a critical task for a wide range of applications, such as medicine [14], cultural heritage documentation [15], multitemporal monitoring [16] or person re-identification [17,18], among others.

Image matching algorithms can be classified in two large groups: (i) traditional hand-crafted methods, and (ii) learning-based methods. The latter group utilises artificial intelligence for the development of new detectors and descriptors learned from the data [19]. While the hand-crafted feature-extraction methods are well-established in photogrammetric processes, they are not able to overcome important geometric, radiometric and spectral changes.

The number of artificial intelligence algorithms that can be used for image matching is rapidly growing. As a consequence, the selection of a suitable combination of detector, descriptor and matching function for a specific case is a complex task [20]. A detailed study must be conducted for each type of data to select the best algorithm from the increasing number of available options. Additionally, it is important not to overlook the manual configuration of certain input parameters, which can be highly theoretical and difficult for end-users to understand. Configuring each option is a time-consuming process, especially when including deep learning methodologies and training processes. Furthermore, there is a lack of tools that facilitate the processing, comparison, and assessment of the different feature-based image matching methodologies.

The purpose of the present study is to try and contribute to the scientific community in this gap. Here, we introduce PhotoMatch, an educational and open-source tool for multi-modal and multi-view feature-based image matching. The tool allows for the use of a wide range of algorithms for keypoint detection, description, and matching. It also provides a method for evaluating and comparing the obtained results among different approaches in a didactic way, including the ability to provide reference data for the evaluation of the tested methodologies. In [21], a first version of the PhotoMatch tool was presented and awarded by ISPRS (ISPRS Scientific Initiatives, 2019). This article presents a new version of the PhotoMatch tool, which includes several improvements and consolidated learning-based methods.

The standard methodology for hand-crafted methods consists of feature detection, feature description, and matching:

- Detectors identify distinctive features (keypoints), localizing meaningful and salient regions of the image, and extracting these regions as patches. These patches are generally normalized in order to achieve invariance to geometric and radiometric transformations. These keypoints are represented by their point representatives, such as the centre of gravity or other distinctive points.
- Descriptors analyse the neighbourhood of the keypoints and create a 2D vector of information based on the different mathematical properties of the point and its neighbourhood. Usually, distance is used to establish the candidate correspondences.
- Matching identifies homologous keypoints between images using the information provided by the descriptors. The most common matching methods are brute-force and Flann [22], and robust matching by means of spatial global or local constraints,

such as those provided by epipolar geometry [23] and RANdom SAmple Consensus (RANSAC) [21–24].

A wide range of detectors and descriptors has been developed in the last decades [25] SIFT [26], and its last version RootSIFT [27], which introduces a slight variation in the descriptor computation; SURF [28]; or MSD [29]. These are just a few examples of the large number of detectors and descriptors available in the scientific community. SIFT has monopolised feature-based matching in the last two decades. SIFT matching relies on keypoints, whose associated patches are normalized to become invariant to scale and rotation changes. Nevertheless, although SIFT is still valid and able to obtain robust results in the SfM pipelines, it is not invariant to considerable scale and rotation changes, and even less invariant to radiometric and/or spectral changes.

Deep learning detectors and descriptors have emerged in recent years as a promising alternative to hand-crafted methods, especially for multi-modal matching [14]. Although learning-based methods are often seen as a replacement of hand-crafted methods, they still face an important number of challenges. In particular, acquiring sufficient data to effectively train and evaluate deep learning algorithms can be challenging in many application fields. Furthermore, the variability in the types of multi-modal combinations complicates the development of tools that can be simultaneously utilized across a wide range of applications [14,30].

The challenge of acquiring the data required for training is being overcome by the development of unsupervised learning approaches. For image matching, unsupervised learning approaches include techniques such as the use of video, where the temporal coherence between frames can be used for model training [31]. Nevertheless, these approaches require a high amount of video data, which are not always available for other applications, such as medical imaging.

In certain complex scenarios, or when dealing with multi-modal datasets, learning-based methods might outperform hand-crafted methods. A high number of deep learning algorithms have been presented for keypoints' detection and description, many of them focused on specific applications [20,30,32,33], and many are fully available and tested. For instance, in the last Image Matching Challenge (IMC) (*Image Matching Challenge—2022 edition*) [34], the best-performing algorithms were ASpanFormer [35], and combinations of SuperGlue [36], SuperPoint [37], LoFTR [38], DKM [39] and DISK [40]. Although the datasets of IMC included images with different positions, cameras, illumination or even filters, they did not include multimodal datasets (i.e., a combination of different sensors or combination of images coming from different wavelenghts). A comparison and evaluation of the best IMC algorithms was also carried out by other authors [41], using multi-view imagery and applied to cultural heritage. However, the obtained results did not show a clear winner, with some algorithms performing better than others under specific conditions. Trying to find specific multi-modal image matching contributions, other authors used TILDE [42], SuperPoint [37], and LF-Net [33]. More recently, an outstanding turning point was the "detect-and-describe" approach, D2-Net, network [43], and the repeatable and reliable detector and descriptor R2D2 [44], which represents a step forward in photogrammetry and computer vision.

Being aware of the pros and cons of the existing learning-based methods, these two methods, D2-Net and R2D2, were included in PhotoMatch.

This paper has been structured as follows: after this introduction, the tool, PhotoMatch, is described in Section 2. Section 3 outlines and analyzes the main results focused on multi-view and multi-modal images. Section 4 is devoted to highlighting the main conclusions and future perspectives.

2. PhotoMatch

PhotoMatch is an educational and open-source tool developed in C++ and Qt, which was awarded by the ISPRS through a Scientific Initiative [45]. It is available at https://github.com/TIDOP-USAL/PhotoMatch/releases (accessed on 20 February 2023). The tool follows a pipeline of six steps: (i) project and session definition, (ii) pre-processing, (iii) feature extraction, (iv) feature matching, (v) quality control, and (vi) export (Figure 1).

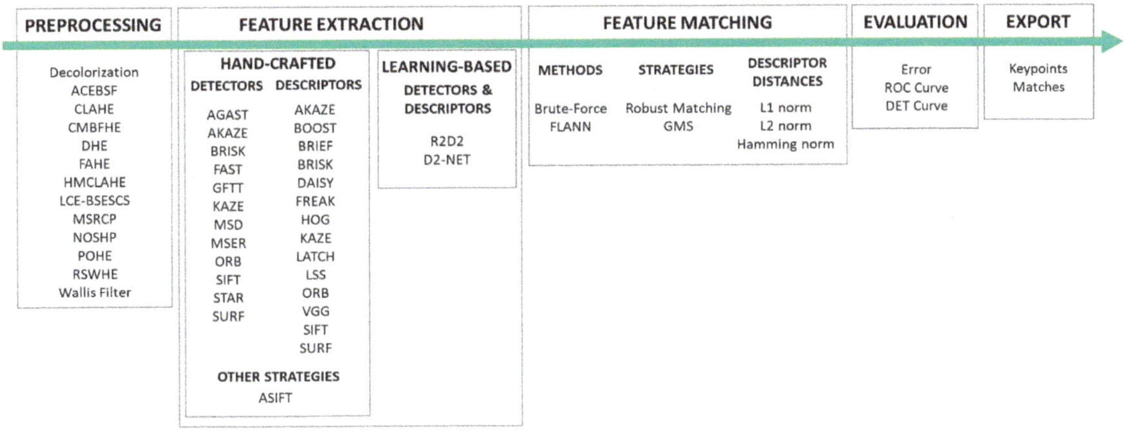

Figure 1. PhotoMatch pipeline, including a list of available algorithms/options for each step.

2.1. Project and Session Definition

This first step allows for the creation of a new project and uploading of the images. Each project can consist of one or several sessions, enabling a comparative assessment of the results. The tool accepts common image formats and an unlimited number of images.

2.2. Image Pre-Processing

Image pre-processing is stated as a fundamental step prior to feature extraction. The goal is to improve the radiometric content of the images, and thus to facilitate the subsequent feature extraction and matching process. This pre-processing is especially useful in cases with unfavourable texture images [46].

PhotoMatch offers different approaches to image pre-processing (Figure 1), including decolorization [47], Adaptive ContrastEnhancement Based on modified Sigmoid Function (ACEBSF) [48], Dynamic Histogram Equalization (DHE) [49], Parametric-Oriented Histogram Equalization (POHE) [50], Recursively Separated and Weighted Histogram Equalization (RSWHE) [51], and Wallis Filtering [52]. Pre-processing is highly recommended to obtain better results in the successive steps.

2.3. Feature Extraction

The feature extraction includes the detection and description of keypoints. The tool includes several alternatives that can be classified as hand-crafted or learning-based feature-extraction methods (Figure 1).

A total of 20 different hand-crafted methods were implemented. These include: SURF [28], SIFT [26], AKAZE [53] or MSD [29]. Most of the hand-crafted algorithms include a detector and a descriptor, which can be combined (e.g., SURF detector and SIFT descriptor). Different advanced parameters can be tuned, providing educational support for each available algorithm. An example of the MSD and SIFT options is provided in Figure 2b.

In addition, the Affine SIFT (ASIFT) [54] algorithm is also available. This algorithm computes a fully affine invariant matching. It is specifically designed to deal with images that present considerable geometric variations in terms of scale and perspective. The algorithm simulates all possible views by modifying the longitude and latitude of the camera orientation parameters. The ASIFT algorithm can also be used, in combination with other similarity invariant-matching methods such as SURF, BRISK [55] or AKAZE.

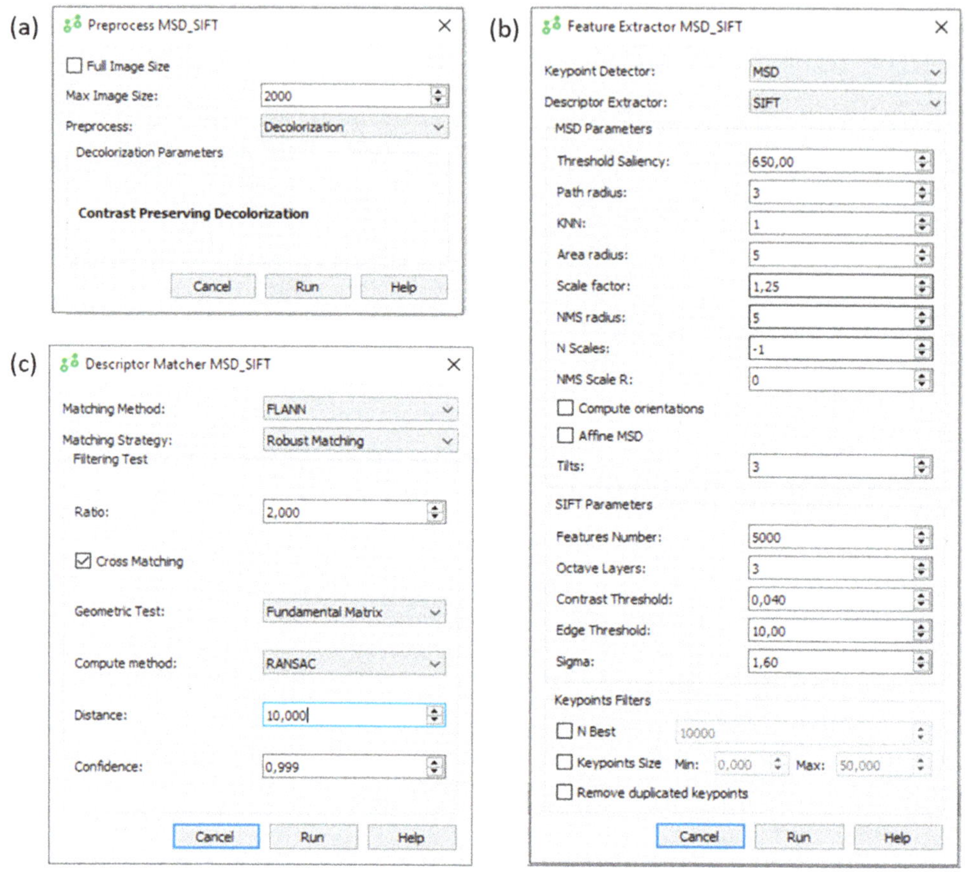

Figure 2. Selection of parameters for preprocessing (**a**), feature extraction (**b**) and matching (**c**) in PhotoMatch. The help menu provides educational support for each advanced parameter.

Regarding the learning-based methods, two deep learning detectors/descriptors were incorporated in PhotoMatch: D2-Net [43] and R2D2 [44]. This selection was made based on their outstanding performance, and considering that these algorithms are freely available and use pretrained models, so they are not designed for a specific type of data.

D2-Net uses a single convolutional neural network for simultaneous feature description and detection. Instead of carrying out the detection of low-level image structures, the process is carried out after the computation of the feature maps, when more reliable information is available. This has been assessed on multi-modal datasets, where it has proven to perform well for the matching of features under challenging illumination or weather conditions.

R2D2 also simultaneously acts as a keypoint detector and descriptor. This includes a local predictor of discriminativeness during learning, to avoid areas with salient features but where accurate matching is not possible due to repetitiveness (e.g., sea waves or canopy). It has been proven to perform especially well for the matching of day and night images.

The included deep learning algorithms were incorporated within PhotoMatch with pretrained models, while the selection of different pretrained models is also an option.

2.4. Matching

The matching process consists of finding the right correspondence between previously detected keypoints. PhotoMatch includes Brute-force and FLANN [22] as classical matching methods, while Robust Matching (RM) and Grid-based Motion Statistics (GMS) [56] are also possible matching strategies. The available descriptors distances are L1, L2, and Hamming Norm [57]. Then, the matching process is filtered using different methods. Homography [37], or Fundamental Matrix [58] can be combined with different computational methods, including RANSAC, all points, Least Median of Squares (LMedS), and Spearman's RHO Correlation Coefficient (Figure 2c).

2.5. Assessment of Results

The main limitation in the analysis of feature-based image matching results is the unavailability of reliable reference data. To overcome this issue, PhotoMatch includes a reference data editor (as shown in Figure 3) that allows for the end-user to manually and accurately introduce a set of matching points. These points are later used to assess different feature-based matching algorithms.

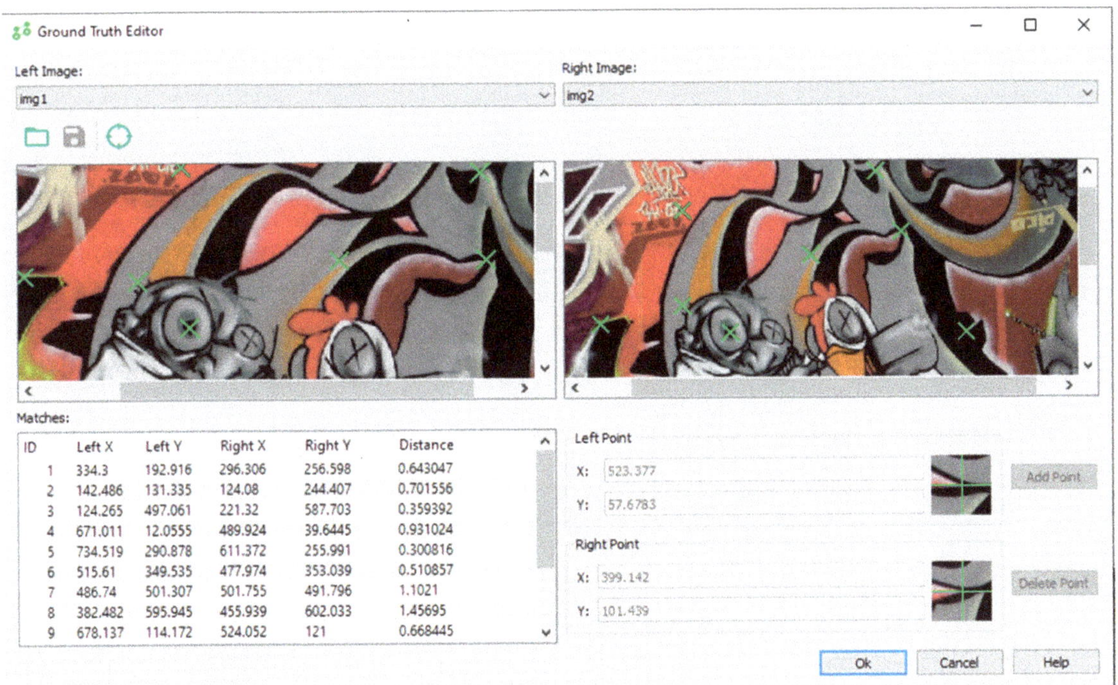

Figure 3. Reference data editor in PhotoMatch with subpixel accuracy.

Once the reference matchings are defined, PhotoMatch calculates and graphically represents the Receiver Operating Characteristic (ROC) curve and the Detection Error Trade-off (DET) curve, which illustrate the error in feature-based image matching. PhotoMatch offers the option to choose between homography or a fundamental matrix to compute these errors. Homography should be used when all points in the image are on the same plane, while the fundamental matrix should be selected when the points are not co-planar.

Furthermore, PhotoMatch provides a user-friendly visualization of the matchings (as shown in Figure 4), allowing for a better interpretation of the results.

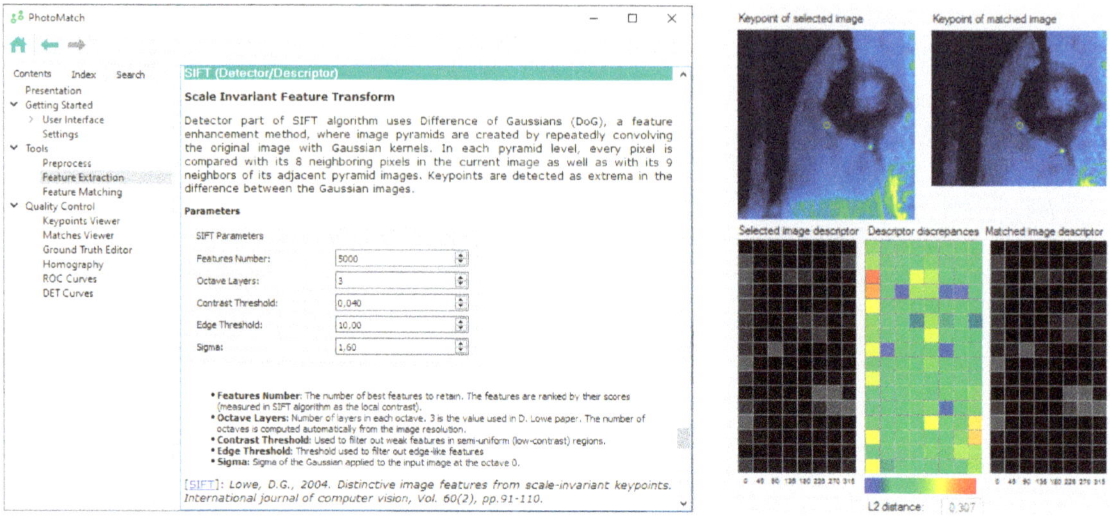

Figure 4. An example of an educational section in PhotoMatch applied to the SIFT algorithm [26].

2.6. Export

Finally, PhotoMatch allows for the exportation of the extracted keypoints and matchings in different formats, including XML and YML for OpenCV and plain text. This allows for end-users to import and use these observations in other tools for image triangulation (bundle adjustment) or photogrammetric reconstruction. This also allows for a more detailed assessment of the results to be carried out, or for the combination of the algorithms presented in PhotoMatch and other approaches.

2.7. Educational Information

PhotoMatch includes educational information with a short introduction to the different algorithms. The scientific references are also included in a more detailed explanation of the process (Figure 4). In this way, the idea is to provide researchers, students, and even end-users, with the information needed to select the optimal parameters and combinations for each algorithm. This also contributes to making PhotoMatch an educational and research resource, far from being a black-box tool. Last but not least, thanks to its exportation capabilities, PhotoMatch offers a solution for SfM tools that cannot correctly solve the matching and, thus, the orientation of the images.

3. Experimental Results and Discussion

Six multi-view and six multi-modal case studies with different characteristics were selected and analysed to show the PhotoMatch capabilities. Different feature detectors, descriptors and matching were used for each dataset and the obtained results were compared and assessed.

3.1. Multi-View

The selected multi-view datasets are related to the close-range photogrammetric applications. Due to the widespread adoption of SfM and MVS tools for 3D modeling, feature-based image matching has become a critical process. As end-users increasingly apply photogrammetry, this method must overcome more challenging conditions than traditional aerial photogrammetry, such as larger geometric and radiometric differences.

To this end, we selected six multi-view datasets, each composed of four images. Four sets of images were obtained from the ETH3D benchmark (https://www.eth3d.net/datasets, accessed on 11 November 2022) and comprised the images of a façade (Figure 5a), a forest (Figure 5b), a playground (Figure 5c) and a boulder (Figure 5d). The façade dataset is characterised by low overlap and repetitive features; the forest dataset is characterised by low resolution and unfavourable lighting conditions; for the playground dataset, the viewpoints between images have large differences; the boulder dataset also has a low overlap, but distinctive features that should help to solve the feature-based image matching.

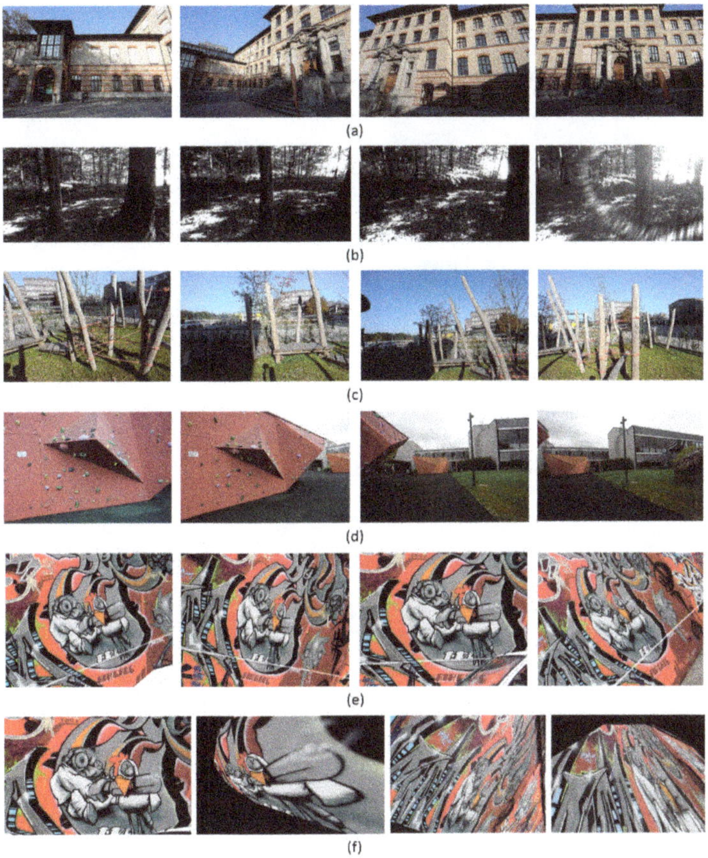

Figure 5. Multi-view datasets: façade (**a**), forest (**b**), playground (**c**), boulder (**d**), graffiti with small geometric differences (**e**) and graffiti with large geometric differences (**f**).

The last two multi-view datasets were obtained from the public benchmark VGG Oxford (Visual Geometry Group—University of Oxford, accessed on 11 November 2022). The images cover a planar wall covered by graffiti. For one of the datasets, the images have good overlap and low geometric differences (Figure 5e). For the last dataset (Figure 5f), two of the images are synthetically derived from the other two and enclose considerable geometric differences, substantially hampering the matching process, even for a human operator.

3.2. Multi-Modal

Considering the increasing popularity of sensors and cameras, the multi-modal matching of images is a growing demand in many applications. We selected some of the most common examples: the combination of thermal and visible imagery for a building (Figure 6a); scanning electron microscopy (SEM) images, including a backscattered electrons and secondary electron images of a mineral surface (Figure 6b); the combination of visible and range imagery from a laser scanner (Figure 6c); the combination of visible and thermal for aerial imagery (Figure 6d); satellite imagery with different wavelengths, where two images were synthetically derived by applying geometric and radiometric distortions to the other two images (Figure 6e); Magnetic Resonance Imaging (MRI) with different visualization parameters, used to highlight different tissues and synthetically derived images (Figure 6f).

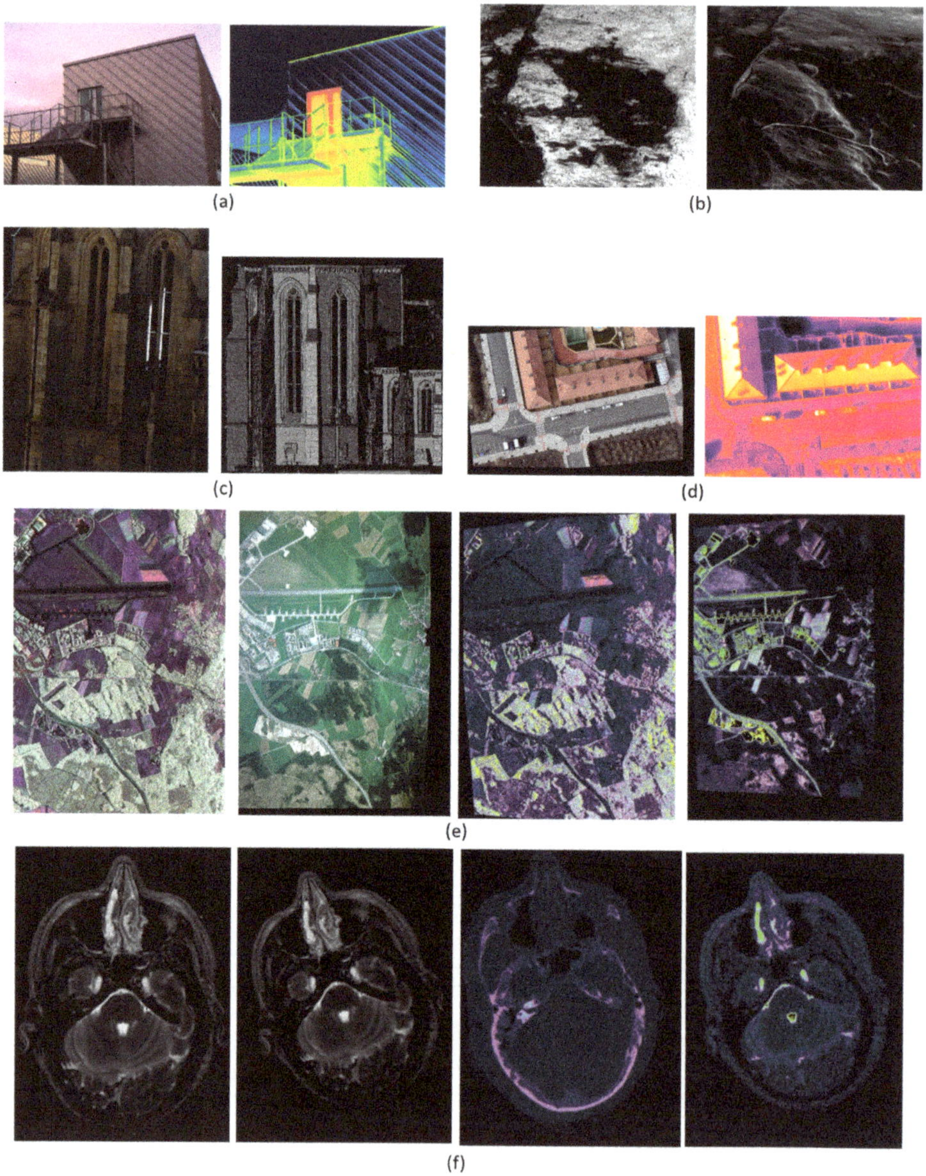

Figure 6. Multi-modal datasets: visible thermal imagery for a building (**a**), SEM images of a bone (**b**), visible range imagery (**c**), visible thermal aerial imagery (**d**), satellite imagery with different wavelengths and synthetically derived images (**e**), magnetic resonance images with different visualization parameters and synthetically derived images (**f**).

For the first case (Figure 6a), the matching was carried out for a visible and a thermal image of a building for its energetic inspection. The second case (Figure 6b) is composed of two SEM images: an imaging technique used to analyse the surface of a sample at very high magnifications, used in various scientific fields such as materials science, geology, archaeology and biology to gain insight into the structure and composition of a sample. The third case (Figure 6c) corresponds to a visible and a range image of a heritage building for its 3D reconstruction and texture mapping. The resulting matching can be used to improve the registration between the camera (visible) and the laser (range), and then to map the high-resolution texture coming from the visible imagery into the 3D point cloud coming from the laser scanner. The fourth case (Figure 6d) combines visible and thermal aerial images of a city area using a drone; this type of aerial image can be used for the estimation of land surface temperature or for the study and mitigation of urban heat islands, among other applications. The fifth case (Figure 6e) is composed of two satellite images taken with different sensors and the other two are synthetically derived from the original ones; in this case, the registration is important for automatic georeferencing. The sixth case (Figure 6f) is also composed of four images: two of them are medical resonance images taken using different parameters and the other two are synthetically derived images. Synthetic images represent possible processing and acquisition modifications by the application of geometric (rotation, scale, perspective) and radiometric (brightness, hue and addition of random noise) changes.

The first (Figure 6a) and third (Figure 6c) multi-modal image pairs contain non-coplanar points; therefore, the evaluation of the matches needs to be carried out using the fundamental matrix. For the rest of the multimodal datasets, the points in the images could be considered totally coplanar, with homography being the best method for their assessment.

3.3. Feature-Based Image Matching Strategies

The process carried out for each dataset consists of three steps: (i) pre-processing, (ii) feature extraction and (iii) matching (Figure 2).

All images were pre-processed by applying decolorization (Figure 2a). Pre-processing is reported as a fundamental step by different authors [21,59]. Decolorization is the simplest pre-processing algorithm provided by PhotoMatch and is commonly used prior to image matching [60].

For the feature extraction step (Figure 2b), many algorithms are provided by PhotoMatch, while a selection of the most representative ones was tested. To this end, hand-crafted methods were identified based on the best results obtained in previous tests [21]. The following combinations of detector and descriptor were assessed: SIFT + SIFT, SURF + SURF, SURF + SIFT, MSD + SIFT and ASIFT. In addition, both deep learning algorithms included in PhotoMatch, R2D2 and D2-Net, were also tested.

For hand-crafted algorithms, the following parameters were selected in PhotoMatch: the maximum number of features was set to 5000; for MSD, the threshold saliency was set to 650, and the number of selected points (KNN) was set to 1. The saliency threshold is linked to the level of dissimilarity between neighboring pixels and should be higher for images with a higher level of detail. KNN refers to the number of salience points considered; in this case, only the points with higher saliency were selected. The reason for selecting these parameters was based on their good performance after different tests, especially for the case of multi-modal images [21].

For the learning-based algorithms, the input parameters were based on choosing among the different pretrained models. Alternatives were tested, and the best pretrained models were chosen. For R2D2, the pretrained model 'r2d2_WASF_N16' was used, while for D2-Net, the pretrained model 'd2_tf' was selected. The choice of these models was based on information provided by the developers of each tool. However, since the models were trained using different datasets, different models may have different outcomes. Therefore, it is recommended to test various options for each specific application.

The selected matching approach (Figure 2c) was the same for all algorithms, since its accuracy and reliability was tested in previous studies [21]. It consisted of FLANN and Robust Matching, supported by ratio test, cross-checking and geometric test (fundamental matrix or homography computed by RANSAC). The RANSAC filtering was carried out using a Lowe ratio test with a value of 2 [61], a distance threshold of 10, and 2000 maximum trials. The homography and fundamental matrices were used to achieve self-supervised validation while supporting the relative orientation backbone. The value of the Lowe ratio test refers to the minimum distance between the two best matches for each keypoint; if the distance is below the threshold, the matches are considered too similar and the keypoint is removed. The distance threshold in RANSAC filtering is used to distinguish inliers from outliers; a higher value would be needed if the dataset is composed of matches with a relatively high error, while more precise algorithms would benefit from lower values. The maximum number of trials controls the trade-off between computational complexity and accuracy.

Detailed information on the different parameters for each algorithm is presented in the help section of the tool (Figure 4).

3.4. Assessment

PhotoMatch provides a reference data editor (Figure 3) that allows for the end-user to select the reference keypoints with subpixel accuracy and compute the error for each point using the homography or fundamental matrix adjustment. Using this reference data editor, each imagery dataset was registered using a set of at least 12 manually selected keypoints and their corresponding matchings. The maximum error for these points was below one pixel for all image pairs.

Once the reference keypoints and matchings are defined, PhotoMatch computes the homography or fundamental matrix transformation between each pair of images. After the keypoints are extracted by each algorithm (detector and descriptor), their coordinates are evaluated through comparison with the reference coordinates obtained via homography, or by computing the distance between each point and the line determined by the collinearity condition in the case of a fundamental matrix transformation (Tables 1 and 2).

Table 1. Number of correct matches (CM) with percentage and mean error (ME) (in px) for the different hand-crafted and learning-based algorithms and the six multi-view datasets. The best results are highlighted in bold.

	Detector Descriptor	SIFT + SIFT	SURF + SURF	SURF + SIFT	MSD + SIFT	ASIFT	R2D2	D2-NET
Facade (Figure 5a)	CM	24 (27.2%)	26 (22.7%)	47 (33.7%)	19 (34.8%)	**190** (27.8%)	68 **(52.5%)**	27 (32.7%)
	ME (px)	175.1	170.6	163.8	123.2	172.2	**110.5**	176.9
Forest (Figure 5b)	CM	108 (79.4%)	89 (71.2%)	139 (80.7%)	77 (80.6%)	**1155** (86.4%)	187 **(94.1%)**	123 (89.1%)
	ME (px)	19.2	21.1	8.9	8.7	13.5	**4.1**	6.3
Playground (Figure 5c)	CM	8 (36.9%)	51 (57.2%)	60 (60.7%)	30 (65.6%)	**214** (74.2%)	47 **(80.5%)**	49 (69.2%)
	ME (px)	607.8	224.6	66.5	40.6	137.5	**26.3**	45.9
Boulder (Figure 5d)	CM	150 (80.1%)	261 (83.2%)	283 (86.4%)	24 (77.8%)	**551 (98.8%)**	322 (96.1%)	533 (91.4%)
	ME (px)	50.9	49.9	30.3	163.5	**12.6**	29.8	20.3
Graffiti low differences (Figure 5e)	CM	681 (99.9%)	613 (99.4%)	602 (96.8%)	62 (93.3%)	**8713 (99.9%)**	241 (98.4%)	165 (94.9%)
	ME (px)	**1.2**	2.4	3.4	14.4	1.3	2.9	8.0
Graffiti high differences (Figure 5f)	CM	91 (94.4%)	62 (90.9%)	50,5 (87.1%)	1 (10.3%)	**2182 (99.8%)**	2 (38.7%)	2 (30%)
	ME (px)	17.0	18.6	31.7	241.0	**1.4**	151.3	171.0

Table 2. Number of correct matches (CM) with percentage and mean error (ME) (in px) for the different hand-crafted and learning-based algorithms in the seven multi-modal datasets. The best results are highlighted in bold.

	Detector Descriptor	SIFT + SIFT	SURF + SURF	SURF + SIFT	MSD + SIFT	ASIFT	R2D2	D2-NET
Visible-Thermal (Figure 6a)	CM	0 (0%)	1 (3,2%)	1 (4,2%)	4 (22,2%)	3 (2,9%)	1 (6,25%)	**53 (81,54%)**
	ME (px)	273.7	143.8	172.2	63.1	202.7	184.3	**13.7**
SEM (Figure 6b)	CM	0 (0%)	0 (0%)	3 (33.3%)	0 (0%)	6 (28.6%)	5 (62.5%)	**16 (76.2%)**
	ME (px)	1470.0	983.4	22.0	860.0	1013.4	10.7	**6.9**
Visible-Range (Figure 6c)	CM	1 (3,3%)	13 (25%)	**154** (53,1%)	3 (23,1%)	47 (32,9%)	5 (31,3%)	77 **(77,8%)**
	ME (px)	216.9	141.9	**34.4**	136.1	140.8	88.2	43.4
Visible-Thermal Aerial (Figure 6d)	CM	0 (0%)	3 (30%)	9 (60%)	0 (0%)	26 (86.7%)	17 **(100%)**	**107** (97.3%)
	ME (px)	383.9	159.2	15.9	259.3	30.7	**3.9**	4.3
Satellite (Figure 6e)	CM	0 (0%)	0 (0%)	7 (41,17%)	9 (75%)	0 (0%)	6 (75%)	**135 (95,7)**
	ME (px)	179.1	154.6	16.5	11.2	178.4	25.3	**4.7**
Magnetic Resonance (Figure 6f)	CM	0 (0%)	0 (0%)	10 (66.7%)	0 (0%)	0 (0%)	5 (16.4%)	**44 (92.6%)**
	ME (px)	194.0	151.9	8.2	122.9	144.1	30.2	**5.3**

3.5. Results

The multi-view and multi-modal datasets were assessed separately. For each dataset, the number of correct matches, percentage of correct matches, and mean error of the matches for the different methodologies (hand-crafted vs. learning-based) are presented. The threshold established for a correct matching was set to 10 px, which is relatively high for precise photogrammetry applications, but can provide a better insight into the approcimate matching ability of the algorithms.

3.5.1. Multi-View

The results for each case and image matching are outlined in Table 1.

For the first four multi-view datasets (Figure 5a–d), all of them corresponding to non-planar environments, R2D2 was the best algorithm in terms of accuracy, while ASIFT was able to obtain a higher number of matches with a lower accuracy. The exception was the boulder multi-view dataset (Figure 5d), where ASIFT achieved the highest accuracy, as the environment does not represent important challenges for matching. For the façade's multi-view dataset (Figure 5a), none of the algorithms (hand-crafted and learning-based) were able to obtain acceptable results as a consequence of the low overlap and repetitive features. Only R2D2 provided the best result, with 52.5% of correct matches (Table 1).

The fifth multi-view dataset (Figure 5e) represents a favourable photogrammetric acquisition with high overlap and low geometric differences. In this case, the hand-crafted algorithms outperform learning-based algorithms, in terms of both accuracy and the number of correct matches (Table 1).

For the last multi-view dataset (Figure 5f), with images with considerable geometric differences covering a wall, ASIFT was the only hand-crafted algorithm capable of computing a high number of accurate matches. Neither the other hand-crafted algorithms, nor the learning-based algorithms, provided acceptable results (Table 1).

Although the performance of hand-crafted methods is guaranteed for multi-view datasets with high overlap and favourable conditions, for challenging environments (i.e., important geometric variations) the image matching is not always successful. The experimental results show that some learning-based algorithms, such as R2D2, are capable of outperforming classical, hand-crafted methods for challenging datasets with low overlap and low-resolution images. Due to its capacity to avoid areas with low reliability, R2D2 outperforms the accuracy of other algorithms for the facade and forest dataset, which are characterized by repetitive features (i.e., windows and canopy). However, ASIFT, which is specifically designed to deal with large perspective distortions, was the only algorithm capable of registering the images in the graffiti dataset with high geometric differences (Figure 5f). This is probably due to the lack of training of the chosen matching learning algorithms for this particular case. It is also worth noting that ASIFT is a technique that simulates different affine distortions to the images, and a similar technique can work with different detectors and descriptors, so the combination of ASFIT and learning-based algorithms would be possible.

In order to evaluate this tool in comparison to other commercial and open-source software, a 3D reconstruction for each multi-view dataset was carried out using Agisoft Metashape 2.0.1 and GRAPHOS [1]. Acceptable results were obtained only for the fifth multi-view dataset (Graffiti with low differences, Figure 5e), while both software failed to compute a 3D reconstruction for the rest of the datasets.

3.5.2. Multi-Modal

The results of the multi-modal dataset are outlined in Table 2. The learning-based algorithms outperform the hand-crafted based algorithms for every dataset. D2-Net is the best-performing algorithm for every case, with the exception of the visible thermal aerial, where R2D2 obtains the highest accuracy. For the visible-range dataset, no acceptable results were obtained using any of the tested algorithms. For the visible-thermal dataset, the mean error was above the threshold, even for the D2-Net algorithm.

The hand-crafted algorithms performed much worse than learning-based algorithms. They were capable of obtaining a mean error below the threshold of ten pixels in only one case (SURF + SIFT for the magnetic resonance dataset).

Learning-based algorithms are shown to be a suitable approach for multimodal image matching for different datasets and applications. An algorithm such as D2-Net has been able to achieve good results for the majority of the presented datasets. Nevertheless, the difference in results for different types of images encourages the study and comparison of different approaches and parameters for any specific application requiring multimodal image matching.

The final matchings obtained for the two best-performing algorithms for each multi-modal dataset can be analysed in Figure 7.

The combination of different hand-crafted algorithms could be useful for some types of multi-modal data [21]. Nevertheless, some learning-based algorithms greatly outperform hand-crafted methods in multi-modal cases, being able to obtain acceptable results when hand-crafted algorithms fail.

In general, the experimental results presented in this paper demonstrate the great variability of results for different approaches and with different case studies. This highlights the importance of offering an educational and open-source tool, PhotoMatch, to compare and assess different algorithms through an experimental evaluation of learning-based and hand-crafted algorithms to better understand their performance across a wide range of scenarios.

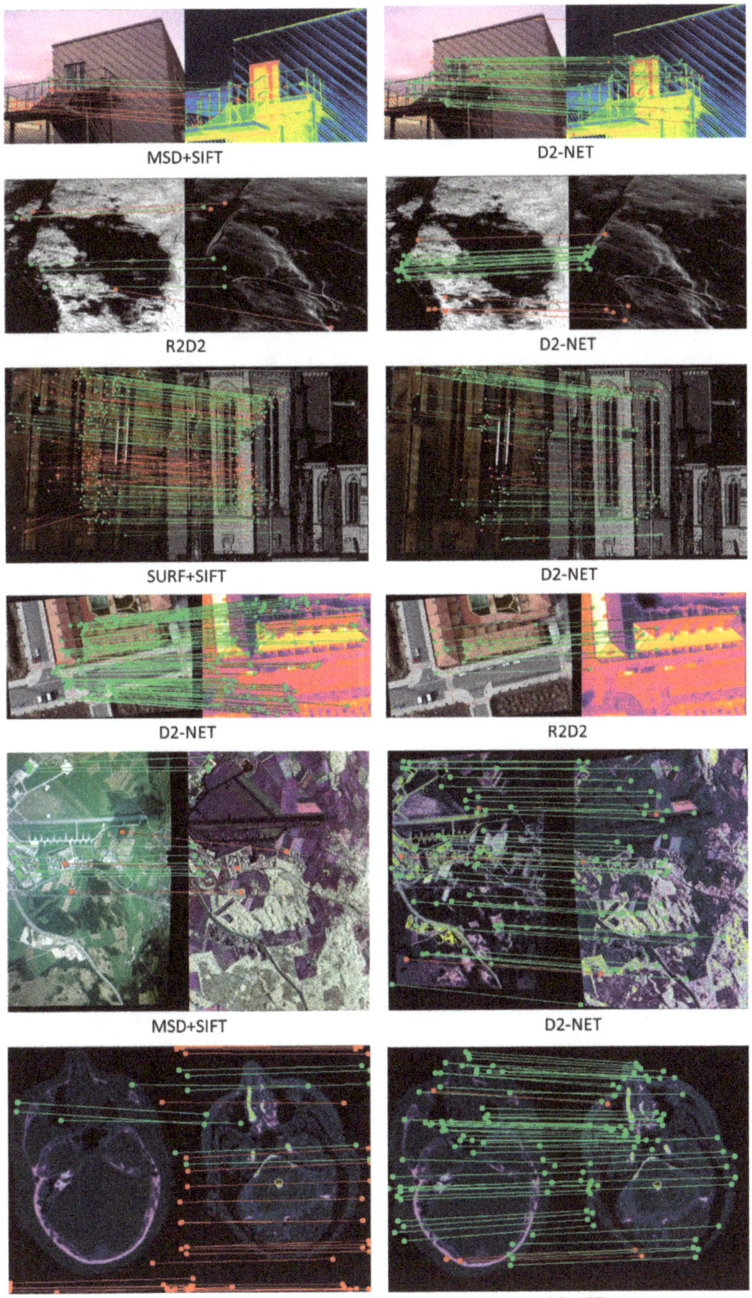

Figure 7. Matchings resulting in the best-performing algorithms for each multi-modal dataset.

4. Conclusions

A growing number of detectors, descriptors, and matching algorithms are available to extract and match keypoints between images. The most important distinction can be made between hand-crafted and learning-based feature-extraction methods. Some of these algorithms for keypoint extraction and matching are well-known and available in different libraries, such as OpenCV, or integrated into SfM tools. Other algorithms require expertise in dealing with source code and programming, and sometimes the use of external libraries. All of them are too abstract to be understood by end-users, requiring the setup of advanced parameters.

Despite the large quantity of available options provided in the scientific community, there are no educational and open-source multi-view and multi-modal image-matching tools to date, which allow for a comparative assessment of hand-crafted and learning-based algorithms.

In real-world problems (e.g., 3D reconstruction, image registration for the analysis of different wavelengths, SLAM or digital correlations between 3D and 2D data for applications such as material deformation analysis), selecting the best-matching algorithm and optimal parameters for a specific application is a time-consuming process requiring very specialised knowledge and is not integrated into the existing tools. This situation can easily lead to the adoption of not-optimal solutions and certainly hampers the adoption of new methodologies.

PhotoMatch provides a solution to this bottleneck, integrating hand-crafted and learning-based algorithms for comparing and assessing feature-based image matching, with special attention to multi-view and multi-modal imagery. PhotoMatch allows for students, researchers, and other end-users to compare and assess different matching methodologies through an educational and friendly environment, and thus to find the best algorithms for different applications. The different case studies exhibit the capabilities of PhotoMatch and its possibility to offer an accurate and reliable input for image orientation and 3D reconstruction. The results also highlight how different combinations of algorithms and setup parameters can lead to significant changes in the validity of the results.

Of course, PhotoMatch was conceived to support future developments, so future work will include the addition of new deep learning algorithms [36,40,61], as well as new detectors and descriptors [62–64]. These additions will be added to new release versions or presented as plugins. This will allow PhotoMatch to present a wider array of algorithm combinations for the assessment of the different approaches, while maintaining its educational goal and ease of use.

Author Contributions: Conceptualization, D.G.-A., F.R., P.R.-G. and D.H.-L.; methodology, E.R.d.O. and I.B.-G.; software, E.R.d.O.; validation, I.B.-G.; investigation, D.G.-A., F.R., P.R.-G. and D.H.-L.; writing—original draft preparation, I.B.-G.; writing—review and editing, E.R.d.O., I.B.-G., D.G.-A., F.R. and P.R.-G.; supervision, D.G.-A. All authors have read and agreed to the published version of the manuscript.

Funding: This research received no external funding.

Institutional Review Board Statement: Not applicable.

Informed Consent Statement: Not applicable.

Data Availability Statement: The PhotoMatch Installer and the source code are available on https://github.com/TIDOP-USAL/PhotoMatch/releases.

Conflicts of Interest: The authors declare no conflict of interest.

References

1. Gonzalez-Aguilera, D.; López-Fernández, L.; Rodriguez-Gonzalvez, P.; Hernandez-Lopez, D.; Guerrero, D.; Remondino, F.; Menna, F.; Nocerino, E.; Toschi, I.; Ballabeni, A.; et al. GRAPHOS—Open-Source Software for Photogrammetric Applications. *Photogramm. Rec.* **2018**, *33*, 11–29. [CrossRef]
2. Dai-Hong, J.; Lei, D.; Dan, L.; San-You, Z. Moving-Object Tracking Algorithm Based on PCA-SIFT and Optimization for Underground Coal Mines. *IEEE Access* **2019**, *7*, 35556–35563. [CrossRef]
3. Dalal, N.; Triggs, B. Histograms of Oriented Gradients for Human Detection. In Proceedings of the 2005 IEEE Computer Society Conference on Computer Vision and Pattern Recognition (CVPR'05), San Diego, CA, USA, 20–25 June 2005; Volume 1, pp. 886–893.

4. Fiaz, M.; Mahmood, A.; Javed, S.; Jung, S.K. Handcrafted and Deep Trackers: Recent Visual Object Tracking Approaches and Trends. *ACM Comput. Surv.* **2020**, *52*, 1–44. [CrossRef]
5. Luo, W.; Xing, J.; Milan, A.; Zhang, X.; Liu, W.; Kim, T.-K. Multiple Object Tracking: A Literature Review. *Artif. Intell.* **2021**, *293*, 103448. [CrossRef]
6. Pal, S.K.; Pramanik, A.; Maiti, J.; Mitra, P. Deep Learning in Multi-Object Detection and Tracking: State of the Art. *Appl. Intell.* **2021**, *51*, 6400–6429. [CrossRef]
7. Wohlhart, P.; Lepetit, V. Learning Descriptors for Object Recognition and 3D Pose Estimation. In Proceedings of the 2015 IEEE Conference on Computer Vision and Pattern Recognition (CVPR), Boston, MA, USA, 7–12 June 2015; pp. 3109–3118.
8. Granshaw, S.I. Editorial: Imaging Technology 1430–2015: Old Masters to Mass Photogrammetry. *Photogramm. Rec.* **2015**, *30*, 255–260. [CrossRef]
9. Morales, A.; González-Aguilera, D.; Gutiérrez, M.A.; López, I. Energy Analysis of Road Accidents Based on Close-Range Photogrammetry. *Remote Sens.* **2015**, *7*, 15161–15178. [CrossRef]
10. Nocerino, E.; Lago, F.; Morabito, D.; Remondino, F.; Porzi, L.; Poiesi, F.; Rota Bulo, S.; Chippendale, P.; Locher, A.; Havlena, M.; et al. A Smartphone-Based 3D Pipeline for the Creative Industry—The Replicate EU Project. *Int. Arch. Photogramm. Remote Sens. Spatial Inf. Sci.* **2017**, *XLII-2-W3*, 535–541. [CrossRef]
11. Ortiz-Sanz, J.; Gil-Docampo, M.; Rego-Sanmartín, T.; Arza-García, M.; Tucci, G. A PBeL for Training Non-Experts in Mobile-Based Photogrammetry and Accurate 3-D Recording of Small-Size/Non-Complex Objects. *Measurement* **2021**, *178*, 109338. [CrossRef]
12. Remondino, F.; Nocerino, E.; Toschi, I.; Menna, F. A Critical Review of Automated Photogrammetric Processing of Large Datasets. *Int. Arch. Photogramm. Remote Sens. Spatial Inf. Sci.* **2017**, *XLII-2-W5*, 591–599. [CrossRef]
13. Rahaman, H.; Champion, E. To 3D or Not 3D: Choosing a Photogrammetry Workflow for Cultural Heritage Groups. *Heritage* **2019**, *2*, 1835–1851. [CrossRef]
14. Jiang, X.; Ma, J.; Xiao, G.; Shao, Z.; Guo, X. A Review of Multimodal Image Matching: Methods and Applications. *Inf. Fusion* **2021**, *73*, 22–71. [CrossRef]
15. Pamart, A.; Morlet, F.; De Luca, L.; Veron, P. A Robust and Versatile Pipeline for Automatic Photogrammetric-Based Registration of Multimodal Cultural Heritage Documentation. *Remote Sens.* **2020**, *12*, 2051. [CrossRef]
16. Wei, Z.; Han, Y.; Li, M.; Yang, K.; Yang, Y.; Luo, Y.; Ong, S.-H. A Small UAV Based Multi-Temporal Image Registration for Dynamic Agricultural Terrace Monitoring. *Remote Sens.* **2017**, *9*, 904. [CrossRef]
17. Kang, J.K.; Hoang, T.M.; Park, K.R. Person Re-Identification Between Visible and Thermal Camera Images Based on Deep Residual CNN Using Single Input. *IEEE Access* **2019**, *7*, 57972–57984. [CrossRef]
18. Kniaz, V.V.; Knyaz, V.A.; Hladuvka, J.; Kropatsch, W.G.; Mizginov, V. *ThermalGAN: Multimodal Color-to-Thermal Image Translation for Person Re-Identification in Multispectral Dataset*; Springer: Cham, Switzerland, 2018; pp. 606–624.
19. Remondino, F.; Menna, F.; Morelli, L. Evaluating Hand-Crafted and Learning-Based Features for Photogrammetric Applications. *Int. Arch. Photogramm. Remote Sens. Spat. Inf. Sci.* **2021**, *XLIII-B2-2021*, 549–556. [CrossRef]
20. Ma, J.; Jiang, X.; Fan, A.; Jiang, J.; Yan, J. Image Matching from Handcrafted to Deep Features: A Survey. *Int. J. Comput. Vis.* **2021**, *129*, 23–79. [CrossRef]
21. González-Aguilera, D.; Ruiz De Oña, E.; López-Fernandez, L.; Farella, E.M.; Stathopoulou, E.K.; Toschi, I.; Remondino, F.; Rodríguez-Gonzálvez, P.; Hernández-López, D.; Fusiello, A.; et al. PHOTOMATCH: An Open-Source Multi-View and Multi-Modal Feature Matching Tool for Photogrammetric Applications. *Int. Arch. Photogramm. Remote Sens. Spat. Inf. Sci.-ISPRS Arch.* **2020**, *43*, 213–219. [CrossRef]
22. Muja, M.; Lowe, D.G. Flann, Fast Library for Approximate Nearest Neighbors. In *International Conference on Computer Vision Theory and Applications (VISAPP'09)*; INSTICC Press: Setúbal, Portugal, 2009; Volume 3.
23. Hartley, R.; Zisserman, A. *Multiple View Geometry in Computer Vision*; Cambridge University Press: Cambridge, UK, 2003; ISBN 978-0-521-54051-3.
24. Zitová, B.; Flusser, J. Image Registration Methods: A Survey. *Image Vis. Comput.* **2003**, *21*, 977–1000. [CrossRef]
25. Chen, L.; Rottensteiner, F.; Heipke, C. Feature Detection and Description for Image Matching: From Hand-Crafted Design to Deep Learning. *Geo-Spat. Inf. Sci.* **2021**, *24*, 58–74. [CrossRef]
26. Lowe, D.G. Distinctive Image Features from Scale-Invariant Keypoints. *Int. J. Comput. Vis.* **2004**, *60*, 91–110. [CrossRef]
27. Arandjelović, R.; Zisserman, A. Three Things Everyone Should Know to Improve Object Retrieval. In Proceedings of the 2012 IEEE Conference on Computer Vision and Pattern Recognition, Providence, RI, USA, 16–21 June 2012; pp. 2911–2918.
28. Bay, H.; Tuytelaars, T.; Van Gool, L. SURF: Speeded up Robust Features. In *Lecture Notes in Computer Science*; Springer: Berlin/Heidelberg, Germany, 2006; Volume 3951, pp. 404–417. [CrossRef]
29. Tombari, F.; Di Stefano, L. Interest Points via Maximal Self-Dissimilarities. In *Lecture Notes in Computer Science*; Springer: Cham, Switzerland, 2015; Volume 9004, pp. 586–600. [CrossRef]
30. Yu, K.; Zheng, X.; Duan, Y.; Fang, B.; An, P.; Ma, J. NCFT: Automatic Matching of Multimodal Image Based on Nonlinear Consistent Feature Transform. *IEEE Geosci. Remote Sens. Lett.* **2022**, *19*, 1–5. [CrossRef]
31. Long, G.; Kneip, L.; Alvarez, J.M.; Li, H.; Zhang, X.; Yu, Q. Learning Image Matching by Simply Watching Video. In Proceedings of the Computer Vision—ECCV 2016, Amsterdam, The Netherlands, 11–14 October 2016; Leibe, B., Matas, J., Sebe, N., Welling, M., Eds.; Springer International Publishing: Cham, Switzerland, 2016; pp. 434–450.

32. Christiansen, P.H.; Kragh, M.F.; Brodskiy, Y.; Karstoft, H. UnsuperPoint: End-to-End Unsupervised Interest Point Detector and Descriptor. *arXiv* **2019**, arXiv:1907.04011.
33. Ono, Y.; Trulls, E.; Fua, P.; Moo Yi, K. LF-Net: Learning Local Features from Images. In Proceedings of the 32nd International Conference on Neural Information Processing Systems, Montréal, QC, Canada, 3–8 December 2018.
34. Image Matching Challenge—2021 Edition. Available online: https://www.cs.ubc.ca/research/image-matching-challenge/current/ (accessed on 11 October 2022).
35. Chen, H.; Luo, Z.; Zhou, L.; Tian, Y.; Zhen, M.; Fang, T.; Mckinnon, D.; Tsin, Y.; Quan, L. ASpanFormer: Detector-Free Image Matching with Adaptive Span Transformer. In Proceedings of the 17th European Conference, Tel Aviv, Israel, 23 October 2022.
36. Sarlin, P.-E.; DeTone, D.; Malisiewicz, T.; Rabinovich, A. SuperGlue: Learning Feature Matching With Graph Neural Networks. In Proceedings of the 2020 IEEE/CVF Conference on Computer Vision and Pattern Recognition (CVPR), Seattle, WA, USA, 13–19 June 2020; pp. 4938–4947.
37. DeTone, D.; Malisiewicz, T.; Rabinovich, A. SuperPoint: Self-Supervised Interest Point Detection and Description. In Proceedings of the 2018 IEEE/CVF Conference on Computer Vision and Pattern Recognition Workshops (CVPRW), Salt Lake City, UT, USA, 18–22 June 2018; pp. 224–236.
38. Sun, J.; Shen, Z.; Wang, Y.; Bao, H.; Zhou, X. LoFTR: Detector-Free Local Feature Matching With Transformers. In Proceedings of the 2021 IEEE/CVF Conference on Computer Vision and Pattern Recognition (CVPR), Nashville, TN, USA, 20–25 June 2021; pp. 8922–8931.
39. Edstedt, J.; Athanasiadis, I.; Wadenbäck, M.; Felsberg, M. DKM: Dense Kernelized Feature Matching for Geometry Estimation. *arXiv* **2022**, arXiv:2202.00667.
40. Tyszkiewicz, M.; Fua, P.; Trulls, E. DISK: Learning Local Features with Policy Gradient. In *Advances in Neural Information Processing Systems*; Curran Associates, Inc.: Red Hook, NY, USA, 2020; Volume 33, pp. 14254–14265.
41. Bellavia, F.; Colombo, C.; Morelli, L.; Remondino, F. Challenges in Image Matching for Cultural Heritage: An Overview and Perspective. In *Image Analysis and Processing*; ICIAP 2022 Workshops; Mazzeo, P.L., Frontoni, E., Sclaroff, S., Distante, C., Eds.; Springer International Publishing: Cham, Switzerland, 2022; pp. 210–222.
42. Verdie, Y.; Yi, K.; Fua, P.; Lepetit, V. TILDE: A Temporally Invariant Learned DEtector. In Proceedings of the IEEE Conference on Computer Vision and Pattern Recognition, Boston, MA, USA, 7–12 June 2015; pp. 5279–5288.
43. Dusmanu, M.; Rocco, I.; Pajdla, T.; Pollefeys, M.; Sivic, J.; Torii, A.; Sattler, T. D2-Net: A Trainable CNN for Joint Description and Detection of Local Features. In Proceedings of the IEEE Computer Society Conference on Computer Vision and Pattern Recognition, Long Beach, CA, USA, 15–20 June 2019; Volume 2019, pp. 8084–8093.
44. Revaud, J.; Weinzaepfel, P.; De Souza, C.; Pion, N.; Csurka, G.; Cabon, Y.; Humenberger, M. R2D2: Repeatable and Reliable Detector and Descriptor. In Proceedings of the Conference on Neural Information Processing Systems (NeurIPS), Vancouver, BC, Canada, 8–14 December 2019; Volume 32. [CrossRef]
45. ISPRS Scientific Initiatives. Available online: https://www.isprs.org/society/si/SI-2019/default.aspx (accessed on 9 August 2022).
46. Gaiani, M.; Apollonio, F.I.; Ballabeni, A.; Remondino, F. Securing Color Fidelity in 3D Architectural Heritage Scenarios. *Sensors* **2017**, *17*, 2437. [CrossRef]
47. Lu, C.; Xu, L.; Jia, J. Contrast Preserving Decolorization with Perception-Based Quality Metrics. *Int. J. Comput. Vis.* **2014**, *110*, 222–239. [CrossRef]
48. Lal, S.; Chandra, M. Efficient Algorithm for Contrast Enhancement of Natural Images. *Int. Arab J. Inf. Technol.* **2014**, *11*, 95–102.
49. Abdullah-Al-Wadud, M.; Kabir, M.D.H.; Akber Dewan, M.A.; Chae, O. A Dynamic Histogram Equalization for Image Contrast Enhancement. *IEEE Trans. Consum. Electron.* **2007**, *53*, 593–600. [CrossRef]
50. Liu, Y.-F.; Guo, J.-M.; Lai, B.-S.; Lee, J.-D. High Efficient Contrast Enhancement Using Parametric Approximation. In Proceedings of the 2013 IEEE International Conference on Acoustics, Speech and Signal Processing, Vancouver, BC, Canada, 26–31 May 2013; pp. 2444–2448.
51. Kim, M.; Chung, M.G. Recursively Separated and Weighted Histogram Equalization for Brightness Preservation and Contrast Enhancement. *IEEE Trans. Consum. Electron.* **2008**, *54*, 1389–1397. [CrossRef]
52. Wallis, K.F. Seasonal Adjustment and Relations between Variables. *J. Am. Stat. Assoc.* **1974**, *69*, 18–31. [CrossRef]
53. Alcantarilla, P.; Nuevo, J.; Bartoli, A. Fast Explicit Diffusion for Accelerated Features in Nonlinear Scale Spaces. In Proceedings of the British Machine Vision Conference 2013; British Machine Vision Association: Bristol, UK, 2013; pp. 13.1–13.11.
54. Morel, J.-M.; Yu, G. ASIFT: A New Framework for Fully Affine Invariant Image Comparison. *SIAM J. Imaging Sci.* **2009**, *2*, 438–469. [CrossRef]
55. Leutenegger, S.; Chli, M.; Siegwart, R.Y. BRISK: Binary Robust Invariant Scalable Keypoints. In Proceedings of the 2011 International Conference on Computer Vision, Barcelona, Spain, 6–13 November 2011; pp. 2548–2555.
56. Bian, J.; Lin, W.-Y.; Matsushita, Y.; Yeung, S.-K.; Nguyen, T.-D.; Cheng, M.-M. GMS: Grid-Based Motion Statistics for Fast, Ultra-Robust Feature Correspondence. In Proceedings of the 2017 IEEE Conference on Computer Vision and Pattern Recognition (CVPR), Honolulu, HI, USA, 21–26 July 2017; pp. 4181–4190.
57. Hamming, R.W. Error Detecting and Error Correcting Codes. *Bell Syst. Tech. J.* **1950**, *29*, 147–160. [CrossRef]
58. Poursaeed, O.; Yang, G.; Prakash, A.; Fang, Q.; Jiang, H.; Hariharan, B.; Belongie, S. Deep Fundamental Matrix Estimation without Correspondences. In Proceedings of the Computer Vision—ECCV 2018 Workshops, Munich, Germany, 8–14 September 2018.

59. Aicardi, I.; Nex, F.; Gerke, M.; Lingua, A.M. An Image-Based Approach for the Co-Registration of Multi-Temporal UAV Image Datasets. *Remote Sens.* **2016**, *8*, 779. [CrossRef]
60. Ancuti, C.O.; Ancuti, C.; Bekaert, P. Decolorizing Images for Robust Matching. In Proceedings of the 2010 IEEE International Conference on Image Processing, Hong Kong, China, 26–29 September 2010; pp. 149–152.
61. Luo, Z.; Zhou, L.; Bai, X.; Chen, H.; Zhang, J.; Yao, Y.; Li, S.; Fang, T.; Quan, L. ASLFeat: Learning Local Features of Accurate Shape and Localization. In Proceedings of the 2020 IEEE/CVF Conference on Computer Vision and Pattern Recognition (CVPR), Los Alamitos, CA, USA, 13–19 June 2020; pp. 6589–6598.
62. Mishchuk, A.; Mishkin, D.; Radenovic, F.; Matas, J. Working Hard to Know Your Neighbor' s Margins: Local Descriptor Learning Loss. In Proceedings of the Advances in Neural Information Processing Systems, Long Beach, CA USA, 4–9 December 2017; Volume 30.
63. Truong, P.; Apostolopoulos, S.; Mosinska, A.; Stucky, S.; Ciller, C.; Zanet, S.D. GLAMpoints: Greedily Learned Accurate Match Points. In Proceedings of the 2019 IEEE/CVF International Conference on Computer Vision (ICCV), Seoul, Republic of Korea, 27 October–2 November 2019; pp. 10732–10741.
64. Singh Parihar, U.; Gujarathi, A.; Mehta, K.; Tourani, S.; Garg, S.; Milford, M.; Krishna, K.M. RoRD: Rotation-Robust Descriptors and Orthographic Views for Local Feature Matching. In Proceedings of the 2021 IEEE/RSJ International Conference on Intelligent Robots and Systems (IROS), Prague, Czech Republic, 27 September–1 October 2021; pp. 1593–1600.

applied sciences

MDPI

Article

Towards Automatic License Plate Recognition in Challenging Conditions

Fahd Sultan [1], Khurram Khan [2], Yasir Ali Shah [3], Mohsin Shahzad [3], Uzair Khan [3] and Zahid Mahmood [3,*]

1 Department of Software Development, Axispoint Technology Solutions Group (ATSG), New York, NY 10119, USA
2 Faculty of Computer Science and Engineering, GIK Institute of Engineering Sciences and Technology, Topi 23460, Pakistan
3 Department of Electrical and Computer Engineering, COMSATS University Islamabad, Abbottabad Campus, Abbottabad 22060, Pakistan
* Correspondence: zahid0987@cuiatd.edu.pk

Abstract: License plate recognition (LPR) is an integral part of the current intelligent systems that are developed to locate and identify various objects. Unfortunately, the LPR is a challenging task due to various factors, such as the numerous shapes and designs of the LPs, the non-following of standard LP templates, irregular outlines, angle variations, and occlusion. These factors drastically influence the LP appearance and significantly challenge the detection and recognition abilities of state-of-the-art detection and recognition algorithms. However, recent rising trends in the development of machine learning algorithms have yielded encouraging solutions. This paper presents a novel LPR method to address the aforesaid issues. The proposed method is composed of three distinct but interconnected steps. First, a vehicle that appears in an input image is detected using the Faster RCNN. Next, the LP area is located within the detected vehicle by using morphological operations. Finally, license plate recognition is accomplished using the deep learning network. Detailed simulations performed on the PKU, AOLP, and CCPD databases indicate that our developed approach produces mean license plate recognition accuracy of 99%, 96.0231%, and 98.7000% on the aforesaid databases.

Keywords: Faster RCNN; license plate recognition; object detection

Citation: Sultan, F.; Khan, K.; Shah, Y.A.; Shahzad, M.; Khan, U.; Mahmood, Z. Towards Automatic License Plate Recognition in Challenging Conditions. *Appl. Sci.* **2023**, *13*, 3956. https://doi.org/10.3390/app13063956

Academic Editor: Sungho Kim

Received: 6 February 2023
Revised: 13 March 2023
Accepted: 16 March 2023
Published: 20 March 2023

1. Introduction

With the growth of big data, object detection and recognition have attracted excellent interest in research communities. This is because it can be used for a wide range of real-world applications, such as medical imaging, augmented reality, sports applications, independent driving, and video surveillance [1–6]. Particularly, the license plate recognition (LPR) is getting more attention due to its widespread applications in various fields, for instance, traffic monitoring, toll collection, and criminal searches [7,8]. Although many of the LPR systems, for instance [9,10], are available in the literature, most of them have been validated and tested on a pre-defined LP specification. Few of these works are also capable of processing multiple LPs. The LPR systems can be categorized into two major categories, which are (i) traditional LPR and (ii) deep learning-based LPR systems. Traditional methods process limited features and utilize hand-crafted features, for example, contours, colors, and edges, to locate the LP. Deep learning-based techniques automatically learn robust features from the data and have recently produced promising results. Deep learning-based techniques consider LP detection as object detection and analyze the recognition as an optical character recognition (OCR) process. Since the number of characters processed by any LPR method is limited, character recognition is also considered an object detection process, so that LP detection and recognition are handled simultaneously. To develop and analyze a LPR algorithm that can deal with multi-style license plate recognition, there are several challenges, as briefly described below.

Lack of Standard LPs: Standardizing the LPs is a significant challenge throughout the world. For instance, license plates in Macao, China, have to meet a strict standard, as shown in Figure 1a [11], whereas the LP of Hong Kong usually has 1–8 characters and Macao LPs are composed of five to six characters, as shown in Figure 1b,c, respectively. General observation in Figure 1a indicates that the first column has an entirely different appearance than the second column. Similarly, the second and third columns have huge contrast and appearance variations. Few of the plates have yellowish and greenish backgrounds, while few have colorless backgrounds. Moreover, the distance between the characters on the plates shown in the first column takes up much less space than those in the third column.

(**a**) (**b**) (**c**)

Figure 1. Various license plate challenges.

Lack of LP Outlines: Many times, the license plates have no outlines, which makes it an extremely difficult task to classify. This task becomes more challenging when the color of the vehicle is the same as that of the LPs. One such example is indicated in Figure 1b. The top image shown in Figure 1b contains a black background and white ground on a black vehicle. Moreover, the bottom vehicle shown in Figure 1b shows a LP case that has a white background due to the white color of the car and a black front ground. These two cases may appear trivial, but for any machine learning algorithm, the aforedescribed task is not easy. The algorithm should be capable enough to distinguish such cases accurately.

The appearance of LPs: Another challenge in LP localization and recognition is to accurately handle the appearance and occlusion challenges of license plates installed at various locations on a vehicle. If a vehicle has more than one license plate, then characters get matched with the background of other plates, which makes it difficult to distinguish LP character regions. A few cases of this challenge are shown in Figure 1c. Figure 1c depicts two license plates with total complement variations on the top vehicle. Moreover, along with the low-intensity light beam, one plate has a white background with black characters on the plate region, while the other plate has a black background with white alphabets on top. In both cases, the vehicle color appears blackish–green. In each of these cases, it poses a significant challenge to any recognition algorithm.

Therefore, one of the aims of this research is to design an accurate LP recognition technique with the capability to handle diverse license plates. Fortunately, numerous research groups have compiled numerous LP datasets. Few of the databases also contain clear vehicle images in different environments and road conditions. We aimed to contribute to the field with this manuscript, as highlighted below.

- Inspired by recent trends in machine learning, a robust LP recognition method is proposed in this paper that accurately recognizes various license plates. Particularly, our developed system uses an intelligent combination of Faster RCNN to detect various vehicles, morphological image processing methods to locate the LP area, and finally, the deep learning-based method to recognize the detected license plate. The systemic application of various modules enables us to achieve a reliable and accurate LP recognition method.

- We consider the license plate as an object for detection and recognition tasks. The output of our created approach for detecting vehicles is a rectangle encompassing the vehicle and license plate region. In contrast, for the license plate recognition task, our system displays the vehicle license plate alphabets and characters above the real license plate once it has been located in the image.
- Our developed technique supports different types of plates from PKU, AOLP, and CCPD datasets. Our obtained results indicate that our developed method effectively recognizes the LPs of these databases. Moreover, our developed technique is intelligent, as it systematically achieves the aforedescribed tasks.

This manuscript is organized as follows: In Section 2, recent license plate recognition techniques are briefly listed. In Section 3, our proposed method is described in detail. Our findings during simulations are listed in Section 4, followed by the conclusions, which also hint at the future extension of this work.

2. Related Work

This section details recent related works on license plate recognition. In [11], a region-based license plate detection method is discussed that initially shifts mean to filter and segment a color vehicle image to get candidate regions. These candidate regions are then analyzed to decide whether a candidate region contains a license plate. Since this method focuses on regions, so it is more robust to interference characters. In [12], the proposed method uses the YOLOv2 to detect vehicles. This work uses a CNN-based method that they refer to as WPOD-NET for LP detection. Meanwhile, a modified YOLO architecture recognizes the LP characters. However, this work also uses character segmentation, which makes this method a bit more complex than the compared methods. In [13], an improved YOLO architecture is deployed for character recognition. This work is tested on the SSIG dataset, which has 2000 images. In [14], a technique is developed that customizes the YOLO network to detect LPs from images that are captured in different conditions, such as different weathers, varying lighting, and other factors. The authors conclude that the YOLO can strike a balance between precision and recall. However, the YOLO is not suitable for detecting angular or small objects. Therefore, its performance in scenarios where the vehicle is far away from the camera needs to be further checked.

In [15], a framework to detect and recognize license plates is discussed for complex scenes, which is based on mask region convolutional neural networks. The evaluation of this framework is further enhanced on four publicly available datasets for different countries. Moreover, this method is tested on diverse range of images, which are captured from multiple scenes, such as varying orientations, poor image quality, blurred images, and complex backgrounds. In [16], a convolutional filter of size 3×3 is used in deep networks to analyze the increasing depth of the architecture by using 16 to 19 layers to process the 24×24 pixel colored image. This work also introduces a pre-processing step by subtracting the average RGB value from each individual pixel. This paper reports significant improvements to ConvNets in the realm of image recognition as a result. In addition, this work also uses a large number of 3×3 convolutional filters that fit well on the investigated datasets only. In [17], initially, candidate regions are selected through a sparse network using winnows classification, followed by filtration through CNN. An interesting novelty introduced in this work is the minimization of training and target domains in an unsupervised manner. However, this work also considers artificially generated synthetic LP images. In [18], the developed method utilizes thin–plate spline transformation and adaptively rectifies a textual LP image. Moreover, a recognition model predicts a character sequence immediately from the rectified image. This work only considers qualitative results on several images. In [19], a 2D attention-based encoder–decoder architecture is developed. This method extracts features by applying the ResNet CNN architecture. The 2D model introduced is capable of accommodating text with different layouts, arbitrary shapes, and different angles. Their reported results are encouraging, and their development reduces data bias and increases model generalization capacity. This method is simple; however, its generalization to standard datasets has not been explored. In a previous study [20], the authors used CycleWGAN to create LP images to improve the performance of recognition. Their work simultaneously generates images of different conditions. Meanwhile, a modified version of the CTC is used to recognize the LP. Their work simultaneously generates images of different conditions. Meanwhile, a modified version of the CTC is used to recognize the LP.

In [21], an end-to-end irregular LPR (EILPR) is proposed using plate-level annotations during training. In the EILPR method, a coarse-to-fine approach is implemented that extracts the LP features for sequence recognition. This work assumes the fact that a LP may generate a perspective bias in the image; therefore, to cater to this fact, an automatic perspective alignment network (APAN) is introduced to extract the fine license plate features. To classify the international license plates, a location–aware 2D attention-based recognition network is used. In [22], a novel ALPR technique, which is referred to as VSNet, is developed. The VSNet contains two CNNs that are combined in a cascading manner. Meanwhile, an integration block is introduced that extracts the spatial features. With vertex supervisory information, authors develop a vertex-evaluation module in VertexNet such that a LP can be repaired as the input images of SCR-Net. A horizontal encoding algorithm is used in the SCR-Net to extract left-to-right features and then recognize a license plate. This work performs well on standard LPs. However, its generalization capability on tilting and rotating LPs has not been explored.

Additionally, ALPRNet is developed to detect and recognize mixed–style license plates [23]. Two fully convolutional object detectors are used in the proposed ALPRNet to classify and recognize LPs. The proposed ALPRNet processes LP and character equally. In this work, object detectors output bound boxes of LPs along with corresponding labels without the application of the RNN branches of the OCR. This is because this is a single–stage network-based method. Therefore, its detection accuracy on challenging datasets has not been explored. In [24], image processing and OCR-based techniques are merged to recognize the LPs. The image processing methods utilize color conversion, Otsu's thresholding, and noise removal. The OCR method uses template matching to predict the characters of LPs. The authors of this work have not examined the scalability of this method and have only used basic tools from signal and image processing. In [25], the proposed LPR method consists of three steps: LP detection, unified character recognition, and multinational LP layout detection. This work is primarily based on the YOLO networks. To extract the correct sequence, a layout detection scheme is introduced, which extracts the sequence of LP numbers from multinational LPs. This study is extensively tested on standard Koran and Taiwan LPs. In [26], the developed LPR method uses a joint combination of adaptive boosting and the LDA to extract features. The CNNC is then used to separate the LP region from irrelevant samples. This work is segmentation-free. However, its recognition capability on real-world images has not been explored. In [27], the developed algorithm uses a distinct, fine-tuned YOLO-v3 platform to extract LP characters from input images. During the training and testing stages, a wide range of LP images have been analyzed. However, this system is fully annotated and consumes over 100 ms to accomplish the task of LP recognition. In addition, an intriguing review article is released that summarizes the many approaches currently utilized to detect various objects [28].

In [29], researchers introduced a robust vehicle detection method using a multi-scale deep convolutional neural network. This work utilizes a standard Gaussian mixture probability hypothesis density filter along with hierarchical data associations (HDA) that isolate detection-to-track and track-to-track associations. Particularly, the cost matrix of various phases is solved using the Hungarian algorithm. For quick execution, detection information, such as bounding boxes and detection scores, is used in the HDA without visual feature information. Although this is an interesting work, the computational difficulty of the approach is not covered. In [30], a region proposal network (RPN) is developed that shares full-image convolutional features with the detection network. The RPN is a fully convolutional network that forecasts object bounds and scores at various positions. The

RPN is trained end-to-end to generate high-quality region proposals that are later used by Fast R-CNN for detection. In this work, the RPN and Fast R-CNN are also merged into a single network by sharing their convolutional features. For the very deep VGG-16 model, this system has a frame rate of 5 fps on a GPU, while achieving encouraging object detection accuracy on several datasets with only 300 proposals per image.

In [31], a wavelet transform based technique to extract license plates from cluttered images is developed. This method comprises of three major stages, which are (i) extracting important contrast features through wavelets. Then, finding a reference line in HL subimage plays an important role to locate the desired license plate region roughly. According, (ii) decrease the searching region of license plate to speed up the execution time, and (iii) localization of license plate through manual adjustments. More importantly, the proposed detection method can locate multiple plates with different orientations in one image. Since the feature extracted is robust to complex backgrounds, the proposed method works well in extracting differently illuminated and oriented license plates. The average accuracy of detection is 92.4%. In [32], authors made use of a combination of the MSER and the stroke width transform (SWT) to detect and isolate the LP character regions. The license plates were finally bordered using the probabilistic Hough transform. The authors discuss that character-based methods are reliable and can lead to a high recall. However, the other text in the image background has a significant impact on performance. This method requires multiple cameras before the system is placed for evaluation. In [33], an interesting license plate recognition system is developed using a sequence of deep CNNs. These CNNs are trained and fine-tuned so that they are robust under different conditions (for instance, lighting, occlusion, or tilt) and work across a variety of license plate templates that include different sizes, backgrounds, or fonts. In [34], a novel line density filter approach was developed that connects regions with high edge density and removes sparse regions in each row and column from a binary edge image. This study indicates that edge-based methods are fast in computation but cannot be applied to complex images because they are too sensitive to unwanted edges.

In [35], the developed LP method consists of three modules for plate detection, character segmentation, and recognition. This method also formulates edge clustering to solve plate detection for the first time. A bilayer classifier, which is improved with an additional null class, is empirically proven to be better than previous methods for character recognition. However, this method is evaluated only on a single dataset, which was also gathered by the authors themselves. In [36], license plate detection and recognition are tackled in standard natural scene images via the development of a segmentation-free method. Inspired by the success of DNNs, these are deployed to learn high-level features in a cascade framework, which leads to improved performance on both detection and recognition. This work also trains 37 CNNs to detect all characters in an image, which results in a high recall. Later, to improve the IoU ratio, bounding box refinement is carried out based on the edge information of the LPs. This method extracts license plates effectively with both high recall and precision. Last, a recurrent neural network with long short-term memory (LSTM) is trained to recognize the sequential features extracted from the whole license plate via CNNs. For scene and lighting variations, this method needs to be further explored. In [37], a unified deep neural network is proposed that localizes license plates and recognizes the letters simultaneously in a single forward pass. This whole network is trained end-to-end and achieves the LP recognition task in a single network, avoiding intermediate error accumulation and resulting in faster processing speed. For performance evaluation, a few datasets that include images captured from various scenes under different conditions are tested. However, this method does not consider the complexity of the developed method.

In [38], researchers use computer graphic scripts and GANs to generate and augment a large number of annotated, synthesized LPs with realistic colors, fonts, and character composition from a small number of real, manually labeled license plate images. In this work, generated and augmented data are mixed and used as training data for the LP recognition network modified from the DenseNet. Simulations reveal that the model trained from the generated mixed training data has much better generalization ability and achieves encouraging detection and recognition accuracy on multiple datasets, even with a very limited number of original real license plates. In [39], a new license plate recognition technique is developed in the wild. This method comprises a tailored CycleGAN model for license plate image generation and an elaborately designed image-to-sequence network for plate recognition. The CycleGAN-based plate generation engine eases the exhausting human annotation work. In this work, huge amounts of training data are obtained with a more balanced character distribution and various shooting conditions that boost the recognition accuracy to a large extent. Moreover, a 2D attentional-based license plate recognizer with an Xception-based CNN encoder is developed that is capable of recognizing various LPs with different patterns under various scenarios accurately.

In [40], a new license plate dataset, to which the authors refer as the CCPD, is developed and tested under different circumstances, for instance, tilt, blur, rotate, or varying weather conditions. This work is novel in the sense that it provides a single platform for researchers to investigate the LP's prevailing issues. In [41], a novel end-to-end method for LP recognition without initial character segmentation is presented as LPRNet. Particularly, this method is inspired by recent breakthroughs in the DNNs and works in real-time with recognition accuracy up to 95% for Chinese license plates: 3 ms/plate on NVIDIA^R GeForce^{TM} GTX 1080 and 1.3 ms/plate on the Intel R Core^{TM} i7-6700K CPU. The LPRNet consists of the lightweight CNN and can be trained end-to-end. The authors of this work recommend that the LPRNet algorithm may be used to create embedded solutions for LPR that feature high levels of accuracy even on challenging Chinese license plates. In [42], a multi-object rectified attention network (MORAN) is proposed for text recognition. The MORAN consists of a multi-object rectification network and an attention-based sequence recognition network. The multi-object rectification network is designed to rectify images that contain irregular text. It decreases the difficulty of recognition and enables the attention-based sequence recognition network to read irregular text. The attention-based sequence recognition network focuses on target characters and sequentially outputs the predictions. Further, to improve the sensitivity of the attention-based sequence recognition network, a fractional pickup algorithm is also developed for an attention-based decoder during the training phase. In [43], a novel decoupled attention network (DAN) is developed that decouples the alignment operation from using historical decoding results. The DAN is an effective, flexible, reliable, and robust end-to-end text recognizer and consists of three components: a feature encoder, a convolutional alignment module, and a decoupled text decoder that generates final predictions by jointly using the feature map and attention maps. Yu et al. [44] used a wavelet transform at first to get the horizontal and vertical details of an image. Meanwhile, empirical mode decomposition (EMD) analysis was employed to deal with the projection data and locate the desired wave crest that indicates the position of a license plate appearing in any corner of the input image. Different versions of YOLO [45–47], which give state-of-the-art accuracy for object detection, have been published in the last few years.

The attempts outlined above are just a few examples of the numerous object detection and recognition algorithms that aim to overcome various LP recognition challenges. The following are a few of the primary reasons that prompted us to create a state-of-the-art license plate recognition algorithm.

- Most of the above-described methods and works have been carried out on standard databases that are gathered by researchers at different times under different conditions. Therefore, it prompted us to develop an algorithm that can reliably handle real-life images in real time while maintaining high recognition accuracy.
- Our study indicates that the methods, which use RNNs as the OCR, are costly in terms of execution time. Similarly, segmentation-based methods are mostly dependent on segmentation performance and highly susceptible to environmental conditions, such as varying illumination conditions, wild weather, or blurring. Therefore, these methods result in low recognition accuracy in such conditions. Even a strong recognizer, if

applied, would produce much lower recognition rates. Therefore, inspired by the aforementioned fact, we aimed to develop a license plate recognition method that could perform well under the scenarios described above.

- The PKU dataset, which is also investigated in this study, contains five prominent classes of vehicles on main highways. These categories cover different day times, varying weather conditions, multiple vehicles and license plates per image, occlusions, and crosswalks on the main highways. The scenarios mentioned are from real life, in which the detection and recognition accuracy of any algorithm might be significantly challenged.
- Many times, the cameras installed on the main highways of various countries in the world capture vehicle images in which license plates appear at an angle, tilted, or partially obscured. This motivates us to develop a system that could facilitate the traffic control and monitoring staff's ability to reliably recognize any suspicious license plate.

3. Methodology

Our developed method has three major modules, which are vehicle detection, license plate detection, and license plate recognition. Figure 2 illustrates the complete flow of our developed method that achieves the aforementioned tasks. The details of each component of the developed method are described below.

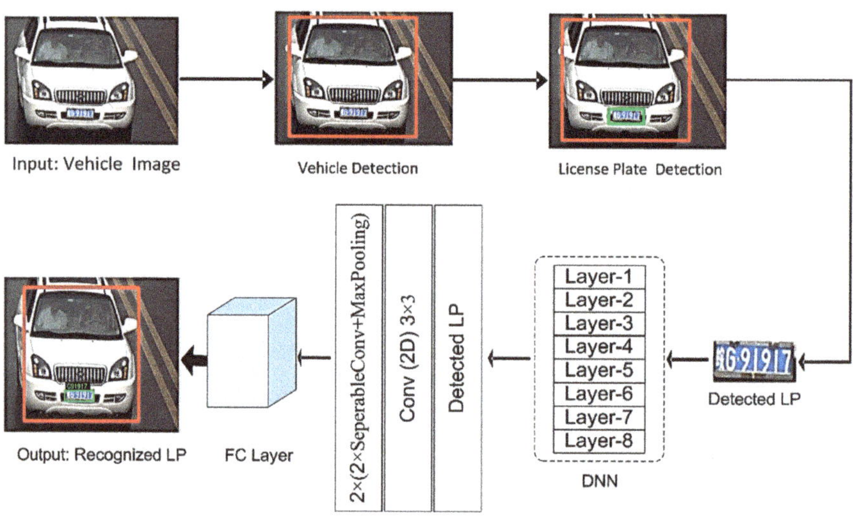

Figure 2. Flow of the proposed license plate recognition procedure.

3.1. Vehicle Detection

To locate objects, for instance, vehicle detection is a critical phase in developing an intelligent traffic monitoring system. In the past few years, the computer vision domain has introduced efficient object detection algorithms. Particularly, Faster RCNN and deep learning-based vehicle detection methods report high detection accuracy in near real-time in different environments [29]. Ultimately, these approaches have become a significant part of autonomous vehicles and self-driving applications. Our research reveals that real-time processing to locate vehicles, as well as good detection accuracy, are essential requirements that any object or vehicle detector should meet. We use a fine-tuned version of the Faster R-CNN [30] to find a vehicle quickly. The reason to detect the vehicle is that it considerably reduces the area to be explored for the existence of the LP in later stages. The purpose of using the Faster R-CNN at this stage is that, during the data training phase, it is at least nine times more rapid than the standard R-CNN. Moreover, it is 213× faster during the test phase and yields higher detection accuracy than its counterpart [30].

Algorithm 1 demonstrates the pseudocode of the employed vehicle detection module. In lines (2) to (17), Faster RCNN is used to locate vehicles' positions. In lines 3–9, Faster R-CNN is fine-tuned to obtain the appropriate region of interests (RoIs) to look for the possible existence of a vehicle in an input image. Therefore, we perform the mini-batch sampling by empirically choosing 128 region proposal networks (RPNs). To generate the RPN, a small network is made to slide over the *conv* feature map, which is output by the last shared *conv* layer. This small network takes as input an $N \times N$ spatial window of the input *conv* feature map. This feature is fed into two siblings' fully connected layers. We use $N = 2$ during our tuning, keeping in mind the fact that the effective receptive field on the input image is large. As a result, 64 RoIs are extracted from an input image. Moreover, to describe the foreground of an object mask, we choose an object proposal with an IoU overlap that contains at least 0.5 ground truth. In lines 10–16 of Algorithm 1, we process an RGB vehicle image with thirteen *conv* layers. As a result, a *conv* feature map (Ψ) is obtained.

Algorithm 1: Pseudocode of the vehicle detection method.

1. **Input:** colored RGB vehicle image
2. **begin** Faster R-CNN
3. initialize fine-tuning
4. **do**
5. extract features ► during training initialization
6. perform mini-batch sampling by $\left(\frac{RPN=128}{N=2}\right)$; 64 RoIs from each image
7. select IoU overlap with ground truth > 0.5
8. back-propagate errors across network layers ► weights optimization for nodes
9. **end**
10. **for** $I \in \{R, G, B\}$ **do**
11. process RGB data with 13 *conv* layers to obtain Ψ
12. generate the RPN by using 3 scales and aspect ratios on Ψ
13. feature map (Ψ) and region proposals are fed to the RoI pooling layer ('I)
14. 'I → (r, c, h, w)
15. for all feature vectors (Ɲ), generate the *FC* layer
16. **end**
17. **end**
18. **Output:** vehicle detection

For every region of the vehicle region proposal network (RPN), we apply nine diverse anchors on Ψ that calculate the probable vehicle regions. Meanwhile, for anchors, three scales are employed, which have resolutions of 128^2, 256^2, and 512^2 pixels along with three aspect ratios, which are 1:1, 1:2, and 2:1, respectively. As shown in line 14, max pooling is performed using five layers on $(\frac{h}{Height} \times \frac{w}{Width})$ with a 7×7 of $Height \times Width$ with h and w being the layer hyper-parameters that are independent of any particular RoI. Each RoI is expressed by a four-tuple (r, c, h, w), which specifies its top-left corner (r, c) and its width (h, w). Every feature vector (Ɲ) is passed as an input into a sequence of fully connected (FC) layers. In between, the RoI pooling layer uses max pooling to transform the features inside a binding region into a small feature map. Moreover, only a few RPN proposals highly overlap with each other. Therefore, to reduce redundancy, we adopt non-maximum suppression (NMS) in the proposal regions.

We fix the IoU threshold for NMS ≥ 0.5, which leaves us about 2000 proposal regions per image with a significant decrease in the number of proposals. After the NMS, the *top-N* ranked proposal regions are estimated to detect vehicles and draw a red rectangle around them. Once the vehicle is located in the input image, we apply our method to locate a LP within the bounding box that contains the vehicle.

3.2. The LP Localization

The detected vehicle, confined by a bounding box that is obtained in the last step of Algorithm 1, is nursed to the LP localization module that aims to detect the LP. Our developed LP localization method has a few interconnected steps. The LP localization method processes the RGB image and transforms it into the HSV components as shown in Equations (1)–(3).

$$H = \cos^{-1}\left[\frac{\frac{1}{2}[(R-G)+(R-B)]}{\sqrt[2]{(R-G)^2+(R-B)(G-B)}}\right] \quad (1)$$

$$S = 1 - \frac{3}{R+G+B}[\min(R,G,B)] \quad (2)$$

$$V = \frac{1}{3}(R+G+B) \quad (3)$$

where H denotes hue, S represents saturation, and V stands for the value components of the transformed image. Our general observation is that an LP in actuality may have diversity and huge color variations. Considering this fact, in Algorithm 2, we introduce colors segmentation from lines 5–13 on each of the HSV components.

During our simulations, we empirically vary the HT_{low} value from 0.02 to 0.40 and HT_{high} from 0.409 to 0.620. Similarly, for the saturation and value channels, their relevant low and high thresholds are ST_{low}, ST_{high}, VT_{low}, and VT_{high}, respectively, as indicated in lines 8–11 of Algorithm 2. For ST_{low}, the values are changed from 0.370 to 0.500, whereas for ST_{high}, they are changed from 0.909 to 1.10. For the V channel in the HSV image, the VT_{low} is set to 0.750 and the VT_{high} is kept at 1.0. After these thresholds are set, the mask images are obtained for each of the H, S, and V channels. For the H channel, the H_{mask} is set to 1 when the H_{Image} obtained is greater than or equal to the low threshold and less than or equal to the high threshold. A similar mechanism is applied to obtain the masks of the S and V channels.

Consequently, a blob image (A) is attained, which is indicated in line (11), which is analyzed by using morphological operations to enhance LP blobs in a sample space (z). Here, dilation $(\oplus)$ is applied using Equation (4), which enlarges the features and adds pixel layers across the regions of associated elements.

$$A \oplus B = \left\{z \middle| \left(\hat{B}\right)z \bigcap A \neq \phi \right. \quad (4)$$

where B indicates a structuring element through which the blob image is dilated. Meanwhile, the closing $(\bullet)$ operation is applied using Equation (5), in which the license plate blob image is first dilated by structuring element B and then eroded by B. The closing operation results in the smoothing of the contour and filling of the holes in the license plate blob.

$$A \bullet B = (A \oplus B) \ominus B \quad (5)$$

When the luminance is unsatisfactory, in Algorithm 2, we suggest illumination rectification as shown in lines 17–21. We use the PCA on the detected input vehicle image to fix the dimming of the image. By applying the PCA, we extract the Luminance and Chrominance channels of the RGB-colored vehicle image. In our work, only the luminance channel is processed further due to the fact that it contains a large amount of energy. After the mean of the luminance vector is calculated, we empirically estimate the low and upper limits of the threshold as shown in lines 19–21 in Algorithm 2. From lines 20–21, the luminance is adjusted to finally obtain the output image (X') with a much better luminance that can be handled later by the developed license plate detection module. We empirically estimate the low and upper limits of the threshold as shown in line (20) in Algorithm 2. We set the value of $threshold_{low}$ to 0.25 and $threshold_{high}$ to 0.95. From lines 21–22, the luminance is adjusted to obtain the final neat and clean enhanced output image (X').

Algorithm 2: License plate detection procedure.

Input: Vehicle image confined by bounding box
For satisfactory luminance, **do;**
begin LP Localization
transform the vehicle-detected image to the HSV domain using Equations (1)–(3)
do segmentation
define HSV threshold limits for every channel
obtain mask images
If $H_{Image} \geq HT_{low}$ and $H_{Image} \leq HT_{high}$ **then** $H_{mask} = 1$
if $S_{Image} \geq ST_{low}$ and $S_{Image} \leq ST_{high}$ **then** $S_{mask} = 1$
If $V_{Image} \geq VT_{low}$ and $V_{Image} \leq VT_{high}$ **then** $V_{mask} = 1$
obtain the LP blob image (A)
if $HSV_{masks} = 1$
end segmentation
use mathematical morphology by Equations (4) and (5)
analyze dimensions through aspect ratio and LP spatial area
else
use PCA and form a luminance vector
calculate the luminance vector mean
approximate $threshold_{high}$ and $threshold_{low}$
If $mean > threshold_{high} \rightarrow$ decrease the luminance,
else If $mean < threshold_{high} \rightarrow$ increase the luminance,
Obtain improved output image (X')
end LP Localization
Output:I' = LP localization

Once the improved luminance image is obtained, the dimensions of the extracted regions are examined to locate the existence of a possible license plate. We analyze the dimensions of the license plate through its spatial area and aspect ratio. Finally, the LP module draws the green bounding box on connected regions, which outlines the existing LP in the image.

3.3. The LP Recognition

After a license plate is detected, normally the conventional LP identification methods segment the plate characters to recognize LP. These steps usually combine image processing techniques or video sequences, and their calculations depend on the true recognition rate and the error recognition rate. As discussed earlier, LP recognition is a difficult task due to the huge variety of plate formats and severely varying outdoor illuminations during the image acquisition phase. Many methods perform well in standard circumstances, for instance, controlled illuminations, restricted vehicle speeds, prespecified roads, and static backgrounds. Several algorithms have been designed to achieve LPR in images. In addition, issues such as processing time, computational complexity, and recognition rate are also important parts of the LPR algorithm. Algorithm 3 shows the pseudocode of the proposed LP recognition algorithm.

As can be seen in Algorithm 3, our developed method contains interconnected steps and performs miscellaneous operations after the LP bounding box is fed to the recognition module. Since the area contained by the LP is normally small, for better visibility, contrast is enhanced using contrast from basic image processing methods. The improved contrast image is binarized and segmented by applying the morphological operations using Equations (4) and (5), respectively.

Algorithm 3: The LP recognition pseudocode.

```
Input: LP bounding box
    begin operations
            Enhance contrast and deblur the image for better visibility
            Binarize the image obtained in the above steps
            Obtain segmented image (S) through dilation and erosion using Equations (4) and (5)
            Get Pre-trained model
              do
                   for S = 1:n
                        Perform prediction on S
                        Build output string
                 end
          end
     end operations
Output: Recognized LP characters
```

On the basic pretrained model, the LP characters are predicted to build the possible LP strings that may appear inside the LP bounding that was processed in the initial stage of the LP recognition module. Algorithm 3 generally depicts the core theme of the LP recognition scenario. All the operations used herein, such as contrast, deblurring, and binarizing the image, are essentially handy for the recognition task.

4. Simulation Results

To simulate, we use a workstation, which has one NVIDIA RTX 2070 GPU along with an Intel CPU-Corel i7-6700. Simulations are done in Python version 3.6.0. Below, we discuss in detail the performance of our proposed LP recognition algorithm.

4.1. Training Data Preparations and Model Training

Before our developed method is executed, we initially prepare the data and make some assumptions to train the model. Algorithm 4 shows the arrangements for preparing the training data. To extract the LP digits from the input image, basic data processing (DP) operations are performed from lines 2–13 of Algorithm 4. Most of these DP operations include desaturating the image through a Gaussian low-pass filter and binarizing the image. Moreover, the erosion and dilation operations described above are also performed. Meanwhile, the LP image is converted to 28×28 pixel image on which random spatial transformations are applied that ultimately result in a 28×28 dataset with prominent characters and their classes.

Since then, we have also performed experiments on the CCPD dataset, which has substantial license plate variations, such as tilted or blurred plates. For the tilted plates, spatial transformations are applied to the 28×28 pixels converted image. This operation essentially corrects the appearance of the license plate and ultimately makes the algorithm easier to process. Similarly, for poor image quality in which characters are not fully visible, characters touch each other due to blur or similar phenomena. In such conditions, mathematical morphological image processing techniques, such as erosion and dilation as described in Equations (4) and (5), respectively, become handy. All the operations listed in Algorithm 4 essentially prepare and result in well-managed, systematic data that is nicely processed by our developed algorithm during the recognition task.

Algorithm 4: Training data preparations.

```
Input: Single digits extracted from LPs
    begin DP
        Extract Single Digit
        Use desaturate
        Use De-Blurring
        Binarize the image
        Erode Image
        Dilate Image
        Convert to 28 × 28 image = (i)
          for s = 1:random (n)
              Perform Random Spatial Transform on (i)
              Save the image (i) and the character class to a dataset
          end
    end DP
Output: Dataset of 28 × 28 resolution with character classes
```

After LP character data is obtained, in the next step, training of the LP recognition model is performed as shown in Algorithm 5, which takes the 28 × 28 LP character image and yields the recognition model with weights. During the first part of the LP training, a 13-layer CNN is used to build a DNN. This DNN is then applied to a 3 × 3 Conv2D layer along with a 2 × 2 MaxPool layer. As shown in Algorithm 5, the next stages also apply a dense layer to perform the 50% dropout to obtain the appropriate model. During the model training, the LP characters are checked and predicted for a small batch of images. Meanwhile, to obtain good accuracy, weights are adjusted at regular intervals after each execution epoch. Once the training data and LP recognition model training are set, in the next section, we demonstrate our detailed observations and findings. Our LP recognition analysis and discussion are based on the PKU, AOLP, and CCPD datasets, which are well-known and widely used in research these days.

Algorithm 5: The LPR model training.

```
Input: Dataset of 28 × 28 Images with character classes
    begin LPR Training
        begin Model Design
            Apply the DNN with 13 Layers of the CNN
            for i = 1:3
                Conv2D 3 × 3
                MaxPool 2 × 2
            end
            Flatten the LP with a dropout of 50%
            for i = 1:3
                Use a dense Layer with a dropout of 50%
            end
        end Model Design
            begin Training
            for epoch = 1:n, get batch of images (i)
                for i = 1:n
                    Provide image to the model and check predicted characters adjust weights
                end
            end
    end LPR Training
Output: Model with set weights
```

4.2. Analysis of the PKU Dataset

During our study, we initiated our experiments on the PKU dataset, which is a well-known publicly available vehicle dataset. Table 1 briefly describes the various vehicle categories in the PKU dataset. Generally, the PKU dataset is a collection of diverse vehicle images that are captured under diverse conditions [31]. As shown in Table 1, this dataset contains a total of 3977 diverse vehicle images. The developers of the PKU dataset divided the vehicles into five distinct categories, which they refer to as G1, G2, G3, G4, and G5. Out of 3977 vehicle images, the PKU dataset also contains a total of 4263 visible license plates, whose pixel resolution varies from 20 to 62 pixels.

Table 1. The PKU dataset description.

Category	Vehicle Conditions	Input Image Resolution (Pixels)	No. of Images	No. of Plates	Plate Height (Pixels)
G1	Cars on roads; ordinary environment at different daytimes; contains only one license plate per image	1082 × 728	810	810	35–57
G2	Cars/trucks on main roads at different daytimes with sunshine; only one license plate in each image	1082 × 728	700	700	30–62
G3	Cars/trucks on highways during the night; one license plate per image	1082 × 728	743	743	29–53
G4	Cars/trucks on main roads; daytimes with reflective glare; one license plate in input images	1600 × 1236	572	572	30–58
G5	Cars/trucks at roads junctions with crosswalks with several plates per image	1600 × 1200	1152	1438	20–60
	Complete PKU dataset		**3977**	**4263**	**20~62**

In Figure 3, we demonstrate a few detection results for both vehicles and license plates for each category of the PKU dataset. We show different vehicles from each category to demonstrate a fair understanding.

Vehicle+LP detection: G1-category: The first row in Figure 3 demonstrates a few images from this category. It is evident for this category that for different-shaped vehicles, the detection module performs well by drawing a red rectangle around the object of interest, which is a vehicle in this case. The detected vehicle image is then analyzed by the LP localization module. In all four of the sample images in Figure 3 from the G1 category, the visible LP is accurately localized by our developed method.

Vehicle+LP detection: G2-category: The second row in Figure 3 demonstrates a few images from the G2 category. As indicated in Table 1, this category mostly contains vehicle images that are captured during different times of the day. In all four images shown for this category, both the vehicle and the LP localization module are in the correct position, thereby indicating the correct position of both of these objects. The first image shown for this category is of the truck, and the rest are the cars. However, the detectors applied to capture these objects are intelligent enough to discriminate between these shapes.

Vehicle+LP detection: G3-category: The third row in Figure 3 demonstrates a few images from the G3 category. Most of the images in this category are nighttime captures of small cars and trucks. It can be observed in the third row of Figure 3 that both objects are accurately located. To fairly discriminate the vehicle and the LP detection for nighttime captured images, we draw the white color bounding box around both the detected vehicle and the LP area.

Vehicle+LP detection: G4-category: The 4th row in Figure 3 demonstrates a few images from the G4 category. This category also contains one license plate in an image, but those are captured in a difficult situation of reflective glare that affects the image quality and the LP area appearance. However, in this case, our applied object detectors handle them efficiently. For each of the different images shown in the fourth row of Figure 3, the good performance of the applied detectors to localize both vehicles and the LP of that vehicle is evident.

Vehicle+LP detection: G5-category: The last row in Figure 3 demonstrates a few images from the G5 category. This category contains a few LPs in an image. As shown in the last row of Figure 3, all the instances of object detection are completely achieved. In particular, the first image in the fifth row shows the object from an angle, which is also correctly spotted by the applied detectors. The rest of the images in this row contain at least two vehicles along with two LPs that are accurately detected.

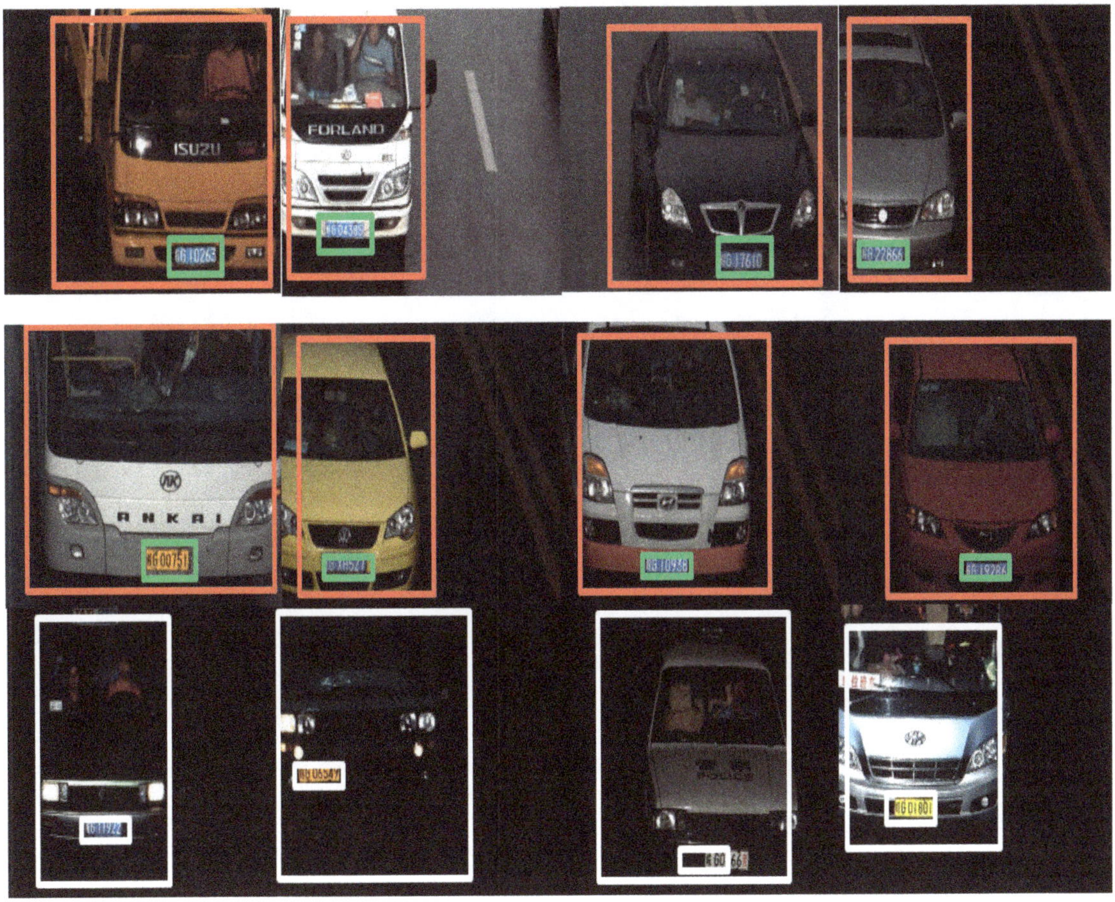

Figure 3. *Cont.*

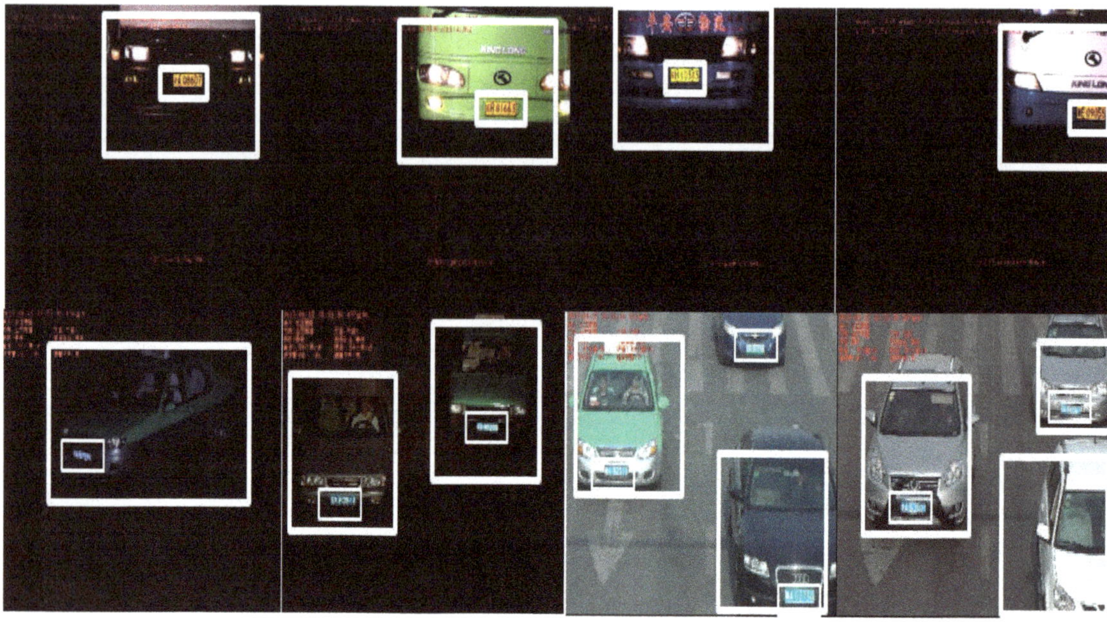

Figure 3. Vehicle+LP detection on PKU dataset.

Table 2 lists the comparison of each category of the PKU dataset for various methods mentioned therein. A few of the important findings from Table 2 are summarized below.

- Each of the compared methods along with our developed method yields 100% vehicle detection accuracy in the G1 and G2 categories, except the work developed in [33], whose vehicle detection accuracy is 99%. Similarly, for the G3 category, the method developed in [33,34] yields 98.20% and 99% vehicle detection accuracy. The remaining approaches all produce 100% vehicle detection results.
- In the G4 category, all of the methods compared can find vehicles with an accuracy of at least 99%. In this category, YOLO-v7-based methods [46,47] yield the highest vehicle detection accuracy of 99.74% and 99.72%, respectively. While for the G5 category, an improved YOLO-v7-based method ranks first, yielding 99.22% vehicle detection accuracy. Our developed method ranks 3rd and yields a vehicle detection result at par with [46] by delivering 99.10% detection accuracy.
- On the PKU dataset to locate vehicles, an improved YOLO-v7-based method ranks first and yields a mean vehicle detection accuracy of 99.79%, followed by standard YOLO-v7 [46], whose accuracy is 99.76%. Our developed method also yields approximately similar results as compared with [46]. Vehicle detection is a prototype in our developed system. Therefore, an accuracy of slightly over 99.75% is very encouraging in the later stages of the algorithm.
- Table 2 also lists the LP detection comparisons for several methods. As can be seen, the improved YOLO-v7 [46] ranks first in all five categories of the PKU dataset in terms of LP detection. The standard YOLO-v7 method [45] ranks 2nd in terms of license plate localization on this dataset. For the G1 and G2 categories, all of the methods compared had a LP detection accuracy of at least 97%, whereas, for the G3 category, the methods listed in Table 2 yielded at least 98% LP detection. For the G4 category, approximately 99% LP detection is achieved. The G5, which is the most challenging category in the PKU dataset, is also addressed nicely. In this category, the methods listed in Table 2 yield at least 98% accurate license plate detection. In addition, our method yields at least 99% LP detection for G1, G2, G3, and G4 categories. In the G5 category,

our finely tuned version of the Faster RCNN achieves 97.30% accurate license plate detection accuracy.
- Our analysis indicates that the mean LP detection accuracies of the works [32–34,45–47] are found to be 98.47%, 98.06%, 98.47%, 99.09%, 99.05%, and 99.13%, respectively. The aforementioned LP detection accuracies are a good indicator that all the compared methods yield at least 98% license plate detection accuracy. YOLO-based methods [45–47] perform well to locate an object, such as a vehicle or license plate. However, from Table 2, we find that our method, which employs a fine-tuned version of the Faster RCNN, yields a mean license plate detection accuracy of 99.04%. The aforedescribed analysis is a good indicator of the application of the various methods to achieve objects, such as vehicles and license plates, in various real-life applications. Vehicle and license plate detection is a prototype of our developed system. Therefore, our deployed detectors also yields at par results with the recently published works.

Table 2. Category-wise Vehicle + License Plate detection comparison (%) on PKU dataset.

		PKU Dataset Categories				
Object	**Ref**	**G1**	**G2**	**G3**	**G4**	**G5**
Vehicle	[32]	100	100	100	99	98.50
	[33]	99	98	98.20	99.10	98
	[34]	100	100	99	99.10	98
	[45]	100	100	100	98.96	99.13
	[46]	100	100	100	99.72	99.10
	[47]	100	100	100	99.74	99.22
	Proposed	**100**	**100**	**100**	**99.70**	**99.10**
License Plate	[32]	99	97.05	98.80	99	98.50
	[33]	97	98.01	98.20	99.10	98
	[34]	98.50	98.22	98.55	99.10	98
	[45]	98.80	99.45	99.15	98.96	99.13
	[46]	99.85	99.50	99.22	99.35	97.35
	[47]	99.87	99.65	99.40	99.40	97.35
	Proposed	**99.81**	**99.50**	**99.20**	**99.40**	**97.30**

With the state-of-the-art method listed in Table 2, detection accuracy is almost at par with that of conventional methods. After the objects, which in our case are vehicles and LPs, are located, in the next phase we process the detected LP area for recognition. It is important to state that in the PKU dataset, all the visible license plate labels are not annotated. Therefore, in the PKU dataset, we labeled the 2250 images. The 1355 images are randomly selected for training, and the other 901 are used for testing. To evaluate license plate recognition accuracy, the license plate was localized by a bounding box as shown in Figure 3 for each category of the PKU dataset. The detected license plate is now fed to our newly developed recognition module.

As shown in Figure 4, the proposed LP recognition technique correctly understands different LPs that appear in each of the five categories of the PKU dataset. The important points noted during the LP recognition task are discussed further below.

LP recognition: G1-category: As shown in the first row in Figure 4, the proposed recognition algorithm precisely identifies the LPs shown therein. Our obtained correct recognition result is shown on top of the original LP on the input vehicle images. The third image in the first row of Figure 4 has a relatively complex background. However, it does not pose any threat to the proposed method of achieving the correct identification result.

LP recognition: G2-category: As shown in the second row in Figure 4, the first three images have different car colors with their own installed LPs. Our method correctly identifies all such cases. However, the fourth image of the bus with visible LP has a relatively complex background. Nevertheless, the proposed method handles this scenario well and achieves the correct result on top of the original LP shown therein.

LP recognition: G3-category: As shown in the third row in Figure 4, the proposed LPR method reliably handles the high-glare images. The LPs on the vehicles in the first two pictures in this row are clear enough to be correctly identified. Similarly, it is clear from this row that our developed method handles low-contrast images in which both the vehicles and the background have blackish appearances. Generally, it is observed in the third row of Figure 4 that our developed method has barely any effect on its recognition performance with blackish objects against a black background.

LP recognition: G4-category: As shown in the fourth row in Figure 4, the area around the vehicles is highly dark. There also appear to be glare and high beams from vehicles. However, in all four images shown for this category in Figure 4, our developed method accurately identifies all the LP numbers and successfully handles the glare situations.

LP recognition: G5-category: As shown in the fifth row in Figure 4, there appear to be multiple vehicles and LPs in the images. For all the images shown, our developed method identifies all the LP that appear in the images. In the second and fourth images, there appear to be three LPs. In the fourth image, our method identifies all three LPs, whereas, in the second image, only two LPs are detected out of three. One reason is the red text that appears in the input image around the LP area, which created a hurdle for our developed method.

Figure 4. *Cont.*

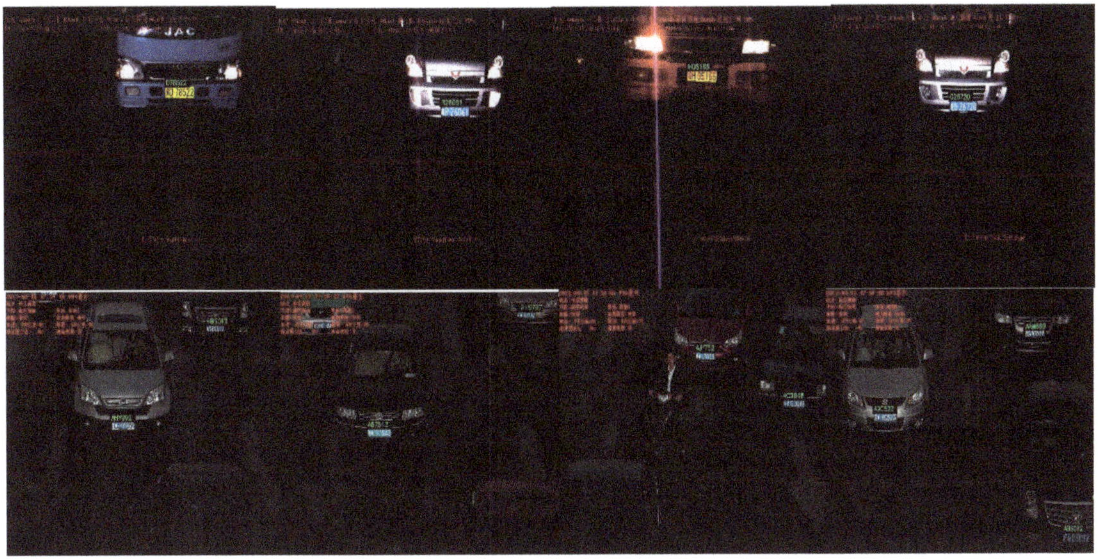

Figure 4. The LP recognition on different images of the PKU dataset.

Table 3 lists the LP recognition rate for each of the PKU categories for works developed in [32–34]. It is important to state here that these methods were chosen for comparison on the PKU dataset because their standard implementation is publicly available. This makes it logical to train these models on the PKU dataset along with our developed method.

Table 3. Category-wise LP recognition accuracy comparison (%) on PKU dataset.

Ref	G1	G2	G3	G4	G5
[32]	96	97.80	92.60	80.00	72.00
[33]	92.00	90.50	90.10	89.60	86.40
[34]	98.00	98.50	90.00	86.01	81.10
Proposed	**100**	**100**	**100**	**99**	**99.63**

From Table 3, it is evident that all the compared methods yield over 90% recognition accuracy for the G1 category. Our developed method yields 100% LP recognition accuracy in this category. The work developed by Zhang et al. [34] ranks second, yielding 98% accurate recognition. For the G2 category, the work developed by Masood et al. [33] yields the lowest LP recognition accuracy of 90.50%. In this category, our developed method ranks first, followed by the work reported in [34]. For the G3 category, the works on [33] and [34] yield almost similar results by producing at least 90% recognition accuracy. In this category, the work developed by Xu et al. [32] also reports 92.60% LP recognition. For the G4 category, we observe that works in [32–34] yield below 90% LP recognition. In this category, our developed method comprehensively outperforms the compared methods. For the G5, which is the most difficult category of the PKU dataset, the work in [32] produces the least accuracy of 72%, followed by [34], whose accuracy is a bit over 80%. In this category, our developed method yields 99.63% recognition accuracy.

In Figure 5, we report the mean license plate recognition accuracy on the PKU dataset. Our proposed LPR method comprehensively beats the compared methods in terms of mean recognition accuracy. As shown in Figure 5, our developed license plate recognition method yields 99.63% accuracy on the PKU dataset. Similarly, the work proposed in [34] ranks second, yielding 90.72% accuracy. On the aforementioned dataset, the work proposed in [32] yields the lowest license plate recognition accuracy of 79.28%. To the best of our knowledge, on the PKU vehicle dataset, the proposed method has almost solved the LPR accuracy problem.

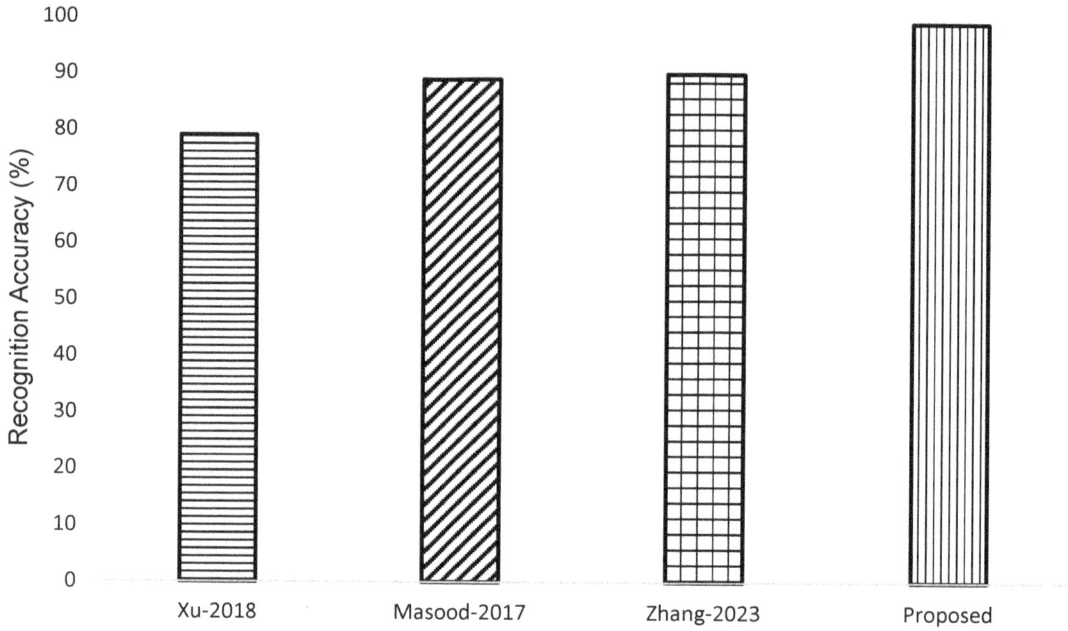

Figure 5. Mean LP recognition comparison on the PKU dataset.

4.3. *Analysis of the AOLP Dataset*

The application-oriented license plate (AOLP) [35] database consists of 2049 images of a Taiwan license plate. This dataset is categorized into three subsets according to complexity levels and photographing conditions, which are access control (AC), traffic law enforcement (TLE), and road patrol (RP). For the readers' information, below we briefly describe the categories contained in the AOLP dataset.

Access Control (AC): The AC refers to the cases in which a vehicle passes a fixed passage at a reduced speed or with a full stop, such as at a toll station or the entrance/exit of a region.

Traffic Law Enforcement (TLE): The TLE refers to cases where a vehicle travels at a regular or higher speed but violates traffic laws, such as a traffic signal or speed limit, and is captured by a roadside camera. Here, 757 images were collected for this application category.

Road Patrol (RP): The RP refers to the cases where the camera is installed or handheld on a patrolling vehicle and takes images of the vehicles from arbitrary viewpoints and distances. Since we do not have any other images with Taiwan license plates, we use any two of these subsets for training and the remaining one for testing, similar to previous practices. Figure 6 shows our obtained results on the AOLP dataset.

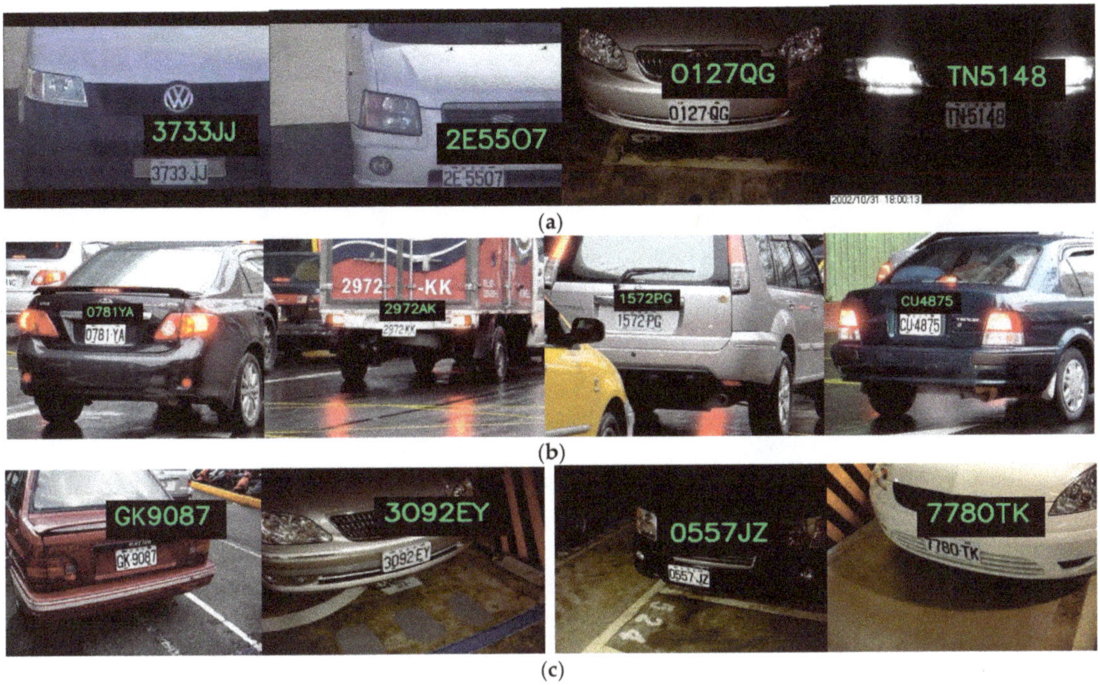

Figure 6. The LPR on the AOLP dataset is organized under (**a**) Access control, (**b**) Traffic law enforcement, and (**c**). Road patrol conditions.

Figure 6 shows the license plate recognition results for each of the aforedescribed categories of the AOLP dataset. The proposed LPR method works well for scenarios where half of the vehicle bonnet is visible along with the license plate location, which is much lower on the horizontal axis. In each of the images in Figure 6a for the AC category, the proposed method accurately identifies the license plates. Similarly, for the TLE category, as shown in Figure 6b, where the license plates appear in the angular view, the proposed method accurately handles this angle variation by correctly identifying all the license plates. The third image in Figure 6b is especially interesting, as here a yellow vehicle appears at the back side of the license plate, which ultimately results in the partial occlusion of the license plate. Although it does not affect the digits of the plate area, the proposed LPR algorithm handles this partial occlusion and accurately identifies the license plate. Figure 6c shows the RP conditions. Clearly, this is a challenging category as there appears to be a large angle deviation of the viewpoint of the license plate, which makes this scenario challenging for most of the machine learning algorithms. However, as can be seen in Figure 6c, the proposed LPR method reliably handles this issue by indicating the correct number on the license plate.

Table 4 lists the LP recognition rate on different classes of the AOLP dataset for works developed in [36–39]. It is important to state here that these methods were chosen for comparison on the AOLP dataset because their evaluations on this dataset, along with standard implementation, are publicly available. This makes a fair reason for us to train these methods on the AOLP dataset along with our developed method. Table 4 also lists the comparison of the proposed LPR with a few recent methods on the AOLP dataset. As can be seen in Table 4, for the AC category, the proposed method yields the highest recognition rate of the license plates in this category.

Table 4. Comparison of the AOLP dataset.

Method	Accuracy % on Each Category			
	AC: No of Images = 681	TLE: No of Images = 757	RP: No of Images = 611	Mean Recognition Accuracy %
[36]	94.9000	94.2000	88.4000	92.5000
[37]	95.3000	96.6000	83.7000	91.8666
[38]	96.6000	97.8000	91.0000	95.1333
[39]	97.3000	98.3000	91.9000	95.8333
Proposed	**97.8970**	**98.2719**	**91.9006**	**96.0231**

Moreover, in the AC category, the work proposed in [39] ranks second among the compared methods. Similarly, for the TLE category, the proposed LPR method ranks second on the AOLP dataset. In this category, the work reported in [39] yields the highest recognition accuracy. However, the work in [36] ranks fourth among all compared methods, yielding slightly over 94% identification accuracy. For the RP category, the method reported in [39] and the proposed method yield almost similar identification accuracy of slightly over 91%, despite the fact that the proposed method is a bit higher. As indicated by the last column in Table 4, the proposed license plate recognition method yields the highest license plate recognition accuracy of 96.0231% on the AOLP dataset. The work listed in [39] ranks second in achieving overall identification accuracy, followed by [38]. In general, and across the whole AOLP dataset, all of the methods compared correctly identify license plates over 91% of the time.

4.4. Analysis of the CCPD Dataset

The CCPD dataset [40] is the largest publicly available LP dataset and has a collection of over 290,000 Chinese LP images. This dataset is separated into several categories according to the difficulty of identification, for instance, the illuminations on the LP area, the distance from the license plate when photographing, and the degree of horizontal and vertical tilts. The CCPD dataset also contains images in different weather conditions, such as rainy, snowy, or foggy. Each category includes 10,000 to 20,000 images. The CCPD-base consists of approximately 200,000 images, of which 100,000 are used for training and the other half are for testing. As listed in Table 5, the other sub-datasets, such as the CCPD-DB, the CCPD-FN, the CCPD-rotate, the CCPD-weather, and the CCPD-challenge, are also used during the test phase.

Table 5. Comparison of the CCPD dataset.

Model	CCPD-Base (100 k)	CCPD-DB (20 k)	CCPD-FN (20 k)	CCPD-Rotate (10 k)	CCPD-Tilt (10 k)	CCPD-Weather (10 k)	CCPD-Challenge (10 k)	Overall Accuracy (%)
[40]	98.5000	96.9000	94.3000	90.8000	92.5000	87.9000	85.1000	95.5000
[41]	99.1000	96.3000	97.3000	95.1000	96.4000	97.1000	83.2000	93.0000
[42]	99.5000	98.1000	98.6000	98.1000	98.6000	97.6000	86.5000	98.3000
[43]	98.9000	96.1000	96.4000	91.9000	93.7000	95.4000	83.1000	96.6000
[44]	99.6000	98.8000	98.8000	96.4000	97.6000	98.5000	88.9000	98.5000
Proposed	**99.8500**	**98.7800**	**98.8000**	**98.1100**	**98.8000**	**98.9000**	**88.8000**	**98.7000**

Figure 7a shows a few samples of the output images on the CCPD-base images. Clearly, the proposed method performs well on all images. Particularly, the left-most image has huge illumination variations with very limited visible contrast in the license plate area. The proposed method handles that scenario well and correctly identifies the license plate. Similarly, the second, third, and fourth images in the top row of Figure 7a are the cases where the license plate appears in the angular view. However, our proposed method handles this scenario and identifies all the license plates. Figure 7b shows the CCPD-blur image output of our developed method. Most of these blurred images were captured in outdoor conditions with strong sunlight and complex backgrounds. Since these images appear blurry, the license plate area has a low resolution. However, it can be seen in the second row of Figure 7b that our developed method performs significantly well and identifies all the license plates shown therein in the second row. Particularly, the first and third images in the second row of Figure 7b are indicative of the good performance of our developed method where the background is complex along with various other objects. Moreover, the third row in Figure 7c is the sample output of our proposed method for the CCPD-FN cases. Clearly, in this case, our developed method is quite accurate and reliably identifies all the license plates shown therein. It is to be noted that the third row in Figure 7 also contains complex backgrounds. However, the good performance of our developed method is unaffected by these factors.

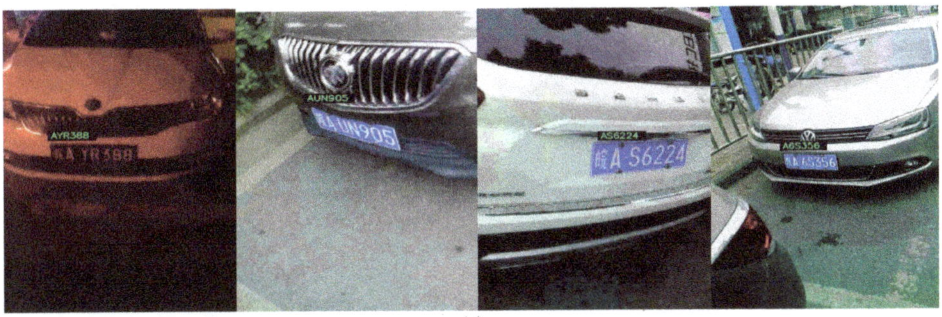

(**a**)

(**b**)

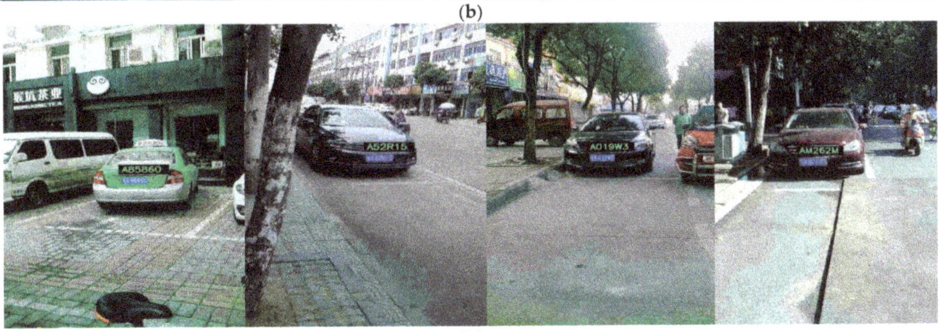

(**c**)

Figure 7. The LPR on the CCPD dataset for (**a**) base, (**b**) blur, and (**c**) FN scenarios.

A more detailed analysis of our developed method is shown in Figure 8. As can be seen, the outputs in the first row of Figure 8 are from the CCPD-rotate category. Particularly, the first image has a rotated license plate along with an overly whitish appearance due to the presence of very strong sunlight. Clearly, the developed method handles such a scenario and accurately identifies the license plate. The fourth image in the first row of Figure 8 has a relative combination of dark and bright contrast. Overall, the proposed method performs well in the CCPD-rotate category and, as seen in Table 5, produces encouraging results. The second row in Figure 8 shows the license plate identification resultant images from the CCPD-tilt category. The first image in the third row of Figure 8 is a low–contrast image example that has severe black contrast. It can be seen that our developed method is unaffected by this situation and accurately identifies the license plate. Similarly, the last image in the third row of Figure 8, which has a slightly misplaced license plate, is highly challenging in the tilt category. However, our developed method also handles this case intelligently and produces accurate output.

More output resultant images from the CCPD dataset are shown in the third row of Figure 8, where a few cases are shown for the different weather conditions. The first three images in the third row of Figure 8 correspond to the snowy weather where our developed method reports accurate results, whereas the fourth image is for the rainy day in which our developed method performs at par and yields accurate results. The fourth row in Figure 8 is for the outputs generated by the algorithm for the CCPD challenge category.

During simulations, we find that this is the most challenging category in the dataset, and it is not easy for every algorithm to handle this. The first image shown in the last row of Figure 8 indicates that both the vehicles and the outside environment are severely dark. However, our developed method handles this scenario and yields accurate recognition results. The same is true for the third image in the last row, where our approach accurately identifies and identifies the license plate. Similarly, for the 2nd image in the last row of Figure 8, there appears to be a shadow on the road and the vehicle, and there is also a bright light in the center of the license plate. However, our developed method passes through this hurdle and yields the correct result. Likewise, the rightmost bottom image in Figure 8 is the case where there are back lights turned on, and half of the license plate has a blue background with white color text on it while the other half has a light grey background with yellow text over it. Consequently, our established approach delivers accurate and encouraging results in this case.

Table 5 lists the LP recognition rate on different classes of the CCPD dataset for works developed in [40–44]. It is important to state here that these methods were chosen for comparison on the AOLP dataset because their evaluations on this dataset, along with standard implementation, are publicly available. This makes a fair reason for us to train these methods on the AOLP dataset along with our developed method. In Table 5, we show the comparison of our developed method with these five techniques on the CCPD dataset for all the categories. It can be seen that, for the CCPD-Base category, our method ranks second out of all the compared methods. In this category, the work reported by [44] has the highest accuracy. In this category, the work conducted in [40] has the least recognition accuracy. For the CCPD-DB category, our method follows [44] and lies in the second position. Here, the work in [41] has the lowest accuracy. For the CCPD-FN category, our method and [44] have the highest license plate recognition accuracy, followed by the work done in [42].

For the CCPD-rotate category, our developed technique beats the compared works and yields the highest identification accuracy of license plates. In this category, the work done in [43] yields the lowest identification rates. Moreover, for the CCPD-Tilt category, our method has the highest recognition accuracy, followed by the work in [40], which has the lowest reported license plate identification rate. For the CCPD-weather category, our method again beats the compared works. Here, the work in [40] has the lowest recognition rates. Furthermore, for the CCPD challenge category, the work presented in [44] has the highest license plate identification rate and [41] has the lowest. In this category, we again rank second out of the compared methods. Our developed method yields the highest overall license plate recognition accuracy, with a 98.7000% correct recognition rate. The work performed in [44] ranks second, and the work reported in [40] lies in the third spot. Overall, the work reported in [41] has the lowest license plate identification rate of 93%.

(**a**)

(**b**)

(**c**)

(**d**)

Figure 8. The LPR on the CCPD includes (**a**) rotate, (**b**) tilt, (**c**) weather, and (**d**) challenge scenarios.

4.5. Computational Complexity

To perform the computational analysis of our developed method and the compared methods, we manually cropped the image resolutions. In our experiments, we selected different image resolutions, which are 700 × 1100, 500 × 800, 400 × 600, 320 × 240, and 300 × 280 pixels. From the compared works in this manuscript, we choose ten methods and executed them on the aforesaid image resolutions. Complete results are detailed in Figure 9. It is evident that the work reported by Yu et al. [41] and Li et al. [36] is computationally complex and consumes more than 3 ms to process the image resolution of 700 × 1100 to yield the final recognition result. Moreover, the works of Yuan et al. [34], Luo et al [42], and Wang et al. [43] also consume more than 2.5 ms to process the aforedescribed image resolution to generate the final resultant image. The works reported in Masood et al. [33] and Wu et al. [38] are computationally efficient and consume nearly 0.5 ms to process the test image for various image resolutions. Therefore, we observe that our developed technique takes slightly over 2 ms to deliver the final result. In terms of the execution time ranking, our developed method ranks fourth out of all of the compared methods. We observe that all the compared methods are near real-time for processing various image resolutions. Once an algorithm is trained on every dataset, our developed method along with other methods can be used in a resource-constrained environment, as we see that all the methods explored in this study work in near real-time in actual living environments with high accuracy.

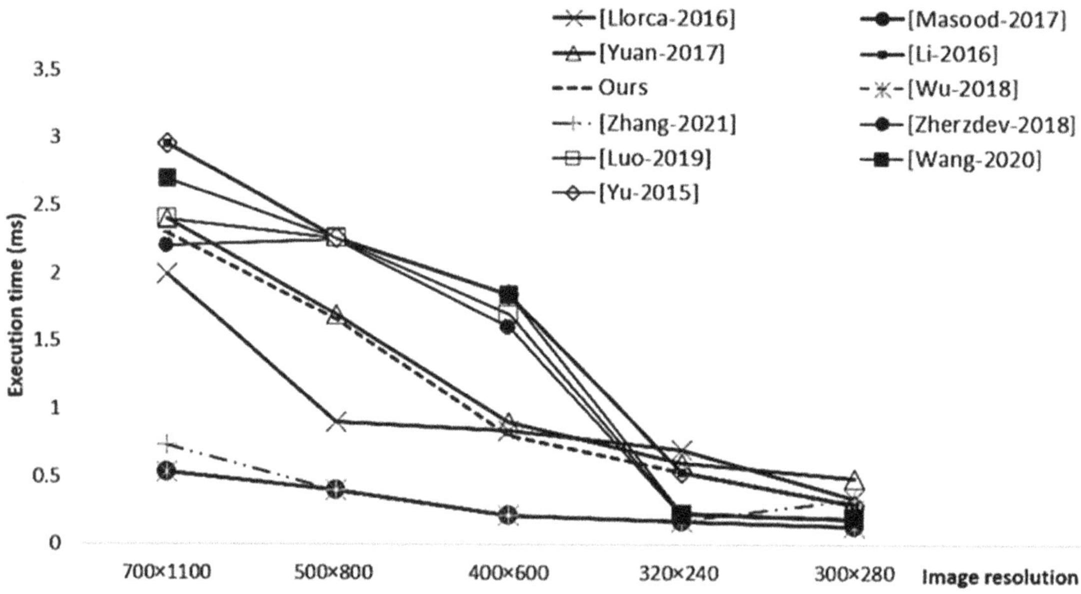

Figure 9. Computational complexity [32–44].

4.6. Discussion

Detailed simulations shown in this paper indicate that object detection, such as vehicle or license plate detection, has been an active research field in recent years. This paper presented a detailed analysis of license plate recognition on three publicly available datasets. For the task of vehicle detection, a Faster RCNN architecture was used. The license plate was located and recognized through our own developed methods. Our findings are indicative of superior outputs on challenging datasets. Moreover, a detailed comparison of our developed method was carried out with several state-of-the-art license plate recognition approaches. We are optimistic that this study will be a fair guideline for beginners and practitioners to modify or use any detector or recognizer for their desired tasks or applications. The outcomes of our developed system for recognizing license plates are summarized below.

PKU Dataset:

On this dataset, our developed method yielded 100% recognition accuracy in the G1, G2, and G3 categories. In the G4 category, our developed method was 99% successful at accurately recognizing the license plate. Finally, in the G5 category, our developed method yielded 99.63% recognition accuracy. Overall, on the PKU dataset, our developed method ranks first out of the three compared methods in terms of license plate recognition accuracy.

AOLP Dataset:

This dataset contains three challenging categories, which are access control, traffic law enforcement, and road patrol. On access control, our developed method yielded 97.8970% accurate recognition accuracy and ranked first herein. On traffic law enforcement, our developed method yielded 98.2719% license plate recognition accuracy and ranked second among the compared methods. On the road patrol category, our developed method generated a mean recognition accuracy of 91.9006% and ranked first among the four compared methods. The whole-mean accuracy on the AOLP dataset by our developed method is 96.0231%.

CCPD Dataset:

This is the largest publicly available license plate dataset and contains challenging scenarios, such as blur, rotation, tilt, and varying weather. On this dataset, our developed method yielded a mean recognition accuracy of 98.7000% and ranked first among all compared methods. In general, for all the other aforementioned categories, our developed method yielded over 98% recognition accuracy. However, for the CCPD challenge category, our developed method yielded slightly over 88% recognition accuracy and ranked second among the five compared methods.

4.7. Limitations

As with any other algorithm for machine learning, we discovered several shortcomings and failures in our method. Figure 10 depicts a handful of these instances with the following observations:

Figure 10. Few failures cases examples.

- It is clear from the rightmost image in the first row of Figure 10 that the input image is extremely blurry with a non-clear license plate. In such a case, our developed method struggles to distinguish the actual words and reads "A" from the license plate as "0".
- Similar is the case for the next two images in the first row of Figure 10. We also observe that there is no specific rule for license plate fonts. Therefore, such cases are very hard to identify correctly. As shown in the first image in the second row of Figure 10, the extreme blur is also a very challenging situation for any algorithm to deal with.
- We observe that occlusion, either partial or full, is also a challenging factor for the machine learning-based license plate identification method. One such case is shown in the third image in the second row of Figure 10, where high intensity light beams have created occlusion in the license plate area and thereby a hurdle for the algorithm to handle with. Therefore, before processing the license plate, such factors should be carefully analyzed.
- We also note that light that falls on the license plate area due to reflection from the vehicle's surface also reduces the recognition ability of the algorithm. One such case is seen in the middle image of the second row in Figure 10. Therefore, before a test license plate is fed to the recognition algorithm, this issue should also be noted. In such cases, an image enhancement or contrast rectification method might be useful to improve the quality of the appearance of the license plate.

5. Conclusions

Accurate detection and recognition of vehicle license plates in natural scene images is an important task to be performed by machine learning algorithms. Nowadays, it is an integral part of modern intelligent traffic control systems. However, this task is quite challenging due to various factors, for instance, the non-uniform patterns of the plates, variations in view angles, such as blurriness, and the occlusions. With such factors, it is always difficult for a single algorithm to handle the aforedescribed issues. This paper discussed methods to detect and identify license plates that appear in an image. The proposed method is composed of three distinct but interconnected steps: (i) vehicle detection, (ii) license plate detection, and (iii) license plate recognition. To locate the vehicles, a fine-tuned version of the Faster RCNN was used, while the license plate area was located through our own developed plate localization module. Finally, the recognition task is achieved using the deep learning network. Simulations were performed on three databases, which are the PKU, the AOLP, and the CCPD license plate dataset. Our proposed method achieves competitive performance and yields 99%, 96.0231%, and 98.7000% recognition rates on the aforesaid datasets. We are optimistic that our findings are promising and will be applicable to a variety of real-world applications, including surveillance and the monitoring of suspicious vehicles.

In the future, the proposed method could be modified to handle extreme blurriness. Similarly, the proposed method could also be improved to handle occlusion. Moreover, the developed algorithm could also be made intelligent by being trained in parallel over various time intervals.

Author Contributions: Conceptualization, F.S. and Z.M.; methodology, F.S.; software, F.S., Y.A.S., K.K., M.S., U.K. and Z.M.; validation, F.S.; formal analysis, F.S., Y.A.S., K.K., M.S., U.K. and Z.M.; investigation, F.S. and Z.M.; resources, F.S., Y.A.S., K.K., M.S., U.K. and Z.M.; data curation, F.S.; writing—original draft preparation, F.S., Y.A.S., K.K., M.S., U.K. and Z.M.; writing—review and editing, F.S., Y.A.S., K.K., M.S., U.K. and Z.M.; visualization, F.S. and Z.M.; supervision, Z.M.; project administration, Z.M. All authors have read and agreed to the published version of the manuscript.

Funding: This research received no external funding.

Institutional Review Board Statement: Not applicable.

Informed Consent Statement: Not applicable.

Data Availability Statement: Not applicable.

Conflicts of Interest: The authors declare no conflict of interest.

References

1. Fan, X.; Zhao, W. Improving robustness of license plates automatic recognition in natural scenes. *IEEE Trans. Intell. Transp. Syst.* **2022**, *23*, 18845–18854. [CrossRef]
2. Xie, L.; Ahmad, T.; Jin, L.; Liu, Y.; Zhang, S. A new CNN-based method for multi-directional car license plate detection. *IEEE Trans. Intell. Transp. Syst.* **2018**, *19*, 507–517. [CrossRef]
3. Soon, F.C.; Khaw, H.Y.; Chuah, J.H.; Kanesan, J. PCANetbased convolutional neural network architecture for a vehicle model recognition system. *IEEE Trans. Intell. Transp. Syst.* **2019**, *20*, 749–759. [CrossRef]
4. Lu, L.; Huang, H. A hierarchical scheme for vehicle make and model recognition from frontal images of vehicles. *IEEE Trans. Intell. Transp. Syst.* **2019**, *20*, 1774–1786. [CrossRef]
5. Farid, A.; Hussain, F.; Khan, K.; Shahzad, M.; Khan, U.; Mahmood, Z. A Fast and Accurate Real-time Vehicle Detection Method Using Deep Learning for Unconstrained Environments. *Appl. Sci.* **2023**, *13*, 3059. [CrossRef]
6. Selmi, Z.; Halima, M.B.; Alimi, A.M. Deep learning system for automatic license plate detection and recognition. In Proceedings of the 2017 14th IAPR International Conference on Document Analysis and Recognition (ICDAR), Kyoto, Japan, 9–15 November 2017; pp. 1132–1138.
7. Shi, J.W.; Zhang, Y. License plate recognition system based on 476 improved YOLOv3 and BGRU. *Comput. Eng. Des.* **2020**, *41*, 2345–2351.
8. Silva, S.M.; Jung, C.R. Real-time license plate detection and recognition using deep convolutional neural networks. *J. Vis. Commun. Image Represent.* **2020**, *71*, 102773. [CrossRef]
9. Lee, Y.; Yun, J.; Hong, Y.; Lee, J.; Jeon, M. Accurate license plate recognition and super-resolution using a generative adversarial networks on traffic surveillance video. In Proceedings of the International Conference on Consumer Electronics-Asia (ICCE-Asia), Jeju, Republic of Korea, 24–26 June 2018; pp. 1–4.
10. Mahmood, Z.; Khan, K.; Khan, U.; Adil, S.H.; Ali, S.S.A.; Shahzad, M. Towards Automatic License Detection. *Sensors* **2022**, *22*, 1245. [CrossRef]
11. Jia, W.; Zhang, H.; He, X. Region-based license plate detection. *J. Netw. Comput. Appl.* **2007**, *30*, 1324–1333. [CrossRef]
12. Silva, S.M.; Jung, C.R. License Plate Detection and Recognition in Unconstrained Scenarios. In *Computer Vision–ECCV 2018 Lecture Notes in Computer Science*; Ferrari, V., Hebert, M., Sminchisescu, C., Weiss, Y., Eds.; Springer: Cham, Switzerland; Volume 11216, pp. 593–609. [CrossRef]
13. Laroca, R.; Severo, E.; Zanlorensi, L.A.; Oliveira, L.S.; Goncalves, G.R.; Schwartsz, W.R.; Menotti, D. A Robust Real-time Automatic License Plate Recognition based on the YOLO Detector. In Proceedings of the International Joint Conference on Neural Network (IJCNN), Rio de Janeiro, Brazil, 8–13 July 2018.
14. Hsu, G.S.; Ambikapathi, A.; Chung, S.L.; Su, C.P. Robust license plate detection in the wild. In Proceedings of the 2017 14th IEEE International Conference on Advanced Video and Signal Based Suveillance (AVSS), Lecce, Italy, 29 August–1 October 2017; pp. 1–6.
15. Selmi, Z.; Halima, M.B.; Pal, U.; Alimi, M.A. DELP-DAR system for license plate detection and recognition. *Pattern Recognit. Lett.* **2020**, *129*, 213–223. [CrossRef]
16. Simonyan, K.; Zisserman, A. Very deep convolutional networks for large-scale image recognition. In Proceedings of the International Conference on Learning Representations (ICLR), San Diego, CA, USA, 7–9 May 2015.
17. Bulan, O.; Kozitsky, V.; Ramesh, P.; Shreve, M. Segmentation and annotation-free license plate recognition with deep localization and failure identification. *IEEE Trans. Intell. Transp. Syst.* **2017**, *18*, 2351–2363. [CrossRef]
18. Zhu, J.Y.; Park, T.; Isola, P.; Efros, A. Unpaired image-to-image translation using cycle-consistent adversarial networks. In Proceedings of the 2017 IEEE International Conference on Computer Vision (ICCV), Vanice, Italy, 22–29 October 2017; pp. 2223–2232.
19. Li, H.; Wang, P.; Shen, C.; Zhang, G. Show, attend and read: A simple and strong baseline for irregular text recognition. In Proceedings of the AAAI Conference on Artificial Intelligence, Honolulu, HI, USA, 27 January–1 February 2019; pp. 8610–8617.
20. Wang, X.; Man, Z.; You, M.; Shen, C. Adversarial generation of training examples: Applications to moving vehicle license plate recognition. *arXiv* **2017**, arXiv:1707.03124.
21. Xu, H.; Zhou, X.-D.; Li, Z.; Liu, L.; Li, C.; Shi, Y. EILPR: Toward End-to-End Irregular License Plate Recognition Based on Automatic Perspective Alignment. *IEEE Trans. Intell. Transp. Syst.* **2021**, *23*, 1–10. [CrossRef]
22. Wang, Y.; Bian, Z.; Zhou, Y.; Chau, L.-P. Rethinking and designing a high-performing automatic license plate recognition approach. *IEEE Trans. Intell. Transp. Syst.* **2022**, *23*, 8868–8880. [CrossRef]
23. Qiuying, H.; Cai, Z.; Lan, T. A single neural network for mixed style license plate detection and recognition. *IEEE Access* **2021**, *9*, 21777–21785.
24. Yogheedha, K.; Nasir, A.; Jaafar, H.; Mamduh, S. Automatic vehicle license plate recognition system based on image processing and template matching approach. In Proceedings of the International Conference on Computational Approach in Smart Systems Design and Applications (ICASSDA), Serawak, Malaysia, 15–17 August 2018; pp. 1–8.
25. Chris, H.; Ahn, S.Y.; Lee, S.W. Multinational license plate recognition using generalized character sequence detection. *IEEE Access* **2020**, *8*, 35185–35199.

26. Khan, A.M.; Awan, S.M.; Arif, M.; Mahmood, Z.; Khan, G.Z. A Robust Segmentation Free License Plate Recognition Method. In Proceedings of the1st International Conference on Electrical, Communication and Computer Engineering (ICECCE), Swat, Pakistan, 24–25 July 2019; pp. 1–6.
27. Tourani, A.; Shahbahrani, A.; Soroori, S.; Khazaee, S.; Suen, C.Y. A robust deep learning approach for automatic iranian vehicle license plate detection and recognition for surveillance systems. *IEEE Access* **2020**, *8*, 201317–201330. [CrossRef]
28. Shashirangana, J.; Padmasiri, H.; Meedeniya, D.; Perera, C. Automated license plate recognition: A survey on methods and techniques. *IEEE Access* **2020**, *9*, 11203–11225. [CrossRef]
29. Hassaballah, M.; Kenk, M.; Muhammad, K.; Minaee, S. Vehicle detection and tracking in adverse weather using a deep learning framework. *IEEE Trans. Intell. Transp. Syst.* **2020**, *22*, 4230–4242. [CrossRef]
30. Ren, S.; He, K.; Ross, G. Faster R-CNN: Towards real-time object detection with region proposal networks. *IEEE Trans. Pattern Anal. Mach. Intell.* **2017**, *39*, 1135–1149. [CrossRef]
31. Hsieh, J.-W.; Yu, S.-H.; Chen, Y.-S. Morphology-based license plate detection from complex scenes. In Proceedings of the 2002 International Conference on Pattern Recognition, Quebec City, QC, Canada, 11–15 August 2002; Volume 3.
32. Llorca, D.F.; Salinas, C.; Jimenez, M.; Parra, I.; Morcillo, A.G.; Izquierdo, R.; Lorenzo, J.; Sotelo, M.A. Two-camera based accurate vehicle speed measurement using average speed at a fixed point. In Proceedings of the 2016 IEEE 19th International Conference on Intelligent Transportation System (ITSC), Rio de Janeiro, Brazil, 1–4 November 2016; pp. 2533–2538.
33. Masood, S.Z.; Shu, G.; Dehghan, A.; Ortiz, E.G. License plate detection and recognition using deeply learned convolutional neural networks. *arXiv* **2017**, arXiv:1703.07330.
34. Yuan, Y.; Zou, W.; Zhao, Y.; Wang, X.; Hu, X.; Komodakis, N. A robust and efficient approach to license plate detection. *IEEE Trans. Image Process.* **2017**, *26*, 1102–1114. [CrossRef] [PubMed]
35. Hsu, G.S.; Chen, J.C.; Chung, Y.Z. Application-Oriented License Plate Recognition. *IEEE Trans. Veh. Technol.* **2013**, *62*, 552–561. [CrossRef]
36. Li, H.; Shen, C. Reading car license plates using deep convolutional neural networks and LSTMs. *arXiv* **2016**, arXiv:1601.05610.
37. Li, H.; Wang, P.; Shen, C. Toward end-to-end car license plate detection and recognition with deep neural networks. *IEEE Trans. Intell. Transp. Syst.* **2019**, *20*, 1126–1136. [CrossRef]
38. Wu, C.; Xu, S.; Song, G.; Zhang, S. How many labeled license plates are needed. In Proceedings of the Chinese Conference on Pattern Recognition and Computer Vision (PRCV), Guangzhou, China, 23–26 November 2018; Springer: Cham, Switzerland, 2018; pp. 334–346.
39. Zhang, L.; Wang, P.; Li, H.; Li, Z.; Shen, C.; Zhang, Y. A robust attentional framework for license plate recognition in the wild. *IEEE Trans. Intell. Transp. Syst.* **2021**, *22*, 6967–6976. [CrossRef]
40. Xu, Z.; Yang, W.; Meng, A.; Lu, N.; Huang, H.; Ying, C.; Huang, L. Towards end-to-end license plate detection and recognition: A large dataset and baseline. In Proceedings of the European Conference on Computer Vision, Munich, Germany, 8–14 September 2018; pp. 255–271.
41. Zherzdev, S.; Gruzdev, A. LPRNet: License plate recognition via deep neural networks. *arXiv* **2018**, arXiv:1806.10447.
42. Luo, C.; Jin, L.; Sun, Z. MORAN: A multi-object rectified attention network for scene text recognition. *Pattern Recognit.* **2019**, *90*, 109–118. [CrossRef]
43. Wang, T.; Zhu, Y.; Jin, L.; Luo, C.; Chen, X.; Wu, Y.; Wang, Q.; Cai, M. Decoupled attention network for text recognition. *arXiv* **2020**, arXiv:1912.10205. [CrossRef]
44. Yu, S.; Li, B.; Zhang, Q.; Liu, C.; Meng, M.-Q.H. A novel license plate location method based on wavelet transform and EMD analysis. *Pattern Recognit.* **2015**, *48*, 114–125. [CrossRef]
45. Zhao, J.; Hao, S.; Dai, C.; Zhang, H.; Zhao, L. Improved Vision-Based Vehicle Detection and Classification by Optimized YOLOv4. *IEEE Access* **2022**, *10*, 8590–8603. [CrossRef]
46. Tran, D.N.N.; Pham, L.H.; Nguyen, H.H.; Jeon, J.W. City-Scale Multi-Camera Vehicle Tracking of Vehicles based on YOLOv7. In Proceedings of the International Conference on Consumer Electronics-Asia (ICCE-Asia), Yeosu, Republic of Korea, 26–28 October 2022; pp. 1–4.
47. Zhao, H.; Zhang, H.; Zhao, Y. Yolov7-sea: Object detection of maritime uav images based on improved yolov7. In Proceedings of the Winter Conference on Applications of Computer Vision, Wailoloa, HI, USA, 3–7 January 2023; pp. 233–238.

applied sciences

MDPI

Article

A Fast and Accurate Real-Time Vehicle Detection Method Using Deep Learning for Unconstrained Environments

Annam Farid [1,*], Farhan Hussain [1,*], Khurram Khan [2], Mohsin Shahzad [3], Uzair Khan [3] and Zahid Mahmood [3,*]

1 Department of Computer and Software Engineering, College of Electrical and Mechanical Engineering (CEME), National University of Sciences and Technology (NUST), Islamabad 44000, Pakistan
2 Faculty of Computer Science and Engineering, GIK Institute of Engineering Sciences and Technology, Topi 23460, Pakistan
3 Department of Electrical and Computer Engineering, COMSATS University Islamabad, Abbottabad Campus, Abbottabad 22060, Pakistan
* Correspondence: annamfarid15@gmail.com (A.F.); farhan.hussain@ceme.nust.edu.pk (F.H.); zahid0987@cuiatd.edu.pk (Z.M.)

Abstract: Deep learning-based classification and detection algorithms have emerged as a powerful tool for vehicle detection in intelligent transportation systems. The limitations of the number of high-quality labeled training samples makes the single vehicle detection methods incapable of accomplishing acceptable accuracy in road vehicle detection. This paper presents detection and classification of vehicles on publicly available datasets by utilizing the YOLO-v5 architecture. This paper's findings utilize the concept of transfer learning through fine tuning the weights of the pre-trained YOLO-v5 architecture. To employ the concept of transfer learning, extensive data sets of images and videos of the congested traffic patterns were collected by the authors. These datasets were made more comprehensive by pointing various attributes, for instance high- and low-density traffic patterns, occlusions, and different weather circumstances. All of these gathered datasets were manually annotated. Ultimately, the improved YOLO-v5 structure becomes accustomed to any difficult traffic patterns. By fine-tuning the pre-trained network through our datasets, our proposed YOLO-v5 has exceeded several other traditional vehicle detection methods in terms of detection accuracy and execution time. Detailed simulations performed on the PKU, COCO, and DAWN datasets demonstrate the effectiveness of the proposed method in various challenging situations.

Keywords: machine learning; object detection; vehicle detection

Citation: Farid, A.; Hussain, F.; Khan, K.; Shahzad, M.; Khan, U.; Mahmood, Z. A Fast and Accurate Real-Time Vehicle Detection Method Using Deep Learning for Unconstrained Environments. *Appl. Sci.* **2023**, *13*, 3059. https://doi.org/10.3390/app13053059

Academic Editor: Yu-Dong Zhang

Received: 25 January 2023
Revised: 22 February 2023
Accepted: 23 February 2023
Published: 27 February 2023

1. Introduction

The Human Vision System (HVS) reliably and accurately performs complex tasks, such as being able to detect and recognize and identify diverse range of objects with little conscious attention. With the recent developments in the Computer Vision (CV) and Machine Learning (ML), and with the availability of capabilities, such as massive data sets, faster GPUs, and better algorithms, it has now become possible for computers to detect, recognize, and classify several items in an image or video with high accuracy [1]. The aim of vehicle detection and classification is to locate vehicles in either images or videos [2]. Efficiency of vehicle localization is a critical step in traffic monitoring or surveillance. Figure 1 shows several detected vehicles from Pakistani traffic images that are achieved using the machine learning algorithms. Therefore, autonomous vehicle detection methods must exactly detect traffic objects, such as cars, vehicles, or police vans or bikes in real-time to gain good control and make right decisions for the public safety [3].

Figure 1. Vehicle detection examples from Pakistani traffic.

With the development of the DNNs, automatic vehicle detection has made substantial progress in recent years, for instance, in Autonomous Driving Systems (ADS) and driver support systems in the context of concerns about traffic congestion and driving safety [4].

To develop intelligent and autonomous systems, for instance, self-directed driving, surveillance, detecting objects, or tracking, vehicle localization is a crucial problem [5]. Automatic driving is a new high technology invention that relies on the ability to only find vehicles [6]. In the metropolitan areas, frequent incidents happen regarding traffic breaches, vehicle mishaps, and thefts that are recorded through the CCTV cameras. Traffic surveillance system detector should be fast, accurate, and reliable enough to detect vehicles in real-time. In the areas of traffic managing systems or surveillance technology, there have been numerous advancements. Two essential conditions are normally considered to rate vehicle detectors, which are its real-time detection ability and whether it has a high detection accuracy of the traffic objects under adverse weather conditions.

One of the interesting efforts to locate vehicles is to detect abnormalities in traffic violations, as well as careless driving on the roads. With the introduction of new technologies in the ITS and the growing demand for automation, the employment of technology in a variety of disciplines has become inevitable. Because of the growing number of cars on the road, automated vehicle traffic monitoring is one of the most important applications being developed for speed or traffic control, offence detection, road tolls, and a variety of other related issues. To manage such issues, large amounts of general budget is consumed. In large and congested cities, traffic surveillance is a major challenge. ITS mobility planning traffic engineering applications have made significant progress in reducing city incidents. Surveillance systems nowadays use traffic flow data, which typically consider crucial factors, such as speed, size, trajectory, and vehicle type. Moreover, vision-based systems are also used nowadays to monitor and record various traffic patterns.

Due to the developments in the DNNs, the ML based models can be reliably used to detect various vehicles, although the training speed of deep learning networks is much slower in the CPU calculations. However, the training time is significantly much less thanks to developing technologies, such as GPUs and the TPUs. When compared to standard ML-based approaches, the DNNs have significantly enhanced performance in various scenarios, such as smart self-governing, self-driving vehicles, intelligent observations, and smart city-based applications. DNNs that are based on neural networks constitute an advanced category of machine learning, which is very handy at resolving difficulties in a variety of complex models that usually are hard to explain through typical statistical techniques.

Moreover, the CNN, which is a form of deep neural network, is extensively used for image recognition and categorization. These are the algorithms that can identify various objects, such as, license plates, cars, people, and a variety of other objects [5]. A primary benefit of the CNN is that it extracts essential features without any human interaction after the training process. Different versions of the CNN, such as R-CNN, Fast-RCNN, and Faster-RCNN are the most popular and commonly utilized CNN approaches [5]. However, the computational load is still too high for devices with limited computing power and space to process photos. The D-based algorithms have long been regarded as effective tools for image recognition. The CNN-based methods have been frequently used in recent approaches among the many vehicle detection algorithms and are divided into region-based and regression-based methods.

The YOLO is a new method to detect diverse vehicles in a single step. The YOLO handles the vehicle perception problem as a regression problem by classifying the image via CNN, which is utilized to achieve robust vehicle detection. The YOLO can retrieve the object's position, category, and confidence score, as well as boost detection speed and detect the motion-blurred vehicles in real-time. A regression-based YOLO technique is one of the most current ways to predict bounding boxes and class probabilities directly in a single neural network. As a result, the YOLO model was created to speed up the process to identify an object and find its location in an image. It uses the CNN to detect several items in an image immediately. To handle vehicles of different shapes and sizes, it integrates predictions from many feature maps with different resolutions. With advancements in the YOLO based methods, such as YOLOv3 and YOLOv5, the YOLO continues to provide higher performance in terms of processing time and accuracy [3].

In this work, we target the detection and classification of vehicles in images using deep learning to explore the feasibility of YOLO based methods. The YOLO family of algorithms is first-order object detection method, which uses an anchor box to integrate various objects localization. Up to now, five versions of YOLO family of algorithms have been released. The YOLOv3 is a milestone in the performance and speed of the YOLO family of algorithms. Our motivation to choose the YOLOv5 detection model is due to its smaller architecture and much fast detection ability than the previous generations of its model families. Recently, researchers in various research domains have enhanced the original YOLOv5 model based on the characteristics of their detection targets, which makes the YOLOv5 algorithm an excellent choice in vehicle detection domain. Our main contributions in this work are listed below.

- We propose a modified version of the YOLO algorithm to achieve vehicle detection in real time. Earlier-developed works have been trained on massive datasets, but still need to be fine-tuned for use in congested traffic environments. However, we augment these datasets with our gathered datasets. We compare the efficiency of our trained version with several recent state-of-the-art methods.
- We detect and classify vehicles in images that are captured in various traffic scenes. We perform detailed study on the PKU, COCO, and DAWN datasets. To achieve higher accuracy on images from our local traffic patterns, we gathered an extensive dataset and applied transfer learning to the YOLOv5. The input to a system is a real-time image, and the output is a bounding box corresponding to all objects in the image, along with the class of object in each box.
- In addition, we employ a transfer learning approach to utilize the knowledge embedded in our local datasets. We believe that the ITS based applications require rapid and precise vehicle identification and classification. It is a challenging task to detect different vehicles abruptly and precisely due to short gaps between vehicles on the road and interference aspects of pictures or video frames containing vehicle images. Therefore, we are optimistic that our developed method provides a good insight into locating vehicles in congested traffic environments.

This paper is organized as follows. Section 2 discusses few recent related works. Section 3 describes in detail the proposed method. Simulation results and discussions are presented in Section 4. Finally, Section 5 concludes the paper and hints towards future research directions. For readers' smooth understanding, Table 1 lists the nomenclature that is used extensively in this paper.

Table 1. Nomenclature.

Acronym	Meaning
CNN	Convolutional Neural Networks
COCO	Common Objects in Context
DLT	Dark Label Tool
DNN	Deep Neural Network
FPN	Feature Pyramid Network
FPS	Frames Per Seconds
HDT/LDT	High Density Traffic/Low Density Traffic
ITS	Intelligent Transportation Systems
LIT	Label Image Tool
mAP	mean Average Precision
MSR	Multi Scale Retinex
PAN	Path Aggregation Network
PKU	Peking University
R-CNN	Region-based Convolutional Neural Networks
RFW	RoboFlow
SSD	Single Stage Detector
TP/TN	True Positives/True Negatives
XAI	Explainable Artificial Intelligence
YOLO	You Only Look Once

2. Related Work

Vehicle detection has gained considerable attention in the research community in the past two decades. In this section, we briefly discuss the recent advances in the vehicle detection domain. For readers' fair understanding, we categorize the literature into two streams as illustrated below.

2.1. Conventional Methods

This section quickly lists a few of the latest conventional vehicle detection approaches. In [6], the developed method detects vehicles in airborne images. In this work, the vehicle localization is attained through the Gaussian Mixture Model (GMM) and background subtraction representations. In [7], an ensemble-based method is developed for various image descriptors, which illustrate the distributions of gradients, color models, and textures. This work reports good results in high resolution aerial images. In [8], a new methodology through the application of the GMM is developed to detect dissimilar complex structures, for example, objects in residential, agricultural, and industrial zones. This work also reflects spectral and spatial constraints. An efficient, GMM-based image segmentation method is utilized in [9]. This method is capable of detecting the frontal view of different vehicles. To locate the vehicles' driving area, lanes are spotted through the application of the Canny edge detector along with Hough transform. To further enhance the efficiency of proposed method, this work uses the HOG features, colors, and the Harr-features of vehicles, and trains the SVM classifier. In [10], the SVM is trained through multi-feature fusion that results in reduced vehicle detection time. In [11], vehicle detection is achieved through integration of the SIFT with the SVM. To further improve classification ability, an integration of pyramids pooling, sliding windows, and NMS is done that substantially enhances the vehicle detection outputs, which are obtained therein.

2.2. YOLO-Based Methods

In [12] a vision-based object detection and recognition framework for autonomous driving was proposed with particular emphasis on: (i) an optimized model based on the structure of YOLOv4 was presented to detect 10 types of objects; (ii) a fine-tuned part affinity fields approach was developed; (iii) eXplainable Artificial Intelligence (XAI) was integrated to assist the approximations in the risk evaluation phase; (iv) an intricate self-driving dataset was developed, which included several different subsets for each relevant task; and (v) an end-to-end system with a high-accuracy model was discussed.

The overall parameters of enhanced YOLOv4 are reduced by 74%, which meets the real-time capacity. Moreover, when evaluated with other methods, the detection precision of the enhanced YOLOv4 improved by 2.6%. In [13], a novel and efficient detector named YOLO-ACN is developed, which is inspired by the high detection accuracy and speed of YOLOv3. This technique is improved by the addition of an attention mechanism, a CIoU (complete intersection over union) loss function, Soft-NMS, and depth wise separable convolution. In this method, initially, the attention mechanism is built in the channel and spatial dimensions in each residual block focus on small targets. Later, CIoU loss is adopted to achieve accurate bounding box regression. Besides, to filter out a more accurate BBox and avoid deleting occluded objects in dense images, the CIoU is applied in the Soft-NMS, and the Gaussian model in the Soft-NMS is employed to suppress the surrounding BBox. Finally, to improve the detection speed, standard convolution is replaced by depth wise separable convolution. Meanwhile, a hard-swish activation function is utilized in deeper layers.

In [14], a multi-stage object detection architecture, which authors refer as Cascade R-CNN, is developed to address objects appearance and detection. The proposed R-CNN is composed of a sequence of detectors that are trained with varying IoU thresholds, to be sequentially more discriminating against close false positives. These detectors are trained stage-to-stage and by leveraging the scrutiny that the output of a detector is a good distribution for training the next higher stage detector. The resampling of improved hypotheses assures that all detectors have a positive set of examples of equivalent size, and thus reducing the overfitting. The same systematic method is applied at inference, enabling a closer match between the hypotheses and the detector quality of each stage. A simple implementation of the Cascade R-CNN is shown to surpass all single-model object detectors on the challenging COCO dataset. Simulations also reveal that the Cascade R-CNN is widely applicable across detector architectures and achieves consistent gains of the baseline detector strength.

A method to detect smoky vehicles with high precision and speed has been proposed in [15] using an enhanced lightweight network based on Yolov5. This work uses Mobilenetv3-small modified Yolov5s' backbone to reduce the number of model parameters and calculations. A vehicle exhaust dataset is collected and created to detect motor vehicle exhaust with high precision. Cutout and saturation transformations were used to enlarge the self-built dataset, which was eventually expanded to 6102 photos, due to the interference of vehicle shadows and occlusion between vehicles. The results demonstrate that applying data augmentation improves detection accuracy by 8.5%. The upgraded network is installed on embedded devices and has a detection speed of 12.5 FPS, which is two times faster than Yolov5. Only 0.48 million network parameters have been improved. This study suggests an effective target detection model as well as a strategy for developing low-cost and quick vehicle exhaust detection equipment. An effective nighttime vehicle detection approach is developed in [16]. First, an optimal MSR algorithm was used to improve the original nighttime photos. The improved photos were then used to fine-tune a pre-trained YOLO v3 network. Finally, the network was employed to distinguish vehicles from each other and outperforming two popular object detection approaches, the Faster R-CNN and SSD, in terms of precision and detection efficiency. The suggested method has an average precision of 93.66%, which is 6.14% and 3.21% higher than the Faster R-CNN and SSD, respectively.

In [17], the proposed work contributed to the field of autonomous driving through the DL techniques to detect objects. This work primarily uses the YOLO to locate numerous objects on the roads and categorized into the type that they belong to with the aid of bounding boxes. The YOLOv4 weights are used to custom train the model to detect the objects, and the data is acquired using the OIDv4 toolkit from the open-source data collection. In [18], an updated YOLOv3 algorithm for vehicle detection is developed. Initially, it clusters the data set using a clustering analysis approach, then optimizes the network structure to raise the number of final output grids and boost the comparatively low vehicle prediction ability. It also optimizes the data set as well as the input image. Its robustness under various external situations is due to its resolution. Experiments demonstrate that the modified YOLOv3 algorithm outperforms the traditional approach in terms of detection accuracy and rate. In [19], researchers proposed the newest YOLOv3 algorithm to detect traffic participants. They trained the network for five different object classes, which are vehicle, truck, pedestrian, traffic signs, and lights. This work also discusses the range of driving scenarios that include bright and overcast sky, snow, fog, and night conditions. In [20], the baseline YOLO is used to detect moving cars. Meanwhile, a modified Kalman filter method is used to dynamically track the detected vehicles, which results in overall competitive performance in both day and night. The testing results reveal that the system is resistant to occluding vehicles or congested highways, with an average vehicle counting accuracy of 92.11% at the rate of 2.55 FPS. In [21], researchers suggested an updated Yolov3 transfer learning-based deep learning algorithm for object detection. In this work, the network is trained on a difficult data set, and the output is fast and precise, which is beneficial for applications that need object detection. In [22], a method is proposed that classifies vehicular traffic on video using a neural network. The necessity to regulate traffic on the roads has emerged as the number of vehicles on the road has increased, resulting in traffic congestion and a high accident rate. Collecting data from video of vehicles on the road will aid in the creation of statistics that can be used to efficiently consider traffic regulation on the roads. The challenge of vehicle categorization on video was solved using the YOLOv5 powerful real-time object classification method. For neural network training 750 images from outdoor surveillance camera were used as a dataset. After testing the model, the recognition accuracy was 89%.

YOLOv2 and YOLO9000 models were discussed in [23]. Their strength in real-time detection and classification of objects in videos made them useful in several applications. The YOLOv2 is very efficient at detecting and classifying simple objects. The GPU features and the Anchor Box approach were used to accomplish the desired speed and precision. Furthermore, YOLOv2 can accurately detect object movement in video recordings. YOLO9000 is a real-time framework that can maximize detection and classification while also bridging the gap. The YOLOv2 model and the YOLO 9000 detection system can detect and classify a wide range of items, from multiple occurrences of a single object to multiple instances of various objects. In [24], an improved YOLOv4-based video stream vehicle target detection system was used to address the problem of slow detection speed. This study first presents a theoretical overview of the YOLOv4 algorithm, then offers an algorithmic technique for increasing detection speed, and lastly conducts real road tests. According to the experimental results, the algorithm in this work can improve detection speed without sacrificing accuracy, which can be used to make decisions for safe vehicle driving.

In [25], the YOLOv5 is used to locate weighty supplies vehicles during cold weather and thus allowed the prediction of parking place slots in real-time. The authors employ infrared network cameras, since snowy conditions and the polar night in the winter pose certain obstacles for image recognition. Authors used the YOLOv5 to analyze if the front cabin and back are adequate features to identify heavy goods vehicles because these photos repeatedly have large overlaps. The trained algorithm reliably distinguishes the front of heavy goods vehicles. However, detecting the back cabin appears to be more difficult, especially when the vehicle is placed far away from the camera. Finally, they show that detecting heavy goods vehicles utilizing their front and rear instead of the entire vehicle improves detection during winters, which mostly experience difficult images with significant objects overlaps and cut-offs.

Recently, some of the learning-based approaches [26] and the CNN based methods [27] also report encouraging results in the vehicle detection domain. In [26], authors developed a box-free instance segmentation method using semi-supervised iterative learning. The iterative learning procedure considered labeling vehicles from the entire scene and then trained the deep learning model for classification. Authors also considered vehicle interiors and borders to isolate instances using a semantic segmentation. In [27], researchers performed a fully convolutional regression network. In this method the training stage uses an input image along with its ground to describe each vehicle as a 2-D Gaussian function distribution. Hence, the vehicle's original format attains a simplified elliptical shape in the ground truth and output images. The vehicle segmentation uses a fixed threshold in the predicted density map to generate a binary mask. This method prevents grouping cars and favors counting. Moreover, vehicles take on a different form that is expressed by the Gaussian function.

In [28], a robust vehicle detection model is developed, which is referred to as YOLOv4_AF. This model introduces an attention mechanism that suppresses the interference features of images through channel length and spatial dimension. In addition, a modification of the Feature Pyramid Network (FPN) part of the Path Aggregation Network (PAN) is also applied to enhance the effective features. This way, the objects are steadily positioned in the 3D space that ultimately improves the vehicle object detection and classification performance. In [29], vehicle detection and tracking are achieved through a multi-scale deep convolution neural network. This work also applies conventional Gaussian mixture probability hypothesis along with hierarchical data association that divides into detection-to-track and track-to-track associations. Moreover, the cost matrix of each stage is resolved using the Hungarian algorithm. Only detection information is used in the previous so as to achieve rapid execution. In [30], Faster-RCNN is tuned to detect vehicles in various scenarios. Moreover, this work also uses basic image processing methods along with morphological operations and multiple thresholding to achieve vehicle exact location in near-real-time. In [31], vehicle and distance detection method is developed in a virtual environment. This work mainly uses the Yolo v5s neural network structure and develops a novel neural network system, which the authors refer as the Yolo v5-Ghost. In the discussed approach, the authors further fine-tuned the network layer structure of the Yolo v5s. Experiments performed therein indicate that this method is suitable to be deployed in real-time environments. The authors of this work also claim that their work is suitable for embedded and edge devices and object detection in general [32]. In [33], a novel bounding box regression loss approach is developed that learns objects bounding box through miscellaneous transformations and variance localizations. The learned localization variance is further merged during non-maximum suppression that increases the localization performance. In [34], a dynamic vehicle detection method, which is based on a likelihood-field-based model and on Coherent Point Drift (CPD), is developed. This study also applies an adaptive thresholding on the distance and grid angular resolutions to detect the moving vehicles. This work also presents the pose estimation that is based on the CPD to estimate the vehicle pose. The scaling series algorithm is also coupled with a Bayesian filter to update the vehicle localization states during various intervals.

In [35], a new Multi-Level Feature Pyramid Network (MLFPN) is proposed that constructs effective feature pyramids to detect objects. This method initially fuses multi-level features and later feeds the base features into a block of alternating joint thinned U-shape networks. Meanwhile, the decoder layers are gathered up with correspondent sizes to build a feature pyramid for object detection. In [36], the proposed method is primarily based on Trident Network (TridentNet), which aims to generate scale-specific feature maps. This scheme also constructs parallel multi-branches in which each branch shares the same transformation parameters. This algorithm also adopts a scale-aware training scheme to specialize each branch by sampling object instances of proper scales for training. The proposed TridentNet achieves significant improvements without any additional parameters. In [37], a single-stage method uses Mask SSD to investigate objects. This work uses a convolutional series to predict pixelwise objects' separation. This work also optimizes the whole network through multitask loss function. Ultimately, the network directly predicts final objects presence results. This work also uses multi-scale and feedback features that perform well on various objects of different scales and aspect ratios. In [38], the developed method uses two classifiers to tackle the problem of failure to locate vehicles that have occlusions or slight interference. It accomplishes vehicle detection through a local binary pattern along with a support vector machine. This method also uses the CNN in the second phase to remove the interference areas between vehicles and any moving object.

In [39], a novel CornerNet is developed achieve accurate object detection. The CornerNet approach detects objects bounding box as a pair of keypoints. The top-left corner and the bottom-right corners are localized through a single CNN. Through an intelligent paired keypoints approach, this method eliminates the need to design a set of anchor boxes that are normally used in prior single-stage detectors. This work also introduces corner pooling, which is a new type of pooling layer and helps the network to better visualize and localize the objects' corners. In [40], a novel approach, which authors refer to as Mask R-CNN, is discussed that extends Faster R-CNN by adding a new branch to predict an object mask. The Mask R-CNN is simple to train and adds only a small overhead to Faster R-CNN. Moreover, Mask R-CNN is easy to generalize to other tasks, for instance to estimate object orientation in the same framework. This method is conceptually simple and flexible and efficiently detects objects in an image. In [41], an anchor-free vehicle detection approach is developed that is capable of detecting arbitrarily oriented vehicles in high-resolution images. This work considers vehicles as a multitask learning problem and predicts high-level vehicle features via a fully convolutional network. In this work, initially, coarse and fine feature maps outputted from different stages of a residual network are integrated through a feature pyramid fusion. Later, four convolutional layers are added to predict possible vehicle features. In [42], a scale-insensitive CNN (SINet) is proposed to locate vehicles with a large variance of scales. Initially, a context-aware RoI pooling is done to maintain the contextual information and original structure of objects. Later, a multi-branch decision mechanism is introduced to minimize the intra-class distance of various features. The proposed techniques can be further equipped with any deep network architectures and keep them trained end-to-end.

The preceding discussion offers a good suggestion that vehicle detection is a crucial step to develop systematic mechanisms, such as an intelligent transportation system. The methods describe above are a few of the efficient and good works that aim to address the vehicle detection problem in various environments. As we will see in Section 4, different datasets are publicly available to address the vehicle detection problem under diverse conditions. We believe that our work is an efficient addition in the vehicle detection domain. In the next section, we discuss our developed vehicle detection method.

3. Proposed Method

In this section, we describe our proposed method in detail. As discussed below, we divide our developed method into the following interconnected steps along with a brief description. Figure 2 shows the flow of the proposed method. In addition, Algorithm 1 shows more details of our developed method.

To test our method, we gather our own dataset from challenging Pakistani traffic environments. This dataset was collected over a period of two months in different cities of Pakistan. As shown in Figure 2 that the gathered data is preprocessed and augmented. Later it is trained by our model. Meanwhile, the YOLO-v5 model is built and trained. Our collected data is from an unknown distribution in Pakistani traffic. Therefore, it is now tested on the YOLO-v5 model. After the YOLO-v5 is applied, we then investigate and analyse our detector. To aid readers' understanding, below we describe the steps and details of our developed method.

Algorithm 1: Pseudo code of the proposed vehicle detection algorithm

```
1.  Input: A test image with one or more visible vehicles.
2.  Execute the algorithm in following order to get the desired result.
3.  begin
4.      Gather data
5.        do
6.          Categorize data into LDT and HDT scene images as indicated by Figure 3a,b.
                                                                        ▶ use the LIT
7.          Annotate data to identify classes of objects as shown by Figure 4a,b.
                                                          ▶ use the DLT for video dataset
8.        end
9.      begin data preprocessing and augmentation
10.       do processing
11.         Preprocess that data and split into train, validation, and test set
                                                                    ▶ use the RFW tool
12.         do augmentation
13.           Standardize the image and video data from step (3) to step (8) up to 416 × 416 pixels.
14.           Crop dataset between 0% and 30% zoom.
15.           Saturate dataset between ±25%.
16.           Vary brightness, such as darken and brighten the images between ±25%.
17.         end
18.     end
19.     begin YOLO-v5
20.         do
21.         Install all Yolov5 repositories to be ready for running object detection training & inference.
22.         Download custom Yolov5 Object detection data.
23.         Configure model and architecture.
24.         begin Training
25.           Train custom YOLO-v5 detector                 ▶ use YOLO-v5 architecture
26.           Use training parameters as:                   ▶ use COCO dataset weights
27.             image size: 416 × 416 pixels,
28.             batch size: vary as 5, 10, and 20,
29.             epochs: vary as 100, 300, and 500,
30.             Configuration: use as per YOLO-v5s, YOLO-v5m, or YOLO-v5L,
31.             Weights: use pre-trained COCO dataset,
32.         end
33.         Run YOLO-v5 inference on test images.
34.           python detect.py --weights runs/train/exp/weights/best.pt,img 416,conf 0.1.
35.     end
36. end
37. Output: An image with detected vehicles through a bounding box around.
```

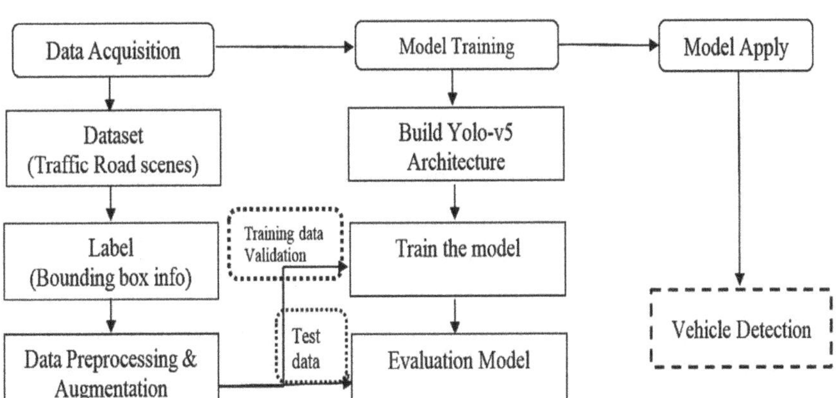

Figure 2. Flow of the proposed method.

3.1. Data Acquisition

To begin with the proposed algorithm, we initially acquire data. First we deal with different conditions on highways. For example, we come across the multi-class objects, such as different types of vehicles, motor bikes, and pedestrians on the roads. Similarly, we also faced severe and crucial challenges, such as massive traffic jams and overlapped vehicles. Therefore, to systematically acquire the data as shown in line (6) of Algorithm 1, we collected the dataset under two different situations, which are (i) High Density Traffic (HDT) scenes that contains multiple objects in an image and (ii) the Low-Density Traffic (LDT) scene that contains only one class per image, with zero overlaps. For improved training, the images of the LDT and the HDT dataset are placed separately.

The LDT Scenes: This dataset was gathered from daily real-time traffic places, for example open parking lots, less crowded roads, and places with fewer crowds. The objective of assembling this dataset is to separately train the model on each class. We collected a total of 600 images from three classes, which are cars, motor cycles, and pedestrians. Example images of the few of the LDT images are shown in Figure 3a.

(**a**)

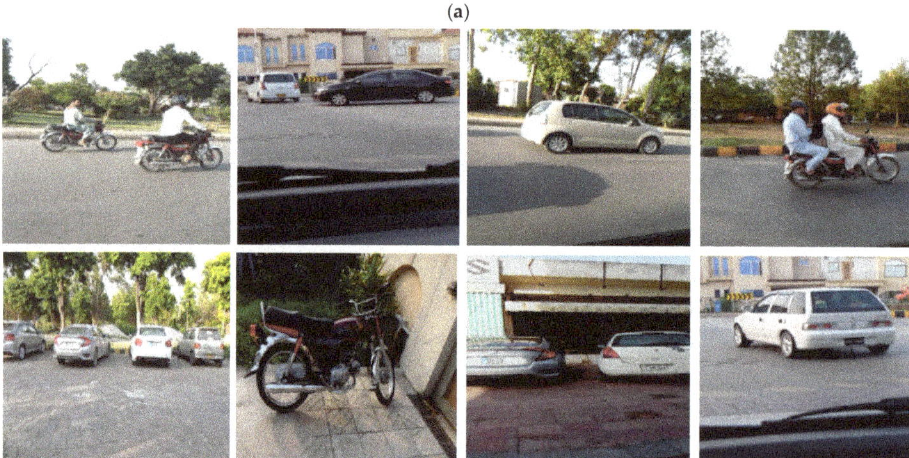

(**b**)

Figure 3. Sample images of our collected dataset: (**a**) Low density traffic scenes and (**b**) High density traffic scenes.

The HDT Scenes: This dataset was collected in congested places, for example public parking lots, big shopping malls, main highways, and places near main traffic sign boards. We gathered a total of 1800 images of the aforementioned classes. We also collected this dataset by thinking about crucial factors, for instance varying illumination, partial/full/long term occlusions, along with collections of objects regardless of size, scale, shape, or appearance. A few such sample images are shown in Figure 3b. The statistics of both the low- and high-density dataset along with each class annotations are described in Table 2.

Table 2. Summary of our collected dataset images.

Dataset		HDT Dataset	LDT Dataset
Source images		1800	600
Annotations		15,618	903
Classes	Car	8457	655
	Motorcycle	4136	136
	Person	3025	112
Total		**3036**	**2406**

Video Dataset: Along with the images, we also gathered a video dataset from the different locations of main highways, such as crossway bridges. A few of the sample images of our collected video dataset are shown in Figure 4a. It can be seen that our collected dataset has different types of vehicles that appear in the image. Moreover, the vehicles' resolution also varies. Collecting such a diverse dataset helps us to develop a robust, reliable, and accurate vehicle detection method, which we believe can be used in any real-time application.

3.2. Data Annotation

It is the proper procedure to label the classes of the datasets to achieve reliable vehicle detection in later stages. This data annotation is an important step for good training of the CNN model so as to get promising results.

As shown in line (7) of Algorithm 1, we have used Label Image Tool (https://github.com/heartexlabs/labelImg (accessed on 12 January 2023)) (LIT) to label and annotate the image dataset. To use the LIT tool, we upload the image dataset to the LIT, which reads the images. Later, we manually assign a bounding box for each object present in the image as shown in Figure 4b. It is evident that for the HDT category, there are several bounding boxes on a single image. However, for the LDT scenes, there are fewer bounding boxes. These bounding boxes specify the label of the respective class, such as vehicle or motorbike. Every object present in the image is manually labeled, which is indicated by bounding boxes. The overall dataset is then divided into three classes, which are cars, motorcycles, and pedestrians. Readers are referred to the LIT link, which is provided at the bottom of this page, which offers detail about the LIT usage.

For the video dataset, the annotation is some way bit extensive. To do it quickly, we used Dark Label Tool (https://github.com/darkpgmr/DarkLabel (accessed on 12 January 2023)) (DLT) as it consumes less time as compared to the LIT module. The DLT automatically divides the uploaded video dataset into frames, for instance frames of 10 s into 360 frames. These frames are now interpolated, in which the first frame draws a bounding box around an object, and the last frame draws the bounding box around the same object. Hence, all the objects in between the 10 s have been annotated and labelled according to the specified classes. Readers are referred to the DLT link, which is provided at the bottom of this page, which detail about the DLT usage, along with more facilities provided therein.

(**a**)

(**b**)

Figure 4. Sample images of our collected dataset: (**a**) video dataset and (**b**) annotated images.

3.3. Data Augmentation

To increase the data features to obtain better results, data preprocessing is the building block of deep learning-based algorithms. We know that real world datasets might be contaminated with noise. Many times, these datasets are inconsistent, or some things may be missing. Sometimes, uneven and unbalanced classes appear. As can be seen in lines (9) to (18) of Algorithm-1, data preprocessing and augmentation is analyzed. As shown in Table 3, we preprocessed our dataset in distinct steps. One is to get the same size of each image of the HDT, the LDT, and video datasets. We make the dataset of 416 × 416 pixels resolution of each image and video. Then the dataset is split into train, test, and validation set. To split the dataset, we used the RoboFlow (https://public.roboflow.com/ (accessed on 12 January 2023)) (RFW) tool as described below.

Table 3. Dataset statistics after augmentation.

Dataset	Classes	Classes	Training	Validation	Testing
HDT data	3	car, motorcycle,	3685	356	177
LDT data	3	and person	1260	120	60
COCO 2017	80	car, motorcycle, person, dog, table, and horse	118,287	5000	40,760
	Total		**123,232**	**5476**	**40,997**

The RFW: this tool hosts free public computer vision datasets in many popular formats. The RFW provides a streamlined workflow to identify edges of various objects in several iterations. With each iteration, the detection models become smarter and more accurate. We used the RFW tool to fragment the entire dataset into train, validation, and test sets. In this study, we keep the split ratio as 7:2:1 that is the image dataset of both categories and the video dataset has been divided into 70% train, 20% validation, and 10% test sets. Training a model on small number of images could result in overfitting [26]. Moreover, it also results in poor generalization despite the fact that the training results are good enough. However, the testing accuracy drops down and the model classifies the samples into one class. In short, the training accuracy is high, but the validation accuracy drops down. To overcome this issue, data augmentation is used, which modifies the data using different techniques and increase the samples of the dataset. Through empirical analysis, we applied the following augmentation techniques on our collected dataset.

Cropping: In this stage, we crop the image dataset of both categories between 0% minimum and 30% maximum zoom.

Saturation: To achieve better results, we change the color ranges of images of both categories and the video as well. In this study, we saturate images ±25%.

Brightness variation: The brightness of the image dataset of both categories has been carefully varied. We darkened and brightened the images ±25%. After applying the data augmentation technique to the image dataset, it is ready to use in a model for object detection. The statistics of the dataset after augmentation are shown in Table 3. As can be seen, we have a total of 123,232 training images, 5476 validation images, and 40,997 test images.

3.4. The YOLO-v5

The YOLO-v5 is one of the latest models to obtain reliable object detection in the YOLO family [26]. YOLO-v5 has four more types, which are, YOLO-v5s, YOLO-v5m, YOLO-v5l, and YOLO-v5×. All of these types differ in size and inference time. The size ranges between 14MB to 168MB. The YOLO-v5 surpasses other conventional object detection procedures mostly in terms of detection accuracy. Moreover, the YOLO-v5 is computationally faster in comparison to its companion YOLO family-based algorithms. As shown in Algorithm 1, the YOLO-v5 is used in this study from lines (19)–(35). There are three main architectural blocks in the YOLO-v5 as discussed below [26].

Backbone: In the YOLO-v5, the Cross Stage Partial (CSP) networks are used as a backbone to extract important features from the given input image. Figure 5 lists the details of the backbone modules that are embedded therein.

Neck: The feature pyramid is constructed with the PAN for features accumulation. The features are then passed to head. Figure 5 lists the details of the Neck module along with the necessary details, which are implanted therein.

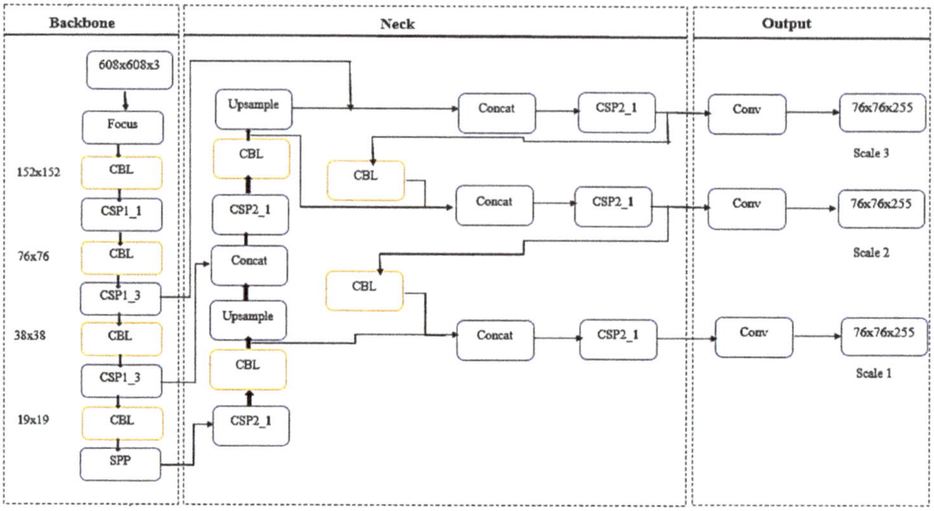

Figure 5. YOLOv5 schematics used in our work.

Head: In this block, the predictions are generated with the help of anchor boxes that ultimately achieves object detection. YOLOv-5 is made more intelligent through a transfer learning mechanism, which is shown in Figure 6, in which an input dataset is processed by the convolutional layers. That in return feeds to the FC layer. Later, our test datasets are processed by pretrained network that yields the final output through the FC layers.

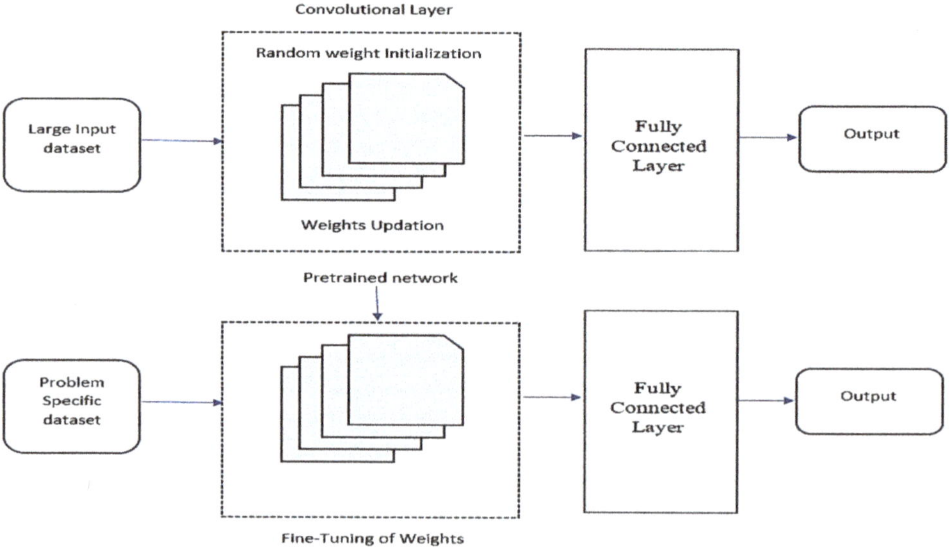

Figure 6. Employed transfer learning in the YOLOv5 architecture.

Therefore, to solve challenging issues in the vehicle detection domain, this above-described strategy introduces a prediction head in the YOLO v5 detection architecture. The introduced transfer learning or the prediction head is generated through random initialization of weights and the fine-tuned network. Moreover, to make the prediction head better detect various vehicles, YOLOv5 architecture also adds a CSP, CBL, and CBS section, and up-samples the fused feature map to generate a larger feature map to detect variations in vehicles appearances.

3.5. Training Strategy

After the data is acquired and processed along with annotations, training and validation datasets are passed to the YOLO-v5s algorithm. For training, we select different parameters, such as batch size, epochs, and image resolution. Training path, testing, and validation dataset are given to the algorithm. We notice that if we train our model from scratch, then we have to initialize it with some random weights. Therefore, we used pre-trained COCO weights for our model training as it saves a considerable amount of time and makes computations easy. Using pre-trained YOLO-v5 model, we get the best weights after transfer learning. Moreover, we used default layers and anchors, as we are utilizing the initial weights of the COCO dataset. We also employed the COCO dataset as a benchmark to train our custom dataset. Furthermore, we have also varied the batch sizes as 5, 10, and 20. We have also changed the epochs to 100, 300, and 500. In this study, the values of confidence $\in [0.4 \sim 0.6]$. After the training phase is done, we use the best weights to detect objects on the dataset. Lastly, we obtain the values of predicted labels and the test images with the bounding boxes with confidence values. The collected dataset and the trained model along with the manuscript will be made publicly available (http://research.cuiatd.edu.pk/secure/ResearchGroups/comsatsresearchgroups.aspx (accessed on 12 January 2023)). In the next section, we present and discuss the simulation results in detail.

3.6. Evaluation Criteria

The following criteria are used to measure the robustness of our developed vehicle detection method:

$$Precision = \frac{True\ Positive\ Cases}{Total\ Positive\ Predictions} \tag{1}$$

$$Recall = \frac{True\ Positive\ Cases}{Total\ Cases} \tag{2}$$

Similarly, mean Average Precision (*mAP*), the average value of Precision, is also computed for the value of Recall over 0 to 1. The mAP is usually applied in object detection algorithms and is shown mathematically below.

$$mAP = \int_0^1 P(R)dR \tag{3}$$

4. Simulation Results

This section presents the detailed simulation results. Extensive experiments are carried out on Google Colaboratory (Colab) platform. The Google Colab provides Intel Xeon CPU with a clock speed of 2.3 GHz and up to 16 GB of RAM. Moreover, the Google Colab also provides NVIDIA K80 or T4 GPU. We use Python V3.6 as a simulation tool for different vehicle datasets as described in subsequent sections. To investigate the performance of vehicle detection methods on different datasets, we select 14 state-of-the-art vehicle detector evaluations and comparisons with the proposed method in terms of accuracy and execution time. All of the compared approaches have been trained on the same training data from each of the PKU, COCO, and DAWN datasets.

4.1. Analysis on the PKU Dataset

The PKU dataset is a collection of diverse vehicle images that are captured under diverse conditions [27]. As shown in Table 4, that this dataset contains a total of 3977 diverse vehicle images. The developers of the PKU dataset divided the vehicles into five distinct and different categories, which they refer as G1, G2, G3, G4, and G5. Out of 3977 vehicle images, the PKU dataset also contains a total of 4263 visible license plates whose pixel resolution varies from 20~62 pixels, which are captured therein. Figure 7 shows a few of the vehicle detection results of our proposed YOLO-based method on all of the five categories of the PKU dataset. As can be seen in the first three rows of Figure 7, for the G1~G3 categories, the proposed method locates all the vehicles in the input images. For the G4 category as shown by the fourth row in the Figure 7, it is obvious that the proposed method is able to locate vehicles that just expose their front bonnet. Moreover, the PKU–G4 category also contains extreme reflective glare. It is always very challenging for any detection algorithm to perform accurately under such circumstances. However, as can be seen, the proposed method handles the aforesaid scenario effectively.

Table 4. The PKU dataset description.

Category	Vehicle Conditions	Input Image Resolution (pixels)	No. of Images	No. of Plates	Plate Height (pixels)
G1	Cars on roads; ordinary environment at different daytimes; contains only one vehicle/license plate per image.	1082 × 728	810	810	35–57
G2	Cars/trucks on main roads at different daytimes with sunshine; only one vehicle in each image.	1082 × 728	700	700	30–62
G3	Cars/trucks on highways during night; one license plate per image.	1082 × 728	743	743	29–53
G4	Cars/trucks on main roads; daytimes with reflective glare; one license plate in input images.	1600 × 1236	572	572	30–58
G5	Cars/trucks at roads junctions with crosswalks; several vehicles per image.	1600 × 1200	1152	1438	20–60
	PKU dataset		**3977**	**4263**	**20~62**

Moreover, the proposed method also performs well on the G5 category, which contains multiple vehicles per image. It can be seen that for different view angles along with the partially occluded vehicles, the proposed method performs well and in most of the instances detects all such vehicles. Figure 7 also reveals that the PKU-G3, G4, and G5 pose a challenge to any detection algorithm due to the fact that the illuminations change abruptly. A few of the images shown in the 3rd, 4th, and 5th rows in Figure 7 have a background that is dark black, or in which the head lights of the vehicles are turned on. In such cases, the proposed method performs well and up to task by locating all the vehicles that appear therein.

Table 5 lists the comparison of the proposed method on the PKU dataset with fourteen other methods. Since we collected most of the data from Pakistani cities, for a fair comparison we tested the methods reported in [28–31] and [33–42] on the whole PKU dataset along with our developed method. Table 5 lists the detailed results with important observations.

- From Table 5, it is evident that all the compared methods except [33,35–37,41] and [42] yield 100% detection accuracy on the PKU–G1~G3 categories.
- On the G4 category, the proposed method ranks 3rd among all the compared fourteen methods in terms of detection accuracy. On the other hand, on the G5 category, our developed method outperforms all the compared methods. The PKU G5 is a challenging category due to fact it contains multiple vehicles per image and also contains several disguising crosswalks that pose a threat to any vehicle detection algorithm.
- From Table 5, we also observe that for G1, G2, and G3 categories, the methods developed in [28–30,38,40] produce 100% vehicle detection result. Our proposed method yields 99.94% vehicle detection accuracy on the G1 category and 100% for the G2 and G3 categories. Therefore, we observe that methods shown in Table 5 have solved the challenge of vehicle detection on these three categories as most of them yield at least 99% accurate vehicle detection.

- Overall, on the PKU dataset, the proposed method ranks 1st at achieving vehicle detection in terms of the mAP as listed in Table 5. The method by [40] ranks 2nd by yielding 99.86% accurate vehicle detection accuracy. The works developed in [30,31] also yield slightly over 99.75% vehicle detection accuracy. In addition, the methods shown in Table 5 report over 97% detection accuracy, which we believe is encouraging in solving real-world traffic problems.
- To best of our knowledge, we observe that vehicle detection challenge is almost solved on the PKU dataset. However, we observe that non-uniform illuminations or high glare at the night could still affect vehicle detection accuracy. Similarly, the researchers who aim to solve the other object detection problems, such as license plate detection or recognition, may need to perform additional preprocessing or postprocessing to achieve reliable detection results.

Figure 7. Vehicle detection results on the PKU dataset from G1 to G5 categories

Table 5. Vehicle detection comparison (%) on PKU dataset.

Ref.	G1	G2	G3	G4	G5	mAP (%)
[28]	100	100	100	98.96	99.13	99.61
[29]	100	100	100	99.73	99.21	99.78
[30]	100	100	100	99.70	99.10	99.76
[31]	100	100	99.40	99.74	98.96	97.74
[33]	99.00	99.00	98.70	98.00	98.90	98.72
[34]	100	100	100	99.00	96.50	99.10
[35]	100	100	99.00	99.64	99.06	98.34
[36]	100	100	99.40	99.74	98.96	97.74
[37]	100	98.50	100	99.50	98.10	99.22
[38]	100	100	100	99.00	98.00	99.40
[39]	99.00	100	100	99.00	98.50	99.30
[40]	100	100	100	99.80	99.50	99.86
[41]	99	99	98.50	98.00	99.00	98.70
[42]	98.90	98.50	98.00	97.50	96.10	97.80
Proposed	**99.94**	**100**	**100**	**99.73**	**99.96**	**99.92**

4.2. Analysis on the COCO Dataset

The COCO dataset is designed to detect and segment various objects that occur in their natural context [32]. As shown in Table 3, the COCO dataset contains various object images, which have been gathered from complex everyday scenes and contains common objects in their natural context. Moreover, objects in this dataset are labeled using per-instance segmentations to aid in precise object localization. Overall, the COCO dataset contains images of 91 object types with a total of two and a half million labeled instances in 328,000 images. Recently, the COCO dataset received extensive attention from researchers investigating various categories of detection including diverse vehicle shapes. Figure 8 shows the vehicle detection results of our proposed method on the COCO dataset. Clearly, Figure 5 depicts the performance of the proposed vehicle detection algorithm on various challenging images of the COCO dataset.

In most of the instances and under huge illumination variations, almost all of the different vehicles are accurately detected by the proposed methodology. We used other objects, such as motorcycles and persons during this phase. Therefore, those are also accurately located in various images in Figure 8. A few such instances can be seen in the 1st image of the 2nd and 3rd rows, respectively. Similarly, the 6th image in the bottom row of Figure 8 also depicts the object detection phenomenon. To further validate the superiority of the proposed method, a comparison with fourteen other methods is listed in Table 6 with some important observations.

- As can be seen for various image resolutions that range from 512 × 512 to 800 × 800 pixels, the proposed method ranks 1st among all the compared methods and reports the highest mAP value of 52.31%. The work reported in [42] ranks 2nd and yields a 50.40% mAp value followed by [41] with a 49.80% mAP value. Our analysis reveals that the work developed in [35,36] are also an encouraging solution for detecting various objects in the challenging COCO dataset.
- On the COCO dataset, the work reported in [31] yields the lowest (27.89%) mAp value followed by [33], whose method yields a mAP value of 29.10%. Moreover, in the current study, work discussed in [40], which uses the ResNet as a backbone, yields a mAP value of 31.80%, which in the context of current study falls on the lower side.

- The crux of this dataset is that the proposed method effectively and reliably detects miscellaneous objects that include vehicles of varying shapes, including motorbikes and jeeps. Furthermore, the proposed method also effectively handles big buses. A few such samples are also in the 1st and 3rd columns of Figure 8.

Figure 8. Vehicle detection results on the COCO dataset.

Table 6. Vehicle detection accuracy comparison on the COCO dataset.

Ref.	Backbone	Data	Input Size	Multi Scale	mAP(%)
[28]	CSPDarkNet53	trainval35K	512 × 512	False	47.62
[29]	CNN	trainval35K	512 × 512	False	48.00
[30]	R-CNN	trainval35K	512 × 512	False	46.20
[31]	BottlenectCSP	trainval35K	512 × 512	False	27.89
[33]	VGGNet-16	trainval35K	512 × 512	False	29.10
[34]	ResNet-101-FPN	trainval35K	512 × 512	False	38.30
[35]	VGGNet-16	trainval35K	800 × 800	False	41.00
[36]	ResNet-101	trainval35K	800 × 800	False	48.40
[37]	ResNet-101	trainval35K	512 × 512	False	39.30
[38]	CNN + SVM	trainval35K	512 × 512	False	49.05
[39]	BN + ReLU	trainval35K	512 × 512	False	32.98
[40]	ResNet-C4-FPN	trainval35K	512 × 512	False	31.80
[41]	ResNet-50	trainval35K	512 × 512	False	49.80
[42]	SiNet	trainval35K	512 × 512	False	50.40
Proposed	**CSP**	**trainval35K**	**512 × 512**	**False**	**52.31**

4.3. Analysis on the DAWN Dataset

The DAWN dataset is designed to investigate the performance of recent vehicle detection methods on a broad range of natural images including adverse weather conditions. The DAWN dataset contains 1000 image of significant variation in terms of vehicle size and category along with pose variation, non-uniform illumination, position, and occlusion from real traffic environments. Additionally, it exhibits a systematic variation for traffic scenes, for instance, bad winter weather, heavy snow, sleet rain, sand, and dust storms. Figure 9 shows detailed results on fog, sand, rain, and snow situations with important observations.

- For the snow category as seen in top row of Figure 9, it is obvious that many times the vehicles are partially visible due to adverse weather conditions, such as fog that is normally experienced in severe winters in areas of various parts of the world. However, our developed method handles all such situations except the 2nd last image of front row in Figure 9, where it is obvious that the vehicle is not visible to the human eye as well.
- For a considerably rainy day as seen in second row of Figure 9, the proposed method accurately locates multiple vehicles that appear therein. In this case, the image scene variations, such as shown in the 2nd and 4th images of the second row in Figure 9 indicates that the proposed method is unaffected by such changes in the image scene. Similarly, the skyscrapers in the vehicle's background as shown in the 5th image of the 2nd row in Figure 9 also do not affect the detection ability of our developed method.
- For a sand situation as indicated in the third row of Figure 9, the proposed method detects all vehicles that appear there in such challenging conditions. In such situations, visibility is normally very low, which poses threats to most of the machine learning algorithms. Particularly, the first two images in the 3rd row of Figure 9 have intra-class scene variations, i.e., both are images effected by sand storms and yet appear differently to the human eye. Even in such cases, our developed method performs well and detects most of the instances that appear in such condition. The 3rd image in this row is quite challenging for human observers as well. However, as indicated there, our developed method handles such situations by successfully locating the vehicles that appear in such scene images.

- The bottom row in Figure 9 is a case when the scene is dominated by snow. In this case, surprisingly, the image appears neat and clean and thus results in a visually pleasing image due to the massive amount of snow which is present in the image. In this case, our developed method accurately detects and labels all the vehicles that appear therein. Particularly, the 3rd image in this row also reveals a red light along with the snow. Yet in this case, the proposed method performs well and successfully locates all the vehicles. Moreover, the last image in this row shows a few vehicles that overlap and result in partial occlusion. However, our developed method performs well in this case as well.

(**a**)

(**b**)

(**c**)

(**d**)

Figure 9. Vehicle detection results on the DAWN dataset row-wise: (**a**) fog; (**b**) rain; (**c**) sand; and (**d**) snow.

In Table 7, we compare our method with the works already described in Tables 5 and 6, respectively. A few of the observations from Table 7 are listed below.

- As can be seen in Table 7, for the fog scenario the work developed in [40] ranks 1st among 14 compared methods by yielding a 29.68% mAP value. Our developed method ranks 2nd out of all compared methods in fog situation and yields a 29.66% mAP value. In the fog situation, the work developed in [38] yields the lowest mAP value (16.50%) followed by [29] whose method yields a mAP value of 24%.
- For the rain scenario on the DAWN dataset, our proposed method and the work developed in [31] yield the highest mAP value of 41.21%. In this category, the work in [34] ranks 2nd and yields an encouraging result of a 41.10% mAP value. For the

aforesaid category, results yielded by [36,37] are also encouraging. For images that are affected by rain, the work in [38] delivers the lowest mAP value of 14.08%.

- For the snow conditions on the DAWN dataset, the work developed in [37] ranks 1st among all compared methods and slightly outperforms the proposed method by yielding a mAP value of 43.02%. For this category, our developed method yields a mAP value of 43.01%. It is important to state here that for this situation, the works in [31,33,36,37] yield almost similar results.
- For the sand condition, our method ranks 1st and outperforms all compared methods by yielding a 24.13% mAP value. On this situation, the works in [28,34,35] yield similar mAP values. For the sand situation, the work in [38] yields the lowest mAP value (10.69%).

Table 7. The mAP (%) comparison on DAWN dataset.

Ref.	Backbone	Image/s	Fog	Rain	Snow	Sand
[28]	CSPDarkNet53	0.085	26.40	31.55	39.95	24.10
[29]	CNN	0.085	24.00	21.10	38.32	23.80
[30]	R-CNN	0.085	27.20	21.30	28.30	18.00
[31]	BottlenectCSP	0.085	29.31	41.21	43.00	24.02
[33]	VGGNet-16	0.085	23.40	24.60	37.90	15.83
[34]	ResNet-101-FPN	0.085	28.95	41.10	43.00	24.09
[35]	VGGNet-16	0.085	23.10	27.65	34.00	24.10
[36]	ResNet-101	0.085	29.70	40.10	43.00	23.99
[37]	ResNet-101	0.085	28.10	40.40	43.02	24.10
[38]	ResNet-101-FPN	0.085	16.50	14.08	15.38	10.69
[39]	Hourglass-104	0.085	25.08	19.14	23.18	17.38
[40]	ResNeXt-101	0.085	29.68	30.32	33.93	24.00
[41]	ResNet-101-FPN	0.085	28.83	27.68	30.19	24.03
[42]	VGGNet-16	0.085	26.45	20.09	27.92	11.31
Proposed	**CSP**	**0.0085**	**29.66**	**41.21**	**43.01**	**24.13**

4.4. Computational Complexity

We evaluate the computational complexity in terms of the time consumed to yield the vehicle detected output image. While evaluating the computational complexity of the methods listed in Figure 10, we manually vary the test image size from 512 × 512 pixels up to 1600 × 1236 pixels on all the three datasets compared in this study. In addition, all the times shown in Figure 10 are the mean execution time on all the three datasets to process a single image and yield the output image.

Moreover, in Figure 10, we compare the execution time of 14 state-of-the-art vehicle detection methods with the proposed method. It can be seen that the work of Liu [34] is computationally more expensive than all of the compared methods. Clearly, the proposed method is computationally most economical and consumes slightly more than 0.50 s to yield a vehicle detected output image. Furthermore, the works reported by Wu et al. [31], Law et al. [39], and He et al. [40] consume nearly 1 s to yield the output image with detected vehicles.

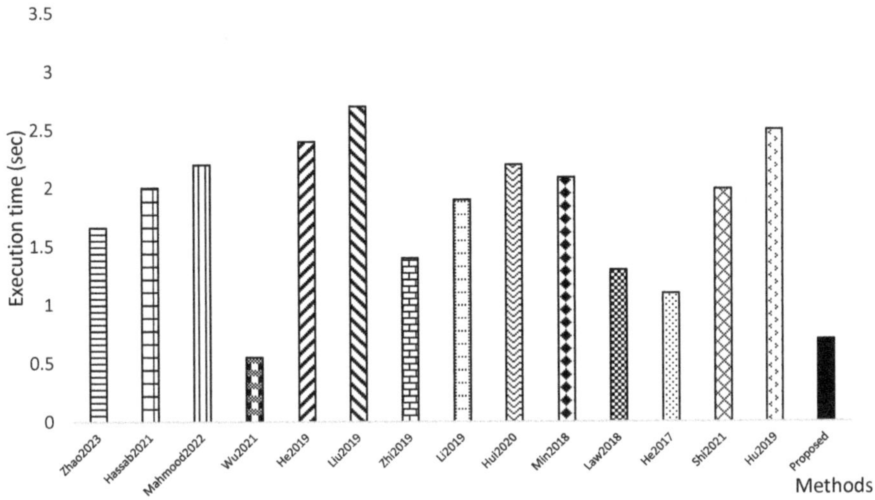

Figure 10. Computational complexity comparison with fourteen other approaches [28–31,33–42].

4.5. Discussion

Although the analysis presented above familiarizes the readers with the feasibility of our developed method to detect diverse vehicles under diverse range of environments, the discussion below will give further insight to the readers.

- Methods compared in this study are state-of-the-art object detectors. We observed that specific method performs well on a specific dataset but are challenged by other datasets. For instance, the work developed in [28] investigates BIT-Vehicle and UA-DETRAC datasets only. These datasets mostly contain high quality frontal view of the different vehicles with image resolution of 1920 × 1080 to 1600 × 1200 pixels. In contrast, the method proposed in current study explores three different datasets that have variations, such as different road conditions, varying weathers, or complex backgrounds. Moreover, the study presented in this manuscript also explores the detection ability of this method on three standard and publicly available datasets.
- The works discussed in [29,34] mostly focus on KITTI and the DAWN datasets that contain the variations as described earlier. However, we also explore their detection ability on five different classes of the PKU dataset that contain huge road and traffic variations along with our proposed method. This will essentially provide a nice baseline to beginners and researchers to develop their specified tasks.
- The work reported in [30] investigates the generic PKU dataset in its five distinct categories. However, this study further explores the detection capability of [30] on the COCO and the DAWN datasets. Moreover, the detection accuracy of the method proposed in this study provides a fair insight into vehicle detection in various scenarios.
- The method developed in [31] examined the CARLA dataset, which we believe is a limited and relatively small vehicle dataset. The findings presented in this study extend the detection capability of this method to three other datasets. In addition, its detection comparison with the proposed method and several other techniques provides much detailed insight about issues in the vehicle detection domain.
- In [33,37], the PASCAL VOC 2007 dataset is explored only. Moreover, work in [33] also analyzes the subdomain of the COCO dataset to show the detection of trains only. In contrast, this study explores the detection capability of [33] on various vehicle classes of the COCO dataset along with the PKU and DAWN datasets. Moreover, the detailed comparison provided in the earlier sections provides a fair baseline to the research community. Furthermore, the work in [37] explored the PASCAL dataset

that already contains annotated images of various objects. This study further expands the detection capability of this method to three different vehicle datasets. Finally, the detailed analysis and comparison provided in earlier section hints towards additional modifications of this algorithm.

- The works in [35,36,39,40] were validated on the MS COCO dataset to detect various objects. The experiments reported in this study extend the detection analysis of the aforementioned approaches to PKU and DAWN datasets as well. Since our method also explores the vehicle detection on these datasets, it will be convenient for researchers and practitioners to choose the appropriate algorithm for their specified applications. Moreover, the work listed in [40] reports the detection of various objects, such as pedestrians, statues, or animals. However, this study reports the detection ability of this algorithm on actual and real-world vehicle images along with several other approaches.
- In [38], the PETS2009 and the changedetection.net 2012 datasets are explored. Results analyzed in their study are mostly standard high quality frontal view images of mono-color cars running on a main highway. In contrast, the analysis presented in this study explores the detection ability this method on different datasets on multiple styles of vehicle and on differently color cars. Moreover, this study also investigates the detection ability of this method on varying illuminations and weathers along with different road conditions.
- The study in [41] analyzed the DLR Munich vehicle and VEDAI datasets. In their study, mostly high-quality aerial vehicle images are analyzed. Few of these are running on roads, while several parked vehicles are shown. However, our study also reports the use of this method on actual daily life vehicle images from three publicly available datasets. We are optimistic that detailed analysis and comparisons provided in this manuscript will be handy for the research community to modify any algorithm for their specified tasks.
- Finally, in [42], the KITTI and the LSVH datasets were explored. Results reported in this study are mostly vans, cars, and bus that run on the main highways. However, our study reports the detection ability of this method on varying illuminations, different weathers, and challenging road conditions from three publicly available datasets. We believe that the analysis provided by our developed method and the detailed comparison listed in this manuscript will provide further insight to the research community.
- All of these are useful efforts to solve and automate the vehicle detection problem under varying conditions. For each of the datasets mentioned above, these methods perform well. One of the objectives of the current study was to test and analyze all of the fourteen methods compared in this paper on standard PKU, COCO, and DAWN datasets. The main reason to choose PKU, COCO, and DAWN datasets is that these datasets contain real world and challenging images. For instance, the PKU dataset has five distinct categories that range from normal images to dark night images along with night glare. Similarly, this dataset also contains multiple images that appear in the input along with partial occlusions and different road conditions. Similarly, as mentioned in Section 4, the COCO dataset is also a huge dataset and contains a diverse range of objects. Moreover, the DAWN dataset also contains various real-world situation, such as fog, rain, snow, and the sand. An evaluation of fourteen different methods on these three datasets will be a fair guideline for researchers and beginners to develop, implement, or modify any algorithm for their specified applications.
- Out of the datasets that are investigated in this study, we find the DAWN dataset a bit more challenging than the others. The main reason is the inclusion of images in challenging conditions, such as fog, rain, contaminated with sand, or snow. Our study indicates that the sandy images reduce the scene visibility and ultimately reduce the detection accuracy of a detector. The 1st image in the top row in Figure 11 depicts such conditions in which very low vehicle detection is achieved. Similarly, as shown in the 2nd image of the top row of Figure 11, low vehicle detection is observed during

a rainy night when the head lights of the vehicle are also turned on. In this case an electricity pole also appears, which results in partial occlusion that ultimately results in reduced object detection.

Figure 11. Sample images of low vehicle detection from compared datasets

- We observe that our proposed method still needs to perform well in different situations, such as when the scene is contaminated with the snow storm or blizzard as shown in the 2nd row of Figure 11. In such cases, background noise dominates results in low visibility. In this scenario, a Retinex-based image enhancement scheme might be useful. For such a scenario, we suggest that an image dehazing-based enhancement could also be effective. We are optimistic that this proposed solution will essentially enhance the object and image scene, which will later make life easier for any of the vehicle detectors deployed. Ultimately, the application of image enhancement technique will significantly increase the detection ability of object detector.
- For images where snow is dominant, image appears overly white, which also decreases the detection accuracy of state-of-the-art object detection methods. In this case, image contrast correction might produce the desirable results. In many cases, the occlusions on the road also pose a threat to the detector, which ultimately results in false detections. In such cases, an occlusion handling method could also be used to reliably detect any object.

- For all of the aforementioned discussion, Figure 11 shows a few of the sample images where our developed method struggles. In images shown in Figure 11, our method either yields a very low vehicle detection rate or produces false detections. Therefore, future research could also focus on few of the cases as shown in Figure 11.

4.6. Final Remarks

Detailed analysis discussed in this paper indicates that vehicle detection has been an active research field in recent years. From providing early warning signals and monitoring up to exercising control, there are several examples of major research in intelligent vehicle detection. This paper presented a detailed analysis on vehicle detection on three publicly available dataset. For the task of vehicle detection, YOLO-v5-based architecture was used. To make the YOLO-v5-based architecture more intelligent and flexible, a transfer learning methodology was introduced. Detailed analysis indicated that the proposed approach performed well on challenging datasets.

In addition, a detailed comparison of the proposed method was carried with fourteen recent state-of-the-art approaches. We are optimistic that this study will be a fair guideline for beginners and practitioners to modify or use any detector for their desired tasks or applications. Below we list the final summary of developed vehicle detection method on three datasets.

PKU: On this dataset, in the G1 category, the proposed method yielded mAP of 99.94%. In the G2 and G3 categories, the proposed method yielded 100% vehicle detection mAP. In the G4 and G5 categories, the proposed method yielded 99.73% and 99.96% mAP, respectively. Overall, on the PKU dataset our method yielded 99.92% vehicle detection accuracy. Out of the fourteen compared methods on the PKU dataset, the proposed method ranked 1st among all compared approaches.

COCO: On this dataset, with image resolution of 512 × 512 pixels, the proposed method yielded 52.31% mAP values and ranked 1st among all the compared methods.

DAWN: This dataset contains four prominent sub classes, which are fog, rain, snow, and sand. On images that were affected by fog, our proposed method yielded a mAP value of 29.66% and ranked 3rd out of fourteen compared methods in this category. Meanwhile, for images that contained rain, our developed method produced a mAP value of 41.21% and ranked 1st along with [31] among all compared works. For images that contained snow, our method yielded a 43.01% mAP value and ranked 2nd among all compared works. In this class, the work developed in [37] ranked 1st by yielding a 43.02% mAP value. For images that contained sand, our developed method yielded a 24.13% mAP value and ranked 1st among all methods. In this class, the work developed in [28] also produced a par result by yielding a mAP value of 24.10%.

5. Conclusions

This paper discussed an accurate, fast, and robust vehicle detection method based on the YOLO-v5 architecture. To develop a robust object detection algorithm, transfer learning was performed. The proposed object detection method was tested on three publicly available datasets, which are the PKU, COCO, and DAWN datasets. Simulation results demonstrated that the proposed method is effective at handling various challenging situations, such as night, rainy, and snow conditions. The proposed method significantly elevated the accuracy and operational efficiency. In addition, the detection technique proposed in this research can additionally be relevant to a large number of real time applications. However, the only caveat is that a giant quantity of data is required for training of the detection model. The YOLO-vs-based vehicle detection method discussed in this paper achieved a 99.92% detection accuracy on the PKU dataset and outperformed five methods compared therein. Similarly, on the COCO dataset, the proposed method yielded a superior mean average precision than several methods compared therein. Furthermore, for highly challenging conditions in the DAWN dataset, the proposed method was superior in terms of detection accuracy for fog, rain, snow, and sandy conditions.

In the future, the proposed method can be further investigated to detect the occluded vehicles. Moreover, for moving objects, motion blur could also be investigated. Furthermore, a cloud computing-based domain can be introduced to handle the resources of complex machine learning algorithms. Our algorithm could also be investigated for haze images in which there is very limited visibility and thus vehicles are barely visible to human eye. Moreover, the impact of changes in the network structure of each type of a YOLO model could also be further explored on the datasets explored in this study. Finally, our developed method could be integrated with deep learning methods to further explore the research of vehicle detection, tracking, or recognition.

Author Contributions: Conceptualization, A.F. and F.H.; methodology, A.F.; software, A.F., K.K., M.S., U.K., Z.M. and F.H.; validation, A.F. and F.H.; formal analysis, A.F., K.K., M.S., U.K., Z.M. and F.H.; investigation, A.F. and F.H.; resources, A.F., K.K., M.S., U.K., Z.M. and F.H.; data curation, A.F.; writing—original draft preparation, A.F., K.K., M.S., U.K., Z.M. and F.H.; writing—review and editing, A.F., K.K., M.S., U.K., Z.M. and F.H.; visualization, A.F. and F.H.; supervision, F.H.; project administration, F.H. All authors have read and agreed to the published version of the manuscript.

Funding: This research received no external funding.

Institutional Review Board Statement: Not applicable.

Informed Consent Statement: Not applicable.

Data Availability Statement: Not applicable.

Conflicts of Interest: The authors declare no conflict of interest.

References

1. Mahmood, Z.; Haneef, O.; Muhammad, N.; Khattak, S. Towards a Fully Automated Car Parking System. *IET Intell. Transp. Syst.* **2018**, *13*, 293–302. [CrossRef]
2. Xiaohong, H.; Chang, J.; Wang, K. Real-time object detection based on YOLO-v2 for tiny vehicle object. *Procedia Comput. Sci.* **2021**, *183*, 61–72.
3. Rani, E. LittleYOLO-SPP: A delicate real-time vehicle detection algorithm. *Optik* **2021**, *225*, 165818. [CrossRef]
4. Tajar, T.; Ramazani, A.; Mansoorizadeh, M. A lightweight Tiny-YOLOv3 vehicle detection approach. *J. Real-Time Image Process.* **2021**, *18*, 2389–2401. [CrossRef]
5. Mahmood, Z.; Bibi, N.; Usman, M.; Khan, U.; Muhammad, N. Mobile Cloud based Framework for Sports Applications. *Multidimens. Syst. Signal Process.* **2019**, *30*, 1991–2019. [CrossRef]
6. Hamsa, S.; Panthakkan, A.; Al Mansoori, S.; Alahamed, H. Automatic Vehicle Detection from Aerial Images using Cascaded Support Vector Machine and Gaussian Mixture Model. In Proceedings of the 2018 International Conference on Signal Processing and Information Security (ICSPIS), Dubai, United Arab Emirates, 7–8 November 2018; pp. 1–4.
7. Mikaty, M.; Stathaki, T. Detection of Cars in HighResolution Aerial Images of Complex Urban Environments. *IEEE Trans. Geosci. Remote Sens.* **2017**, *55*, 5913–5924.
8. Arı, Ç.; Aksoy, S. Detection of Compound Structures Using a Gaussian Mixture Model With Spectral and Spatial Constraints. *IEEE Trans. Geosci. Remote Sens.* **2014**, *52*, 6627–6638. [CrossRef]
9. Hbaieb, A.; Rezgui, J.; Chaari, L. Pedestrian Detection for Autonomous Driving within Cooperative Communication System. In Proceedings of the 2019 IEEE Wireless Communications and Networking Conference (WCNC), Marrakesh, Morocco, 15–18 April 2019; pp. 1–6.
10. Xiong, L.; Yue, W.; Xu, Q.; Zhu, Z.; Chen, Z. High Speed Front-Vehicle Detection Based on Video Multi-feature Fusion. In Proceedings of the 2020 IEEE 10th International Conference on Electronics Information and Emergency Communication (ICEIEC), Beijing, China, 17–19 July 2020; pp. 348–351.
11. Yawen, T.; Jinxu, G. Research on Vehicle Detection Technology Based on SIFT Feature. In Proceedings of the 8th International Conf on Electronics Info. and Emergency Communication (ICEIEC), Beijing, China, 15–17 June 2018; pp. 274–278.
12. Li, Y.; Wang, H.; Dang, L.M.; Nguyen, T.N.; Han, D.; Lee, A.; Jang, I.; Moon, H. A Deep Learning-Based Hybrid Framework for Object Detection and Recognition in Autonomous Driving. *IEEE Access* **2020**, *8*, 194228–194239. [CrossRef]
13. Li, Y.; Li, S.; Du, H.; Chen, L.; Zhang, D.; Li, Y. YOLO-ACN: Focusing on small target and occluded object detection. *IEEE Access* **2020**, *8*, 227288–227303. [CrossRef]
14. Cai, Z.; Vasconcelos, N. Cascade R-CNN: Delving into High Quality Object Detection. In Proceedings of the IEEE Conference on Computer Vision and Pattern Recognition, Salt Lake City, UT, USA, 18–23 June 2018; pp. 6154–6162.

15. Wang, C.; Wang, H.; Yu, F.; Xia, W. A High-Precision Fast Smoky Vehicle Detection Method Based on Improved Yolov5 Network. In Proceedings of the 2021 IEEE International Conference on Artificial Intelligence and Industrial Design (AIID), Guangzhou, China, 28–30 May 2021; pp. 255–259. [CrossRef]
16. Miao, Y.; Liu, F.; Hou, T.; Liu, L.; Liu, Y. A Nighttime Vehicle Detection Method Based on YOLO v3. In Proceedings of the 2020 Chinese Automation Congress (CAC), Shanghai, China, 6–8 November 2020; pp. 6617–6621. [CrossRef]
17. Sarda, A.; Dixit, S.; Bhan, A. Object Detection for Autonomous Driving using YOLO [You Only Look Once] algorithm. In Proceedings of the 2021 Third International Conference on Intelligent Communication Technologies and Virtual Mobile Networks (ICICV), Tirunelveli, India, 4–6 February 2021; pp. 1370–1374. [CrossRef]
18. Zhao, S.; You, F. Vehicle Detection Based on Improved Yolov3 Algorithm. In Proceedings of the 2020 International Conference on Intelligent Transportation, Big Data & Smart City (ICITBS), Vientiane, Laos, 11–12 January 2020; pp. 76–79. [CrossRef]
19. Ćorović, A.; Ilić, V.; Đurić, S.; Marijan, M.; Pavković, B. The Real-Time Detection of Traffic Participants Using YOLO Algorithm. In Proceedings of the 2018 26th Telecommunications Forum (TELFOR), Belgrade, Serbia, 20–21 November 2018; pp. 1–4. [CrossRef]
20. Lou, L.; Zhang, Q.; Liu, C.; Sheng, M.; Zheng, Y.; Liu, X. Vehicles Detection of Traffic Flow Video Using Deep Learning. In Proceedings of the 2019 IEEE 8th Data Driven Control and Learning Systems Conference (DDCLS), Dali, China, 24–27 May 2019; pp. 1012–1017. [CrossRef]
21. Machiraju, G.S.R.; Kumari, K.A.; Sharif, S.K. Object Detection and Tracking for Community Surveillance using Transfer Learning. In Proceedings of the 2021 6th International Conference on Inventive Computation Technologies (ICICT), Coimbatore, India, 20–22 January 2021; pp. 1035–1042. [CrossRef]
22. Snegireva, D.; Kataev, G. Vehicle Classification Application on Video Using Yolov5 Architecture. In Proceedings of the 2021 International Russian Automation Conference (RusAutoCon), Sochi, Russia, 5–11 September 2021; pp. 1008–1013. [CrossRef]
23. Jana, A.P.; Biswas, A.; Mohana. YOLO based Detection and Classification of Objects in video records. In Proceedings of the 2018 3rd IEEE International Conference on Recent Trends in Electronics, Information & Communication Technology (RTEICT), Bangalore, India, 18–19 May 2018; pp. 2448–2452. [CrossRef]
24. Hu, X.; Wei, Z.; Zhou, W. A video streaming vehicle detection algorithm based on YOLOv4. In Proceedings of the 2021 IEEE 5th Advanced Information Technology, Electronic and Automation Control Conference (IAEAC), Chongqing, China, 12–14 March 2021; pp. 2081–2086. [CrossRef]
25. Kasper-Eulaers, M.; Hahn, N.; Berger, S.; Sebulonsen, T.; Myrland; Kummervold, P.E. Short Communication: Detecting Heavy Goods Vehicles in Rest Areas in Winter Conditions Using YOLOv5. *Algorithms* **2021**, *14*, 114. [CrossRef]
26. De Carvalho, O.L.F.; de Carvalho Júnior, O.A.; de Albuquerque, A.O.; Santana, N.C.; Guimarães, R.F.; Gomes, R.A.T.; Borges, D.L. Bounding box-free instance segmentation using semi-supervised iterative learning for vehicle detection. *IEEE J. Sel. Top. Appl. Earth Obs. Remote Sens.* **2022**, *15*, 3403–3420. [CrossRef]
27. Tayara, H.; Soo, K.G.; Chong, K.T. Vehicle detection and counting in high-resolution aerial images using convolutional regression neural network. *IEEE Access* **2018**, *6*, 2220–2230. [CrossRef]
28. Zhao, J.; Hao, S.; Dai, C.; Zhang, H.; Zhao, L. Improved Vision-Based Vehicle Detection and Classification by Optimized YOLOv4. *IEEE Access* **2022**, *10*, 8590–8603. [CrossRef]
29. Hassaballah, M.; Mahmoud; Kenk, M.; Muhammad, K.; Minaee, S. Vehicle detection and tracking in adverse weather using a deep learning framework. *IEEE Trans. Intell. Transp. Syst.* **2020**, *22*, 4230–4242. [CrossRef]
30. Mahmood, Z.; Khan, K.; Khan, U.; Adil, S.H.; Ali, S.S.A.; Shahzad, M. Towards Automatic License Plate Detection. *Sensors* **2022**, *22*, 1245. [CrossRef]
31. Wu, T.H.; Wang, W.T.; Liu, Y.Q. Real-time vehicle and distance detection based on improved yolo v5 network. In Proceedings of the 2021 3rd World Symposium on Artificial Intelligence (WSAI), Guangzhou, China, 18–20 June 2021; pp. 24–28.
32. Lin, T.; Maire, M.; Belongie, S.; Hays, J.; Perona, P.; Romanan, D.; Dollar, P.; Zitnick, C. Microsoft COCO: Common objects in context. In Proceedings of the Computer Vision–ECCV 2014: 13th European Conference, Zurich, Switzerland, 6–12 September 2014; pp. 740–755.
33. He, Y.; Zhu, C.; Wang, J.; Savvides, M.; Zhang, X. Bounding box regression with uncertainty for accurate object detection. In *Proceedings of the In Proceedings of the Ieee/Cvf Conference on Computer Vision and Pattern Recognition, Long Beach, CA, USA, 15–20 June 2019 pp.*; pp. 2888–2897.
34. Liu, K.; Wang, W.; Tharmarasa, R.; Wang, J. Dynamic vehicle detection with sparse point clouds based on PE-CPD. *IEEE Trans. Intell. Transp. Syst.* **2019**, *20*, 1964–1977. [CrossRef]
35. Zhao, Q.; Sheng, T.; Wang, Y.; Tang, Z.; Chen, Y.; Cai, L.; Ling, H. M2Det: A single-shot object detector based on multi-level feature pyramid network. In Proceedings of the AAAI Conference on Artificial Intelligence, Montréal, QC, Canada, 17 July 2019; Volume 33, pp. 9259–9266.
36. Li, Y.; Chen, Y.; Wang, N.; Zhang, Z. Scale-aware trident networks for object detection. *arXiv* **2019**, arXiv:1901.01892.
37. Zhang, H.; Tian, Y.; Wang, K.; Zhang, W.; Wang, F.-Y. Mask SSD: An effective single-stage approach to object instance segmentation. *IEEE Trans. Image Process.* **2020**, *29*, 2078–2093. [CrossRef]
38. Min, W.; Fan, M.; Guo, X.; Han, Q. A new approach to track multiple vehicles with the combination of robust detection and two classifiers. *IEEE Trans. Intell. Transp. Syst.* **2018**, *19*, 174–186. [CrossRef]
39. Law, H.; Deng, J. CornerNet: Detecting objects as paired keypoints. In Proceedings of the European Conference on Computer Vision (ECCV), Glasgow, UK, 17–24 May 2018; pp. 734–750.

40. He, K.; Gkioxari, G.; Dollár, P.; Girshick, R. Mask R-CNN. In Proceedings of the IEEE International Conference on Computer Vision, Venice, Italy, 22–29 October 2017; pp. 2961–2969.
41. Shi, F.; Zhang, T.; Zhang, T. Orientation-Aware Vehicle Detection in Aerial Images via an Anchor-Free Object Detection Approach. *IEEE Trans. Geosci. Remote Sens.* **2021**, *59*, 5221–5233. [CrossRef]
42. Hu, X.; Xu, X.; Xiao, Y.; Chen, H.; He, S.; Qin, J.; Heng, P.-A. SINet: A scale-insensitive convolutional neural network for fast vehicle detection. *IEEE Trans. Intell. Transp. Syst.* **2019**, *20*, 1010–1019. [CrossRef]

applied sciences

MDPI

Article

Recognition and Classification of Handwritten Urdu Numerals Using Deep Learning Techniques

Aamna Bhatti [1], Ameera Arif [1], Waqar Khalid [2], Baber Khan [3], Ahmad Ali [4], Shehzad Khalid [2] and Atiq ur Rehman [5,*]

1 School of Electrical Engineering and Computer Science, National University of Sciences and Technology, Islamabad 24090, Pakistan
2 Computer Engineering Department, Bahria University, Islamabad 44000, Pakistan
3 Department of Electrical and Computer Engineering, International Islamic University, Islamabad 04436, Pakistan
4 Department of Software Engineering, Bahria University, Islamabad 44000, Pakistan
5 Artificial Intelligence and Intelligent Systems Research Group, School of Innovation, Design and Engineering, Mälardalen University, Högskoleplan 1, 722 20 Västerås, Sweden
* Correspondence: atiq.ur.rehman@mdu.se or atiqjadoon@gmail.com

Abstract: Urdu is a complex language as it is an amalgam of many South Asian and East Asian languages; hence, its character recognition is a huge and difficult task. It is a bidirectional language with its numerals written from left to right while script is written in opposite direction which induces complexities in the recognition process. This paper presents the recognition and classification of a novel Urdu numeral dataset using convolutional neural network (CNN) and its variants. We propose custom CNN model to extract features which are used by Softmax activation function and support vector machine (SVM) classifier. We compare it with GoogLeNet and the residual network (ResNet) in terms of performance. Our proposed CNN gives an accuracy of 98.41% with the Softmax classifier and 99.0% with the SVM classifier. For GoogLeNet, we achieve an accuracy of 95.61% and 96.4% on ResNet. Moreover, we develop datasets for handwritten Urdu numbers and numbers of Pakistani currency to incorporate real-life problems. Our models achieve best accuracies as compared to previous models in the literature for optical character recognition (OCR).

Keywords: urdu numeral recognition; convolutional neural network; SVM; GoogLeNet; ResNet

Citation: Bhatti, A.; Arif, A.; Khalid, W.; Khan, B.; Ali, A.; Khalid, S.; Rehman, A.u. Recognition and Classification of Handwritten Urdu Numerals Using Deep Learning Techniques. *Appl. Sci.* **2023**, *13*, 1624. https://doi.org/10.3390/app13031624

Academic Editor: José Salvador Sánchez Garreta

Received: 5 December 2022
Revised: 11 January 2023
Accepted: 20 January 2023
Published: 27 January 2023

1. Introduction

OCR technology scans printed characters to determine their shape by recognizing edge information, and then translates them into characters by process of character recognition [1]. In recent years, one of the most fascinating and difficult research areas in the fields of image processing and pattern recognition has been handwriting recognition. It has many applications such as OCR, pattern classification, postal mail sorting, bank cheque processing, form data entry, etc. Such character recognizers prove to be fruitful for humans because of their speed and accuracy. Mostly, they are based on deep learning models and solve the problem efficiently.

In the study of languages, Urdu is one of those cursive languages that is hugely popular in South Asian countries such as Pakistan, India, Bangladesh, Bhutan, and Nepal. It is the national language of Pakistan and is widely spoken in urban areas, while its adoption as a second language in rural areas is in progress. It is an amalgamation of many languages; hence, its script contains loanwords and is written in different variants all around the globe. Another important characteristic is that it is bidirectional with its numerals written from left to right while its script is written in the opposite direction [2]; therefore, this becomes a problem for OCR. It has up to 40 letters in its script and 10 numerals [3]. Urdu, Persian, and Eastern Arabic numerals are written on similar patterns; however, some of the digits differ, as shown in Figure 1, which is yet again another concern. Some other challenges that are encountered during its OCR is blurred text, torn paper, and spacing between letters when the data is found in written form [3].

Urdu	۰	۱	۲	۳	۴	۵	۶	۷	۸	۹
Persian	۰	۱	۲	۳	۴	۵	۶	۷	۸	۹
Eastern Arabic	٠	١	٢	٣	٤	٥	٦	٧	٨	٩

Figure 1. Urdu, Persian, and Eastern Arabic numerals.

Our aim is to develop a classifier for our novel dataset. The motivation behind developing an Urdu numeral classifier is that majority of the work has been done on English numerals; however, no such work has been done on Urdu numerals which are distinct in features as compared to English. Data unavailability is a major obstacle in the development of Urdu handwritten character recognition [4,5]. As no dataset of Urdu numerals is available publicly for research purposes, we, therefore, present a novel dataset gathered specifically for this study. In this way, problems related to Urdu character recognition, OCR, and intelligent character recognition (ICR) can be solved as efficiently as they are solved in other languages. The previous datasets are of Latin, Arabic handwritten numerals [6], text lines of Urdu [7], integration of MNIST with Urdu numerals [8], Persian numerals [9], Bengali numerals [10], Urdu Nastalique Handwritten Dataset (UNHD) [11], Urdu Printed Text Image Database (UPTI) [12], and Urdu spoken and text words [13]. We did come across an Urdu numerals' dataset by Husnain et al. [14], but that did not incorporate variations of crumbled, torn, and ink spots on paper. So, we sought to create our novel dataset of 9800 images of handwritten Urdu numerals written by over 200 individuals with their left and right hands. The difficulty of accurately classifying handwritten characters is increased by variances in writing style, character size and form, and resemblances to other characters [5]. Hence these papers were then crumbled and torn, and some had ink spots added on them for variety. The pictures of these images were preprocessed by employing a Gaussian filter and connected component labeling. These images were then fed to machine learning and deep learning models for classification purpose. Deep learning models find out complex structures in massive datasets by using the backpropagation algorithm to indicate how a machine can change its internal parameters in each layer from representation in the previous layer [15]. We applied our proposed CNN, GoogLeNet, ResNet, and SVM on the dataset. Our proposed solution is powerful, yet simple, and results in a performance which is comparable or higher than the state-of-the-art when evaluated on our novel dataset. Since the model is deployed for real-world applications, we tested its reliability on practical applications, i.e., Pakistani currency. The chosen currency notes were 10, 20, 50, 100, 1000, and 5000. A sample of test images is shown in Figure 2.

Figure 2. Test images of Pakistani currency notes.

The main contributions of this paper are:

- Proposition of new Urdu numerals dataset that contains variations because of crumbled, torn, and ink spotted paper.
- After CNN extracts the features, we use two different activation functions: SVM and Softmax.

- The proposed CNN is compared with GoogLeNet and ResNet. The conducted experiments suggest our models' better accuracy.

So, the major advantage of our proposed work is that our data includes noise added by environment as compared to the previous datasets that were collected in simple situations. Now, our models also learn these distinctions and, hence, perform relatively well on real life examples, which was yet again missing in previous work.

The paper is divided into five sections. Section 2 elaborates the state-of-the-art techniques, our dataset collection, and other datasets available for Urdu language classifiers. The proposed model and technical details are discussed in Section 3. Section 4 provides a review of the classification results. In Section 5, the details of the experiments and their corresponding results are explained, while last section concludes the paper and presents a direction for future work.

2. Literature Review

Handwriting recognition has been around in the field of computer science for almost half a century now. In [1], the oldest techniques that have been in use for character recognition since 1959 are discussed. It originates from the work suggested by Eden in 1968 known as analysis-by-synthesis. This is the basis for syntactic approaches in character recognition. K. Gaurav and Bhatia P. K. [2] discuss the advancements in preprocessing techniques when input data ranges from simple handwritten documents to deformed images, or images having viewpoint variation and background clutter. They concluded that applying only one preprocessing technique is never enough to obtain high accuracies, but rather a mixture of preprocessing techniques is applied to obtain reliable results. Basically, two types of recognition systems exist; online systems and offline systems [16,17]. Online character recognition works when user writes on a special writing surface, computer recognizes it and converts it into codes with respect to time, while offline recognition systems are images or documents fed as input with text written on them and are converted into digital form. Offline recognition works in phases where images are first segmented, cropped, and resized, their features are extracted, and then they are classified [18]. This paper presents work on Urdu numerals using offline recognition. Table 1 presents a summary of the accuracies achieved for other algorithms using Urdu datasets. All these papers used different datasets and since those datasets are not available publicly, it was difficult for us to compare them with our techniques. Additionally, these datasets were collected on different lines and solved various problems, but underlying concept is same that they work on variations of Urdu language datasets; hence, their comparison is necessary.

Table 1. Accuracy reported on different techniques.

Article Reference	Techniques Applied	Accuracy Achieved
[19]	Back propagation neural network	90%
[8]	Kohonen self-organization maps	91%
[20]	Shape context-based digit recognition computation	93%
[21]	Fuzzy rule	97.4%
[22]	Capsule-Net	98.5%

In [23], N. Gautam, R. S. Sharma, and G. Hazrati state work done on Eastern Arabic numerals through OCR. S. Abdelazeem et al. [6] compared the problems encountered in Latin and Arabic handwritten numerals by using the Arabic Handwritten Digits Database (ADBase) [24]. H. Kour and N. K. Gondhi proposed a recognition system [19] based on approaches of segmentation for feature extraction, slant analysis for slant removal, and dictionary search for classification. It resulted in a recognition rate of 93% for isolated characters and the same for numerals. In another study [25], J. Memon, M. Sami, and R. A. Khan provided an in-depth review of statistical, kernel, artificial neural network (ANN), template matching, and structural methods for classification of OCR. In a very interesting work presented by Ahmed, S.B., Naz, S., Swati, S. et al. [7], a 6.04–7.93% error rate on

700 unique text lines was achieved (including Urdu numerals and Urdu handwritten samples) after applying 1-D bidirectional long short-term memory (BLSTM) networks. In [8], L. Javed, M. Shafi, M. I. Khattak, and N. Ullah presented the utilization of Kohonen self-organization maps on 6000 handwritten Urdu numerals and obtained an efficiency of 91%. A work similar to this paper was conducted by Saad Ahmad in [7], where Urdu text was integrated with the modified National Institute of Standards and Technology dataset (MNIST) to learn the similar nature of patterns. They used CNN and multidimensional long short-term memory (MDLSTM) on UNHD samples by pretraining the network on MNIST. Their results showed the highest recall of 0.84 and 0.93 for precision. With their roots in statistical learning theory, SVMs have been used widely for image classification and character recognition tasks, so we studied their uses in different languages. In [26], Ebrahimzadeh, R. et al. employed a linear SVM as a classifier for the MNIST dataset to obtain a 97.25% accuracy rate. Das et al. [27] extracted local features of a handwritten Bangla digits dataset using a genetic algorithm which were then fed to an SVM. It gave promising results of 96.7%. Abu et al., in [10], discussed a task-oriented model that make use of densely connected neural networks termed Bengali handwriting digit classification (BDNet). The ISI Bengali handwritten numeral dataset was used to train it. In [28], Duddela et al. discussed the task of image classification by employing NN and CNN on Devangari script. Fatemeh et al. [29] proposed a novel approach that stacks ensemble classifiers to identify handwritten numbers. They employed CNN and BLSTM that takes the probability vector of the image class as an input. The model has been tested on Arabic and Persian numerals. In [9], Savita et al. proposed a hybrid model that combines CNN that serves as a feature extractor and SVM that acts as a binary classifier.

Finally, the recent notable technique MetaQNN discussed in [30] which was put forward in 2018. It relies on reinforcement learning for the design of CNN architectures and has its roots in the neuroevolution of committees of CNN. It has an error rate of 0.44% and 0.32% when using an ensemble of the most appropriate found neural networks. Lastly, data collection covers a huge portion of this paper and major work was performed on these preprocessing of images. Ahmed, R. et al., in [31], provided the insight on how to go forward with the data collection and preprocessing. They implemented the algorithms of binarization, dots removing, and thinning which are used for our feature extraction phase.

In deep learning, the next step for an OCR-based problem is the dataset where standardization is essential to obtain exemplary results. During exploration and definition of the problem statement, no standard data set was found. We did come across a dataset containing Urdu spoken and text words by EMILLE (Enabling Minority Language Engineering) [13], which is a collection of 67 million word corpus of South Asian languages. Another ligature corpus by the Centre for Language Engineering (CLE) [18] in Pakistan has been extracted from a 19.3 million corpus gathered from different domains such as sports, news, finance, culture, and consumer information. A similar ligature corpus presented by Sabbour and Shaifat [12] is called Urdu Printed Text Image Database (UPTI) which was created along similar lines as the Arabic Printed Text Image (APTI), proposed by Slimane et al. [32]. UPTI contains 10,063 synthetically generated text lines and ligature images. Lastly, an offline dataset by Ahmed et al. [11] for Urdu text by the name of Urdu Nastalique Handwritten Dataset (UNHD) was found. It was created in 2013 by collecting samples of 8 Urdu text lines having few Urdu numerals to produce 312,000 words with 10,000 text lines.

3. Proposed Model

In this section, we present our CNN model which is trained to learn Urdu numerals. The network is trained on raw image pixels having preprocessed and cropped images of Urdu numerals. It classifies the dataset into 10 feature-mapped classes. Figure 3 elaborates these steps in the form of a block diagram and the following subsections give an insight into these steps. We start with preprocessing the set of images to get them into desired shape for all models. Our custom built CNN is applied to the final set of images. The features extracted from our CNN are fed to the Softmax activation function and SVM classifier in parallel to obtain an in-depth review of our results. GoogLeNet and ResNet architectures are also applied on the set of images using transfer learning to compare results with our base architecture.

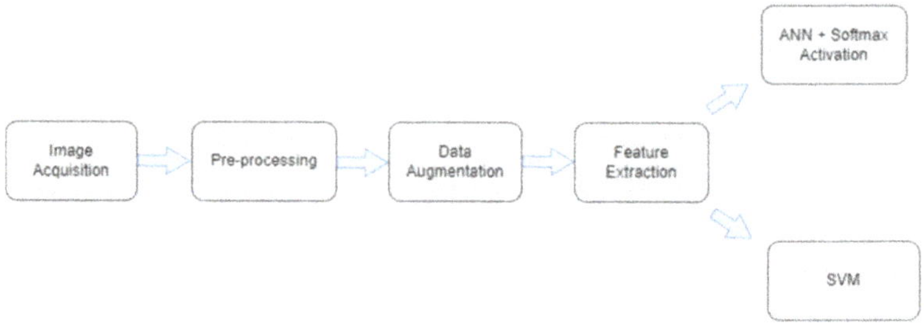

Figure 3. Block diagram.

3.1. *Image Acquisition*

In order to continue with our research work, we sought to make our own unique dataset as all the available datasets did not match our paper's requirements. Our dataset contains a total of 9800 images of 10 Urdu numerals written with left and right hands to create diversity. Each person wrote 0 to 9 numerals four times. These were people belonging to various age groups and different fields of life. This was done subconsciously to include people with diverse writing styles so as to bring variety to our dataset.

3.2. *Preprocessing*

The scanned pages of handwritten data are used for preprocessing to remove any redundant information that could be misclassified by employing connected component labeling. In order to contain different types of noise in our dataset, we crumbled some pages as shown in Figure 4, added additional dots on the page, and dropped ink spots so the classifier does not have a simple version of the dataset but instead has complex samples. These modified pages are then filtered and thresholded to maintain their maximum information. Firstly, the noise is suppressed by applying a Gaussian filter which not only subdues the effect of noise but also maintains sharp edges. A Gaussian filter with different sigma values is applied. The ideal sigma value is found to be 3, which is checked in accordance with thresholding as shown in Figure 5.

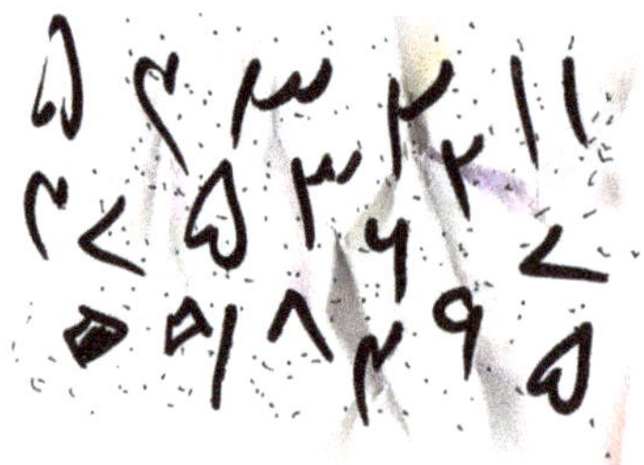

Figure 4. Actual crumbled paper.

Finally, the images are thresholded to obtain binary images. With different experiments, we found out that if the threshold value is kept low, i.e., 50, it removes the background noise completely. However, the problem is that it also lightens the boundaries of the handwritten numbers which is a great failure as shown in Figure 6. On the other hand, if a large threshold value is set, i.e., 150, it incorporates all noise in the image. The noisy dots which are closer to handwritten numbers when joined with them got mixed. This is illustrated in Figure 7 where it is difficult to distinguish the two 2s. This can result in a loss for our data because the model would be unable to classify such numbers. So, after intense experimentation, we found that the appropriate thresholding level is 120 which retains an accurate amount of information according to Figure 8. The images are cropped in a way so as to remove maximum background and obtain images similar to MNIST. MNIST [33] is a benchmark dataset for numerals, and its images are normalized and centered in a fixed-size image. Finally, the images are resized to 32 by 32 pixels and saved in their respective folders for ease of labeling.

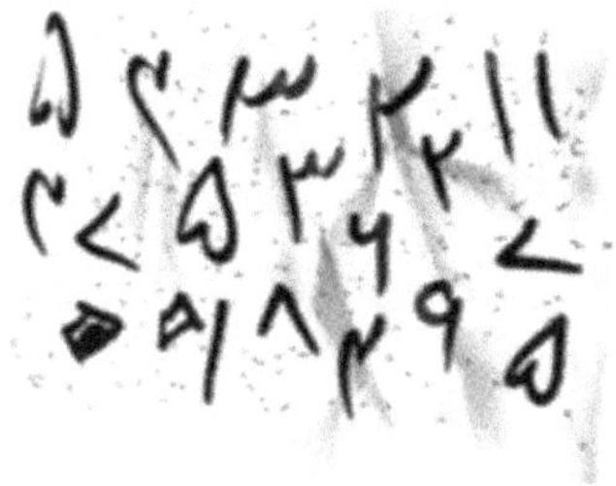

Figure 5. Application of Gaussian filter.

Figure 6. Low threshold level.

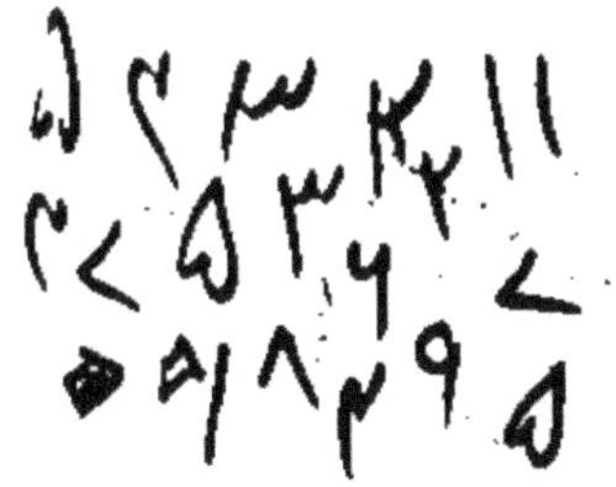

Figure 7. High threshold level.

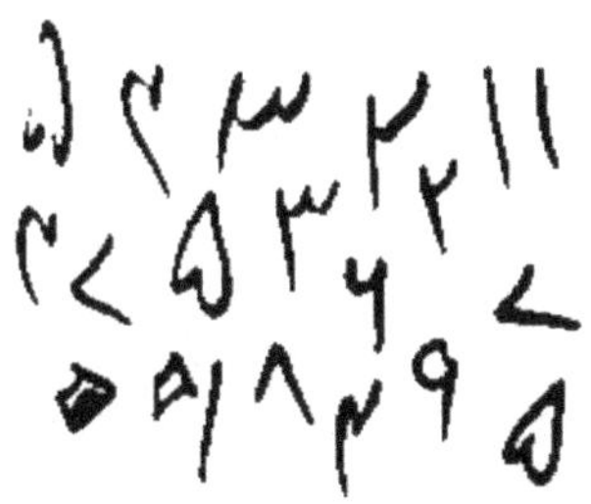

Figure 8. Ideal threshold level.

3.3. Augmentation

Data augmentation is a technique to create new data artificially so as to bring variation into the original dataset. Different techniques of augmentation (such as flipping, rotating, and adding noise) are applied on separate numbers because some numerals upon these transformations change into another number and could not preserve their respective labels [34]. For instance, the numbers zero, one, and five are flipped and labelled as is because they are unaffected by flipping.

However, when digit eight is rotated in a 45° counterclockwise direction it becomes Urdu digit seven. Thus, it is then labelled as seven. Similarly, digit seven is rotated 45° in clockwise direction to resemble digit eight as shown in Figure 9a. As Urdu digit two and six are similar in terms of shape and are counter flips of each other, the digit two is flipped and labelled as six, while digit six is flipped to be labelled as two. Deep research is performed regarding these transformations as a slight error could change the class of data. For example, digit six could change into digit two upon a vertical flip as shown in Figure 9b. For rest of the numbers, three, four, and nine, noise is added to them which is depicted in Figure 9c. As a result of augmentation, our dataset increased to almost 13,000.

(a) (b) (c)

Figure 9. Sample of (**a**) rotated image, (**b**) flipped image, and (**c**) noisy image.

3.4. Feature Extraction

3.4.1. Convolutional Neural Network

We propose a dataset of Urdu handwritten numerals with 10 labels and 13,000 images in all. Initially, all the images are standardized by dividing current pixel value by sum of all pixels. This represents image pixels in range of 0 to 1. Then, each image is resized to new dimensions while ensuring that no information is lost. The CNN model consists of four core substructures which are used repeatedly with different nonlinearities to bring the best results.

1. The input layer contains raw pixel values, and, in this case, each image of the size $32 \times 32 \times 3$ pixels is fed to the CNN. Here, 32 represents the width and height of the image while 3 is the color channels—red, green, and blue.
2. The convolution layer connects local receptive field of the input with neurons in the next layer. This is achieved through a simple dot product of kernel and input image. A kernel size of 3×3 is maintained throughout the model, whereas padding is set to 1. It is followed by batch normalization of convolution layer. Each output of convolution layer uses the ReLU activation function followed by pooling layer. ReLU activations work better than sigmoid function in terms of gradient vanishing problems. ReLU was picked out of other nonlinearities (e.g., tanh, sigmoid) after comparing their results in our CNN model. Batch normalization is applied after each convolution layer to improve generalization [35].
3. The pooling layer down samples input along spatial dimensions. One of the most famous pooling layers is 'Max Pooling' which is used here to extract the highest pixel value in the current space. These extracted features are then fed to the classifiers which are discussed further. Figure 10 depicts the architecture of our proposed model and Table 2 provides an analysis of required computation resources and learning parameters.

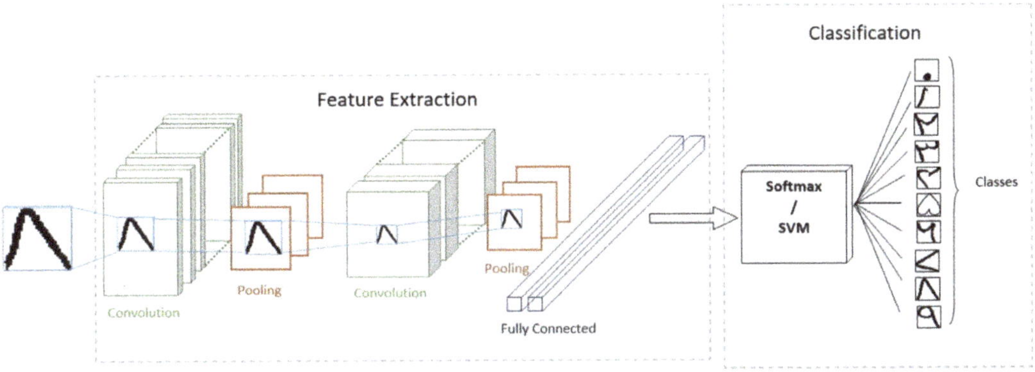

Figure 10. Architecture of CNN for Urdu handwritten numeral recognition.

Table 2. Summary of proposed CNN model.

Layer	Output Shape	Number of Parameters
Conv	32 × 32 × 256	7168
Batch Norm	32 × 32 × 256	1024
MaxPool	16 × 16 × 256	0
Conv	16 × 16 × 128	295,040
Batch Norm	16 × 16 × 128	512
MaxPool	8 × 8 × 128	0
Flatten	8192	0
Dense	90	737,370
Dense	64	5824
Dense	10	650
Total parameters		1,047,588
Trainable parameters		1,046,820
Nontrainable parameters		768

3.4.2. GoogLeNet

GoogLeNet [36] is based on the idea of an inception layer that covers a large area but maintains fine resolution on a dataset for small information. Because GoogLeNet achieved the top 5 error rate of 6.67%, we used it to train the Urdu numeral dataset. A major task was to tune the three main parameters learning rate, number of epochs, and batch size. A batch size of 32 with a learning rate of 0.001 gave us the best results for 30 epochs. The learning curve for GoogLeNet is shown in Figure 11 which does not show underfitting or overfitting as both losses reached a point of stability. This is an exceptionally good result for a novel dataset like ours.

3.4.3. ResNet

The intuition behind ResNet is that deep neural networks are hard to train due to their huge number of layers, especially where the problem of vanishing gradient occurs [37]. ResNet50 is the variant that is used in this paper. It is 50 layers deep as the name indicates and its pretrained version from the ImageNet dataset is used. Images of size 224 × 224 with 3 color channels were used. For ResNet, batch size and epochs were used as tunable parameters. Here, a batch size of 16 with a learning rate of 0.001 and 40 epochs achieved the best results. The learning curves show a good generalization between training and validation data on the Urdu numeral dataset. They are plotted in Figure 12.

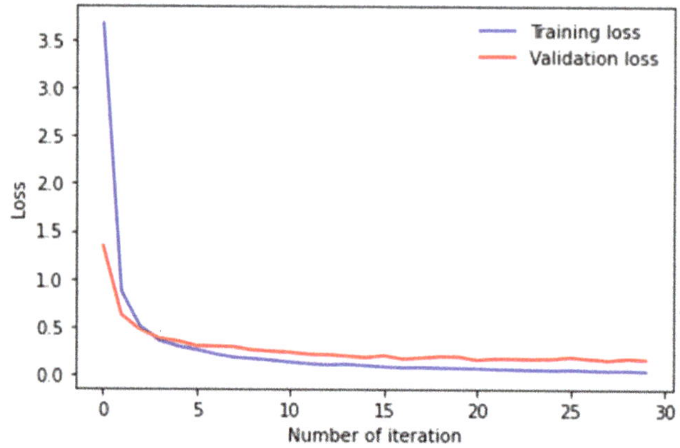

Figure 11. Training and validation learning curves on GoogLeNet.

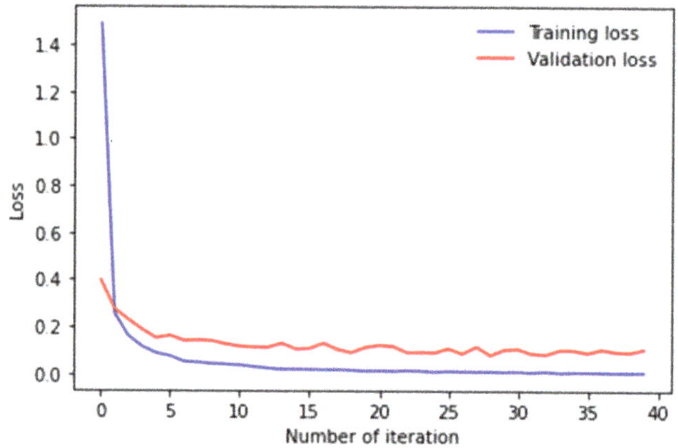

Figure 12. Training and validation learning curves on ResNet.

4. Classification

For classification purposes, SVM and Softmax activation functions are used. The features extracted from CNN are fed to these classifiers separately to manipulate their different results. First, we apply the Softmax activation function on the features extracted from the CNN which classifies the output into a probability distribution of 10 classes. Its function is given as follows:

$$softmax_j = \frac{\exp(x_i)}{\sum_j \exp(x_j)} \tag{1}$$

It computes the exponential of the input parameter and the sum of the exponential parameters of all existing values in the inputs while giving output in the ratio of the exponential of the parameter and the sum of the exponential parameter. The learning curve for our CNN model in Figure 13 does not show either underfitting or overfitting. The training curve shows how well the model is learning while the validation curve shows its rate of generalization. The loss is lower on the training set as compared to the validation set. It can be concluded that it is a good fit as both losses decrease to a point of stability and the gap between them is negligible.

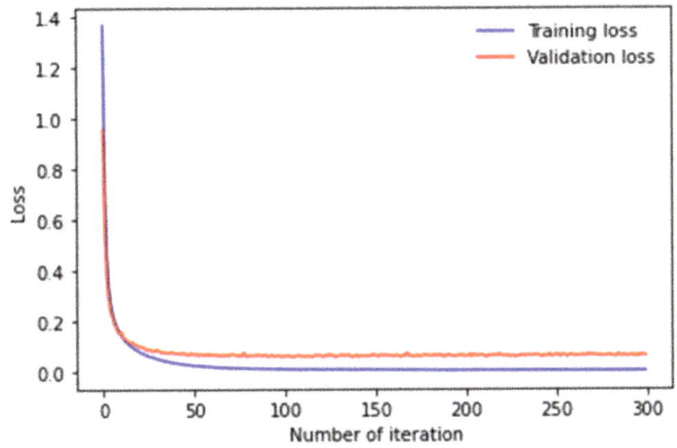

Figure 13. Training and validation learning curves for Softmax activation function.

Then, the SVM is applied to the same extracted features. The SVM is a supervised classification algorithm that works on features extracted from images rather than raw images. Its training equation is given as: set of attributes–label pairs (xi,yi), $i = 1, 2, \ldots\ldots, l$:

$$minimize \quad w^T w + C \sum_{i=1}^{m} \xi_i^2$$

$$subject\ to \quad y_i\left(x_i^T w + b\right) \geq 1 - \xi_i, \quad (i=1, \ldots, m) \tag{2}$$

Studies have shown that it works better for classification as compared to the traditional Softmax function but fails for multiclass problems. This is validated in our results as shown in Figure 14 which is a good generalization curve.

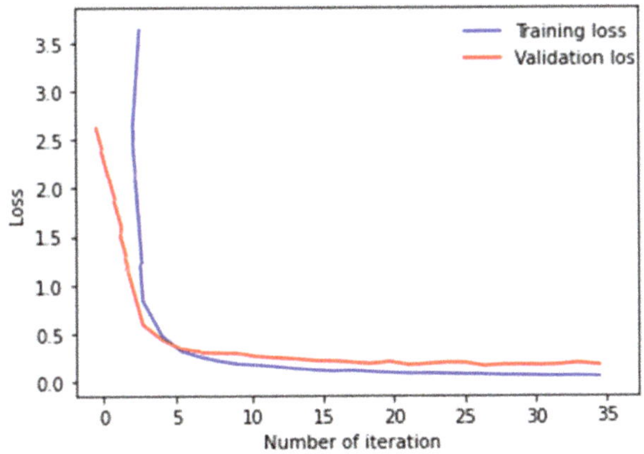

Figure 14. Training and validation learning curves on SVM.

5. Experiment and Discussion

To validate the accuracy of our proposed model and two architectures (GoogLeNet and ResNet), we used our novel dataset of Urdu numerals. They were compared for performance on different datasets and a different number of labels. For all our models, we used two types of datasets.

1. One is the original one that we proposed initially, and we evaluated it by splitting it into 85–15 ratio.

2. The other one is made as a separate set consisting of Pakistani currency note images which are used for testing only but are trained using the Urdu dataset. This is done to test our models on real-life scenarios.

Batch size, number of epochs, and learning rate are used for hyperparameter tuning. Different batch sizes of 16, 32, 64, and 128 are checked for each of the three models with different types of regularization. For our own CNN and ResNet, batch sizes of 16 gave the best results, while for GoogLeNet, 32 gave optimal results. A learning rate of 0.001 is used for CNN, SVM, and ResNet, while a 0.003 learning rate is used for GoogLeNet. Stochastic gradient descent (SGD) with momentum 0.9 worked best as compared to other types of optimizations such as Adamax, Adam. In CNN, it is observed that training deep neural networks for more layers brought sensitivity to weights and settings of the learning algorithm. To solve this, batch normalization is employed to standardize the layers for each batch. This also caused the number of epochs to reduce which in turn stabilized the learning process [36]. However, it also brought an increase in training time because of the use of an optimizer—SGD. Both batch normalization and dropout are used in the CNN after much hyperparameter tuning. It is concluded that combinations of the above values of hyperparameters produced maximum accuracy with the Urdu numeral dataset [38]. We tried to use dropout too, but it reduced the accuracy, so we relied on batch normalization only. On the other hand, a dropout rate of 0.5 was used for SVM to obtain the best results.

The performance graphs indicate that the increased accuracy is because of more neurons in more layers. These neurons helped in choosing the features for a dataset in a deep manner. It is quite noteworthy that our proposed CNN gave the best results equivalent to ResNet and GoogLeNet as shown in Figure 15. The combination of CNN with SVM gave the best accuracy of 99.96% with a promising validation accuracy of 99%. These are ground breaking results for a dataset that is unique. In its comparison training, the accuracy of the CNN with the Softmax classifier is 98.89% which is equally promising. GoogLeNet showed a training accuracy of 99.06%, ResNet showed a training accuracy of 99.88%. The validation accuracy of CNN with softmax classifier is 98.41% and 96.4% for ResNet, while that of GoogLeNet is comparatively lower at 95.61%. These accuracies conclude that our custom built CNN with SVM outperformed all the available models.

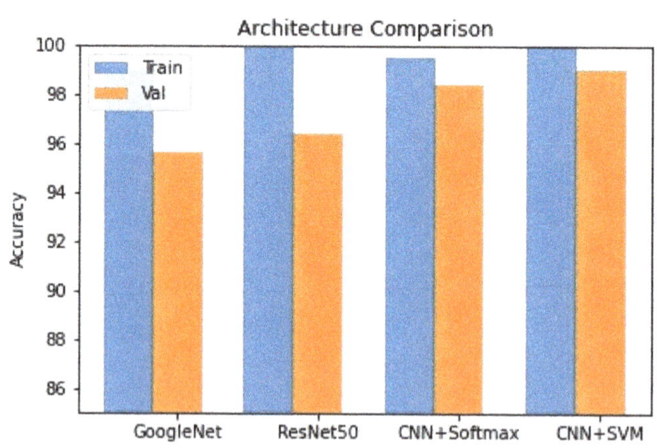

Figure 15. Comparison of accuracies on 4 proposed architectures.

5.1. Comparison with Existing Methods

In order to authenticate our results, we compared the performance of our data with the previously published model by Husnain, M. and Saad Missen et al. [39]. We tried 16, 32, 40, 60, and 80 neurons in fully connected layers with SGD and batch sizes of 32, 64, and 128. Momentum was varied between 0.7 and 0.99. Adamax with batch sizes of 32, 64, and 128 was checked to obtain the best results. We achieved an accuracy of 95.7% on their model and its variants. However, we achieved the final test accuracy of 99.0% using SVM as a classifier on our dataset as compared to their test accuracy of 98.41% on their Urdu numeral dataset. This proves that classification using SVM results in better accuracies for the features extracted for Urdu numerals. A brief comparison of accuracies of previously published papers is given in Table 3. As is evident from the results, our models beat the previous techniques with remarkable accuracies.

Table 3. Comparison of handwritten Urdu numerals recognition on different classifiers.

Systems	Dataset	Classifier	Accuracy Achieved
[11]	UNHD(Urdu characters and ligatures)	BLSTM	93.96%
[40]	Sindhi handwritten numbers	Self-organizing map neural network	86.89%
[21]	Handwritten Urdu numerals	Rule based technique, HMM	97.4% (Rule based technique), 96.2% (HMM)
[41]	Handwritten Urdu numerals	Daubechies wavelet	92.05%
[42]	Urdu handwritten characters and numerals	Convolutional neural network	98.3%
[26]	Urdu handwritten characters and numerals	Convolutional neural network	98.3%
Our Approach	Handwritten Urdu numerals	GoogLeNet and ResNet	95.7%
Our Approach	Handwritten Urdu numerals	Feature extraction using convolution layer and Softmax Activation for classi cation	98.41%
		Feature extraction using convolution layer and SVM for classi cation	**99.3%**

Additionally, in comparison to the previous paper [33], in terms of test accuracy, our approach achieved 99.0% as compared to their accuracy of 98.3%, which is optimal considering that the Urdu numeral dataset is novel. This dataset shows promising training and validation accuracies on GoogLeNet, ResNet, and the proposed approach as shown in Figure 15.

Additionally, we experimented with our model on a novel Pakistani currency dataset and achieved a test accuracy of 89.41%, which further validates the performance and robustness of our model on real-world problems.

5.2. Expanded Testing Set

As our model is tested on real-life examples thus, it is deemed necessary to validate our results with a separate dataset. This dataset consists of 90 images captured with the same tools as were used for the training data. The preprocessing techniques similar to training data were applied to it. The only difference here is that it contained 4 classes of zero, one, two, and five instead of 10 classes. So, during training and testing of the currency dataset, the training data for Urdu numerals was minimized to the same 4 classes. The same CNN network was used as proposed in Figure 10. An accuracy of 97.97% was achieved on the training set along with 89.41% on the test set. To obtain the best results, we retuned the hyperparameters of the model. We chose different batch sizes ranging from 16 to 128 to find the one where we achieved the best results while keeping the learning rate at 0.001. Furthermore, the number of epochs was varied accordingly which resulted in getting momentum 0.7 with SGD.

Along with CNN, ResNet, GoogLeNet, and SVM were also applied to this expanded test set to obtain an insight into their accuracies and losses. The parameter batch size was kept to 32 for both ResNet and GoogLeNet. The bar graph in Figure 16 shows that our proposed CNN model with the SVM worked best on a real-life dataset with a validation accuracy of 91%. On the other hand, the validation accuracies of GoogLeNet (53.47%), ResNet50 (64.24%), and Softmax (89.41%) are worth considering. As we worked on particular classes instead of all the 10 classes and the data within these 4 classes was also less in quantity, so this caused a reduction in accuracies for real-life data.

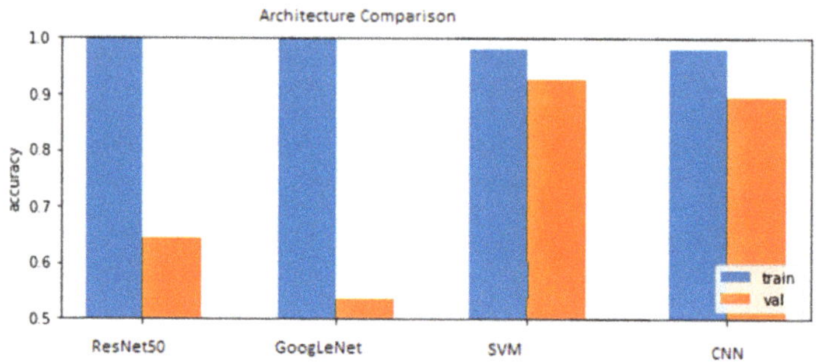

Figure 16. Bar plot for comparison of accuracies on real-life example.

6. Conclusions and Future Work

In this paper, we proposed two approaches to classify novel datasets of Urdu numerals. The first approach extracts features using convolution layers and uses Softmax activation followed by fully connected layers for classfication. The second approach applies an SVM classifier to the features extracted from the convolution layer. All the models give best results in terms of accuracy where the first approach provides a validation accuracy of 98.41%, while an accuracy of 99.0% is achieved by the second approach. The accuracy of 96.4% on ResNet and 95.61% on GoogLeNet is achieved on this novel dataset. We tested these models on Pakistani currency to see their reliability in real-world application after being trained on our dataset. To implement this, we developed another dataset from Pakistani currency notes and evaluated our proposed models with it. Our handwritten Urdu numerals dataset is unique and any such dataset is not available publicly. This hinders research in the domain of the Urdu language. Moreover, our dataset is refined and is collected along the lines of the MNIST dataset, so it provides the best results with real-life problems as shown in our paper.

In the future, we plan on increasing the Urdu numeral dataset and then making it publicly available so as to motivate researchers to work in this field. Increasing this dataset will also increase the accuracies of all models. Additionally, our dataset and CNN can help develop a system to identify and count currency notes. Since the performance of deep learning algorithms in real-world applications is of utmost importance, we plan on testing it on other applications such as recognizing the Surah numbers of The Holy Quran and numbers on Pakistani postage stamps. The sole motivation of this paper is to bring our mother language Urdu to a competitive level with all the latest research.

Author Contributions: Conceptualization, W.K. and S.K.; methodology, A.A. (Ameera Arif) and A.B.; software, A.B.; validation, W.K., A.u.R., B.K. and A.A. (Ahmed Ali).; data curation, A.B., A.A. (Ameera Arif) and W.K.; writing—original draft preparation, A.B.; writing—review and editing, A.A (Ameera Arif)., W.K., A.u.R. and B.K.; visualization, A.A. (Ameera Arif) and S.K.; supervision, A.A. (Ahmed Ali) and S.K.; funding acquisition, A.u.R. All authors have read and agreed to the published version of the manuscript.

Funding: This research received no external funding.

Conflicts of Interest: The authors declare no conflict of interest.

References

1. Singh, R.; Mishra, R.K.; Bedi, S.; Kumar, S.; Shukla, A.K. A Literature Review on Handwritten Character Recognition based on Artificial Neural Network. *Int. J. Comput. Sci. Eng.* **2018**, *6*, 753–758. [CrossRef]
2. The Online Encyclopedia of Writing Systems and Languages. Available online: https://www.omniglot.com/writing/urdu.htm (accessed on 9 June 2020).
3. Spitz, A.L.; Andreas, D. Document Analysis Systems. In *Proceedings of the International Association for Pattern Recognition Workshop*; World Scientific: Singapore, 1995; pp. 237–292.
4. Sharif, M.; Ul-Hasan, A.; Shafait, F. Urdu Handwritten Ligature Generation Using Generative Adversarial Networks (GANs). In Proceedings of the Frontiers in Handwriting Recognition: 18th International Conference, ICFHR 2022, Hyderabad, India, 4–7 December 2022; Springer-Verlag: Berlin/Heidelberg, Germany, 2022; pp. 421–435. [CrossRef]
5. Misgar, M.M.; Mushtaq, F.; Khurana, S.S.; Kumar, M. Recognition of offline handwritten Urdu characters using RNN and LSTM models. *Multimed. Tools Appl.* **2022**, *82*, 2053–2076. [CrossRef]
6. Gautam, N.; Sharma, R.S.; Hazrati, G. Eastern Arabic Numerals: A Stand out from Other Jargons. In Proceedings of the International Conference on Computational Intelligence and Communication Networks (CICN), Jabalpur, India, 12–14 December 2015; pp. 337–338. [CrossRef]
7. Memon, J.; Sami, M.; Khan, R.; Uddin, M. Handwritten Optical Character Recognition (OCR): A Comprehensive Systematic Literature Review (SLR). *IEEE Access* **2020**, *8*, 142642–142668. [CrossRef]
8. Khan, S. A Mechanism for Offline Character Recognition. *Int. J. Res. Appl. Sci. Eng. Technol.* **2019**, *7*, 1086–1090. [CrossRef]
9. Haghighi, F.; Omranpour, H. Stacking ensemble model of deep learning and its application to Persian/Arabic handwritten digits recognition. *Knowl. Based Syst.* **2021**, *220*, 106940. [CrossRef]
10. Das, N.; Sarkar, R.; Basu, S.; Kundu, M.; Nasipuri, M.; Basu, D.K. A genetic algorithm based region sampling for selection of local features in handwritten digit recognition application. *Appl. Soft Comput.* **2012**, *12*, 1592–1606. [CrossRef]
11. Slimane, F.; Kanoun, S.; Hennebert, J.; Alimi, A.; Ingold, R. A study on font-family and font-size recognition applied to Arabic word images at ultra-low resolution. *Pattern Recognit. Lett.* **2013**, *34*, 209–218. [CrossRef]
12. Center for Language Engineering Urdu Ligatures from Corpus Page. Available online: http://www.cle.org.pk/software/ling_resources/UrduLigaturesfromCorpus.htm (accessed on 11 June 2020).
13. Ahmed, R.; Musa, M. Preprocessing Phase for Offline Arabic Handwritten Character Recognition. *Int. J. Comput. Appl. Technol. Res.* **2016**, *5*, 760–763. [CrossRef]
14. Borse, R.; Ansari, I.A. *Offline Handwritten and Printed Urdu Digits Recognition using Daubechies Wavelet*; ER Publication: New Delhi, India, 2015.
15. Kumar, G.; Bhatia, P.K. Analytical Review of Preprocessing Techniques for Offline Handwritten Character Recognition. In Proceedings of the 2nd International Conference on Emerging Trends in Engineering Trends in Engineering and Management ICETEM, Rohtak India, 21–22 July 2013.
16. Akhtar, P. An Online and Offline Character Recognition Using Image Processing Methods—A Survey. *Int. J. Commun. Comput. Technol.* **2016**, *4*, 102. [CrossRef]
17. Liu, C.; Yin, F.; Wang, D.; Wang, Q. Online and offline handwritten Chinese character recognition: Benchmarking on new databases. *Pattern Recognit.* **2012**, *46*, 155–162. [CrossRef]
18. Baker, P.; Hardie, A.; McEnery, T.; Cunningham, H.; Gaizauskas, R.J. EMILLE, A 67-Million Word Corpus of Indic Languages: Data Collection, Mark-up and Harmonisation. In Proceedings of the Third International Conference on Language Resources and Evaluation (LREC'02) LREC, Las Palmas, Canary Islands Spain, 29–31 May 2002.
19. Javed, L.; Shafi, M.; Khattak, M.I.; Ullah, N. Hand-written Urdu Numerals Recognition Using Kohonen Self Organizing Maps. *Sindh Univ. Res. J. SURJ* **2015**, *47*, 403–406.
20. Razzak, M.I.; Hussain, S.A.; Belaid, A.; Sher, M. Multi-font Numerals Recognition for Urdu Script based Languages. *Int. J. Recent Trends Eng.* 2009.
21. Kour, H.; Gondhi, N.K. Machine Learning approaches for Nastaliq style Urdu handwritten recognition: A survey. In Proceedings of the 6th Communication International Systems Conference (ICACCSon) Advanced, Coimbatore, India, 23 April 2020; pp. 50–54. [CrossRef]
22. Yusuf, M.; Haider, T. Recognition of Handwritten Urdu Digits using Shape Context. *INMIC* **2004**. [CrossRef]
23. Iqbal, T.; Ali, H.; Saad, M.M.; Khan, S.; Tanougast, C. CapsuleNet for Urdu Digits Recognition. In Proceedings of the 10th IEEE International Conference on Intelligent Data Acquisition and Advanced Computing Systems: Technology and Applications, Metz, France, 18–21 September 2019.
24. Abdelazeem, S. Comparing Arabic and Latin Handwritten Digits Recognition Problems. *Int. J. Comput. Inf. Eng.* **2009**, *3*, 1583–1587. [CrossRef]
25. Abdelazeem, S.; El-Sherif, E. The Arabic Handwritten Digits Databases ADBase & MADBase. Available online: http://datacenter.aucegypt.edu/shazeem/ (accessed on 14 May 2020).
26. Ahmed, S.B.; Hameed, I.A.; Naz, S.; Razzak, M.I.; Yusof, R. Evaluation of Handwritten Urdu Text by Integration of MNIST Dataset Learning Experience. *IEEE Access* **2019**, *7*, 153566–153578. [CrossRef]
27. Ebrahimzadeh, R.; Jampour, M. Efficient Handwritten Digit Recognition based on Histogram of Oriented Gradients and SVM. *Int. J. Comput. Appl.* **2014**, *104*, 10–13. [CrossRef]

28. Sufian, A.; Ghosh, A.; Naskar, A.; Sultana, F.; Sil, J.; Hafizur Rahman, M.M. BDNet: Bengali Handwritten Numeral Digit Recognition based on Densely connected Convolutional Neural Networks. *J. King Saud Univ. Comput. Inf. Sci.* **2020**, *34*, 2610–2620. [CrossRef]
29. Prashanth, D.S.; Mehta, R.V.K.; Sharma, N. Classification of Handwritten Devanagari Number An analysis of Pattern Recognition Tool using Neural Network and CNN. *Procedia Comput. Sci.* **2020**, *167*, 2445–2457. [CrossRef]
30. Ahlawat, S.; Choudhary, A. Hybrid CNN-SVM Classifier for Handwritten Digit Recognition. *Procedia Comput. Sci.* **2020**, *167*, 2554–2560. [CrossRef]
31. Baldominos, A.; Saez, Y.; Isasi, P. Evolutionary Convolutional Neural Networks: An Application to Handwriting Recognition. *Neurocomputing* **2018**, *283*, 38–52. [CrossRef]
32. Sabbour, N.; Shafait, F. A segmentation-free approach to Arabic and Urdu OCR. In Proceedings of the SPIE 8658, Document Recognition and Retrieval XX, 86580N, Burlingame, CA, USA, 3–7 February 2013. [CrossRef]
33. Ahmed, S.; Naz, S.; Swati, S.; Razzak, M. Handwritten Urdu character recognition using one-dimensional BLSTM classifier. *Neural Comput. Appl.* **2017**, *31*, 1143–1151. [CrossRef]
34. LeCun, Y. The MNIST DATABASE of handwritten digits. Available online: http://yann.lecun.com/exdb/mnist/ (accessed on 14 June 2020).
35. Shorten, C.; Khoshgoftaar, T. A survey on Image Data Augmentation for Deep Learning. *J. Big Data* **2019**, *6*, 60. [CrossRef]
36. He, K.; Zhang, X.; Ren, S.; Sun, J. Deep residual learning for image recognition. In Proceedings of the IEEE Conference on Computer Vision and Pattern Recognition (CVPR), Las Vegas, NV, USA, 27–30 June 2016; pp. 770–778. [CrossRef]
37. Szegedy, C.; Liu, W.; Jia, Y.; Sermanet, P.; Reed, S.; Anguelov, D.; Erhan, D.; Vanhoucke, V.; Rabinovich, A. Going deeper with convolutions. In Proceedings of the IEEE Conference on Computer Vision and Pattern Recognition (CVPR), Boston, MA, USA, 7–12 June 2015; pp. 1–9. [CrossRef]
38. Prabhu, R. CNN Architectures—LeNet, AlexNet, VGG, GoogLeNet and ResNet. Available online: https://medium.com/@RaghavPrabhu/cnn-architectures-lenet-alexnet-vgg-googlenet-and-resnet-7c81c017b84 (accessed on 14 June 2020).
39. Garbin, C.; Zhu, X.; Marques, O. Dropout vs. batch normalization: An empirical study of their impact to deep learning. *Multimed. Tools Appl.* **2020**, *79*, 12777–12815. [CrossRef]
40. Husnain, M.; Missen, M.M.S.; Mumtaz, S.; Jhanidr, M.Z.; Coustaty, M.; Muzzamil Luqman, M.; Ogier, J.M.; Choi, G.S. Recognition of Urdu Handwritten Characters Using Convolutional Neural Network. *Appl. Sci.* **2019**, *9*, 2758. [CrossRef]
41. Chandio, A.A.; Jalbani, A.H.; Leghari, M.; Awan, S.A. Multi-Digit Handwritten Sindhi Numerals Recognition using SOM Neural Network. *Mehran Univ. Res. J. Eng. Technol.* **2017**, *36*, 8. [CrossRef]
42. Malik, S.; Khan, S.A. Urdu online handwriting recognition. In Proceedings of the IEEE Symposium on Emerging Technologies, Islamabad, Pakistan, 18 September 2005. [CrossRef]

applied sciences

MDPI

Article

Coarse-to-Fine Structure-Aware Artistic Style Transfer

Kunxiao Liu, Guowu Yuan *, Hao Wu and Wenhua Qian

School of Information Science and Engineering, Yunnan University, Kunming 650504, China
* Correspondence: gwyuan@ynu.edu.cn

Abstract: Artistic style transfer aims to use a style image and a content image to synthesize a target image that retains the same artistic expression as the style image while preserving the basic content of the content image. Many recently proposed style transfer methods have a common problem; that is, they simply transfer the texture and color of the style image to the global structure of the content image. As a result, the content image has a local structure that is not similar to the local structure of the style image. In this paper, we present an effective method that can be used to transfer style patterns while fusing the local style structure to the local content structure. In our method, different levels of coarse stylized features are first reconstructed at low resolution using a coarse network, in which style color distribution is roughly transferred, and the content structure is combined with the style structure. Then, the reconstructed features and the content features are adopted to synthesize high-quality structure-aware stylized images with high resolution using a fine network with three structural selective fusion (SSF) modules. The effectiveness of our method is demonstrated through the generation of appealing high-quality stylization results and a comparison with some state-of-the-art style transfer methods.

Keywords: image processing; nonphotorealistic rendering (NPR); style transfer; structure-aware; deep learning

Citation: Liu, K.; Yuan, G.; Wu, H.; Qian, W. Coarse-to-Fine Structure-Aware Artistic Style Transfer. *Appl. Sci.* **2023**, *13*, 952. https://doi.org/10.3390/app13020952

Academic Editors: Zahid Mehmood Jehangiri, Mohsin Shahzad and Uzair Khan

Received: 16 December 2022
Revised: 5 January 2023
Accepted: 8 January 2023
Published: 10 January 2023

1. Introduction

Artistic style transfer is an attractive image-processing technique that is used to generate a new image that preserves the structure of a content image but carries the pattern of a style image. Recently, the seminal image-optimization method proposed by Gatys et al. [1] was used to achieve style transfer by adopting the correlation of features extracted from a pretrained deep neural network and the iterative optimization process. Like the method presented by Gatys et al. [1], style transfer by relaxed optimal transport and self-similarity (STROTSS) [2] is also an image-optimization style transfer method; this method has achieved superior stylization results by adopting the relaxed earth mover's distance (rEMD) loss in a multiscale optimization process. However, the expensive computational cost of these image-optimization methods restricts their use in practice applications in industry. To speed up the optimization procedure, Johnson et al. [3] and Ulyanov et al. [4] proposed model-optimization style transfer methods. They train a feed-forward neural network that can be used to synthesize images with a single given style image in real time. Both adaptive instance normalization (AdaIN) [5] and whitening and coloring transforms (WCTs) [6] are model-optimization methods but are also arbitrary style transfer methods, in which style patterns of arbitrary style images are transferred by adopting some feature transforms. After reviewing these methods, we have found that although local style texture and content structures can generally be combined, some key structures of the style image are not accurately learned. For example, the color blocks and brushstrokes that constitute the main objects in style images are not transferred very well. Meanwhile, in some cases, these methods produce distorted objects and incongruous artistic effects in stylized images. Therefore, our main task is to transfer the local structure of the style image to the content image and adopt a coarse-to-fine strategy to enhance the artistic details of the stylization results.

We propose a novel artistic style transfer network for fusing an essential style structure to a content structure and synthesizing a structure-aware stylized image. In our model, a coarse network is designed to obtain reconstructed coarse stylized features in the first stage. Because the coarse network works only at a low resolution, the coarse stylized features can discard the trivial structure details of the content image and combine the global content structure with the style patterns. Then, the task of a fine network is to adopt these reconstructed coarse stylized features obtained at a low resolution and the original content image with a high resolution to synthesize the final high-resolution stylized image in the second stage. By adopting some SSF modules to fuse the coarse stylized features into the fine network, the final high-resolution stylized images can selectively integrate structural information at different scales. Our main contributions are as follows:

1. We introduce a novel style transfer model that can be used to synthesize appealing structure-aware stylization results. This model consists of a coarse network and a fine network. The former roughly transfers style patterns that include holistic structural information and color distribution information, and then the latter enhances the details of the style patterns by fusing multiscale features.
2. We propose a SSF module for fusing the reconstructed features to the content features in a fine network. This module can help the fine network select essential structural information for feature fusion on the basis of the channel attention mechanism. As a result, the color distribution of the style images can be accurately transferred.
3. It is demonstrated through experiments that our method can be used to synthesize high-quality stylizations, where the main structures of the content image are preserved and the local structures of the style image are transferred. These stylization results can maintain the same artistic expression as style images by discarding trivial content details and injecting key local style structures.

The rest of the paper Is organized as follows. In Section 2, the works related to different style transfer methods are reviewed. In Section 3, the pipeline of our framework and the details of our two networks are described. Moreover, the different loss functions are introduced. Different experimental results are shown and discussed in Section 4. The conclusion is summarized in Section 5.

2. Related Work

2.1. Style Transfer

The goal of style transfer is to combine the texture of a style image with the structure of a content image. Gatys et al. [1] proposed a seminal iterative method that was based on a pretrained visual geometry group (VGG) network [7]. In this method, the content structure and the style texture can be used to synthesize a new image, but it is expensive, and a stylized image is generated only after the training process has been completed. Inspired by Gatys et al. [1], Johnson et al. [3] proposed a feed-forward method, which can be used to synthesize arbitrary images with a fixed style by an encoder-decoder architecture; the time and computation costs are reduced when using this method. Numerous methods have been developed to speed up the style transfer process [4,8] and improve the visual quality [9–11]. Sanakoyeu et al. [12] also improved the stylization quality by proposing a style-aware loss, but they trained a network with a set of style images instead of a style image. This approach aimed to combine many style images created by one artist to synthesize a stylized image with the overall style of this artist. The dual style generative adversarial network (DualStyleGAN) [13] is proposed to characterize the content and style of a portrait by retaining an intrinsic style path to control the style of the original domain and an extrinsic path to model the style of the target extended domain. Peking Opera face makeup (POFMakeup) [14] also is a portrait style transfer method that can transfer the style of a portrait with a Peking Opera face to a target portrait. Lin et al. [15] combined a universal style transfer method with image fusion and color enhancement methods to solve the problems of the color scheme, the strength of style strokes, and the adjustment of image contrast.

To simultaneously handle multiple styles, [16] proposed a flexible conditional instance normalization approach embedded in style transfer networks to learn multiple styles, and [17] achieved multistyle generation in a generative network architecture with a learnable inspiration layer. Ye et al. [18] adopted a mechanism and instance segmentation to achieve a regional multistyle style transfer model, which can solve the problem of unnatural connections between regions. Alexandru et al. [19] combined various existing style transfer frameworks to propose a novel framework that can generate intriguing artistic stylization results by performing geometric deformation and using different styles from multiple artists.

In AdaIN [5], adaptive instance normalization is implemented to train a network with various styles, providing the ability to transfer arbitrary styles after the training process. In WCT [6], the whitening and coloring transforms are adopted to synthesize arbitrary styles with a pretrained VGG network and a series of pretrained image restructuring decoders. Based on WCT, Wang et al. [20] achieved the diversity of style transfer by adopting a deep feature perturbation (DFP) operation while preserving the quality of stylization results, and Wang et al. [21] synthesized ultraresolution stylized images and reduced the convolutional filters by using a knowledge-distillation method. A style-attentional network (SANet) [22] is also an arbitrary style transfer method that can be used to efficiently generate stylized images by injecting local style patterns into content features on the basis of using the style attention mechanism.

2.2. Style Transfer Based on Multiscale Learning

Recently, some style transfer methods have been used to transfer style patterns on the basis of multiscale learning. Multiscale holistic style transfer is achieved in Avatar-Net [23] on the basis of the use of an hourglass with multiple skip connections and a style decorator. STROTSS [2] is an image-optimization method that adopts multiscale learning to update the content image and generate high-quality stylized images. Yang et al. [24] proposed a novel video style transfer framework that can render high-quality artistic portraits on the basis of the multiscale content features and preserve the frame details. A Laplacian pyramid style network (LapStyle) [25] also exhibits high visual quality and is based on a drafting network and a revision network. First, the former transfers the global style patterns, and then, the latter enhances local style details. However, too many content structure details are preserved in these methods. Key local style structures are not fused into stylized images in any of these methods. In contrast, our method transfers global style patterns at low resolution using a coarse network, which needs to be trained only once to reconstruct coarse stylized features. Our fine network enhances local style details with multiscale features from the coarse network and the high-resolution content image. As a result, our method can discard trivial local content structures and synthesize high-quality structure-aware stylized images by using a coarse-to-fine process. The differences between our method and the methods in previous studies are shown in Table 1.

Table 1. The differences between our method and those in previous studies.

Methods	Image-Optimization	Model-Optimization	Single Style	Multiple Style	Arbitrary Style
Ours		√	√		
[1,2]	√		√		
[3,4,8–15,24,25]		√	√		
[16–19]		√		√	
[5,6,20–23]		√			√

3. Proposed Method

3.1. Framework Overview

Inspired by the painting process of artists, in which the coarse structure and color distribution are first constructed and then fine details are added, our framework employs a coarse network and a fine network to simulate the coarse-to-fine process. As shown in Figure 1, given a content image $x_c \in \mathbb{R}^{3\times h_c \times w_c}$ and a style image $x_s \in \mathbb{R}^{3\times h_s \times w_s}$, our model eventually generates a stylized image $x_{cs} \in \mathbb{R}^{3\times h_{cs} \times w_{cs}}$. In the first stage, the coarse network takes $\overline{x}_c$ and $\overline{x}_s$ as inputs, where $\overline{x}_c$ and $\overline{x}_s$ are the results of downsampling x_c and x_s by 2, respectively. Then three restructured coarse stylized features $\overline{f}_r^{(i)} \in \mathbb{R}^{c_r^{(i)} \times h_r^{(i)} \times w_r^{(i)}}$ (i = 1, 2, 3) are generated by the coarse network, where $c_r^{(i)}$, $h_r^{(i)}$, and $w_r^{(i)}$ are the number of channels, height, and width of the i restructured feature, respectively. In the second stage, the fine network takes x_c and $\overline{f}_r^{(i)}$ as inputs and then generates the final stylized image x_{cs} by adopting SSF modules for feature fusion.

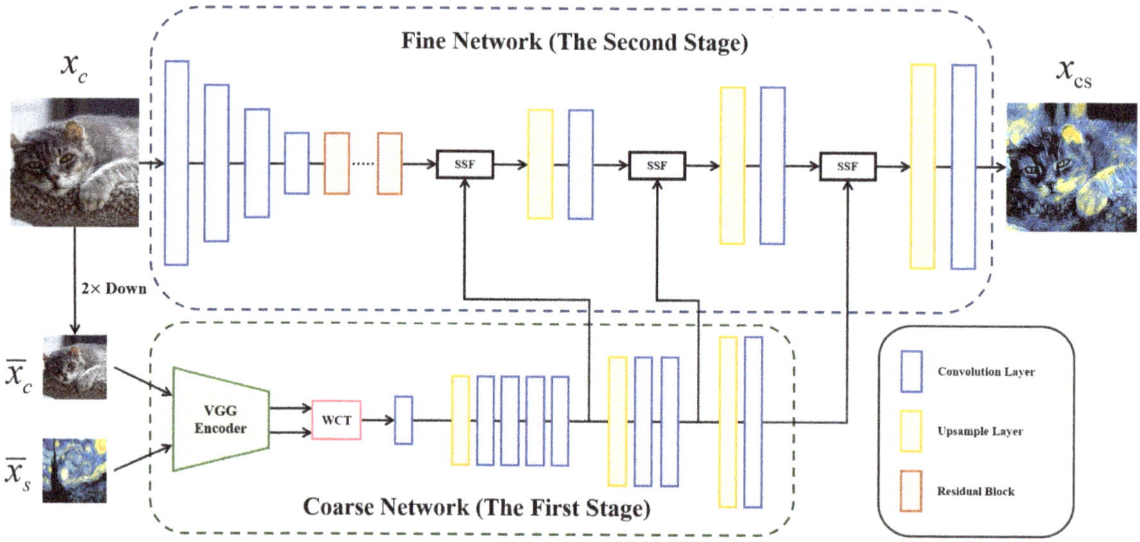

Figure 1. Overview of our framework.

As shown in Figure 2, different stylized images are generated by our method. In Figure 2b, we adopt the last restructured coarse stylized features only to directly restructure the coarse stylized image by the coarse network in the first stage. The coarse stylized image discards the unnecessary local structures of the content image and transfers the global color distribution of the style image. Then, the fine network is employed to encode the high-resolution content image to obtain the content features, and these content features and three coarse reconstructed features from the coarse network are decoded to generate the high-quality structure-aware stylized image in the second stage. As shown in Figure 2c, the final appealing stylized image is synthesized by adopting our full model with a coarse network and a fine network. Moreover, to more clearly show the local style structure of the final stylized image, we use the color control method [26] to keep the color of the final stylized image consistent with the color of the original content image. As illustrated in Figure 2d, although the color distribution of the stylized image remains the same as that of the content image, the local structure of the stylized image is similar to that of the style image.

(a) (b) (c) (d)

Figure 2. Different stylization results from the same content image and style image: (**a**) the content image is a cat, and the style image is Starry Night by Vincent van Gogh; (**b**) this stylized image is generated directly by our coarse network in the first stage; (**c**) the final stylized image is generated by our full model in the second stage; (**d**) this stylized image maintains the same color as the content image using color control.

3.2. Coarse Network

One problem with recent style transfer methods is that too many structural details of the content image are retained during the transfer of style patterns. In the stylized image, there are some small structures from the content image that do not change; they simply transfer the color and texture of the style image. These local structures that do not exist in the style image appear in the stylized image, resulting in a stylized image that fails to show the spirit of the artistic expression of the style image. The reason is that these methods directly extract features from high-resolution images and cannot decide which details to discard from the content image. Contrary to previous work, our coarse network transfers rough style patterns at low resolution. As a result, there is a larger receptive field to learn low-frequency information to determine the overall structure of the image. Then, some unnecessary high-frequency information is ignored during training. As shown in Figure 3, the coarse network can transfer more details that are unnecessary in the coarse stylized image at high resolution. At low resolution, the coarse network can discard some trivial details of the structure and keep the objects smooth in the stylized image.

(a) (b) (c)

Figure 3. Comparison of two stylized images generated by the coarse network at different resolutions: (**a**) the original content image with resolution of 512 × 512; (**b**) the stylized image with resolution of 256 × 256; (**c**) the stylized image with resolution of 512 × 512.

3.2.1. WCT Module

Inspired by WCT [6], our coarse network adopts whitening and coloring transforms to transfer coarse style patterns at low resolution. The whitening transform can remove inessential information related to style while preserving the global structure of the content. Then, the coloring transform can capture the salient visual style and fuse some style structures to content structures. WCT is a multilevel stylization process that uses different rectified linear unit (ReLU) layers of VGG features ReLU_X_1 (X = 1, 2, ... , 5) and transfers style patterns in a coarse-to-fine pipeline. The higher-layer features are adopted to capture complex local structures, while lower-layer features carry low-level color and texture information. The difference between our coarse network and WCT is that we use only a single-level whitening and coloring transform for stylization. Moreover, we do not directly reconstruct the stylized features to generate an image; however, we utilize the reconstructed features at different layers during reconstruction. As a result, our coarse network, which has the ability to capture the multilevel information by reconstructing the coarse stylized features at different levels, can save computing resources.

3.2.2. Architecture of Coarse Network

The architecture of coarse network, which is shown in Figure 1, includes an encoder, a WCT module, and a decoder. (1) The encoder is a pretrained VGG-19 network, which is fixed during training. Given $\overline{x}_c$ and $\overline{x}_s$, the VGG encoder extracts the content feature $\overline{f}_c$ and the style feature $\overline{f}_s$ at ReLU_4_1. (2) Then, we apply a WCT module for whitening and coloring transformation. As shown in Figure 4a, the whitening transform is adopted to linearly transform $\overline{f}_c$ to obtain $\overline{f}_c'$. Next, the coloring transform is carried out to obtain $\overline{f}_{cs}$ by using $\overline{f}_c'$ and $\overline{f}_s$. (3) Finally, we adopt a reconstruction decoder to reconstruct the coarse stylized feature $\overline{f}_{cs}$. The decoder is designed to be symmetrical to the VGG-19 network, where the nearest neighbor upsampling layer is used for enlarging the feature map. We take $\overline{f}_{cs}$ as input for reconstruction and then generate these restructured stylized features $\overline{f}_r^{(i)}$ as outputs. In this reconstruction decoder, these outputs are output before the second upsampling layer, before the third upsampling layer, and after the last convolution layer. These $\overline{f}_r^{(i)}$ will become a part of the input of the fine network.

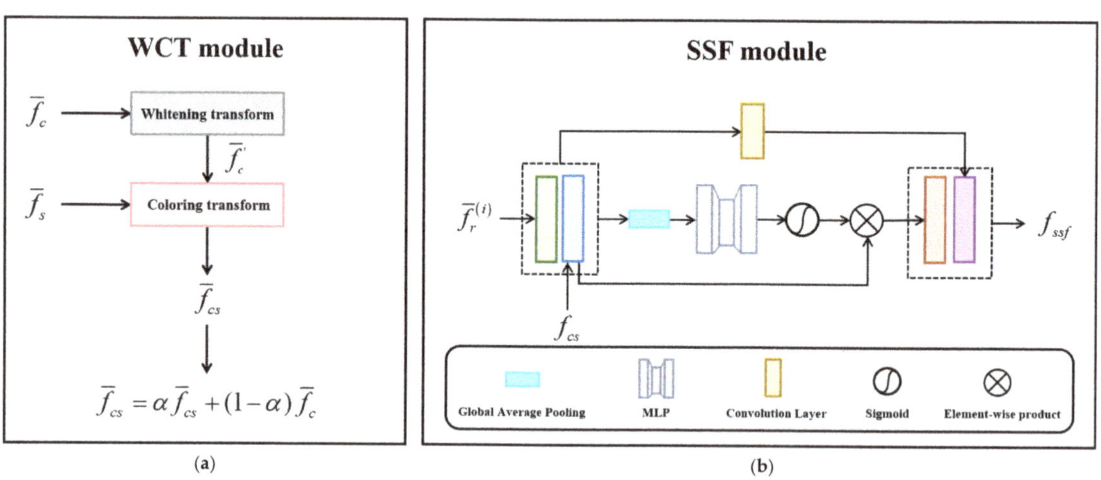

Figure 4. The schematics for two modules: (**a**) WCT module; (**b**) SSF module.

3.3. Fine Network

The fine network aims to synthesize high-resolution stylized images by fusing the reconstructed coarse stylized features to the reconstructed content features. The reconstructed content features are from the high-resolution content image and contains the global semantic information and local detail information. Contrary to the reconstructed content features at high resolution, the reconstructed stylized features generated from the coarse network preserve only the main content structure while blending some local structural style information. By fusing multiscale information, the fine network can pay more attention to the holistic structure of the content and ignore some trivial details by using our SSF modules. Then, some significant details can be added to the structure, and appealing artistic effects in the stylized image can be enhanced. In addition, fusing the reconstructed coarse stylized information can greatly reduce the time cost of the training process of the fine network, and the desired stylization results can be achieved at an earlier point in time.

3.3.1. SSF Module

The structural selective fusion (SSF) module is designed to fuse the reconstructed coarse stylized features from the coarse network to the reconstructed content features in the decoder of the fine network. Inspired by the attention mechanism [27,28], we employ a weight matrix to select the key structural information of the reconstructed content features, which is learned by adopting the merged features. The merged features are obtained by concatenating reconstructed coarse stylized features and the reconstructed content features. The matrix can help the SSF module obtain the selective features that focus on meaningful structural information, and the selective feature is one part of the output of the SSF module. Another part of the output is the refined merged features, which include different scale information, such as some crucial local textures or global structures.

The architecture of the SSF module is shown in Figure 4b. First, we concatenate the reconstructed coarse stylized features $\overline{f}_r^{(i)}$ and the reconstructed content features f_{cs} as input $f_{csr} \in \mathbb{R}^{(c_{cs}+c_r)\times w_r \times h_r}$. The reconstructed content features f_{cs} are the output of the convolution layer in the decoder of the fine network (except that the first SSF module uses the content features f_c from the encoder of the fine network as f_{cs}). We adopt an average-pooling operation to aggregate the spatial information of f_{csr} to generate the input of the multilayer perceptron, which is adopted to produce an attention map $M_{cs} \in \mathbb{R}^{c_{cs}\times 1\times 1}$ as the weight matrix. In summary, the attention map is calculated as follows:

$$M_{cs}(f_{csr}) = \sigma(MLP(AvgPool(f_{csr}))) \tag{1}$$

where σ denotes the sigmoid function. Then the selective feature f'_{cs} is calculated as follows:

$$f'_{cs} = M_{cs}(f_{csr}) \otimes f_{cs} \tag{2}$$

where $\otimes$ denotes element-wise multiplication. Meanwhile, f_{csr} is fed into a convolutional layer to produce a refined merged feature $f'_{csr} \in \mathbb{R}^{c_r\times w_r \times h_r}$. Eventually, the SSF module generates the final output $f_{ssf} \in \mathbb{R}^{(c_{cs}+c_r)\times w_r \times h_r}$ as the fused feature by directly concatenating f'_{cs} and f'_{csr}.

3.3.2. Architecture of Fine Network

As shown in Figure 1, fine network is designed as a flexible encoder-decoder architecture, with an encoder, a series of residual blocks, and a decoder. The encoder contains a convolutional layer with a stride of 1 and three convolutional layers with strides of 2, followed by several residual blocks. The decoder contains three upsampling layers, three convolutional layers with strides of 1, and three SSF modules. We use an SSF module before each upsampling layer. Given the content image x_c as the input of fine network, the encoder and several residual blocks generate the content feature f_c. Then, SSF modules generate the fused features f_{ssf} by taking $\overline{f}_r^{(i)}$ and f_{cs} as inputs, where f_{cs} is the output of these convolution layers in the decoder (except the first SSF module, which takes f_c as f_{cs}). These fused features f_{ssf} are fed into an upsampling layer and a convolution layer. Finally, the decoder generates the final stylized image x_{cs} after the last convolution layer.

3.4. Loss Function

Our coarse network needs to train only once, and it is fixed during the training of the fine network. Compared with WCT [6], we train only one reconstruction decoder network to reconstruct the coarse stylized feature. Our coarse network can reconstruct the stylized features at three levels or directly generate a coarse stylized image by taking advantage of the reconstruction decoder. Following WCT, we adopt pixel reconstruction and perceptual loss [3] to train our decoder for image reconstruction:

$$l_{re} = \left|\left|I_o - I_i\right|\right|_2^2 + \lambda\left|\left|\Phi(I_o) - \Phi(I_i)\right|\right|_2^2 \quad (3)$$

where I_i and I_o are the input image and output image, respectively, and Φ is the VGG encoder that extracts features at ReLU_X_1 (X = 1, 2, 3, 4). In addition, λ is the weight to balance the two losses.

The fine network is optimized with content and style loss during training. As shown in Figure 5, we keep a single x_s and a set of x_c from a content dataset, then x_{cs} is a stylized image generated by the fine network. For x_s, x_c, and x_{cs}, we can use a pretrained VGG-19 encoder to extract their features $F_c^{(t)} \in \mathbb{R}^{c_t \times h_t \times w_t}$, $F_s^{(t)} \in \mathbb{R}^{c_t \times h_t \times w_t}$, and $F_{cs}^{(t)} \in \mathbb{R}^{c_t \times h_t \times w_t}$, where t denotes the features extracted at ReLU_t (t = 1_1, 1_2, 2_1, 2_2, 3_1, 3_3, 4_1).

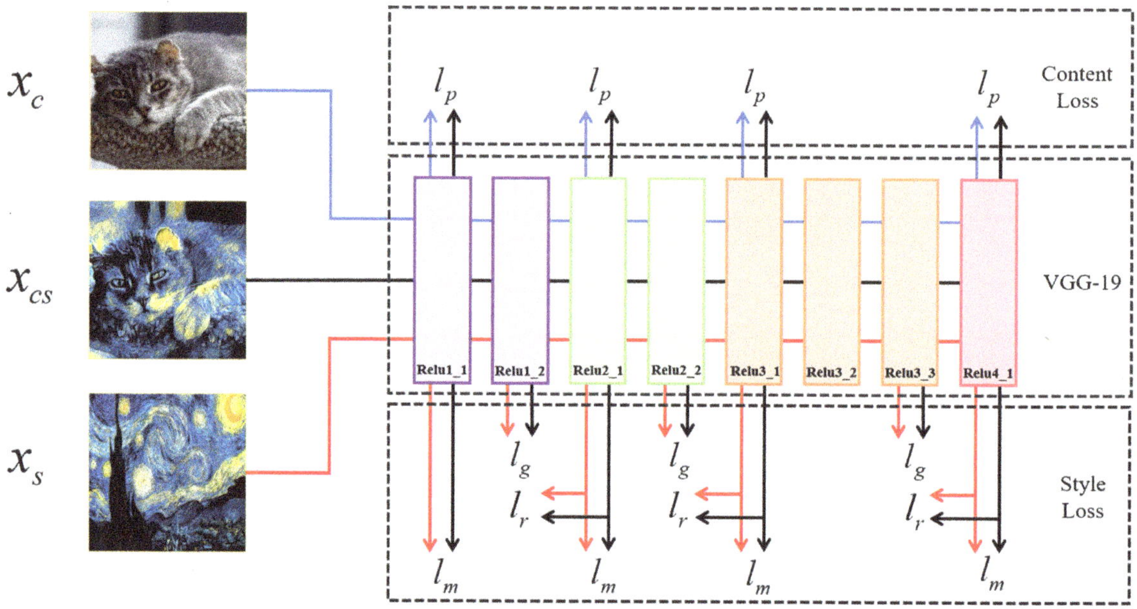

Figure 5. The schematic for loss network.

For content loss, we adopt the commonly used perceptual loss between $F_c^{(t)}$ and $F_{cs}^{(t)}$ proposed in [3]. The perceptual loss can measure high-level perceptual and semantic differences between images, and it is defined as follows:

$$l_p = \frac{1}{c_t h_t w_t}\left|\left|F_c^{(t)} - F_{cs}^{(t)}\right|\right|_2^2 \quad (4)$$

For style loss, we adopt three style losses. The first and most significant style loss is the relaxed earth mover's distance (rEMD) loss [2], which helps the fine network generate visual effects with minimum distortion to the layout of the content image. This loss plays a key role in migrating the structural forms of the style image to the target image. The rEMD loss between $F_s^{(t)}$ and $F_{cs}^{(t)}$ can be calculated as follows:

$$l_r = \max\left(\frac{1}{h_t w_t}\sum_{i=1}^{h_t w_t} \min_j C_{ij}, \frac{1}{h_t w_t}\sum_{j=1}^{h_t w_t} \min_i C_{ij}\right) \quad (5)$$

where C is the cost matrix, which can be calculated as the cosine distance between $F_s^{(t)}$ and $F_{cs}^{(t)}$:

$$C_{ij} = D_{cos}\left(F_{s,i}^{(t)}, F_{cs,j}^{(t)}\right) = 1 - \frac{F_{s,i}^{(t)} \cdot F_{cs,j}^{(t)}}{\left|\left|F_{s,i}^{(t)}\right|\right|\left|\left|F_{cs,j}^{(t)}\right|\right|} \quad (6)$$

The second style loss is the commonly used style reconstruction loss proposed by Gatys et al. [1], which is the difference between the Gram matrices of $F_s^{(t)}$ and $F_{cs}^{(t)}$:

$$l_g = \left|\left|G\left(F_s^{(t)}\right), G\left(F_{cs}^{(t)}\right)\right|\right|_2^2 \quad (7)$$

where G denotes the calculation of the Gram matrix of the feature vectors. Finally, we use the mean-variance loss as the third style loss, which is similar to the style reconstruction loss. We can use this type of loss to reduce unnecessary visual effects in the stylized image and keep the magnitude of the stylized feature the same as that of the style feature:

$$l_m = \left|\left|\mu\left(F_s^{(t)}\right) - \mu\left(F_{cs}^{(t)}\right)\right|\right|_2^2 + \left|\left|\sigma\left(F_s^{(t)}\right) - \sigma\left(F_{cs}^{(t)}\right)\right|\right|_2^2 \quad (8)$$

where μ and σ denote the mean and covariance of the feature vectors, respectively.

The overall optimization objective is defined as follows:

$$L = \alpha l_p + \lambda_1 l_r + \lambda_2 l_g + \lambda_3 l_m \quad (9)$$

where α, λ_1, λ_2, and λ_3 are weight terms. By adjusting α, we can control the degree of stylization. Specifically, l_p and l_m both work on ReLU_1_1, ReLU_2_1, ReLU_3_1, and ReLU_4_1; then, l_r works on ReLU_2_1, ReLU_3_1, and ReLU_4_1. Following Johnson et al. [3], l_g works on ReLU_1_2, ReLU_2_2, and ReLU_3_3.

4. Experimental Results and Analysis

4.1. Experimental Dataset and Implementation Details

During training, we use the MS-COCO [29] dataset as the set of content images and select some famous art paintings as style images. To show the experimental results of our method, we also select some copyright-free images as content images, from Pexels.com.

In our experiment, the coarse network is trained on the MS-COCO dataset only once for image reconstruction, and the weight λ in Equation (1) is set as 1. In the experiments, we use the content images and the style image with a resolution of 512 × 512. Then these images are downsampled by 2. Each image that is input into the coarse network has a resolution of 256 × 256. During the training of the fine network, we use the Adam [30] optimizer with a learning rate of 1×10^{-4}, and the batch size is set as 1 because of the limitation of the graphics processing unit (GPU) memory. To train a style, a training process consists of 15,000 iterations. The loss weight terms α, λ_1, λ_2, and λ_3 are set to 1, 20, 1000, and 5, respectively. The experimental environment configuration is shown in Table 2.

Table 2. Experimental environment configuration.

Designation	Information
Operating system	Windows 10
System configuration	CPU: AMD Ryzen 9 5900X
	GPU: NVIDIA GeForce RTX 3090
Software	PyCharm 2021.3.1 (Community Edition)
	Python 3.8.12
Python library	Cuda 11.7
	Pytorch 1.8
	Torchvision 0.9
	Numpy 1.21 Matplotlib 3.5.1

4.2. Qualitative Comparisons with Methods in Prior Works

Inspired by the recent WCT [6] and STROTSS [2] methods, our method adopts the whiting and coloring transformation proposed in WCT and the rEMD loss proposed in STROTSS. In Figure 6, we compare our method with WCT and STROTSS. WCT can transfer the color distribution and simple texture of arbitrary style images; however, some context local structure is discarded, resulting in messy and disordered stylized images (e.g., rows 1, 2, and 3). STROTSS is an image-optimization style transfer method that transfers the visual attributes from the style image to the content image with minimum semantic distortion. Nevertheless, too many structural details are preserved, and the overall palette of the style image is not accurately transferred (e.g., rows 2 and 3). In contrast to these two methods, our method can transfer the main structure and discard some trivial details of the content image. Moreover, some notable local structures of the style image, such as brushstrokes, can be fused into the global structure of the content image, and the overall palette of the stylized image remains the same as that of the style image. For example, in the second and fourth rows, the color blocks of mountains and the brushstrokes of vegetation in our stylized images are explicitly similar to those in the style images. Our model can learn some key style structures while ignoring some unimportant content details.

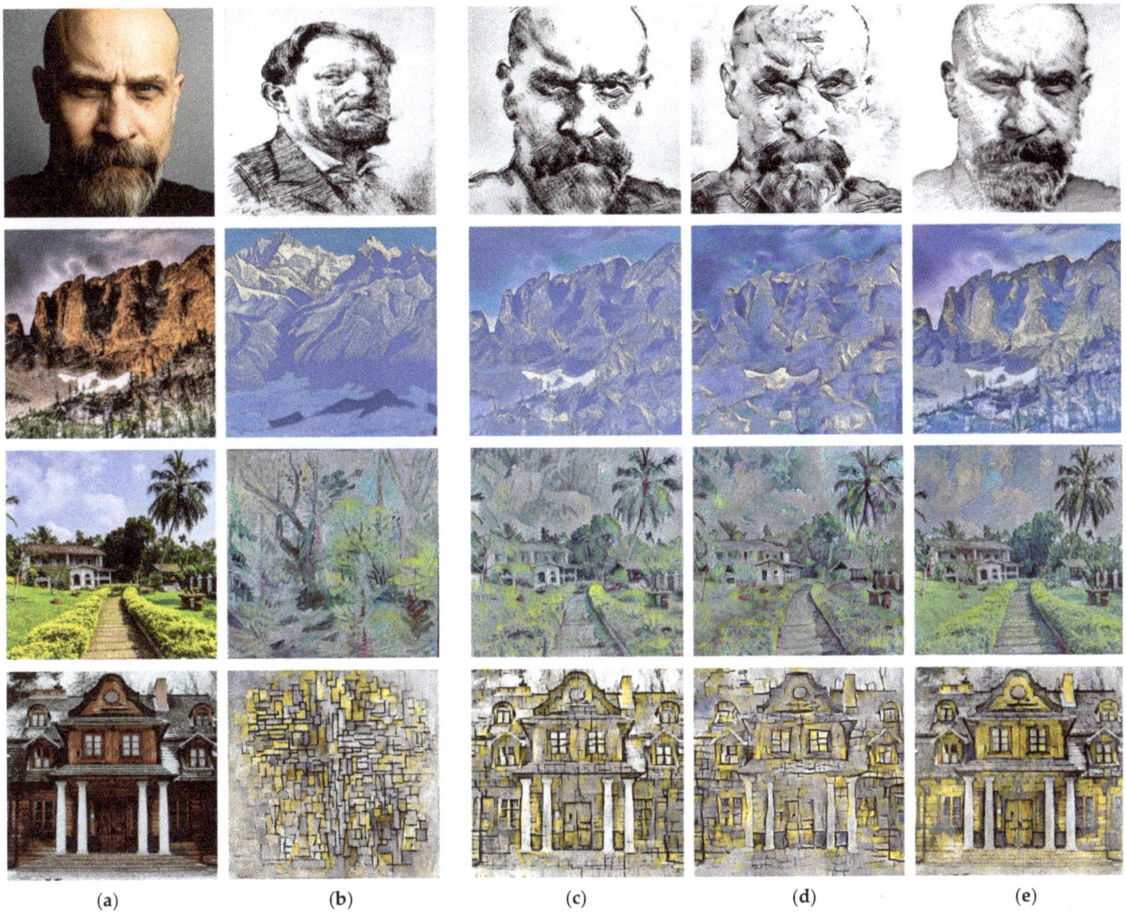

Figure 6. Qualitative comparisons between our method, WCT [6] and STROTSS [2]: (a) The content images; (b) The style images; (c) The stylized images are generated by our method; (d) The stylized images are generated by WCT; (e) The stylized images are generated by STROTSS.

As shown in Figure 7, we compare our method with other state-of-the-art style transfer methods. Gatys et al. [1] proposed the original optimization-based style transfer algorithm, which can transfer the overall style texture and the color distribution. However, some incongruous textures appear in the stylized images, leading to the stylizations' looking unnatural (e.g., rows 4, 5, and 6). Similar to our method, the method proposed by Johnson et al. [3] is also a feed-forward method. It can combine the local color and texture of style images with the structure of the content but often maintains too many content structures and may play a role in shifting the color histogram only in some cases (e.g., rows 1, 2, and 3). AdaIN [5] and SANet [22] are both arbitrary style transfer models, which mainly transfer simple style patterns. AdaIN often fails to transfer the color distribution of style images, and SANet has the severe problem of messy texture and disordered structure (e.g., rows 4, 5, and 6). All of the methods mentioned above maintain some unnecessary small local structures of the content images, and the essential local structures of style images are not integrated into the target image. In contrast to these methods, our model can simultaneously transfer the style color distribution accurately and combine the local style structure with the global content structure. For example, in the fourth row, the image of the rabbits generated by our method looks more harmonious and natural in the stylized image.

It seems as though the style image consists of ink dots; the same artistic expression can be exhibited by our method.

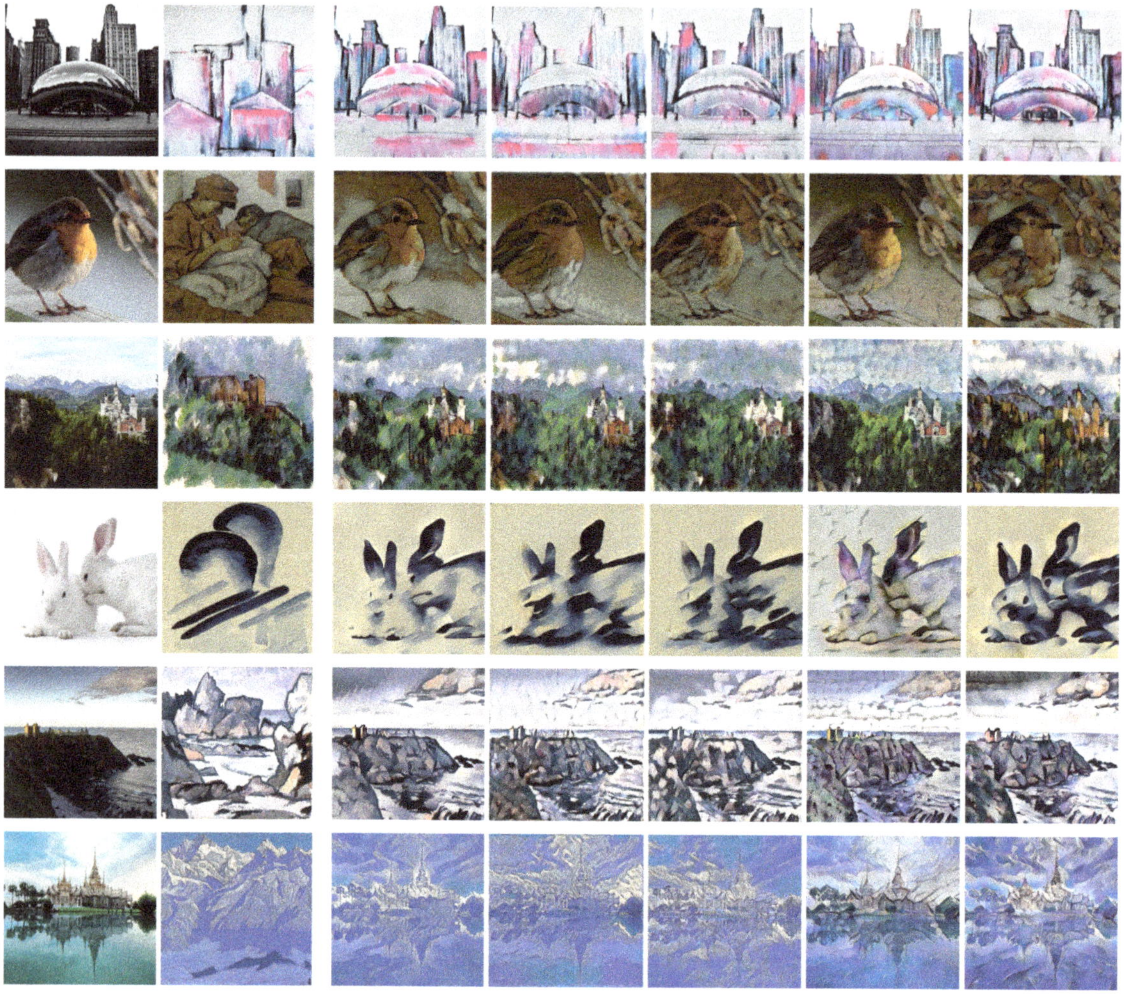

Figure 7. Qualitative comparisons between our method and other state-of-the-art methods: (**a**) The content images; (**b**) The style images; (**c**) The stylized images are generated by our method; (**d**) The stylized images are generated by Gatys et al. [1]; (**e**) The stylized images are generated by Johnson et al. [3]; (**f**) The stylized images are generated by AdaIN [5]; (**g**) The stylized images are generated by SANet [22].

4.3. Quantitative Comparisons with Methods in Prior Works

In the experiment of quantitative comparisons, we use the learned perceptual image patch similarity (LPIPS) proposed in [31] and the structural similarity index measurement (SSIM) proposed in [32] to compute the difference in style structure between the stylized image and the style image. In each method, 1500 pairs of stylized and style images that include 10 styles are used to compute the average distance. As shown in Table 3, lower values indicate the higher similarity of human perceptual judgments when we use LPIPS as the metric, and higher values indicate the higher structural similarity when we use SSIM as the metric. For both evaluation metrics, our proposed method achieves the highest similarity in style structure. The experimental results show that our method can synthesize structure-aware stylized images that have a higher structural similarity to the style images.

Table 3. Quantitative comparisons of LPIPS and SSIM between our method and six state-of-the-art methods.

Method	Our	WCT [6]	STROTSS [2]	Gatys et al. [1]	Johnson et al. [3]	AdaIN [5]	SANet [22]
LPIPS	0.6287	0.6393	0.6516	0.6477	0.6452	0.6445	0.6408
SSIM	0.2135	0.1975	0.2108	0.2022	0.2068	0.1933	0.1893

4.4. Comparisons of Time Efficiency with Methods in Prior Works

We further compare the time efficiency of our proposed method with other state-of-art methods. In each method, we synthesize 100 stylized images with a resolution of 512×512. All experiments are conducted on the same environment configuration. As shown in Table 4, Johnson et al. [3] achieve the highest time efficiency because they use only a simple encoder-decoder architecture to generate stylized images. Like [3], AdaIN [5] and SANet [22] also use the simple encoder-decoder network to generate stylized images. However, they apply some feature transform modules in their networks to integrate content features and style features. As a result, their time efficiencies are lower than [3] but are still satisfactory. Different from these three methods that work at the same image scale, our model includes two networks and works in two stages. Although our model can capture richer multiscale information and synthesize higher-quality stylized images, the time efficiency of our method is only slightly lower than that of AdaIN and SANet. We traded a small increase in time cost for a promising improvement in the quality of stylized images. WCT [6] has low time efficiency because it uses five encoders and decoders to generate a stylized image. The time efficiencies of STROTSS [2] and Gatys et al. [1] are far lower than other methods because they are image-optimization methods that generate only one stylized image after a training process.

Table 4. Running time comparison between our method and six state-of-the-art methods (in seconds).

Method	Our	WCT [6]	STROTSS [2]	Gatys et al. [1]	Johnson et al. [3]	AdaIN [5]	SANet [22]
Time (s)	0.829	2.816	40.157	20.418	0.075	0.105	0.291

4.5. User Study

The user study is conducted on social media, and all participants are anonymous and voluntary. We choose 10 content images and 10 style images to synthesize 10 stylized images in each method and then ask subjects to select their favorite one. By the end of this user study, we had collected 341 votes from these anonymous participants. As shown in Figure 8, we show the percentage of votes for each method. The result shows that the stylization results obtained by our method are more appealing than those of other methods.

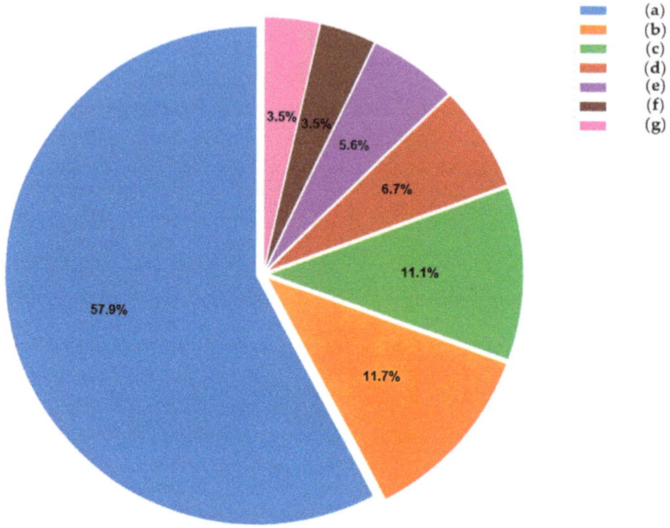

Figure 8. User preference results of our method and six state-of-the-art methods: (**a**) Our method; (**b**) SANet [22]; (**c**) STROTSS [2]; (**d**) AdaIN [5]; (**e**) WCT [6]; (**f**) Gatys et al. [1]; (**g**) Johnson et al. [3].

4.6. Ablation Study on Loss Function

We conduct ablation experiments to verify the effectiveness of each loss term used for training our model, and the results are shown in Figure 9. (1) Without perceptual loss l_p, too many structures of the content image are discarded; for example, the basic structure of the dog disappears in the stylized image. (2) Without Gram matrix loss l_g, the stylization result is acceptable because mean-variance loss l_m has a similar effect to l_g, but the color distribution of the stylized image is slightly different from that of the style image. Moreover, the textures of the dog in the stylized image are increasingly denser and smaller. (3) Without rEMD loss l_r, the texture distribution is chaotic, and some visual artifacts occur in the stylized image. (4) Without mean-variance loss l_m, the global color distribution of the stylized image is not exactly the same as that of the style image; for example, the dark color of the dog in the stylized image is more similar to that in the content image. This dark black color is completely absent in the style image.

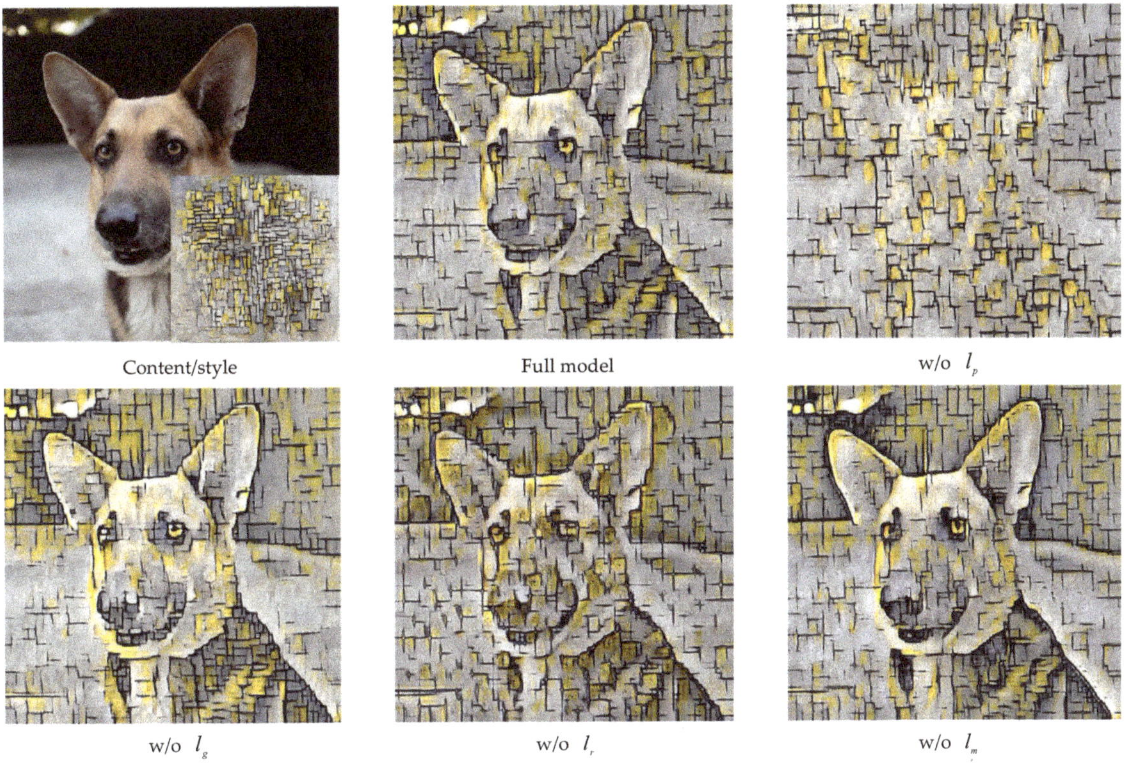

Figure 9. Ablation study of the effects of different loss functions used during training.

4.7. Effectiveness of Coarse Network

During training, we compare our full model with the model without the coarse network. As shown in Figure 10, our full model is trained faster than the model without the coarse network. The preliminary stylization result can be obtained with fewer iterations. Moreover, the stylized images of the comparison during the training phase are shown in Figure 11. At 3000 iterations, our full model can generate a stylized image with a basic structure, while the model without the coarse network generates a completely unstructured image. At 10,000 iterations, the stylization result of our full model is substantially acceptable. However, the stylized result of the model without the coarse network is less than satisfactory because the main structure has not been generated. At 30,000 iterations, the model without the coarse network finally synthesizes the final stylized image, but some messy textures and unnatural structures appear in the stylized image. Compared with this compromised stylized result, our full model can generate an enhanced promising stylized result with more-refined details, such as the brushstrokes of the cat's fur and eyes at 30,000 iterations, which are more delicate and finer than those at 10,000 iterations.

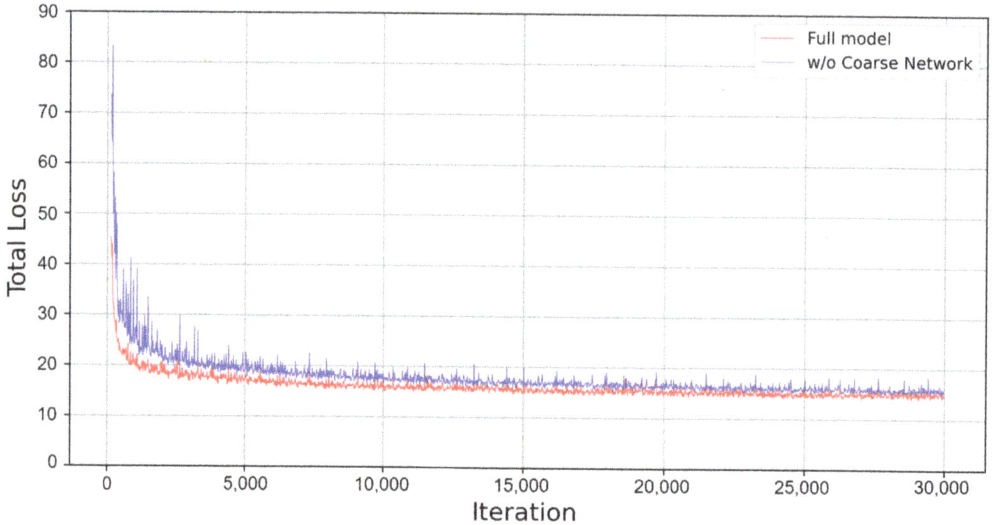

Figure 10. Comparison of the full model and the model without the coarse network in terms of total loss.

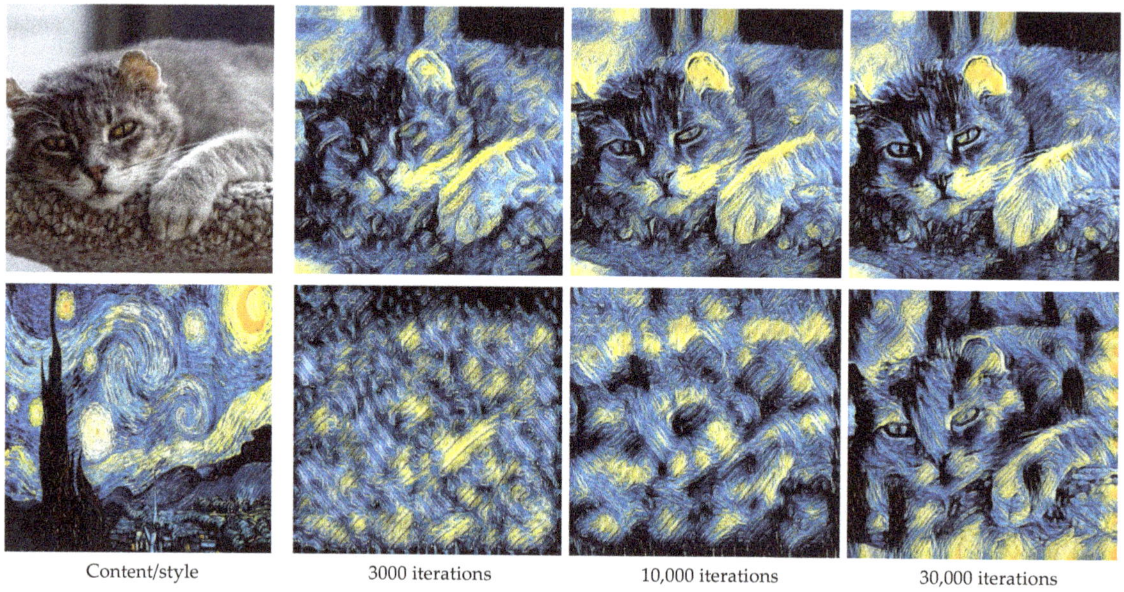

Figure 11. Comparison of stylized images using the full model and the model without the coarse network during training. In the first row, the stylized results are generated by our full model. In the second row, the stylized results are generated by the model without the coarse network.

4.8. *Effectiveness of Fine Network*

As shown in Figure 12, we demonstrate the effectiveness of the fine network. Without the fine network, the coarse network can transfer the color and texture of style images, but the local details and global structure are worse than when our full model is utilized. The stylized image generated directly by the coarse network resembles an unfinished work in progress.

Figure 12. Comparison of stylized images of the full model and the model without the fine network: (**a**) the content images; (**b**) the style images; (**c**) the stylized images generated by the model without fine network; (**d**) the stylized images generated by full model.

4.9. Effectiveness of the SSF Modules

We compare two feature fusion methods through some experiments. In the first method, the reconstructed coarse stylized features from the coarse network are fused to the reconstructed content features in the fine network on the basis of our SSF modules. In the second method, we directly concatenate these two features for feature fusion. As Figure 13 shows, the stylization results that are based on the second method are transferred to the wrong color distribution in some regions. According to the first method, our model can accurately transfer the color distribution, and more-natural textures in the stylized images can be generated by selecting more-critical information.

(**a**) Style (**b**) Content (**c**) Full model (**d**) w/o SSF modules

Figure 13. Comparison of stylized images of our models with different feature fusion methods: (**a**) the content images; (**b**) the style images; (**c**) the stylized images generated by full model; (**d**) the stylized images generated by the model without SSF modules.

4.10. Additional Experiments

In Figure 14, we zoom in on some details in style images, content images, and stylized images. The local structures of these style images are transferred to the content image, and the object of the stylized images looks like a reasonable combination that is composed of the style structures rather than a simple mixture of the content structure and the style texture.

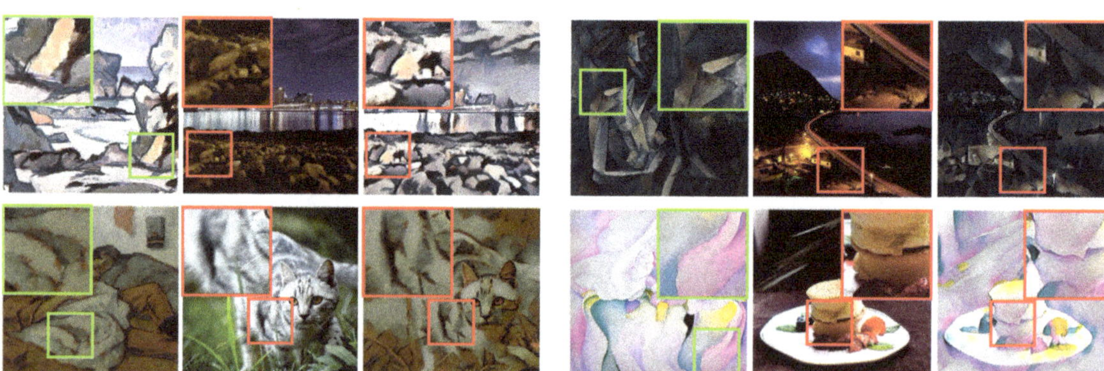

Style I Content I Stylization I Style II Content II Stylization II

Figure 14. Comparison of local style details.

As shown in Figure 15, we can control the stylization degree by adjusting the weight term α in the training phase. These experiments demonstrate that the main content structure can be preserved even though the stylization degree is large. Some local style structures, such as lines or color blocks, can be fused to the global content structure.

Content $\alpha = 4$ $\alpha = 1$ $\alpha = 0.1$ Style

Figure 15. Trade-off of content-style losses.

Following Gatys et al. [26], we incorporate color control and spatial control into our method. In Figure 16b, the color distribution and the local structure of the stylized image are consistent with those of the style image. Then we use color control to make the stylized image preserve the global color of the content image. In Figure 16c, although the color is similar to the content image, the local structure and texture are the same as those of the style image. In Figure 17, we use spatial control to transfer different regions of the content image to different styles. The stylization result is appealing as the local style structures and color distribution are greatly maintained. Both experiments demonstrate that our model can synthesize high-quality structure-aware stylized images by fusing key local structures from the style image to the main content structure while discarding some trivial details from the content image.

(**a**) Content/style (**b**) Stylization (**c**) Stylization by color control

Figure 16. Color control.

(**a**) Content (**b**) Style I/Mask I (**c**) Style II/Mask II (**d**) Stylization by spatial control

Figure 17. Spatial control.

5. Conclusions

The conclusions are summarized as follows:

1. We proposed a novel feed-forward style transfer algorithm that fuses the local style structure into the global content structure. Different from most style transfer methods that work at the same scale, our model can integrate richer information from features from different scales and then synthesize high-quality structure-aware stylized images.

2. We first proposed a coarse network to generate reconstructed coarse stylized features at low resolution, which can capture the main structure of the content image and transfer the holistic color distribution of the style image. Then, we proposed a fine network to enhance local style patterns and three SSF modules to selectively fuse the reconstructed stylized features to reconstructed content features at different levels.
3. Through comparative experiments, it was demonstrated that our method was effective in synthesizing appealing high-quality stylized images, and these stylization results outperformed the results generated by current state-of-the-art style transfer methods. The experimental results also demonstrated the effectiveness of the coarse network, the fine network, and the SSF module.

Although the high-quality stylization results can be synthesized by our method, our model generated the stylized images with a single style only after a training process. In future studies, we will achieve a novel arbitrary style transfer framework that is based on our full model in this paper. Appealing high-quality structure-aware stylized images with an arbitrary style can be generated by this framework after a training process. In addition, we will try to use more feature transform methods to replace the whitening and coloring transforms for achieving higher running time efficiency.

Author Contributions: K.L. proposed the style transfer method and designed the framework. K.L. conducted the experiments. K.L. analyzed and discussed the experimental results. K.L. and G.Y. wrote the article. G.Y., H.W. and W.Q. revised the article and provided valuable advice for ablation experiments. All authors have read and agreed to the published version of the manuscript.

Funding: This research was funded by the Natural Science Foundation of China (grant no. 62162065, 62061049, 12263008), the Application and Foundation Project of Yunnan Province (grant no. 202001BB050032), the Department of Science and Technology of Yunnan Province–Yunnan University Joint Special Project for Double-Class Construction (grant no. 202201BF070001-005), the Expert Workstation of Yunnan Province (202105AF150011), and the Postgraduate Practice and Innovation Project of Yunnan University (grant no. 2021Y177).

Data Availability Statement: The dataset of content images adopted during training is openly available online. MS-COCO: https://cocodataset.org/#home (accessed on 25 March 2022).

Conflicts of Interest: The authors declare no conflict of interest.

Abbreviations

SSF	Structural selective fusion
NPR	Nonphotorealistic rendering
STROTSS	Style transfer by relaxed optimal transport and self-similarity
rEMD	Relaxed earth mover's distance
AdaIN	Adaptive instance normalization
WCT	Whitening and coloring transforms
VGG	Visual geometry group
DualStyleGAN	Dual style generative adversarial network
POFMakeup	Peking Opera face makeup
SANet	Style-attentional network
LapStyle	Laplacian pyramid style network
ReLU	Rectified linear unit
GPU	Graphics processing unit
LPIPS	Learned perceptual image patch similarity
SSIM	Structural similarity index measurement

Symbol

x_c	Content image
x_s	Style image
x_{cs}	Stylized image
$\overline{x}_c$	The result of downsampling x_c by 2
$\overline{x}_s$	The result of downsampling x_s by 2
$\overline{f}_r^{(i)}$	Restructured coarse stylized features
$c_r^{(i)}$	Channels of $\overline{f}_r^{(i)}$
$h_r^{(i)}$	Height of $\overline{f}_r^{(i)}$
$w_r^{(i)}$	Width of $\overline{f}_r^{(i)}$
$\overline{f}_c$	Content feature extracted from VGG network
$\overline{f}_s$	Style feature extracted from VGG network
$\overline{f}'_c$	The result of linearly transforming $\overline{f}_c$
$\overline{f}_{cs}$	Stylized feature generated by WCT module
f_{cs}	Reconstructed content features
f_{csr}	The input of SSF module
M_{cs}	Attention map of f_{csr}
f'_{cs}	The result of refining f_{cs}
f'_{csr}	The result of refining f_{csr}
f_{ssf}	The output of SSF module
l_{re}	Reconstruction loss
I_i	Input image
I_o	Output image
Φ	VGG encoder that extracts features at ReLU_X_1
λ	Weight term of l_{re}
$F_c^{(t)}$	Content feature extracted at ReLU_t
$F_s^{(t)}$	Style feature extracted at ReLU_t
$F_{cs}^{(t)}$	Stylized feature extracted at ReLU_t
l_p	Perceptual loss
l_r	Relaxed earth mover's distance (rEMD) loss
C	Cost matrix
D_{cos}	Cosine distance
l_g	Gram matrix loss
G	Calculation of the Gram matrix
l_m	Mean-variance loss
μ	Mean
σ	Covariance
L	Overall optimization objective
α	Weight term of L
λ_1	Weight term of L
λ_2	Weight term of L
λ_3	Weight term of L

References

1. Gatys, L.A.; Ecker, A.S.; Bethge, M. Image Style Transfer Using Convolutional Neural Networks. In Proceedings of the 2016 IEEE Conference on Computer Vision and Pattern Recognition (CVPR), Seattle, WA, USA, 27–30 June 2016; pp. 2414–2423. [CrossRef]
2. Kolkin, N.; Salavon, J.; Shakhnarovich, G. Style Transfer by Relaxed Optimal Transport and Self-Similarity. In Proceedings of the IEEE/CVF Conference on Computer Vision and Pattern Recognition (CVPR), Long Beach, CA, USA, 16–20 June 2019; pp. 10043–10052. [CrossRef]
3. Johnson, J.; Alahi, A.; Fei-Fei, L. Perceptual Losses for Real-Time Style Transfer and Super-Resolution. In Proceedings of the Computer Vision—ECCV 2016, 14th European Conference, Amsterdam, The Netherlands, 11–14 October 2016; Springer: Cham, Switzerland, 2016; pp. 694–711. [CrossRef]
4. Ulyanov, D.; Lebedev, V.; Vedaldi, A.; Lempitsky, V. Texture Networks: Feed-forward Synthesis of Textures and Stylized Images. In Proceedings of the 33rd International Conference on Machine Learning, New York, NY, USA, 20–22 June 2016. [CrossRef]

5. Huang, X.; Belongie, S. Arbitrary Style Transfer in Real-time with Adaptive Instance Normalization. In Proceedings of the 16th IEEE International Conference on Computer Vision (ICCV), Venice, Italy, 22–29 October 2017; pp. 1510–1519. [CrossRef]
6. Li, Y.J.; Fang, C.; Yang, J.M.; Wang, Z.W.; Lu, X.; Yang, M.H. Universal Style Transfer via Feature Transforms. In Proceedings of the 31st Annual Conference on Neural Information Processing Systems (NIPS), Long Beach, CA, USA, 4–9 December 2017.
7. Simonyan, K.; Zisserman, A. Very deep convolutional networks for large-scale image recognition. *arXiv* **2014**. [CrossRef]
8. Li, C.; Wand, M. Precomputed Real-Time Texture Synthesis with Markovian Generative Adversarial Networks. In Proceedings of the 14th European Conference on Computer Vision (ECCV), Amsterdam, Netherlands, 8–16 October 2016; pp. 702–716. [CrossRef]
9. Li, C.; Wand, M. Combining Markov Random Fields and Convolutional Neural Networks for Image Synthesis. In Proceedings of the 2016 IEEE Conference on Computer Vision and Pattern Recognition (CVPR), Seattle, WA, USA, 27–30 June 2016; pp. 2479–2486. [CrossRef]
10. Wang, X.; Oxholm, G.; Zhang, D.; Wang, Y.F. Multimodal Transfer: A Hierarchical Deep Convolutional Neural Network for Fast Artistic Style Transfer. In Proceedings of the 30th IEEE/CVF Conference on Computer Vision and Pattern Recognition (CVPR), Honolulu, HI, USA, 21–26 July 2017; pp. 7178–7186. [CrossRef]
11. Ulyanov, D.; Vedaldi, A.; Lempitsky, V. Improved Texture Networks: Maximizing Quality and Diversity in Feed-forward Stylization and Texture Synthesis. In Proceedings of the 30th IEEE/CVF Conference on Computer Vision and Pattern Recognition (CVPR), Honolulu, HI, USA, 21–26 July 2017; pp. 4105–4113. [CrossRef]
12. Sanakoyeu, A.; Kotovenko, D.; Lang, S.; Ommer, B. A Style-Aware Content Loss for Real-Time HD Style Transfer. In Proceedings of the 15th European Conference on Computer Vision (ECCV), Munich, Germany, 8–14 September 2018; pp. 715–731. [CrossRef]
13. Yang, S.; Jiang, L.M.; Liu, Z.W.; Loy, C.C. Pastiche Master: Exemplar-Based High-Resolution Portrait Style Transfer. In Proceedings of the IEEE/CVF Conference on Computer Vision and Pattern Recognition (CVPR), New Orleans, LA, USA, 18–24 June 2022; pp. 7683–7692. [CrossRef]
14. Zhang, F.C.; Liang, X.M.; Sun, Y.Q.; Lin, M.G.; Xiang, J.; Zhao, H.H. POFMakeup: A style transfer method for Peking Opera makeup. *Comput. Electr. Eng.* **2022**, *104*, 108459. [CrossRef]
15. Lin, C.C.; Hsu, C.B.; Lee, J.C.; Chen, C.H.; Tu, T.M.; Huang, H.C. A Variety of Choice Methods for Image-Based Artistic Rendering. *Appl. Sci.* **2022**, *12*, 6710. [CrossRef]
16. Dumoulin, V.; Shlens, J.; Kudlur, M. A learned representation for artistic style. *arXiv* **2016**. [CrossRef]
17. Zhang, H.; Dana, K. Multi-style Generative Network for Real-Time Transfer. In Proceedings of the 15th European Conference on Computer Vision (ECCV), Munich, Germany, 8–14 September 2018; pp. 349–365. [CrossRef]
18. Ye, W.J.; Liu, C.J.; Chen, Y.H.; Liu, Y.J.; Liu, C.M.; Zhou, H.H. Multi-style transfer and fusion of image's regions based on attention mechanism and instance segmentation. *Signal Process.-Image Commun.* **2023**, *110*, 116871. [CrossRef]
19. Alexandru, I.; Nicula, C.; Prodan, C.; Rotaru, R.P.; Tarba, N.; Boiangiu, C.A. Image Style Transfer via Multi-Style Geometry Warping. *Appl. Sci.* **2022**, *12*, 6055. [CrossRef]
20. Wang, Z.; Zhao, L.; Chen, H.; Qiu, L.; Mo, Q.; Lin, S.; Xing, W.; Lu, D. Diversified arbitrary style transfer via deep feature perturbation. In Proceedings of the IEEE/CVF Conference on Computer Vision and Pattern Recognition, Nashville, TN, USA, 20–25 June 2021; pp. 7789–7798.
21. Wang, H.; Li, Y.J.; Wang, Y.H.; Hu, H.J.; Yang, M.H. Collaborative Distillation for Ultra-Resolution Universal Style Transfer. In Proceedings of the IEEE/CVF Conference on Computer Vision and Pattern Recognition (CVPR), Electro Network, Seattle, WA, USA, 14–19 June 2020; pp. 1857–1866. [CrossRef]
22. Park, D.Y.; Lee, K.H. Arbitrary Style Transfer with Style-Attentional Networks. In Proceedings of the 32nd IEEE/CVF Conference on Computer Vision and Pattern Recognition (CVPR), Long Beach, CA, USA, 16–20 June 2019; pp. 5873–5881. [CrossRef]
23. Sheng, L.; Lin, Z.Y.; Shao, J.; Wang, X.G. Avatar-Net: Multi-scale Zero-shot Style Transfer by Feature Decoration. In Proceedings of the 31st IEEE/CVF Conference on Computer Vision and Pattern Recognition (CVPR), Salt Lake City, UT, USA, 18–23 June 2018; pp. 8242–8250. [CrossRef]
24. Yang, S.; Jiang, L.M.; Liu, Z.W.; Loy, C.C. VToonify: Controllable High-Resolution Portrait Video Style Transfer. *ACM Trans. Graph.* **2022**, *41*, 15. [CrossRef]
25. Lin, T.W.; Ma, Z.Q.; Li, F.; He, D.L.; Li, X.; Ding, E.R.; Wang, N.N.; Li, J.; Gao, X.B. Drafting and Revision: Laplacian Pyramid Network for Fast High-Quality Artistic Style Transfer. In Proceedings of the IEEE/CVF Conference on Computer Vision and Pattern Recognition (CVPR), Electro Network, Nashville, TN, USA, 19–25 June 2021; pp. 5137–5146. [CrossRef]
26. Gatys, L.A.; Ecker, A.S.; Bethge, M.; Hertzmann, A.; Shechtman, E. Controlling Perceptual Factors in Neural Style Transfer. In Proceedings of the 30th IEEE/CVF Conference on Computer Vision and Pattern Recognition (CVPR), Honolulu, HI, USA, 21–26 July 2017; pp. 3730–3738. [CrossRef]
27. Hu, J.; Shen, L.; Sun, G. Squeeze-and-Excitation Networks. In Proceedings of the 31st IEEE/CVF Conference on Computer Vision and Pattern Recognition (CVPR), Salt Lake City, UT, USA, 18–23 June 2018; pp. 7132–7141. [CrossRef]
28. Woo, S.H.; Park, J.; Lee, J.Y.; Kweon, I.S. CBAM: Convolutional Block Attention Module. In Proceedings of the 15th European Conference on Computer Vision (ECCV), Munich, Germany, 8–14 September 2018; pp. 3–19. [CrossRef]
29. Lin, T.Y.; Maire, M.; Belongie, S.; Hays, J.; Perona, P.; Ramanan, D.; Dollar, P.; Zitnick, C.L. Microsoft COCO: Common Objects in Context. In Proceedings of the 13th European Conference on Computer Vision (ECCV), Zurich, Switzerland, 6–12 September 2014; pp. 740–755. [CrossRef]
30. Kingma, D.P.; Ba, J. Adam: A method for stochastic optimization. *arXiv* **2014**. [CrossRef]

31. Zhang, R.; Isola, P.; Efros, A.A.; Shechtman, E.; Wang, O. The Unreasonable Effectiveness of Deep Features as a Perceptual Metric. In Proceedings of the 31st IEEE/CVF Conference on Computer Vision and Pattern Recognition (CVPR), Salt Lake City, UT, USA, 18–23 June 2018; pp. 586–595. [CrossRef]
32. Wang, Z.; Bovik, A.C.; Sheikh, H.R.; Simoncelli, E.P. Image quality assessment: From error visibility to structural similarity. *IEEE Trans. Image Process.* **2004**, *13*, 600–612. [CrossRef] [PubMed]

applied sciences

MDPI

Review

A Survey of OCR in Arabic Language: Applications, Techniques, and Challenges

Safiullah Faizullah [1,*], Muhammad Sohaib Ayub [2], Sajid Hussain [2] and Muhammad Asad Khan [3]

1 Department of Computer Science, Islamic University, Madinah 42351, Saudi Arabia
2 Department of Computer Science, Lahore University of Management Sciences, Lahore 54792, Pakistan
3 Department of Telecommunication, Hazara University, Mansehra 21120, Pakistan
* Correspondence: safi@iu.edu.sa; Tel.: +1-848-239-7700

Abstract: Optical character recognition (OCR) is the process of extracting handwritten or printed text from a scanned or printed image and converting it to a machine-readable form for further data processing, such as searching or editing. Automatic text extraction using OCR helps to digitize documents for improved productivity and accessibility and for preservation of historical documents. This paper provides a survey of the current state-of-the-art applications, techniques, and challenges in Arabic OCR. We present the existing methods for each step of the complete OCR process to identify the best-performing approach for improved results. This paper follows the keyword-search method for reviewing the articles related to Arabic OCR, including the backward and forward citations of the article. In addition to state-of-art techniques, this paper identifies research gaps and presents future directions for Arabic OCR.

Keywords: optical character recognition; Arabic OCR; preprocessing; segmentation; classification; postprocessing

Citation: Faizullah, S.; Ayub, M.S.; Hussain, S.; Khan, M.A. A Survey of OCR in Arabic Language: Applications, Techniques, and Challenges. *Appl. Sci.* **2023**, *13*, 4584. https://doi.org/10.3390/app13074584

Academic Editor: Antonio Fernández-Caballero

Received: 26 February 2023
Revised: 28 March 2023
Accepted: 3 April 2023
Published: 4 April 2023

1. Introduction

Optical character recognition (OCR) enables the recognition of text characters from digital images, scanned documents, and video streams. OCR software analyses the image of text and converts it into machine-encoded text, which can then be edited, searched, and indexed. OCR can be used for a wide range of applications, including document scanning, automated indexing, and form processing. Further, OCR software can be integrated into various systems, such as document management systems, workflow systems, and mobile apps. There are some challenges to OCR systems, such as the writing style, text size, and quality of the document (handwritten, printed, or scanned), which cause challenges while implementing OCR [1], and a big challenge also comes while implementing OCR in hardware systems, which helps in many regards, such as a 'Quran Read Pen' that helps blind and illiterate people to read Quran [2].

1.1. Types of OCR

There are different types of OCR systems depending on the language and writing mode of the images. For example, the documents can be handwritten, printed, or scanned, and can contain one or more languages. Therefore, OCR systems can be categorized as unilingual or multilingual based on language. A unilingual OCR system can recognize only one language, and the Arabic OCR model is an example of a unilingual OCR system. On the other hand, some OCR systems perform recognition and extraction tasks for multiple languages; these are called multilingual OCR systems.

OCR systems can be categorized into offline and online OCR systems, as shown in Figure 1. An offline OCR system is a type of OCR system where the input documents are presented in scanned, printed, and handwritten formats [3]. These OCR systems provide online services that can be used for various purposes such as mail sorting, bank cheque reading, signature verification, utility bill processing, and insurance applications. Digital pens help blind or illiterate people by reading text in audio form. Many online recognition systems are implemented in different fields such as number-plate recognition [4]. Similarly, an online OCR system is capable of receiving and processing real-time input images. For offline recognition, multiple models are used with different datasets for different algorithms to get better recognition accuracy [5].

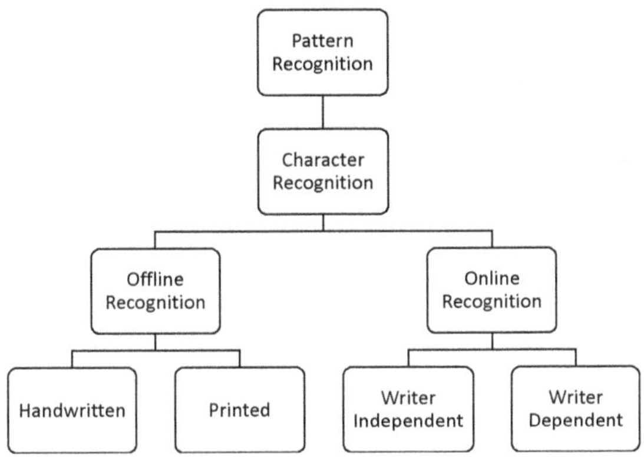

Figure 1. Types of OCR systems in Arabic and their modes of processing.

1.2. Language vs. Script

As we work on the Arabic OCR system, getting basic information about Arabic is essential. For this purpose, the concepts of language, script, and writing styles are crucial. Thus, language refers to the communication system humans use, which includes the grammar, vocabulary, and pronunciation used to convey meaning. On the other hand, the script refers to the written representation of a language, such as an alphabet or letters used to write words and sentences. A language can be written in multiple scripts and can use a script to write various languages. For example, the English letters are written in Latin script, while the Arabic language can be written in Arabic. In Arabic, the most-used styles are Naskh and Nastaleeq. Script similarities can be used to compensate for lack of availability of large amounts of training data for deep-learning-based OCR models [6].

In [5], the authors explain the basics of the Arabic language, i.e., Arabic is written from right to left and from top to bottom. It has 28 letters, which include three vowels, i.e., ا , و, and ي. These letters change their shapes according to their usage in different words. Upper and lower case annotation does not exist. A total of 15 letters out of 28 have a point or dot above or under the letter. Arabic letters are connected from the right or left sides or both sides. However, six letters cannot be connected to their successors in a word; those letters are ا،د،ذ،ر،ز،و. There is another term called Tanween, which produces sound at the end of the words; these symbols are "ً ٌ ٍ" in written form. Punctuation marks have their own format in Arabic; e.g., the question mark in English is written as '?', but its shape in Arabic is '؟'.

1.3. Challenges

Some challenges are faced while designing an OCR system for the Arabic language. Arabic script uses diacritics and ligatures to indicate short vowels and certain consonant combinations, and OCR systems need to recognize and process these diacritics and ligatures correctly. The Arabic language has several types of two- and three-letter consonant combinations, i.e., shadda, sukoon, and tashkeel. An OCR system needs to recognize around 70 to 80 symbols in total for the Arabic language, including basic letters, diacritic marks, and other symbols used in the Arabic script. The Arabic script has 28 basic letters and several complex characters formed by joining multiple basic letters, and OCR systems need to recognize and separate these complex characters. Arabic handwriting can vary greatly, making it more difficult for OCR systems to recognize the characters correctly. Arabic OCR datasets are usually smaller than those of other languages, leading to difficulty with training and fine-tuning the model. The images for OCR can be in multiple forms, i.e., computer-rendered images, scanned images, photographed images, and handwritten scans. These image types have challenges regarding the recognition rate for the OCR process.

Bafjaish et al. [7] also discussed some challenges of the Arabic OCR system. Dots come in Arabic in different places, sometimes above or below the baseline. These dots have much importance in the Arabic language; if you miss any dot somehow or during skew detection/correction, it will change the meaning of the letter or word, reducing the accuracy of the OCR model. Many of the scripts have a non-cursive style, meaning the letters present in a word have some gaps, making them easy to recognize, reducing the challenge, and making the task easy. However, the Arabic language has a cursive style , and the connectivity of letters makes text recognition more complicated. As Arabic letters are compounded to form a word, every font style shows a different level of ligature in words.

1.4. Applications

OCR systems are used now in many fields to make the workflow fast and accurate, so for this kind of digitization, OCR is used. In [8], the authors present a survey of the application of OCR and perform experiments for some applications. The OCR applications discussed are as follows:

- Invoice Imaging: Used in many businesses to track business records.
- Legal Industry: To digitize documents and enter the data directly into the databases, OCR is used.
- Banking: OCR is also widely used in banking services. For example, to process check payments, cheques are scanned and transferred in seconds.
- Healthcare: In healthcare, many forms, reports, and insurance applications are processed into databases and for other purposes; OCR helps to transfer all kinds of patient data.
- Captcha: Captcha is used to secure systems. A few letters, numbers, or both are used in a captcha, and the image is distorted. Humans can easily read this captcha, but not an average computer program.
- Automatic Number Recognition: It is used for surveillance systems to track vehicles' records by getting their number plates. OCR is also used to recognize the characters and numbers from the number plates.
- Handwriting Recognition: in this application of OCR, the text is extracted from handwritten documents and photographs. For this purpose, the model learns and identifies fonts and languages for better results.
- Scanned Receipts: some challenges comes while scanning receipts for extracting information from them, i.e., variations in receipt layout, noise, and distortion [9].

1.5. Brief OCR Process

Arabic OCR systems involve several complex steps, and Figure 2 shows the brief overflow of specific OCR steps to be followed. The image is first preprocessed to improve its quality and make recognization easier, which includes operations such as skew correction, noise reduction, and contrast enhancement. The text area is then broken down into individual segments of characters or words. The segmented characters are then recognized, and the best match is selected using a database of known characters. The recognized text then undergoes further processing to correct errors and improve accuracy. The final result is a machine-readable text document that software applications can edit, search, and analyze.

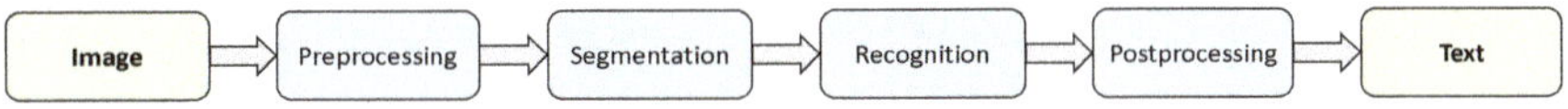

Figure 2. Brief overview of OCR process.

1.6. Goals and Outlines

This paper aims to provide an overview of the current state-of-the-art in Arabic OCR technology. It presents the main challenges and limitations of existing Arabic OCR systems. We highlight the main research trends and future directions for Arabic OCR development. We conclude by identifying the main research gaps and areas that need further study.

The rest of the paper is organized as follows: In Section 2, we review the datasets available to evaluate the Arabic OCR. In Section 3, we summarize the existing literature for each step of Arabic OCR, highlighting the main research trends and advances in the field. We conclude by summarizing the survey's main findings and highlighting the main research gaps and areas that need further study.

2. Datasets

The dataset is an important part of any OCR system to validate the results of OCR. It is especially challenging for the Arabic language because of the cursive nature of the Arabic language, diacritics, different writing styles in which each word's overall shape changes, text sizes, and other reasons. The collection of the Arabic dataset is also very limited due to the low-resource nature of the Arabic language. Previously, refs. [10,11] shared some commonly used datasets, as shown in Table 1. They presented the shared datasets for Arabic, Urdu, and Persian, both publicly available and otherwise.

2.1. Handwritten Text

Urdu and Arabic have many similarities. Writing styles are identical, and both have cursive nature as well; both start from right to left, and Urdu has about 39 to 40 letters; Arabic is similar to Urdu but has fewer characters. Urdu borrows a large vocabulary from Arabic (almost 30%). Most Urdu speakers can read Quran because of Urdu and Arabic similarities. Thus, their datasets and trained models are commonly used as well.

Table 1. Available datasets with their stats, dataset type, and mode of availability.

Dataset	Type of Content	Availability	Size of Dataset
ACTIV2 [12]	Embedded words	Public	10,415 text images
QTID [13]	Synthetic words	Private	309,720 words and 249,428 characters
IFN/ENIT [14]	Handwritten words	Public	115,000 words and 212,000 characters
AHDB [15]	Handwritten words and digits	Private	30,000 words
APTI [16]	Printed words	Public	113,284 words and 648,280 characters
HACDB [17]	Handwritten characters	Public	6600 characters and 50 writers
UPTI [18]	Printed text lines	Public	10,000 text lines
Digital Jawi [19]	Jawi paleography images	Public	168 words and 1524 characters
KHATT [20]	Handwritten text lines	Public	9327 lines, 165,890 words and 589,924 characters
ALIF [21]	Embedded text lines	Upon request	1804 words and 89,819 characters
ACTIV [22]	Embedded text lines	Public	4824 lines and 21,520 words
SmartATID [23]	Printed and handwritten pages	Public	9088 pages
Degraded historical [24]	Handwritten documents	Public	10 handwritten images and 10 printed images
Printed PAW [25]	Printed subwords	Upon request	415,280 unique words and 550,000 sub words
Checks [26]	Handwritten subwords and digits	Private	29,498 subwords and 15,148 digits
Numeral [27]	Handwritten digits	Public	21,120 digits and 44 writers
Forms [28]	Handwritten characters	Private	15,800 characters and 500 writers
KAFD [29]	Printed pages and lines	Public	28,767 pages and 644,006 lines
AHDBIFTR [30]	Handwritten images	Public	497 word images and 5 writers
ARABASE [31]	Handwritten text	Public	47,000 words and 500 free Arabic sentences
CEDAR [32]	Handwritten pages	Private	20,000 words, 10 writers, and 100 documents
CENPARMI [26]	Handwritten subwords and digits	Public	6000 digit images

Shafi and Zia [33] surveyed automatic Urdu text recognition techniques and described the algorithms, techniques, datasets, challenges, and future directions for Urdu OCR. Additionally, [34] reviewed the availability of datasets and suggested more training data to address the unique challenges of OCR systems.

Due to their similarities, both languages have some datasets available. The authors of [35] presented a dataset of handwritten Urdu numerals. In [11], the authors proposed an Urdu Nastaliq Handwritten Dataset (UNHD), which is written by 500 writers on A4-size paper and is available on request (https://www.kaggle.com/datasets/drsaadbinahmed/unhd-dataset, accessed on 28 March 2023). Khosrobeigi et al. [36] also presented a Persian language dataset; this dataset is collected from different Persian-language new websites, and the description of the dataset is shown in Table 2; this dataset is split into 80% for training and 20% for testing purpose.

Table 2. Example Persian dataset collected from different news websites.

Description	Stats
Total text lines of dataset	4,000,000
Total words	15,000,000
Unique words	200,000
Text lines per image	70
Total used fonts (with sizes)	11 fonts (sizes:12, 14, and 18)

There are some datasets available that are used for handwritten text recognition of Urdu, and, as we know, Urdu and Arabic use the same vocabulary and alphabet as well. Therefore, we can use Urdu datasets as well and achieve good results. For this purpose, ref. [37] presents some datasets of handwritten Urdu text recognition, which give outstanding results; the dataset descriptions and their availability are also shown in Table 3.

Table 3. Sample handwritten Urdu datasets.

	UPTI	CALAM	UNHD
Total writers	250	725	500
Text lines	60,000	3043	10,000
Words	240,000	46,664	187,200
Characters	970,650	101,181	312,000
Availability	Private	Private	Public

Naz et al. [38] summarized the state-of-the-art in OCR research for Urdu-like cursive scripts, concentrating on Nastaliq and Naskh scripts in the Urdu, Pushto, and Sindhi languages. The study discusses the quirks of these scripts as well as the text-picture databases that are readily accessible. Three categories have been established: printed, handwritten, and internet character recognition. The database is discussed, which includes 60,329 isolated digits, 12,914 strings, 1705 symbols, 14,890 isolated characters, and 318 different patterns of dates.

Alghamdi and Teahan [39] discussed the most commonly used datasets for training and evaluation of OCR systems for printed Arabic script, including the IFN/ENIT Arabic handwritten dataset, the "Handwriting Arabic Corpus" (HAC) dataset, and the RIMES dataset containing a large collection of printed and handwritten documents. The authors provide an overview of the available datasets and emphasize the importance of high-quality datasets for improving the accuracy of OCR systems.

Publicly available scanned image datasets are tested by [40], e.g., the WATAN and APTI datasets with extensive vocabularies. The datasets are split into a training set and a testing set, where training data contain 282,000 word images and 1,200,000 characters images while testing 5500 words, and 100,500 characters are used. The trained model achieves an overall accuracy of 97.94%. As there are many challenges in Arabic optical character recognition (AOCR), [41] surveys various approaches and methods to detect and reduce errors.

In [42], the authors proposed a system synthesizing Arabic handwritten words and text pages to generate comprehensive databases for training and validating OCR systems. In the database, vocabulary of the 50,000 most-common Arabic words are used for error correction.

2.2. Printed Arabic

In Arabic OCR, printed, handwritten, and historical documents are used. To process printed Arabic documents, [43] presents top-down, bottom-up, and hybrid approaches and discusses the phases of preprocessing, segmentation, feature extraction, and classification.

An efficient, font-independent word and character segmentation algorithm for printed Arabic documents is proposed in [44]. Profile projection is used for the font-independent technique. Interquartile Range (IQR) is used for word segmentation. For character segmentation, two approaches are used, i.e., the holistic approach (segmentation-free approach) and the analytical approach, which is a segmentation-based approach, and the process followed for this purpose is shown in Figure 3. For this purpose, ATPI is the dataset used to make a font-independent OCR system that achieves 97.51% accuracy.

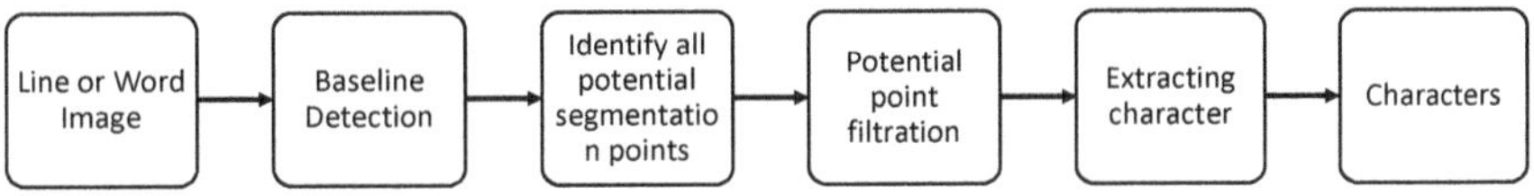

Figure 3. Character segmentation stages in order to recognize characters with maximum accuracy.

In [45], the authors propose a thinning algorithm in the preprocessing stage. A new chain code representation technique is proposed using an agent-based model for feature extraction from non-dotted Arabic text images. A character segmentation technique based on the extracted features is also introduced. A compression-based method is applied to recognize Arabic text in the classification stage. The system was tested on a public dataset and produced an accuracy of 77.3%.

The authors of [46] demonstrate the effective use of unsupervised algorithms for writer attribution of historical scanned documents and forensic document analysis. Some distinct handwriting styles differ in various ways, including character size, stroke width, loops, ductus, slant angles, and cursive ligatures. Additionally covered are prior efforts on labeled data that provide excellent accuracy rates utilizing the Hidden Markov Model (HMM), Support Vector Machine (SVM), and semi-supervised Recurrent Neural Networks (RNN).

Transformer-based models are a type of deep learning method to deal with sequential data [47]. Several metrics are used to evaluate the performance of the proposed method; those metrics are character error rate (CER) and word error rate (WER). Furthermore, results show that the proposed method improves the recognition rate of historical documents.

The data generated by IoT devices such as the Quran Read Pen, which helps to read Quran specifically to illiterate or blind people [48], is shared via the Quranic Text Image Dataset (QTID). It contains 309,720 images of words and 2,494,428 characters taken from the Quran, which uses the sequence-to-sequence model and CNN and achieves a high recognition rate. The character recognition rate (CRR) with and without diacritics is about 97.60% and 97.05%, respectively, and the overall recognition rate of this model is 99.48%, while the CNN model gives the CRR with and without diacritics of about 98.90% and 98.51%, respectively.

Feature extraction and classification techniques are used for character segmentation in ancient manuscripts for their preservation and information extraction [49].

2.3. Scanned Documents/Receipts

Information extraction from scanned documents is difficult compared to regular documents because of the rough layout and low resolution. Preprocessing involves processing the scanned document successfully, as information extraction from the scanned documents/receipts is the key perspective. ICDAR [50] presents a competition wherein 1000 scanned receipts are used to extract information; this competition includes some tasks such as text recognition, layout analysis, and information extraction.

2.4. Quranic Text

As Quran is a Holy Book, it is recited worldwide, and everyone wants to recite it correctly without any mistakes. Bashir et al. [51] review the Quranic NLP techniques, approaches used, tools, and datasets, and recitation via speech-recognition method. The techniques used in the paper are text preprocessing, text matching, clustering, classification, and speech processing. Quranic NLP work includes grammatical NLP analysis and semantic- and ontology-based technologies using BLSTM. The model took recitation of different reciters for training purposes, and a feature widely used for speech recognition, named mel-frequency cepstral coefficients (MFCCs), gives a 99.89% recognition rate for 3 s of recitation, which is far better than all of the other techniques.

3. OCR Process

The OCR process refers to identifying and converting printed or handwritten text characters into machine-encoded text. It typically involves several steps, including preprocessing, segmentation, recognition, and postprocessing. During preprocessing, the input image is cleaned up and enhanced to improve the quality of recognition. Segmentation involves breaking the image into individual or groups of letters or characters. Feature extraction is the process of identifying and extracting the relevant features of each character, such as its shape, size, and orientation. In recognition, the characters are classified by comparing them to a set of known characters, and the best match is selected as the recognized character. Finally, postprocessing of recognized text is performed to remove errors from text and improve accuracy and overall results of OCR. OCR recognition accuracy can vary depending on several factors, such as the quality of the input image, the font type and size, and the language being recognized. The flow of the overall OCR process is shown in Figure 4. We have provided a high-level description of the various techniques and methods involved in the OCR process in Table 4.

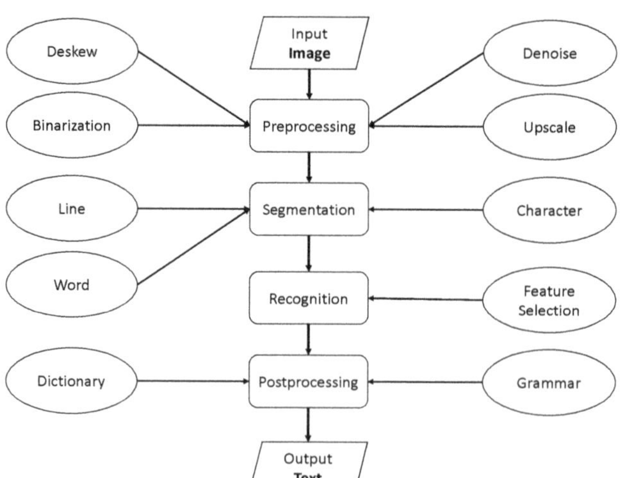

Figure 4. The flow of the OCR process along with OCR phases and methods involved.

Table 4. Comparison of techniques applicable in various OCR methods.

Technique	Method	Brief Description
Preprocessing	Binarization	Transforms the input image into a binary format
	Keystone Correction	Aligns the distortion on the edges of the image
	Skew correction	Corrects the angle of the rotated text
	Denoising	Filters-out the extra-noisy pixels from the image
	Dilation	Restores an eroded image by cleaning it up
	Erosion	Removes object boundaries and unwanted parts in images
	Thinning	Reduces thickness of objects by removing boundary pixels
	Upscaling	Enhances the resolution of the image
Segmentation	Line	Image is divided into lines for line-by-line processing
	Word	Each line is divided into words using spacing methods
	Character	Image of each word is divided into individual characters
Recognition	Template Matching	Matches an input image with predefined characters
	Feature Extraction	Extracts features and classifies image using learning algorithm
	Neural Networks	Uses interconnected neurons to predict text from image
	Deep Learning	Uses neural networks with many layers to learn patterns
	Decision Trees	Builds tree-like structure with decisions and consequences
	SVM	Constructs hyperplane separating image into different classes
	Naive Bayes	Uses Bayes' theorem to classify an input image
	Random Forest	Builds multiple decision trees, combining their outputs
	CNN	Uses deep learning with convolutional layers to classify image
	RNN	Neural network for processing sequences (characters in OCR)
	*k*NN	Classifies image based on *k*-nearest neighbors' majority class
	HT	Detects lines, circles, and edges from image for text extraction
	HOG	Computes image gradients in histograms and extract features
	HMM	Models transition probabilities of text for accurate recognition
	Profile Projection	Extracts character features using projection onto 1D axis
Postprocessing	Spell-check	Error correction, text enhancement, and restoration
	Contextual Analysis	Analyses the surrounding words based on specific context
	Confidence Scoring	Assigns scores to words—higher score means more accurate
	Language Model	Uses large corpus of text to guess best word in context
Evaluation	Character Error Rate	Percentage of characters incorrectly predicted
	Word Error Rate	Percentage of words incorrectly predicted
	Recognition Rate	Percentage of characters/words correctly recognized

3.1. Preprocessing

The formatting issues in images can have a negative impact on the accuracy of OCR models. Examples of these issues include problems related to image orientation or color correction. To improve the accuracy of these models during the training phase, image preprocessing techniques are commonly used. These techniques may involve resizing, grayscale conversion, skew correction, and/or enhancing the resolution of the image.

3.1.1. Binarization and Thinning

Binarization converts a grayscale or color image into a binary image, representing each pixel as either black or white. It is an essential preprocessing step that helps to segment the text from the background and increase the contrast between the characters and the background. The objective of binarization is to transform the input image into a binary format that enhances the visibility of the characters and makes them more easily recognized by the OCR system. Various binarization techniques are used in OCR, including thresholding, adaptive thresholding, and Otsu's method.

A method for preprocessing images of historical documents for OCR and search includes image binarization, skew correction, and line segmentation. The method was tested on a dataset of historical documents. The results show that it improves the accuracy of OCR by reducing errors caused by skew and noise and can be effectively applied to historical documents of various types. The binarization step of the method converts the image into a black-and-white image to make it easier for OCR software to recognize the text [52].

An approach for binarization of non-uniformly illuminated document images to accurately recognize alphanumerical characters is presented in [53]. The proposed method combines local and global thresholding methods, i.e., Sauvola and Otsu methods to achieve robust binarization and improved performance compared to existing binarization methods. In the Sauvola binarization method, the local threshold is calculated using a Sauvola algorithm, which takes into account the local mean and standard deviation of the pixel intensities. In the Otsu binarization method, the global threshold is calculated using an Otsu algorithm, which maximizes the variance between the two classes of pixel intensities.

Thinning, also known as skeletonization, reduces the thickness of the image by deleting the boundary pixels while preserving the shape and structure. The goal of thinning is to obtain the structure of the objects in the image.

Tellache et al. [54] propose and compare different thinning algorithms for improving the performance of OCR for Arabic script. The results show that the Hybrid algorithm performed the best and improved the OCR accuracy by reducing the errors caused by variations in line thickness. The method can be effectively applied to different types of Arabic text, including handwritten and printed text. Results indicate that the Hybrid algorithm improved the OCR accuracy by reducing the errors caused by variations in line thickness. The results also show that the Hybrid algorithm can be effectively applied to different types of Arabic text, including handwritten and printed text.

3.1.2. Denoising

Unwanted changes in the intensity of an image that cover up the underlying image structure are known as noise. Noise can appear in images or documents that contain text in different ways, i.e., scanning documents, compressing files, printing documents, and noise during text recognition in the form of errors. For scanning documents, the noise introduced is in the form of changes to document quality, exposure, lighting, and blurring of the text. Noise can also be introduced during file compression, as it is used to reduce the actual size of the files, so it can add noise in the form of quality loss. Noise can also be introduced while printing the document due to the lack of printing quality of that particular machine or due to variations in ink or toner density or blurring.

Noise and image distortions significantly degrade OCR performance [55]. Noise removal is necessary for every image-processing task, and filters are used to remove unwanted variations in the image while preserving the essential details. Filters are used according to the filter behavior [56]; for example, a Gaussian filter is used to smooth an image by reducing high-frequency noise. A median filter works by replacing pixel values with the median value of the neighbouring pixels to remove impulse noise.

The authors of [57] proposed a deep learning architecture based on a convolutional neural network (CNN) for detecting and recognizing text in distorted document images of different languages. The proposed approach combines two specialized modules for text detection and recognition for automatically learning discriminative features for character recognition; it achieves outstanding performance, surpassing the best competing models by at least 13% for text detection and 7.5% for text recognition. The developed global model demonstrates a high level of robustness and significantly outperforms all other schemes in comprehensive benchmarks.

Denoising is also performed using morphological operations. Morphological operations process images according to their shapes; each pixel corresponds to its neighboring pixels. Salt-and-pepper noise is a common type of noise that appears due to random black-and-white pixels in an image, and morphological operations are used to remove such noise. Erosion and dilation are commonly used morphological operations for denoising [58]. Erosion eliminates the isolated noise pixels, and dilation fills up small, empty holes around the image caused by noise [59]. Opening removes small noisy pixels, whereas closing operations fill empty gaps in the image. The combination of opening and closing is generally used to denoise the image in the preprocessing step, as shown in Figure 5.

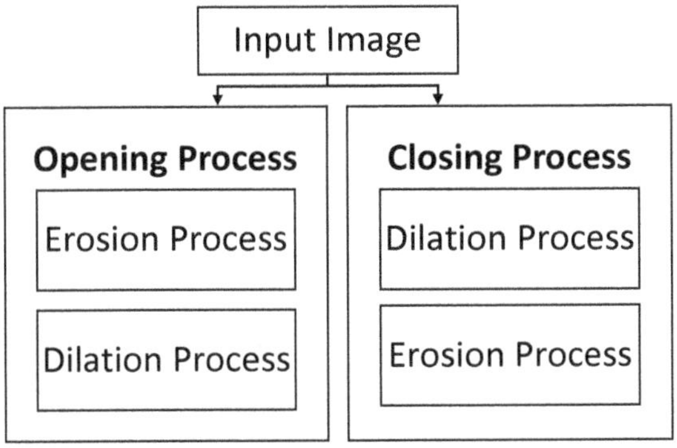

Figure 5. Opening and closing of an image.

3.1.3. Deskewing

Deskewing is the detection of rotated text in an image and computing its angle to correct its rotation [60]. It involves text-block detection, computing the angle of the rotated text, rotating the text, and correcting the image's skewness.

An adaptive deskewing method for document pictures that recognizes the image type and selects an appropriate correction technique based on image type is proposed by [61]. The text direction of the document picture is determined by the method and is used as a parameter to pick a more appropriate projection direction. The research provides many approaches for repairing various sorts of document photographs, as well as a layout-based image categorization system. The results of the experiments suggest that the algorithm is accurate and resilient, although its complexity may restrict its capacity to predict skew over a specific threshold. A voting-based deskewing method is proposed by [62]; it chooses the best deskewing algorithm based on the accuracy of skew correction for large digitization projects.

The Probabilistic Hough Transformation (PHT) method for skew detection and correction in OCR systems for scanned documents is presented in [63]. The method works in two steps: detecting lines of text and clustering them. Factors that affect OCR performance, such as skew, blur, image distortion, and noise, are addressed in the preprocessing phase, with skew being the main focus; an example is shown in Figure 6. The proposed method was tested on different datasets and showed better results than other methods used by researchers. The method calculates skew angles using the following equations:

$$n_{hieght} = (n_{width} * h)/w \quad (1)$$

where n_{height} represents normalized height, n_{width} represents normalized width, and w and h represent width and height, respectively.

$$\begin{aligned} m &= (y_2 - y_1)/(x_2 - x_1) \\ A &= arctan[(y_2 - y_1)/(x_2 - x_1)] \end{aligned} \quad (2)$$

where m represents slope/gradient of a line, and A represents the angle of each line. The proposed skew detection and correction method is used on different datasets and achieves good results compared to other methods used by researchers.

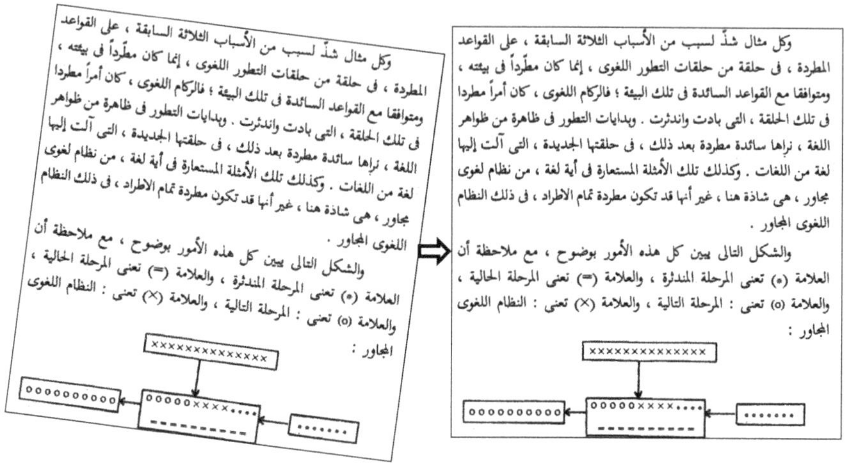

Figure 6. A skewed document (on the left) is deskewed (on the right) to achieve better OCR results.

Hough transformation (HT) is a technique to detect lines in an image and deskew the text. The authors of [7] use the HT method for skew correction. HT creates a parameter space (Hough space) in which each point in the space represents a possible line in the image. By detecting lines in the image, the method can detect the skew of the text by finding the angle at which the lines are inclined. The proposed method can detect skew angles with an accuracy of about 97% and can correct them with an average error of about 0.8°. The method was evaluated through experiments on a dataset of Quran images. The method is robust to noise, which means it can still detect and correct skew even when images have noise present. It can also be applied to images of different quality, showing consistent performance regardless of the image quality. The method was tested on a diverse set of Quran images, including images with different text sizes and levels of quality, and the results were consistent.

3.1.4. Keystone Correction

Keystone correction is used to correct slanted images. OCR algorithms work best when the text in the image is aligned with the x-axis. However, the text may be slanted due to how the original document was scanned or photographed. Keystone correction involves applying mathematical transformations to the image to correct for the slant of the text, which can be done using various algorithms, such as Hough transform or RANSAC. Distortion is aligned at the top/bottom using vertical keystone correction and left/right of the image with horizontal keystone correction. After capturing the image, there will be many technical challenges with the image that make it difficult to read the text. The morphological process and Bézier curve method resolve these challenges [64].

3.1.5. Upscaling

The process of upscaling, also known as Super Resolution, involves enhancing the resolution of the image. Random forests can also be used to upscale images [65]. They use the standard benchmarks for super-resolution and present the training and evaluation accuracy. A deep learning method for upscaling binary document images using super-resolution is the generative adversarial network (SRGAN) [66], which improves readability and OCR performance compared to traditional interpolation methods, as per evaluation metrics. The method involves training a CNN on low-resolution and high-resolution binary document images to generate high-resolution images.

3.2. Segmentation

Segmentation is an important step in OCR involving separation of an image into its constituent parts, such as lines, words, and characters, for recognition. Three types of segmentation techniques are mainly used, i.e., line, word, and character segmentation. Word segmentation, specifically, is challenging in cursive languages such as Arabic due to a lack of clear separation between characters. Traditional segmentation methods in Arabic OCR rely on rules and heuristics based on character features, but deep learning techniques such as convolutional and recurrent neural networks have shown promising results in automatic segmentation. Accurate segmentation is crucial for recognition, and improving Arabic OCR segmentation can have significant implications for document digitization, text-to-speech conversion, and language translation.

The techniques used in [67] include image processing techniques such as thresholding to convert the image into a binary image, morphological operations to perform operations such as erosion and dilation on the binary image, and connected component analysis to identify and label connected regions in the image. The authors use two databases for evaluating the performance of the proposed method. The first database is the publicly available RDI-Arabic dataset, which consists of 1000 images of Arabic text documents. The second database is a new dataset created by the authors and consists of 1000 images of Arabic text documents. The research results show that the proposed image-processing techniques can accurately segment Arabic text documents into text lines, words, and characters with a high degree of accuracy. The researchers use several evaluation metrics, such as F-score, precision, and recall, to evaluate the performance of the proposed method. The results show that the proposed method outperforms traditional methods regarding segmentation accuracy.

Urdu language characters are the super-set of Arabic language characters, and certain challenges are faced when performing segmentation of Urdu-like cursive languages [68]. Arabic is mainly written in Naskh, while Urdu is written in Nastaliq style. There are some challenges in the segmentation of Urdu script, i.e., cursive nature, difficult fonts such as Nastaliq, and letters changing their shape into different forms as required in the word; and cue points are hard to find in the Naskh or Nastaliq style. That is why segmentation is challenging, and character segmentation has been considered difficult in previous research. Researchers mostly used the projection profile method for segmentation, which can perform a vertical projection of the given text. However, in Arabic or Urdu, text writing starts from right to left and top to bottom, so vertical and horizontal projection is required. Analytical approaches are difficult and give the wrong character recognition results, but explicit and implicit recognition systems also give better accuracy. At the same time, holistic approaches are considered best by researchers for better accuracy with the correct recognition. Furthermore, there are better approaches than segmentation-free approaches for large vocabularies. For a reasonable accuracy rate, segmentation should perform well using the approaches that are correct and appropriate for segmentation.

Thorat et al. [69] presented a survey to discuss the methods used by previous researchers. It discussed OCR systems, tools, applications, phases, and methods. There are two types of documents, i.e., unilingual and multilingual documents. Some OCR systems are discussed, i.e., Google Docs OCR, Tesseract, ABBYY FineReader, Transym, and I2OCR, which help to provide services and help to extract text from different types of documents with different languages, whether they are unilingual or multilingual. Multilingual documents contain text from multiple languages, and the techniques used are binarization, layout analysis, page segmentation, preprocessing, feature extraction, classification, and recognition. Some approaches are used for the segmentation-free approach and HMM. There are some applications where OCR systems are used to ease use and increase work productivity in healthcare, education, banking, insurance, automatic exam paper checking, bills and invoices, newspapers, and comics. Some phases are discussed to process the document to get better accuracy, including image acquisition, preprocessing, segmentation, classification and recognition, and postprocessing. Moreover, methods used by previous researchers are matrix matching, fuzzy logic, structural analysis, and neural networks.

3.2.1. Line Segmentation

In line segmentation, the skew-corrected image is divided into lines. Line segmentation is an important step in OCR as it allows the separation of the image into individual lines so that the OCR system can process them one-by-one and improve the recognition of the text in the image. Connected component analysis, project profile, and machine learning-based approaches can be used to perform line segmentation. Once the lines of text have been segmented, the OCR system can process each line individually, which can improve the accuracy of the character recognition process.

A method for line segmentation of printed Arabic text with diacritics using a divide-and-conquer algorithm is presented in [70]. It breaks the image of printed text into smaller blocks, applies image processing techniques to extract the text lines, and then applies a set of heuristic rules to remove false positives and adjust the segments as necessary. It uses image processing techniques such as thresholding, morphological operations, and connected component analysis. The research results show that the proposed method can accurately segment printed Arabic text with diacritics into text lines with a high degree of accuracy. The research uses several evaluation metrics such as F-score, precision, recall, and F-Measure to evaluate the performance of the proposed method.

Brodic et al. [71,72] proposed a basic standardized test framework for evaluating the quality of text line segmentation algorithms in OCR systems for accurate handwritten text recognition. Their proposed framework includes experiments for measuring the accuracy of text line segmentation, skew rate, and reference text line evaluation.

3.2.2. Word segmentation

After line segmentation, each word from the line is segmented by dividing the line of the text into individual words. Several techniques are used for word segmentation, i.e., the Spacing method involves using the spaces between words to segment the text into words. The dictionary-based method uses a dictionary of words to match against the text and segments the text into words based on the matches. Character-based methods use a combination of known character patterns, such as word breaks and punctuation, to segment the text into words. Deep-learning approaches use supervised learning on annotated datasets to learn the complex relationships between adjacent characters in order to infer word boundaries [73,74].

The authors of [75] present word segmentation in Arabic handwritten images using a convolutional recurrent neural network (CRNN) architecture. The authors employ a sliding-window approach for word segmentation, where each window is classified as either a word or non-word using a support vector machine (SVM) classifier. The experimental results show that the proposed CRNN architecture achieves state-of-the-art performance in Arabic handwriting word recognition, with an accuracy of 86.95% on the IFN/ENIT dataset.

Patil et al. [76] propose a semantic segmentation approach for images containing mixed text to segment the image into different regions based on their content. The segmented regions are then processed using different OCR methods that are specifically tailored to the type of text in each region.

3.2.3. Character Segmentation

Character-level segmentation is a technique used to segment an image of a single word into individual letters and characters. It is an optional step depending on the context of the OCR system that is being used. It may be unnecessary if the text has separate letters within a word, as the letters and characters can be segmented in the previous step using a threshold. However, character-level segmentation must be performed if the text has cursive handwriting or a nature where letters are joined.

A method for recognizing and transcribing text from a visual Arabic scripting news ticker from a broadcast stream is presented in [77]. The technique used in this research includes image processing techniques such as thresholding, morphological operations, and connected component analysis to segment the text from the background. It uses machine learning algorithms such as CNNs and long short-term memory (LSTM) to transcribe text. The research results show that the proposed method can accurately recognize and transcribe text from a visual Arabic scripting news ticker from a broadcast stream. The research uses several evaluation metrics, such as character error rate (CER) and word error rate (WER), to evaluate the performance of the proposed method. The results show that the proposed method outperforms traditional methods in recognition accuracy, achieving a lower CER and WER than traditional methods on the dataset used in the research.

Alginahi [78] discusses Arabic character segmentation approaches, including traditional methods such as vertical and horizontal projection, contour tracing and thinning, template matching, neural networks and HMM, holistic approaches and segmentation-free approaches, projection profile, baseline, contour tracing, graph theory, and morphology. The paper also discusses the challenges and limitations of each approach and suggests areas for future research. It also discusses the benefits and problems with character segmentation, especially in Arabic; problems are due to its cursive nature and the different shapes of each character depending on the word's appearance. Problems also occur because of datasets. Therefore, the Arabic Language Technology Center (ALTEC) have provided limited free access to a reliable dataset.

A method for segmenting characters, letters, and digits from Arabic handwritten document images using a hybrid approach is presented in [79]. The method uses image processing techniques, such as thresholding, morphological operations, and connected component analysis, to segment the text from the background and to separate the characters. In addition, machine learning algorithms, including *k*-means clustering and a Random Forest Classifier, are used to classify the segments into individual characters. The research demonstrates that the proposed method can accurately segment characters from Arabic handwritten document images with a high degree of accuracy. The method outperforms traditional methods regarding segmentation accuracy, as evidenced by several evaluation metrics, including F-score, precision, and recall. The proposed method achieves a higher F-score, precision, and recall than traditional methods on the dataset used in the research.

Morphological operators are also used for segmenting Arabic handwritten words [80]. The process involves using morphological operations, such as erosion and dilation, to extract the text from the background and segment the words. The technique used in this research is based on morphological operators, which perform operations such as erosion and dilation on binary images, to extract the text and separate the words. The method also uses image processing techniques, such as thresholding, to convert the image into a binary image and connected component analysis to identify and label connected regions corresponding to words. The research results show that the proposed method can accurately segment Arabic handwritten words with a high degree of accuracy. The research used several evaluation metrics, such as F-score, precision, and recall, to evaluate the performance of the proposed method. The results show that the proposed method outperforms traditional methods in terms of segmentation accuracy.

Some previous works have proposed segmentation-free approaches for Arabic and Urdu OCR, but they do not produce accurate results on clean text. The authors of [18] apply a machine learning model on clean Urdu and Arabic datasets, producing 91% and 86% accuracy on clean UPTI datasets. The authors of [34] review current approaches and challenges unique to Urdu OCR and suggest that future research should focus on developing more-sophisticated algorithms, improving training data, and addressing the unique challenges of the Urdu script. They also propose that integrating Urdu OCR with other technologies, such as machine learning and computer vision, would provide new opportunities for research in the field.

Handwritten digit recognition is discussed in [81]. It has various applications and is used in different fields such as postal mail sorting, bank check numbering, amount processing, and number entries of various forms such as taxes, insurance, and utility bills. There are handwritten images of 10 digits in the dataset from 0 to 9, and this dataset is taken from MNIST, which contains 60,000 and 10,000 images for training and testing purposes, respectively. Then, some phases and approaches are used to process the images to train and test the model, which helps to get better accuracy and gives a maximum recognition rate. Discussed phases and approaches are preprocessing and feature extraction, and then classification is used. In classification, machine learning approaches, i.e., Decision Tree, Support Vector Machine (SVM), and Artificial Neural Network (ANN), are used. The deep learning approach implements a Visual Geometry Group model with 16 layers (VGG16 model); this model is used for deep-learning image-classification problems. By using the proposed techniques, we get better recognition and a high accuracy rate; i.e., in decision tree 86%, SVM 91%, ANN 97%, and CNN 98.84% accuracy is achieved.

3.3. Recognition

Recognition, also called classification in some previous works, is the process of identifying and assigning a specific character or set of characters to a given input image. After preprocessing and segmentation, the OCR system compares the extracted features of each character or group of characters with a set of predefined templates or models. The OCR system creates these templates during a training phase, where a large dataset of sample images is used to teach the system how to recognize each character.

The classification/recognition process can involve several algorithms, including template matching, neural networks, and support vector machines. Template matching involves comparing the features of each character to predefined templates and selecting the template with the closest match to the recognized character. Neural networks and support vector machines use machine learning algorithms to learn and classify patterns in the data and can often achieve higher accuracy than template matching.

A survey of feature extraction and classification techniques is presented in [82]. The techniques include digitization, preprocessing, segmentation, feature extraction, and post-processing. Statistical, structural, template matching, artificial neural network, and kernel classification methods are also discussed.

An efficient feature-descriptor selection for improved Arabic handwritten word recognition is presented in [83]. The approach uses three image features, Histogram Oriented Gradient (HOG), Gabor Filter (GF), and Local Binary Pattern (LBP), for feature extraction, and trains a *k*NN algorithm to build models. The best model achieved an accuracy of 99.88%. A publicly available IFN/ENIT Arabic dataset is also introduced. The researcher use the global approach in the research, which is considered successful and in many cases is used more than the analytical approach.

Various methods and techniques are used in multilingual OCR [84], including preprocessing, binarization, segmentation, feature extraction, and recognition. Segmentation uses three different approaches, i.e., top-down, bottom-up, and hybrid approach, including page, line, and word/character level. The research also explores the challenges to and limitations of current multilingual OCR systems, such as dealing with different scripts and languages and the need for large amounts of annotated training data. The paper highlights the importance of multilingual OCR for applications such as digital libraries, document archiving, and machine translation. Overall, the paper provides a comprehensive overview of the current state of multilingual OCR research and its potential applications. The paper highlights the importance of multilingual OCR for various applications such as digital libraries, document archiving, and machine translation.The authors of [85] discuss the ongoing design of a tool for automatically extracting knowledge and cataloging documents in Arabic, Persian, and Azerbaijani using OCR, text processing, and information-extraction techniques.

A method using a combination of CNN and RNN for recognizing Arabic text in natural scene images is presented in [86]. The method utilizes an attention mechanism to focus on the most relevant parts of the image and improve text recognition accuracy. Two datasets are used in this research for training purposes, i.e., ACTIV and ALIF. ACTIV contains 21520 images of text lines, while ALIF contains 6532 images of text lines; both datasets are extracted from different news channels. Testing is done on the Arabic natural scene text dataset (ANST) the authors created. The proposed method is evaluated on this dataset, which shows that it outperformed the state-of-the-art methods for Arabic text recognition in natural scene images with an accuracy of 92.4%. The model uses CNN to extract features from the image and RNN to process the features and generate the text recognition output. At the same time, the attention mechanism improves the accuracy of recognition (https://tc11.cvc.uab.es/datasets/type/11, accessed on 18 March 2023).

Language detection, document categorization, and region of interest (RoI) identification with KERAS and TensorFlow are used to perform manuscript analysis and recognition in OCR systems [87]. The RoI includes tables, titles, paragraphs, figures, and lists. The deep-learning-trained model uses bounding-box regression to identify the target and serves as a reference point for object detection. The system integrates the fast gradient sign method (FGSM) and uses deep learning to recognize multilingual systems. It also investigates page segmentation methods to enhance accuracy. The system performs well against adversarial attacks on Arabic manuscripts and achieves an accuracy of 99%. It uses Hierarchical Agglomerate Clustering (HAC) to group objects in clusters based on similarity/relation. The research aims to improve preprocessing and identify parameters to enhance the accuracy of page segmentation methods.

A survey of various methods and techniques used for recognizing text in natural images and videos is presented in [88]. The authors discuss various challenges and propose approaches for recognizing text in the wild due to font, color, and background variations. The paper covers traditional methods as well as more-recent methods based on deep learning, such as CNN and RNN. The authors also discuss evaluation metrics and datasets used for text recognition in the wild. Overall, the paper provides an overview of the state-of-the-art in text recognition in the wild and highlights areas that need further research. Additionally, the authors provide a comprehensive review of publicly available resources on their Github repository (https://github.com/HCIILAB/Scene-Text-Recognition, accessed on 22 March 2023).

Bouchakour et al. [89] use the CNN classifier for printed Arabic OCR using a combination of texture, shape, and statistical features. Evaluation of the proposed method achieves an accuracy of 97.23%, which shows the effectiveness of combining image features with a CNN classifier. Similarly, the authors of [90] conduct experiments to analyze the impact of different hyper-parameters and network architectures on the performance of a CNN model for OCR of handwritten digits.

The authors of [91] survey OCR methodologies for Urdu fonts, such as Nastaliq and Naskh, and other similar languages having Urdu-like scripts, such as Arabic, Pashtu, and Sindhi. Moreover, the main focus of this survey is to compare all of the phases involved in OCR, i.e., Image Acquisition, Preprocessing, Segmentation, Feature Extraction, Classification, and Recognition. The Urdu Printed Text Images (UPTI) dataset is divided into training, testing, and validation. It contains about 10,000 text images; so for implementation, multidimensional LSTM-RNN is used, and it achieves an accuracy of 98%. Many stages in the phases make the OCR system better, and it is important to follow all of the stages to make a perfect system, as shown in Figure 7. In the past, researchers used multiple datasets and found a specific output such as character recognition or ligatures recognition and achieved good accuracy.

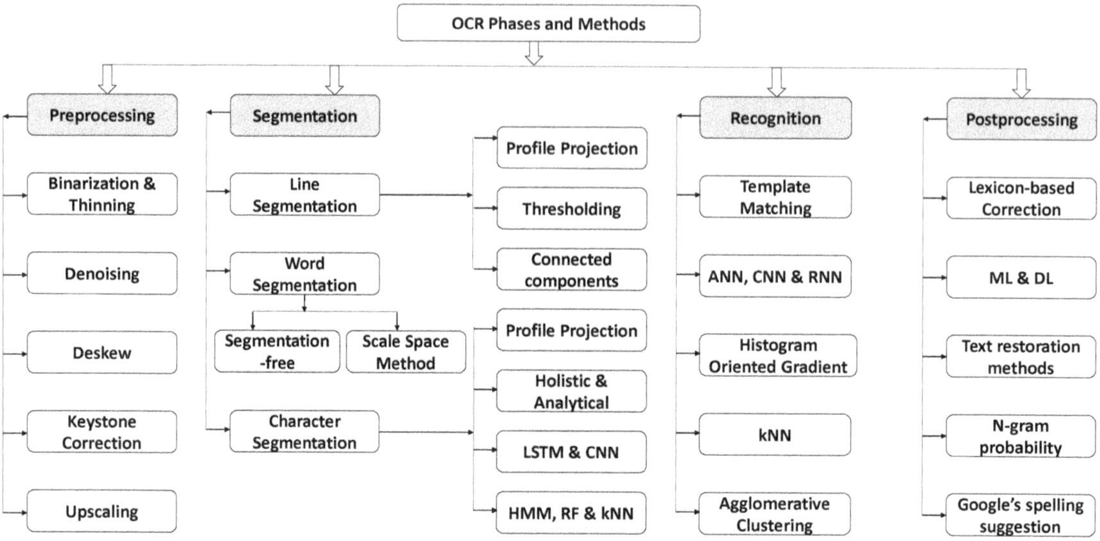

Figure 7. Processes and techniques in each phase of the OCR system.

An artificial neural network (ANN) classifier to identify the characters in a text is presented in [92]. The dataset is generated from different documents with different qualities of the result. A preprocessing step is used to eliminate noise; then, the segmentation phase is done using multiple steps, i.e., line, word, and character segmentation. A feature-extraction step for the character's image is performed to obtain features for all the pixels. The authors trained the ANN classifier on a dataset of printed Arabic text and then evaluated its performance on a separate test set. The results show that the system can accurately recognize the characters in the test set with high recognition rates. A KERAS model is used with a three-layer ANN classifier for character classification. Noise density and multi-spatial resolution matrices are used for evaluation, showing a fast, efficient, high-performance OCR system.

Mittal and Garg [93] review the various techniques used for text extraction from images and documents using OCR; they also discuss the challenges and limitations of text extraction. Reviewed papers are grouped on the basis of the types of OCR techniques that are used. The authors found that the most-used OCR techniques are based on machine learning, specifically deep learning. They also found that the OCR techniques that use multiple-recognition engines and preprocessing techniques perform better than single-recognition-engine-based techniques.

Deep learning models such as CNN, RNN, and attention-based models are discussed in [94]. The paper discusses the performance of models on different Arabic handwritten datasets, and these models show much-improved character recognition rate, word recognition rate, and overall recognition rate. The researchers also discussed challenges dealing with different handwriting styles and sizes.

The authors of [95] present a method for recognizing Arabic handwritten characters using different Holistic techniques, CNN, and deep learning models. At the same time, all of the techniques are well reviewed in this research, i.e., preprocessing, segmentation, feature extraction, recognition, and postprocessing. The authors found that using these models and techniques properly improves the recognition rate of the system by a significant margin.

In [96], automatically extracting and processing text from images is discussed. The challenges that may arise in OCR stages are also explored, as well as the general phases of an OCR system, including preprocessing, segmentation, normalization, feature extraction, classification, and postprocessing. Additionally, the paper highlights OCR's main applications and provides a review of the state-of-the-art at the time of its publication.

The approaches to processing text from documents are reviewed in [97]. In this review, text detection and text transcription are discussed. Text detection is a task whereby a process detects or finds text in a document or image. It is a difficult task but can be detected easily by box bounding and text detection as object detection. Text detection as object detection is challenging but is solved using computer vision tasks such as single-shot multi-box detectors and fast R-CNN models. Meanwhile, in text transcription, the text is extracted in editable form from the document or image of interest. Document layout analysis is the dominant part of selecting the region of interest. Then, the authors identify some datasets used in past research: ICDAR, Total-Text, CYW1500, SynthText, and the updated dataset FUNSD.

3.4. Postprocessing

Postprocessing is the final step in the OCR process; it involves improving the accuracy and quality of the recognized text. Postprocessing techniques can include various methods, such as spell checking, contextual analysis, confidence scoring, and language-model integration. Spell checking involves comparing the recognized text against a dictionary of words to identify and correct spelling errors. Contextual analysis analyses the recognized text within the context of the surrounding text to identify and correct errors that may be caused by confusion with other words. Confidence scoring assigns a confidence score to each recognized character based on the certainty of the OCR system's recognition, and characters with lower confidence scores are flagged for review or correction. The language model is also used to analyze the recognized text in the language context. Postprocessing significantly improves the accuracy and quality of the recognized text.

An overview of the different postprocessing techniques developed and applied to OCR output, including methods for correction of errors, text enhancement, and text restoration, is discussed in [98]. Various methods are used in postprocessing, including spell checking, grammar checking, lexicon-based correction, machine learning, and deep-learning-based approaches. The paper also discusses using different types of features, such as character-level, word-level, and document-level features, as well as preprocessing techniques, such as segmentation and normalization.

Several approaches have been used in the postprocessing of OCR output to improve the accuracy and completeness of the recognized text:

- Spell checking: checks the spelling of the recognized text and corrects any errors by comparing it to a dictionary [99].
- Grammar checking: checks the grammar of the recognized text and corrects any errors by comparing it to a set of grammar rules.
- Lexicon-based correction: uses a lexicon (a list of words and their possible variations) to correct errors in the recognized text by comparing it to the lexicon and suggesting alternative words where there are errors.
- Machine-learning-based approaches: uses machine learning algorithms, such as decision trees, random forests, and support vector machines, to correct errors in the recognized text.
- Deep-learning-based approaches: uses deep learning algorithms, such as CNNs and RNNs, to correct errors in the recognized text.
- Text enhancement: includes techniques to improve the recognized text's visibility, legibility, and readability, such as binarization, deskewing and smoothing of text.
- Text restoration: includes techniques to recover missing or degraded text, such as text in-painting, completion, and restoration.

The authors of [98] also present the results of various experiments and evaluations conducted to assess the performance of postprocessing techniques. These include comparisons of different methods and systems, evaluations of the effects of different types of features and preprocessing techniques, and evaluations of the performance of systems on different types of OCR output and languages.

Doush et al. [100] present a word-context and rule-based technique for OCR postprocessing. OCR is used to obtain text from scanned documents, and the output text is not always 100% accurate. Thus, after obtaining the text, postprocessing step is required to recognize and minimize the errors. The presented research is about printed documents, it lies in an offline OCR system. Cursive nature also causes problems in this step because characters in Arabic are connected, and there are other additional things such as diacritics and different shapes of each character in different words. These things bring more complications to this step. A very small amount of work has been done on the Arabic postprocessing technique because it shows a very high character and word error rate after recognition. Therefore, it is a less attractive side for researchers. In the proposed research, an Arabic text database is prepared, available in three formats, i.e., HTML, PDF, and scanned-document images. The database has 4581 files, and there are about 8994 scanned images. Thus, from the database, 1000 images are used for training by the rule-based method, which reduces the word error rate.

Bassil and Alwani propose an algorithm using Google's online spelling suggestions [101], which helps to improve the accuracy and suggest what should come after each character to make a meaningful word according to the sentence. All of the suggestions given by Google's spelling suggestion algorithm are based on N-gram probability. The authors hybridize this method to make the proposed postprocessing technique. Therefore, the proposed hybrid postprocessing system starts with the generated OCR file (token). The language model checks each token, and if the language model does not find this token, then the error model takes this token and suggests the correct word. This token/word again goes into the language model in an attempt to find matches for the token. If it matches, then the model moves to the next word; otherwise, the error model again suggests the new token provided by Google's spelling suggestion algorithm, as shown in Figure 8.

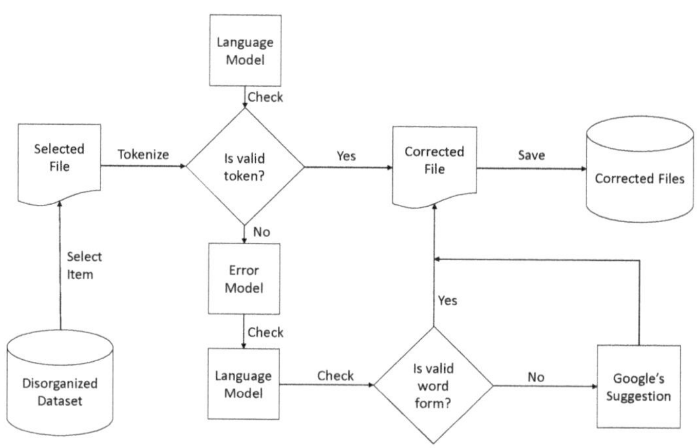

Figure 8. Hybrid postprocessing technique based on Google's spelling suggestion algorithm.

The authors of [102] present a corpus-based technique for improving the performance of an Arabic OCR system. The method involves using a large corpus of texts in Arabic to train the OCR system and improve its recognition accuracy. The corpus of texts is preprocessed to ensure that it is suitable for OCR training, and then it is used to train the OCR system. The performance of the OCR system is then evaluated using a set of test images, and the recognition accuracy is compared to that of traditional methods. The study also shows that using a larger corpus of texts leads to better performance of the OCR system. This phase helps to provide better recognition or reduce the word error rate and character error rate. The collection of a large corpus of texts in Arabic is used to train the OCR system and then preprocess the corpus to ensure that it is suitable for OCR training. The OCR system is trained using the preprocessed corpus, and the OCR system's performance is evaluated using a set of test images. The research shows that the corpus-based technique improves the recognition accuracy of the OCR system by a significant margin compared to traditional methods. It also shows that using a larger corpus of texts leads to better performance of the OCR system.

3.5. Evaluation

Evaluation measures the accuracy and quality of the recognized text output. OCR evaluation typically involves comparing the recognized text to the original input image or document and calculating various performance metrics to assess the quality of the OCR output. Some common metrics used in OCR evaluation include character error rate (CER), word error rate (WER), recall, precision, and F1 score. OCR evaluation can be performed using various methods depending on the goal of the application, such as manual inspection, crowdsourcing, or automated evaluation software. Evaluation is an important step in OCR development and deployment, as it allows developers and users to assess the accuracy and quality of the OCR output and make improvements or adjustments to the OCR system as needed.

An Arabic OCR evaluation tool is discussed in [103]. The Arabic language is difficult to recognize due to its characters'/alphabets' behavior in different words. Arabic OCR's accuracy could be higher because of improper evaluation of performance metrics. Different tools and software have been introduced to help find the performance and accuracy of the applied algorithms. The introduced software is built specially to check Arabic OCR; it checks the performance based on objectives, i.e., accuracy and evaluation metrics. These tools briefly describe the text in characters with or without dots, baseline, and diacritics and the class in which the text lies. These tools include Tesseract, easy OCR, Paddle-Paddle OCR, and PyMuPdf. Recognition rate (RR), character error rate (CER), and word error rate (WER) are the evaluation measures used by these programs to evaluate OCR output.

Kiessling et al. [104] discuss various open-source tools for Arabic OCR systems containing Tesseract (https://github.com/tesseract-ocr/tesseract, accessed on 24 February 2023), OCRad (https://www.gnu.org/software/ocrad/, accessed on 24 February 2023), and GOCR (https://jocr.sourceforge.net/, accessed on 24 February 2023). These tools are evaluated using an Arabic dataset, and Tesseract gives better recognition results. However, it is slower than other evaluated tools. These tools give better accuracy for high-resolution documents, and the accuracy gets worse as the quality of documents decreases.

The authors of [105] provide an overview of existing tools and metrics used to evaluate the OCR system in previous research. Their paper covers traditional evaluation techniques and discusses their pros and cons. It also discusses evaluation metrics such as character error rate, word error rate, and recognition rate. The detailed review also discusses the many challenges involved in evaluating the performance of the OCR system.

The performances of generative and discriminative recognition models for offline Arabic handwritten recognition are compared using generatively trained hidden Markov modeling (HMM), discriminatively trained conditional random fields (CRF), and discriminatively trained hidden-state CRF (HCRF) in [106]. The study presents recognition outcomes for words and letters and assesses the efficiency of all three strategies using the Arabic IFN/ENIT dataset.

Singh et al. [107] discussed offline handwritten word recognition in Devanagari. A holistic-based approach is used in this approach, wherein a word is considered a single entity, and the approach processes it further for extraction and recognition. A class of 50 words recognizes every word based on a feature vector set, uniform zoning, diagonal, centroid, and feature-based. The proposed system uses gradient-boosted algorithms to enhance the performance; furthermore, some other classifiers are used for this purpose, i.e., *k*NN and Random Forest Classifier. Thus, the overall achieved accuracy is 94.53%. The authors generate the dataset during the implementation, which is available on request. The paper also provided some information on previous research in this area, where researchers used Hidden Markov Model, Support Vector Machine, and Multi-Layer perceptron classifiers. The authors also highlighted the importance of the availability of quality datasets for the improved performance of OCR techniques.

The impact of OCR quality on the accuracy of short text classification tasks is presented in [108]. A multi-class classification of short text is introduced. For this, the authors propose a dataset of beauty product images that contains 27,500 entries of labeled brand data and generate results based on targeting specific brands. The authors also show that preprocessing techniques such as text normalization and noise reduction can improve the performance of the classification model on low-quality OCR text.

In addition to papers improving specific processes involved in OCR, some previous papers present a combination of various OCR techniques for overall improved process accuracy. OCR4all [109] is an open-source OCR software that combines state-of-the-art OCR components and continuous model training into a comprehensive workflow for processing historical printings and provides a user-friendly GUI and extensive configuration capabilities. The software outperforms commercial tools on moderate layouts. It achieves excellent character error rates (CER) on very complex early printed books, making it a valuable tool for non-technical users and providing an architecture allowing easy integration of newly developed tools.

3.6. *Summary of Presented Techniques*

This survey aims to provide a comprehensive literature review of the various techniques involved in Arabic OCR and to provide valuable insights into the current state-of-the-art in Arabic OCR. To achieve this goal, we analyzed a range of research papers and articles that focused on different Arabic OCR techniques and methods. Our findings are summarized in Table 5, which presents a concise overview of the methods employed in the reviewed papers. The table outlines the different OCR techniques, including preprocessing, segmentation, recognition, and postprocessing, along with their performance evaluations in terms of accuracy using various types of printed, scanned, and handwritten datasets. This survey is useful for researchers and practitioners who are interested in Arabic OCR systems. By comparing and contrasting the different techniques in the complete pipeline of the Arabic OCR, they can choose the most appropriate one for their specific task. To further aid researchers and practitioners, Table 6 presents various methods commonly employed in OCR systems and their respective advantages and disadvantages.

Table 5. A brief tabular outline of the described papers with the proposed OCR techniques and their performance evaluations.

OCR Techniques	OCR Tasks					Accuracy
	Preprocessing	Segmentation	Recognition	Postprocessing	Evaluation	
Ahmad et al. [63]	✓	✓	✓			99.3% (Scanned)
Bafjaish et al. [7]	✓	✓	✓			90% (Scanned)
Karthick et al. [59]	✓	✓	✓	✓		87.4% (Handwritten), 90% (Scanned)
Abdo et al. [67]	✓	✓	✓		✓	94.1% (Printed)
Qaroush et al. [70]	✓	✓			✓	11% (Segmentation)
Tayyab et al. [77]	✓	✓	✓		✓	98.36% (Scanned)
Alginahi [78]		✓	✓			93.65% (Handwritten), 86.14% (Scanned)
Verma and Ali [82]	✓	✓				No Recognition
Hamida et al. [83]	✓		✓		✓	99.88% (Handwritten)
Butt et al. [86]	✓		✓		✓	87% (Scanned)
Nguyen et al. [98]				✓	✓	No Recognition
Doush et al. [100]				✓	✓	No Recognition
Neudecker et al. [105]					✓	No Recognition
Vitman et al. [108]	✓		✓		✓	83.5% (High-quality), 58.4%(Low-quality)

Table 6. Pros and cons of various recognition methods for Arabic OCR.

Method	Pros	Cons
Template matching	Simple and easy to implement	Limited accuracy, sensitive to noise and variations in text
Deep learning	High accuracy, can handle variations in text	Requires large amounts of training data, computationally expensive
*k*NN	For small datasets, takes less training time and make predictions quickly	Sensitive to noisy or irrelevant features
RNN	For processing large sequential data and can learn term dependencies	Computationally expensive and sensitive to overfitting
Hough Transformation	Robust to noise and can detect lines and circles at any orientation	Computationally expensive when dealing with large images
Histogram Oriented Gradient	Extracts features such as edge orientation and texture, and is computed quickly	Ineffective at detecting finer details and is sensitive to variations in lighting and contrast
Hidden Markov Model	Models complex patterns and can be trained on large/sequential datasets	Computationally expensive to train and sensitive to model parameters
Profile Projection	Extracts features from images, such as character width and spacing	Sensitive to variations in lighting and contrast.
Random Forest	Relatively easy to train and can handle noisy or missing data	Does not perform well on highly imbalanced or sparse datasets
SVM	Used for classification tasks and can handle high-dimensional data	Computationally expensive non-linear kernels require hyperparameter tuning
Hybrid approaches	Combines the strengths of multiple methods	More complex and difficult to implement

4. Discussion and Conclusions

We surveyed that researchers use multiple approaches and datasets to get better recognition rates. The literature review suggests that a proper process needs to be followed, which includes preprocessing, segmentation (text-area detection and line, word, and character segmentation), recognition, and postprocessing. We discussed the pros and cons of each technique discussed in this survey. For example, segmentation-based approaches give better results than segmentation-free approaches, and vertical/horizontal projection produces good results for word and character segmentation. However, OCR results depend upon a good dataset as well. Some datasets are available for the Arabic OCR, but only a few are publicly available. Postprocessing and dataset availability require more attention from researchers. For postprocessing, if Google spelling checker-like algorithms are implemented and improved, then this stage can perform very well, enhancing the overall result of the OCR system. We need a publicly available dataset with an extensive vocabulary of printed and handwritten text (characters and words) for the dataset.

In conclusion, we presented a survey of the state-of-the-art Arabic OCR, which has come a long way in recent years, with several approaches and techniques developed to improve its accuracy and performance. However, many challenges still need to be addressed, including dealing with the variability and complexity of the Arabic script and the large number of dialects and variations in the language. Despite these challenges, the potential benefits of Arabic OCR are clear, and researchers and developers are working hard to continue to improve and refine the technology. We will likely see even more accurate and reliable Arabic OCR systems with continued research and development.

Author Contributions: Conceptualization, S.F., M.S.A. and S.H.; methodology, S.H., M.S.A. and M.A.K.; investigation, M.S.A. and S.H.; resources, M.S.A. and S.H.; writing—original draft preparation, M.S.A. and S.H.; writing—review and editing, S.F., M.S.A., S.H. and M.A.K.; visualization, M.S.A., S.H. and M.A.K.; supervision, S.F. and M.A.K.; project administration, S.F. and M.A.K.; funding acquisition, S.F. All authors have read and agreed to the published version of the manuscript.

Funding: This work is funded by the Deputyship of Research & Innovation, Ministry of Education, Saudi Arabia, through project number 964. In addition, the authors would like to express their appreciation for the support provided by the Islamic University of Madinah.

Institutional Review Board Statement: Not applicable.

Informed Consent Statement: Not applicable.

Data Availability Statement: Publicly available data sources were used for this work.

Conflicts of Interest: The authors declare no conflict of interest.

References

1. Alhomed, L.S.; Jambi, K.M. A survey on the existing arabic optical character recognition and future trends. *Int. J. Adv. Res. Comput. Commun. Eng. (IJARCCE)* **2018**, *7*, 78–88.
2. Beg, A.; Ahmed, F.; Campbell, P. Hybrid OCR techniques for cursive script languages-a review and applications. In Proceedings of the International Conference on Computational Intelligence, Communication Systems and Networks, Liverpool, UK, 28–30 July 2010; pp. 101–105.
3. Djaghbellou, S.; Bouziane, A.; Attia, A.; Akhtar, Z. A Survey on Arabic Handwritten Script Recognition Systems. *Int. J. Artif. Intell. Mach. Learn. (IJAIML)* **2021**, *11*, 1–17. [CrossRef]
4. Islam, N.; Islam, Z.; Noor, N. A survey on optical character recognition system. *arXiv* **2017**, arXiv:1710.05703.
5. Rashid, D.; Kumar Gondhi, N. Scrutinization of Urdu Handwritten Text Recognition with Machine Learning Approach. In Proceedings of the International Conference on Emerging Technologies in Computer Engineering, Xiamen, China, 21–23 October 2022; Springer: Berlin/Heidelberg, Germany, 2022; pp. 383–394.
6. Idrees, S.; Hassani, H. Exploiting Script Similarities to Compensate for the Large Amount of Data in Training Tesseract LSTM: Towards Kurdish OCR. *Appl. Sci.* **2021**, *11*, 9752. [CrossRef]
7. Bafjaish, S.S.; Azmi, M.S.; Al-Mhiqani, M.N.; Radzid, A.R.; Mahdin, H. Skew detection and correction of Mushaf Al-Quran script using hough transform. *Int. J. Adv. Comput. Sci. Appl.* **2018**, *9*. [CrossRef]
8. Singh, A.; Bacchuwar, K.; Bhasin, A. A survey of OCR applications. *Int. J. Mach. Learn. Comput.* **2012**, *2*, 314. [CrossRef]
9. Antonio, J.; Putra, A.R.; Abdurrohman, H.; Tsalasa, M.S. A Survey on Scanned Receipts OCR and Information Extraction. In Proceedings of the International Conference on Document Analysis and Recognit, Jerusalem, Israel, 29–30 November 2022.
10. Al-Sheikh, I.S.; Mohd, M.; Warlina, L. A review of arabic text recognition dataset. *Asia-Pac. J. Inf. Technol. Multimed. (APJITM)* **2020**, *9*, 69–81. [CrossRef]
11. Ahmed, S.B.; Naz, S.; Swati, S.; Razzak, M.I. Handwritten Urdu character recognition using one-dimensional BLSTM classifier. *Neural Comput. Appl.* **2019**, *31*, 1143–1151. [CrossRef]
12. Zayene, O.; Masmoudi Touj, S.; Hennebert, J.; Ingold, R.; Essoukri Ben Amara, N. Open datasets and tools for arabic text detection and recognition in news video frames. *J. Imaging* **2018**, *4*, 32. [CrossRef]
13. Badry, M.; Hassan, H.; Bayomi, H.; Oakasha, H. QTID: Quran Text Image Dataset. *Int. J. Adv. Comput. Sci. Appl.* **2018**, *9*, 385–391. [CrossRef]
14. Pechwitz, M.; Maddouri, S.S.; Märgner, V.; Ellouze, N.; Amiri, H. *IFN/ENIT-Database of Handwritten Arabic Words;* CIFED: Hammamet, Tunis, 2002; Volume 2, pp. 127–136.
15. Al-Ma'adeed, S.; Elliman, D.; Higgins, C.A. A data base for Arabic handwritten text recognition research. In Proceedings of the International workshop on frontiers in handwriting recognition, Niagara-on-the-Lake, ON, Canada, 6–8 August 2002; pp. 485–489.
16. Slimane, F.; Ingold, R.; Kanoun, S.; Alimi, A.M.; Hennebert, J. *Database and Evaluation Protocols for Arabic Printed Text Recognition;* DIUF-University of Fribourg: Fribourg, Switzerland, 2009; p. 1.
17. Lawgali, A.; Angelova, M.; Bouridane, A. HACDB: Handwritten Arabic characters database for automatic character recognition. In Proceedings of the European Workshop on Visual Information Processing (EUVIP), Paris, France, 10–12 June 2013; pp. 255–259.
18. Sabbour, N.; Shafait, F. A segmentation-free approach to Arabic and Urdu OCR. In Proceedings of the Document Recognition and Retrieval, San Jose, CA, USA, 16–20 January 2005; Volume 8658, pp. 215–226.
19. Saddami, K.; Munadi, K.; Arnia, F. A database of printed Jawi character image. In Proceedings of the International Conference on Image Information Processing (ICIIP), Waknaghat, India, 21–24 December 2015; pp. 56–59.
20. Mahmoud, S.A.; Ahmad, I.; Al-Khatib, W.G.; Alshayeb, M.; Parvez, M.T.; Märgner, V.; Fink, G.A. KHATT: An open Arabic offline handwritten text database. *Pattern Recognit.* **2014**, *47*, 1096–1112. [CrossRef]
21. Yousfi, S.; Berrani, S.A.; Garcia, C. ALIF: A dataset for Arabic embedded text recognition in TV broadcast. In Proceedings of the International Conference on Document Analysis and Recognition (ICDAR), Tunis, Tunisia, 23–26 August 2015; pp. 1221–1225.
22. Zayene, O.; Hennebert, J.; Touj, S.M.; Ingold, R.; Amara, N.E.B. A dataset for Arabic text detection, tracking and recognition in news videos-AcTiV. In Proceedings of the International Conference on Document Analysis and Recognition (ICDAR), Tunis, Tunisia, 23–26 August 2015; pp. 996–1000.
23. Chabchoub, F.; Kessentini, Y.; Kanoun, S.; Eglin, V.; Lebourgeois, F. SmartATID: A mobile captured Arabic Text Images Dataset for multi-purpose recognition tasks. In Proceedings of the International Conference on Frontiers in Handwriting Recognition (ICFHR), Hyderabad, India, 4–7 December 2016; pp. 120–125.
24. Sulaiman, A.; Omar, K.; Nasrudin, M.F. A database for degraded Arabic historical manuscripts. In Proceedings of the International Conference on Electrical Engineering and Informatics (ICEEI), Langkawi, Malaysia, 25–27 November 2017; pp. 1–6.
25. Bataineh, B. A Printed PAW Image Database of Arabic Language for Document Analysis and Recognition. *J. ICT Res. Appl.* **2017**, *11*, 200–212. [CrossRef]

26. Al-Ohali, Y.; Cheriet, M.; Suen, C. Databases for recognition of handwritten Arabic cheques. *Pattern Recognit.* **2003**, *36*, 111–121. [CrossRef]
27. Awaidah, S.M.; Mahmoud, S.A. A multiple feature/resolution scheme to Arabic (Indian) numerals recognition using hidden Markov models. *Signal Process.* **2009**, *89*, 1176–1184. [CrossRef]
28. Asiri, A.M.; Khorsheed, M.S. Automatic Processing of Handwritten Arabic Forms using Neural Networks. In Proceedings of the IEC (Prague), Prague, Czech Republic, 26–28 August 2005; pp. 313–317.
29. Luqman, H.; Mahmoud, S.A.; Awaida, S. KAFD Arabic font database. *Pattern Recognit.* **2014**, *47*, 2231–2240. [CrossRef]
30. Ramdan, J.; Omar, K.; Faidzul, M.; Mady, A. Arabic handwriting data base for text recognition. *Procedia Technol.* **2013**, *11*, 580–584. [CrossRef]
31. Amara, N.E.B.; Mazhoud, O.; Bouzrara, N.; Ellouze, N. ARABASE: A Relational Database for Arabic OCR Systems. *Int. Arab J. Inf. Technol.* **2005**, *2*, 259–266.
32. Srihari, S.; Srinivasan, H.; Babu, P.; Bhole, C. Handwritten arabic word spotting using the cedarabic document analysis system. In Proceedings of the Symposium on Document Image Understanding Technology (SDIUT-05), College Park, MD, USA, 2–4 November 2005; pp. 123–132.
33. Shafi, M.; Zia, K. Urdu character recognition: A systematic literature review. *Int. J. Appl. Pattern Recognit.* **2021**, *6*, 283–307. [CrossRef]
34. Khan, N.H.; Adnan, A. Urdu optical character recognition systems: Present contributions and future directions. *IEEE Access* **2018**, *6*, 46019–46046. [CrossRef]
35. Bhatti, A.; Arif, A.; Khalid, W.; Khan, B.; Ali, A.; Khalid, S.; Rehman, A.u. Recognition and Classification of Handwritten Urdu Numerals Using Deep Learning Techniques. *Appl. Sci.* **2023**, *13*, 1624. [CrossRef]
36. Khosrobeigi, Z.; Veisi, H.; Hoseinzade, E.; Shabanian, H. Persian Optical Character Recognition Using Deep Bidirectional Long Short-Term Memory. *Appl. Sci.* **2022**, *12*, 11760. [CrossRef]
37. Husnain, M.; Saad Missen, M.M.; Mumtaz, S.; Coustaty, M.; Luqman, M.; Ogier, J.M. Urdu handwritten text recognition: A survey. *IET Image Process.* **2020**, *14*, 2291–2300. [CrossRef]
38. Naz, S.; Hayat, K.; Razzak, M.I.; Anwar, M.W.; Madani, S.A.; Khan, S.U. The optical character recognition of Urdu-like cursive scripts. *Pattern Recognit.* **2014**, *47*, 1229–1248. [CrossRef]
39. Alghamdi, M.; Teahan, W. Printed Arabic script recognition: A survey. *Int. J. Adv. Comput. Sci. Appl.* **2018**, *9*, 415–427. [CrossRef]
40. Osman, H.; Zaghw, K.; Hazem, M.; Elsehely, S. An Efficient Language-Independent Multi-Font OCR for Arabic Script. *arXiv* **2020**, arXiv:2009.09115.
41. Muhammad, M.; ElGhazaly, T. Handling OCR-degraded arabic text: A comprehensive survey. In Proceedings of the ISSR Conference, Turku, Finland, 27–30 June 2013.
42. Dinges, L.; Al-Hamadi, A.; Elzobi, M.; El-Etriby, S. Synthesis of common Arabic handwritings to aid optical character recognition research. *Sensors* **2016**, *16*, 346. [CrossRef]
43. Bouressace, H. A Review of Arabic Document Analysis Methods. In Proceedings of the International Conference on Pattern Analysis and Intelligent Systems (PAIS), Oum El Bouaghi, Algeria, 12–13 October 2022; pp. 1–7.
44. Qaroush, A.; Jaber, B.; Mohammad, K.; Washaha, M.; Maali, E.; Nayef, N. An efficient, font independent word and character segmentation algorithm for printed Arabic text. *J. King Saud-Univ.-Comput. Inf. Sci.* **2022**, *34*, 1330–1344. [CrossRef]
45. Al Ghamdi, M.A. A Novel Approach to Printed Arabic Optical Character Recognition. *Arab. J. Sci. Eng.* **2022**, *47*, 2219–2237. [CrossRef]
46. Majumdar, S.; Brick, A. Recognizing Handwriting Styles in a Historical Scanned Document Using Scikit-Fuzzy c-means Clustering. *arXiv* **2022**, arXiv:2210.16780.
47. Mostafa, A.; Mohamed, O.; Ashraf, A.; Elbehery, A.; Jamal, S.; Khoriba, G.; Ghoneim, A.S. OCFormer: A Transformer-Based Model For Arabic Handwritten Text Recognition. In Proceedings of the International Mobile, Intelligent, and Ubiquitous Computing Conference (MIUCC), Cairo, Egypt, 26–27 May 2021; pp. 182–186.
48. Badry, M.; Hassanin, M.; Chandio, A.; Moustafa, N. Quranic script optical text recognition using deep learning in IoT systems. *CMC-Comput. Mater. Contin.* **2021**, *68*, 1847–1858. [CrossRef]
49. Moudgil, A.; Singh, S.; Gautam, V. An Overview of Recent Trends in OCR Systems for Manuscripts. In *Cyber Intelligence and Information Retrieval*; Springer: Berlin, Germany, 2022; pp. 525–533.
50. Huang, Z.; Chen, K.; He, J.; Bai, X.; Karatzas, D.; Lu, S.; Jawahar, C. Icdar2019 competition on scanned receipt ocr and information extraction. In Proceedings of the International Conference on Document Analysis and Recognition (ICDAR), Sydney, Australia, 20–25 September 2019; pp. 1516–1520.
51. Bashir, M.H.; Azmi, A.M.; Nawaz, H.; Zaghouani, W.; Diab, M.; Al-Fuqaha, A.; Qadir, J. Arabic natural language processing for Qur'anic research: A systematic review. *Artif. Intell. Rev.* **2022**. [CrossRef]
52. Gupta, M.R.; Jacobson, N.P.; Garcia, E.K. OCR binarization and image pre-processing for searching historical documents. *Pattern Recognit.* **2007**, *40*, 389–397. [CrossRef]
53. Michalak, H.; Okarma, K. Robust combined binarization method of non-uniformly illuminated document images for alphanumerical character recognition. *Sensors* **2020**, *20*, 2914. [CrossRef] [PubMed]
54. Tellache, M.; Sid-Ahmed, M.; Abaza, B. Thinning algorithms for Arabic OCR. In Proceedings of the Pacific Rim Conference on Communications Computers and Signal Processing, Victoria, BC, Canada, 19–21 May 1993; Volume 1, pp. 248–251.

55. Mohsenzadegan, K.; Tavakkoli, V.; Kyamakya, K. Deep Neural Network Concept for a Blind Enhancement of Document-Images in the Presence of Multiple Distortions. *Appl. Sci.* **2022**, *12*, 9601. [CrossRef]
56. Mahmud, J.U.; Raihan, M.F.; Rahman, C.M. A complete OCR system for continuous Bengali characters. In Proceedings of the Conference on Convergent Technologies for Asia-Pacific Region (TENCON), Bangalore, India, 15–17 October 2003; Volume 4, pp. 1372–1376.
57. Mohsenzadegan, K.; Tavakkoli, V.; Kyamakya, K. A Smart Visual Sensing Concept Involving Deep Learning for a Robust Optical Character Recognition under Hard Real-World Conditions. *Sensors* **2022**, *22*, 6025. [CrossRef]
58. Nashwan, F.M.; Rashwan, M.A.; Al-Barhamtoshy, H.M.; Abdou, S.M.; Moussa, A.M. A holistic technique for an Arabic OCR system. *J. Imaging* **2017**, *4*, 6. [CrossRef]
59. Karthick, K.; Ravindrakumar, K.; Francis, R.; Ilankannan, S. Steps involved in text recognition and recent research in OCR; a study. *Int. J. Recent Technol. Eng.* **2019**, *8*, 2277–3878.
60. Cao, Y.; Wang, S.; Li, H. Skew detection and correction in document images based on straight-line fitting. *Pattern Recognit. Lett.* **2003**, *24*, 1871–1879. [CrossRef]
61. Bao, W.; Yang, C.; Wen, S.; Zeng, M.; Guo, J.; Zhong, J.; Xu, X. A Novel Adaptive Deskewing Algorithm for Document Images. *Sensors* **2022**, *22*, 7944. [CrossRef]
62. Boiangiu, C.A.; Dinu, O.A.; Popescu, C.; Constantin, N.; Petrescu, C. Voting-based document image skew detection. *Appl. Sci.* **2020**, *10*, 2236. [CrossRef]
63. Ahmad, R.; Naz, S.; Razzak, I. Efficient skew detection and correction in scanned document images through clustering of probabilistic hough transforms. *Pattern Recognit. Lett.* **2021**, *152*, 93–99. [CrossRef]
64. Li, Y.; Zou, F.; Yang, S.; Liu, H.; Ding, Y.; Zhu, K. Research on Improving OCR Recognition Based on Bending Correction. In Proceedings of the International Information Technology and Artificial Intelligence Conference (ITAIC), Chongqing, China, 11–13 December 2020; Volume 9, pp. 833–837.
65. Schulter, S.; Leistner, C.; Bischof, H. Fast and accurate image upscaling with super-resolution forests. In Proceedings of the Conference on Computer Vision and Pattern Recognition, Boston, MA, USA, 7–12 June 2015; pp. 3791–3799.
66. Pandey, R.K.; Vignesh, K.; Ramakrishnan, A. Binary document image super resolution for improved readability and OCR performance. *arXiv* **2018**, arXiv:1812.02475.
67. Abdo, H.A.; Abdu, A.; Manza, R.R.; Bawiskar, S. An approach to analysis of Arabic text documents into text lines, words, and characters. *Indones. J. Electr. Eng. Comput. Sci.* **2022**, *26*, 754–763. [CrossRef]
68. Naz, S.; Umar, A.I.; Shirazi, S.H.; Ahmed, S.B.; Razzak, M.I.; Siddiqi, I. Segmentation techniques for recognition of Arabic-like scripts: A comprehensive survey. *Educ. Inf. Technol.* **2016**, *21*, 1225–1241. [CrossRef]
69. Thorat, C.; Bhat, A.; Sawant, P.; Bartakke, I.; Shirsath, S. A Detailed Review on Text Extraction Using Optical Character Recognition. *ICT Anal. Appl.* **2022**, 719–728.. [CrossRef]
70. Qaroush, A.; Awad, A.; Hanani, A.; Mohammad, K.; Jaber, B.; Hasheesh, A. Learning-free, divide and conquer text-line extraction algorithm for printed Arabic text with diacritics. *J. King Saud-Univ.-Comput. Inf. Sci.* **2022**, *34*, 7699–7709. [CrossRef]
71. Brodic, D.; Milivojevic, D.R.; Milivojevic, Z.N. An approach to a comprehensive test framework for analysis and evaluation of text line segmentation algorithms. *Sensors* **2011**, *11*, 8782–8812. [CrossRef]
72. Brodić, D.; Milivojević, D.R.; Milivojević, Z. Basic test framework for the evaluation of text line segmentation and text parameter extraction. *Sensors* **2010**, *10*, 5263–5279. [CrossRef]
73. Reisswig, C.; Katti, A.R.; Spinaci, M.; Höhne, J. Chargrid-OCR: End-to-end trainable optical character recognition through semantic segmentation and object detection. In Proceedings of the Workshop on Document Intelligence at NeurIPS 2019, Vancouver, BC, Canada, 14 December 2019.
74. Agarwal, M.; Hassan, F.; Pandey, G.; Ghosh, S. Handwriting recognition using deep learning. In *Emerging Trends in Data Driven Computing and Communications: Proceedings of DDCIoT 2021*; Springer: Berlin/Heidelberg, Germany, 2021; pp. 67–81.
75. Boualam, M.; Elfakir, Y.; Khaissidi, G.; Mrabti, M. Arabic handwriting word recognition based on convolutional recurrent neural network. In Proceedings of the 6th International Conference on Wireless Technologies, Embedded, and Intelligent Systems (WITS 2020), Fez, Morocco, 14–16 October 2020; Springer: Berlin/Heidelberg, Germany, 2022; pp. 877–885.
76. Patil, S.; Varadarajan, V.; Mahadevkar, S.; Athawade, R.; Maheshwari, L.; Kumbhare, S.; Garg, Y.; Dharrao, D.; Kamat, P.; Kotecha, K. Enhancing Optical Character Recognition on Images with Mixed Text Using Semantic Segmentation. *J. Sens. Actuator Netw.* **2022**, *11*, 63. [CrossRef]
77. Tayyab, M.; Hussain, A.; Alshara, M.A.; Khan, S.; Alotaibi, R.M.; Baig, A.R. Recognition of Visual Arabic Scripting News Ticker from Broadcast Stream. *IEEE Access* **2022**, *10*, 59189–59204. [CrossRef]
78. Alginahi, Y.M. A survey on Arabic character segmentation. *Int. J. Doc. Anal. Recognit. (IJDAR)* **2013**, *16*, 105–126. [CrossRef]
79. Boraik, O.A.; Ravikumar, M.; Saif, M.A.N. Characters Segmentation from Arabic Handwritten Document Images: Hybrid Approach. *Int. J. Adv. Comput. Sci. Appl.* **2022**, *13*, 395–403. [CrossRef]
80. AbdAllah, N.; Viriri, S. Off-Line Arabic Handwritten Words Segmentation using Morphological Operators. *arXiv* **2021**, arXiv:2101.02797.
81. Jabde, M.; Patil, C.; Mali, S.; Vibhute, A. Comparative Study of Machine Learning and Deep Learning Classifiers on Handwritten Numeral Recognition. In Proceedings of the International Symposium on Intelligent Informatics, Trivandrum, India, 31 August–2 September 2022.

82. Verma, R.; Ali, J. A-survey of feature extraction and classification techniques in OCR systems. *Int. J. Comput. Appl. Inf. Technol.* **2012**, *1*, 1–3.
83. Hamida, S.; El Gannour, O.; Cherradi, B.; Ouajji, H.; Raihani, A. Efficient feature descriptor selection for improved Arabic handwritten words recognition. *Int. J. Electr. Comput. Eng.* **2022**, *12*. [CrossRef]
84. Peng, X.; Cao, H.; Setlur, S.; Govindaraju, V.; Natarajan, P. Multilingual OCR research and applications: An overview. In Proceedings of the International Workshop on Multilingual OCR, Washington, DC, USA, 24 August 2013; pp. 1–8.
85. Bergamaschi, S.; De Nardis, S.; Martoglia, R.; Ruozzi, F.; Sala, L.; Vanzini, M.; Vigliermo, R.A. Novel perspectives for the management of multilingual and multialphabetic heritages through automatic knowledge extraction: The digitalmaktaba approach. *Sensors* **2022**, *22*, 3995. [CrossRef]
86. Butt, H.; Raza, M.R.; Ramzan, M.J.; Ali, M.J.; Haris, M. Attention-based CNN-RNN Arabic text recognition from natural scene images. *Forecasting* **2021**, *3*, 520–540. [CrossRef]
87. Al-Barhamtoshy, H.M.; Jambi, K.M.; Rashwan, M.A.; Abdou, S.M. An Arabic Manuscript Regions Detection, Recognition and Its Applications for OCRing. *Trans. Asian-Low-Resour. Lang. Inf. Process.* **2023**, *22*, 1–28. [CrossRef]
88. Chen, X.; Jin, L.; Zhu, Y.; Luo, C.; Wang, T. Text recognition in the wild: A survey. *ACM Comput. Surv. (CSUR)* **2021**, *54*, 1–35. [CrossRef]
89. Bouchakour, L.; Meziani, F.; Latrache, H.; Ghribi, K.; Yahiaoui, M. Printed Arabic Characters Recognition Using Combined Features and CNN classifier. In Proceedings of the International Conference on Recent Advances in Mathematics and Informatics (ICRAMI), Tebessa, Algeria, 21–22 September 2021; pp. 1–5.
90. Ahlawat, S.; Choudhary, A.; Nayyar, A.; Singh, S.; Yoon, B. Improved handwritten digit recognition using convolutional neural networks (CNN). *Sensors* **2020**, *20*, 3344. [CrossRef]
91. Ashraf, N.; Arafat, S.Y.; Iqbal, M.J. An Analysis of Optical Character Recognition (OCR) Methods. *Int. J. Comput. Linguist. Res.* **2019**, *10*, 81. [CrossRef]
92. Al-Sadawi, B.; Hussain, A.; Ali, N.S. High-Performance Printed Arabic Optical Character Recognition System Using ANN Classifier. In Proceedings of the Palestinian International Conference on Information and Communication Technology, Gaza, Palestine, 28–29 September 2021; IEEE Computer Society: Colombia, DC, USA, 2021; pp. 1–6.
93. Mittal, R.; Garg, A. Text extraction using OCR: A systematic review. In Proceedings of the International Conference on Inventive Research in Computing Applications, Coimbatore, India, 15–17 July 2020; pp. 357–362.
94. Alrobah, N.; Albahli, S. Arabic handwritten recognition using deep learning: A survey. *Arab. J. Sci. Eng.* **2022**, *47*, 9943–9963. [CrossRef]
95. Alwaqfi, Y.M.; Mohamad, M.; Al-Taani, A.T. Generative Adversarial Network for an Improved Arabic Handwritten Characters Recognition. *Int. J. Adv. Soft Comput. Its Appl.* **2022**, *14*, 176–195. [CrossRef]
96. Hamad, K.; Mehmet, K. A detailed analysis of optical character recognition technology. *Int. J. Appl. Math. Electron. Comput.* **2016**, *1*, 244–249. [CrossRef]
97. Subramani, N.; Matton, A.; Greaves, M.; Lam, A. A survey of deep learning approaches for ocr and document understanding. *arXiv* **2020**, arXiv:2011.13534.
98. Nguyen, T.T.H.; Jatowt, A.; Coustaty, M.; Doucet, A. Survey of post-ocr processing approaches. *ACM Comput. Surv. (CSUR)* **2021**, *54*, 1–37. [CrossRef]
99. Neto, A.F.d.S.; Bezerra, B.L.D.; Toselli, A.H. Towards the natural language processing as spelling correction for offline handwritten text recognition systems. *Appl. Sci.* **2020**, *10*, 7711. [CrossRef]
100. Doush, I.A.; Alkhateeb, F.; Gharaibeh, A.H. A novel Arabic OCR post-processing using rule-based and word context techniques. *Int. J. Doc. Anal. Recognit. (IJDAR)* **2018**, *21*, 77–89. [CrossRef]
101. Bassil, Y.; Alwani, M. Ocr post-processing error correction algorithm using google online spelling suggestion. *arXiv* **2012**, arXiv:1204.0191.
102. Aliwy, A.H.; Al-Sadawi, B. Corpus-based technique for improving Arabic OCR system. *Indones. J. Electr. Eng. Comput. Sci.* **2021**, *21*, 233–241. [CrossRef]
103. Alghamdi, M.A.; Alkhazi, I.S.; Teahan, W.J. Arabic OCR evaluation tool. In Proceedings of the International conference on computer science and information technology (CSIT), Amman, Jordan, 13–14 July 2016; pp. 1–6.
104. Kiessling, B.; Kurin, G.; Miller, M.T.; Smail, K.; Miller, M. Advances and Limitations in Open Source Arabic-Script OCR: A Case Study. *Digit. Stud. Champ NumÉRique* **2021**, *11*. [CrossRef]
105. Neudecker, C.; Baierer, K.; Gerber, M.; Clausner, C.; Antonacopoulos, A.; Pletschacher, S. A survey of OCR evaluation tools and metrics. In Proceedings of the International Workshop on Historical Document Imaging and Processing, Lausanne, Switzerland, 5–10 September 2021; pp. 13–18.
106. Elzobi, M.; Al-Hamadi, A. Generative vs. Discriminative Recognition Models for Off-Line Arabic Handwriting. *Sensors* **2018**, *18*, 2786. [CrossRef]
107. Singh, S.; Garg, N.K.; Kumar, M. On the performance analysis of various features and classifiers for handwritten devanagari word recognition. *Neural Comput. Appl.* **2023**, *35*, 7509–7527. [CrossRef]
108. Vitman, O.; Kostiuk, Y.; Plachinda, P.; Zhila, A.; Sidorov, G.; Gelbukh, A. Evaluating the Impact of OCR Quality on Short Texts Classification Task. In Proceedings of the Mexican International Conference on Artificial Intelligence, Monterrey, Mexico, 24–29 October 2022; Springer: Berlin/Heidelberg, Germany, 2022; pp. 163–177.

109. Reul, C.; Christ, D.; Hartelt, A.; Balbach, N.; Wehner, M.; Springmann, U.; Wick, C.; Grundig, C.; Büttner, A.; Puppe, F. OCR4all—An open-source tool providing a (semi-) automatic OCR workflow for historical printings. *Appl. Sci.* **2019**, *9*, 4853. [CrossRef]

applied sciences

MDPI

Review

On the Use of Deep Learning for Video Classification

Atiq ur Rehman [1,2,*], Samir Brahim Belhaouari [3], Md Alamgir Kabir [1] and Adnan Khan [3]

1 Artificial Intelligence and Intelligent Systems Research Group, School of Innovation, Design and Technology, Mälardalen University, Högskoleplan 1, 722 20 Västerås, Sweden
2 Department of Electrical and Computer Engineering, Pak Austria Fachhochschule, Institute of Applied Sciences and Technology, Haripur 22621, Pakistan
3 Division of Information and Computing Technology, College of Science and Engineering, Hamad Bin Khalifa University, Doha 34110, Qatar
* Correspondence: atiq.ur.rehman@mdu.se or atiqjadoon@gmail.com

Abstract: The video classification task has gained significant success in the recent years. Specifically, the topic has gained more attention after the emergence of deep learning models as a successful tool for automatically classifying videos. In recognition of the importance of the video classification task and to summarize the success of deep learning models for this task, this paper presents a very comprehensive and concise review on the topic. There are several existing reviews and survey papers related to video classification in the scientific literature. However, the existing review papers do not include the recent state-of-art works, and they also have some limitations. To provide an updated and concise review, this paper highlights the key findings based on the existing deep learning models. The key findings are also discussed in a way to provide future research directions. This review mainly focuses on the type of network architecture used, the evaluation criteria to measure the success, and the datasets used. To make the review self-contained, the emergence of deep learning methods towards automatic video classification and the state-of-art deep learning methods are well explained and summarized. Moreover, a clear insight of the newly developed deep learning architectures and the traditional approaches is provided. The critical challenges based on the benchmarks are highlighted for evaluating the technical progress of these methods. The paper also summarizes the benchmark datasets and the performance evaluation matrices for video classification. Based on the compact, complete, and concise review, the paper proposes new research directions to solve the challenging video classification problem.

Keywords: automatic video classification; deep learning; handcrafted features; video processing

Citation: ur Rehman, A.; Belhaouari, S.B.; Kabir, M.A.; Khan, A. On the Use of Deep Learning for Video Classification. *Appl. Sci.* **2023**, *13*, 2007. https://doi.org/10.3390/app13032007

Academic Editors: Zahid Mehmood Jehangiri, Mohsin Shahzad, Uzair Khan and Yu-Dong Zhang

Received: 2 December 2022
Revised: 21 January 2023
Accepted: 30 January 2023
Published: 3 February 2023

1. Introduction

The task of automatically classifying videos has become very successful recently. Particularly, the subject has drawn increased interest since deep learning models became an effective method for automatically classifying videos. The importance of the accurate video classification task can be realized by the large amount of video data available online. People around the world generate and consume a huge amount of video content. Currently, on YouTube only, over 1 billion hours of video are being watched by different people every single day. In recognition to the importance of the video classification task, a combined effort is being made by researchers for proposing an accurate video classification framework. Companies such as Google AI are investing in different competitions to solve the challenging problem under constrained conditions. To further advance the progress of the automatic video classification task, Google AI has released a public dataset called YouTube-8M with millions of video features and more than 3700 labels. All these efforts being made demonstrate the need for a powerful video classification model.

An artificial neural network (ANN) is an algorithm based on interconnected nodes to recognize the relationships in a set of data. Algorithms based on ANNs have shown a great success in modeling both the lineßar and the non-linear relationships in the underlying data. Due to the huge success rate of these algorithms, they are extensively being used for different real-time applications [1–4]. Moreover, with an increase in the availability of huge datasets, the deep learning models have specifically shown a significant improvement in the classification of videos. This paper reviews studies based on deep learning approaches for video classification.

Contribution

There are several existing reviews and survey papers related to video classification in the scientific literature. Some of the recent works are summarized here in Table 1. However, these review papers do not include the recent state-of-art works, and they have some limitations. In the following text, the limitations and highlights of these works are discussed.

Table 1. Summary of recent related works.

Reference	Year	Coverage	Highlights	Drawbacks
A. Anusya [5]	2020	2014–2019	Video classification, tagging, and clustering.	Not comprehensive and lacks concise information.
Rani et al. [6]	2020	2001–2016	Text, audio, and visual modalities for video classification.	Missing analysis of recent state-of-art approaches.
Y. Li et al. [7]	2020	2012–2019	Live sport video classification.	More specific to live sport video classification.
Md Islam et al. [8]	2021	2004–2020	Machine learning approaches for video classification.	Focus of review is not on deep learning approaches.
Ullah. H. et al. [9]	2021	2015–2020	Human activity recognition using deep learning.	Focus only on the human activity recognition.
This study	2022	2000–2022	Comprehensive deep learning review for video classification.	-

1. A more recent review was done by A. Anusya [5]; this review covers very few methods for video classification, clustering, and tagging. However, the review provided is not comprehensive and lacks concise information, coverage of topic, datasets, analysis of state-of-art approaches, and research limitations;
2. Rani et al. [6] also conducted a recent review on video classification methods, and their review covered some recent video classification approaches and summary-based description of some recent works. This review also had some limitations including the missing analysis of recent state-of-art approaches and a very limited description of topics covered;
3. Y. Li et al. [7] recently conducted a systematic and good review on live sport video classification. This review covers most of the recent works in live sport video classification, including the tools, video interaction features, and feature extraction methods. This is a comprehensive review, but the findings are not summarized in tables for research gaps and advantages and disadvantages of existing methods for a quick review. Moreover, this review is more specific to live sport video classification;
4. A recent review was also done by Md Islam et al. [8]; in this review, they included all the methods for video classification, including deep learning. However, as the focus of review is not on deep learning approaches, these methods are therefore not completely covered in this review;
5. Ullah. H. et al. [9] also conducted a recent systematic review; however, the focus of their review remained on human activity recognition;
6. Z. Wu. [10] presented a concise review on video classification specific to deep learning methods. This review provides a good description on deep learning models, feature extraction tools, benchmark dataset, and comparison of existing methods for video classification. However, this review was conducted in the year 2016, and it does not cover the recent state-of-art deep learning methods;

7. Q. Ren [11] conducted a simple review on video classification methods; however, the techniques covered in this review are not well described, and the review also lacks in the description of research gaps, benchmark datasets, limitations of existing methods, and performance metrics.

In contrast to the existing reviews on classification of videos, this paper provides a more comprehensive, concise, and up-to-date review of deep learning approaches for video classification. In this current review, most of the recent state-of-art contributions related to the topic are analyzed and critically summarized. Deep learning is an emerging and vibrant field for the analysis of videos; therefore, we hope this review will help in stimulating future research along the line. The following are the key contributions to this review paper:

1. A summary of state-of-art, CNN-based deep learning models for image analysis;
2. An in-depth review of deep learning approaches for video classification highlighting the notable findings;
3. A summary of breakthroughs in the automatic video classification task;
4. Analysis of research trends from past towards future;
5. Description of benchmark datasets, evaluations metrics, and comparison of recent state-of-art deep learning approaches in terms of performance.

The rest of the paper is organized as follows: Section 2 reviews some existing CNNs for images; Section 3 provides an in-depth review on deep learning models for video classification; Section 4 provides a summary for benchmark datasets, evaluation metrics, and comparison of existing state-of-art methods for the video classification task; and Section 5 provides conclusion and future research directions.

2. Convolutional Neural Networks (CNN) for Image Analysis

Deep learning models, specifically convolutional neural networks (CNNs), are well known for understanding images. A number of CNN architectures are proposed and developed in the scientific literature for image analysis. Among these, the most popular architectures are LeNet-5 [12], AlexNet [13], VGGNet [14], GoogleNet [15], ResNet [16], and DenseNet [17]. The trend that follows from the formerly proposed architectures towards the recently proposed architectures is to deepen the network. A summary of these popular CNN architectures along with trend of deepening the network is shown in Figure 1, where the depth of network increases from left-most (LeNet-5) to right-most (DenseNet). Deep networks are believed to better approximate the target function and to generate better feature representation with more powerful discriminatory powers [18]. Although deeper networks are better in terms of having more discriminatory powers, the deeper networks require more data for training and more parameters to tune [19]. Finding a professionally labeled, huge dataset is still a big challenge faced by the research community, and therefore, it limits the development of deeper neural networks.

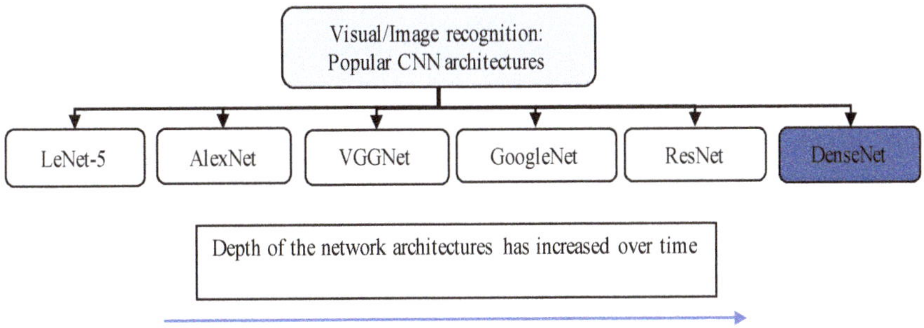

Figure 1. State-of-art image recognition CNN networks. The trend is that the depth and discriminatory powers of network architectures increases from formerly proposed architectures towards the recently proposed architectures.

3. Video Classification

In this section, a very comprehensive and concise review for deep learning models employed in the video classification task is provided. This section covers a description on video data modalities, traditional handcrafted approaches, breakthroughs in video classification, and recent state-of-art deep learning models for video classification.

3.1. Video Data Modalities

As compared to images, videos are more challenging to understand and classify due to the complex nature of the temporal content. However, three different modalities, i.e., visual information, audio information, and text information, might be available to classify videos in contrast to image classification, where only a single visual modality can be utilized. Based on the availability of different modalities in videos, the task of classification can be categorized as a uni-modal video classification or a multi-modal video classification, as summarized in Figure 2. The existing literature has utilized both of these models for the video classification task, and it is generally believed that models utilizing multi-modal data perform better than the models based on uni-modal data [20,21]. Moreover, the visual description [22] of a video works better than the text [23] and the audio [24,25] description for the classification purpose of a video.

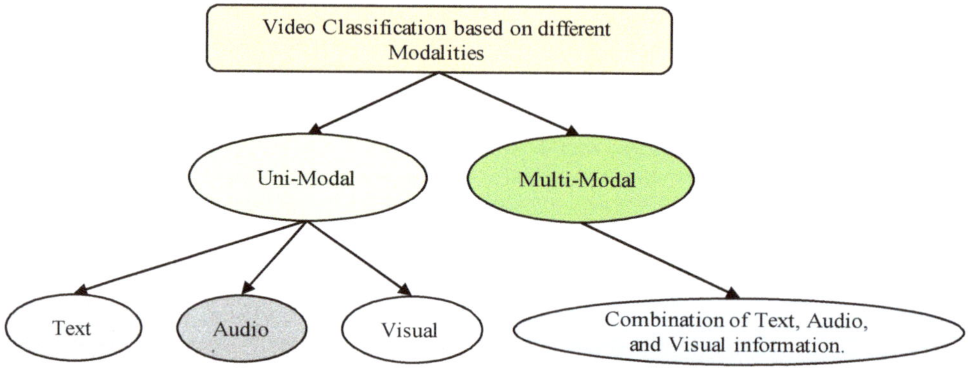

Figure 2. Different modalities used for classification of videos.

3.2. Traditional Handcrafted Features

During the earlier developments of the video classification task, the traditional handcrafted features were combined with state-of-art machine learning algorithms to classify the videos. Some of the most popular handcrafted feature representation techniques used in the literature are spatiotemporal interest points (STIPs) [26], improved dense trajectories (iDT) [27], SIFT-3D [28], HOG3D [29], motion boundary histogram [30], action-bank [31], cuboids [32], 3D SURF [33], and dynamic-poselets [34]. These hand-designed representations use different feature encoding schemes such as the ones based on pyramids and histograms. iDT is one of these handcrafted representations that is widely considered the state-of-the-art. Many recent competitive studies demonstrated that handcrafted features [35–38] and high-level [39,40] and mid-level [41,42] video representations have contributed towards the task of video classification with deep neural networks.

3.3. Deep Learning Frameworks

Along with the development of more powerful deep learning architectures in the recent years, the trend for the video classification task has followed a shift from traditional handcrafted approaches to the fully automated deep learning approaches. Among the very common deep learning architectures used for video classification is a 3D-CNN model. An example of 3D-CNN architecture used for video classification is given in Figure 3 [43]. In this architecture, 3D blocks are utilized to capture the video information necessary to classify the video content. One more very common architecture is a multi-stream architecture, where the spatial and temporal information is separately processed, and the features extracted from different streams are then fused to make a decision. To process the temporal information, different methods are used, and the two most common methods are based on (i) RNN (mainly LSTM) and (ii) optical flow. An example of a multi-stream network model [44], where the temporal stream is processed using optical flow, is shown in Figure 4. A high-level overview of the video classification process is shown in Figure 5, where the stages of feature extraction and prediction are shown with the most common type of strategies used in the literature. In the upcoming sections, the breakthroughs in video classification and studies related to classification of videos, specifically using deep learning frameworks, are summarized, describing the success rate of utilizing deep learning architectures and the associated limitations.

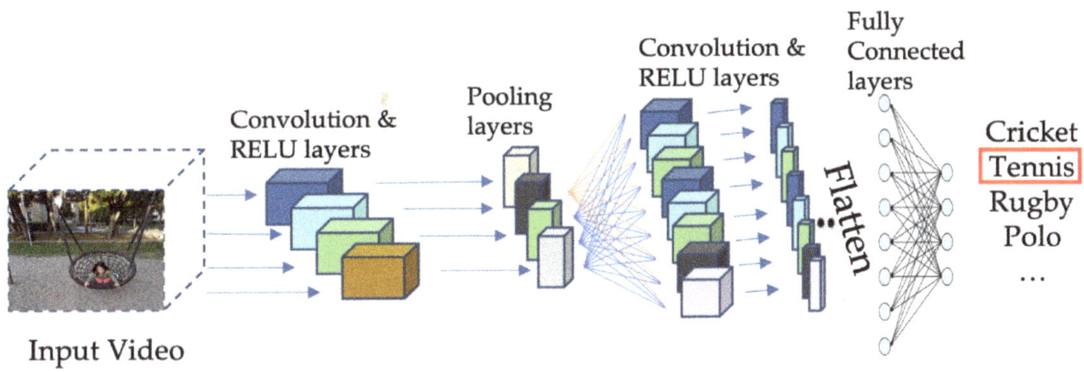

Figure 3. An example of 3D-CNN architecture to classify videos.

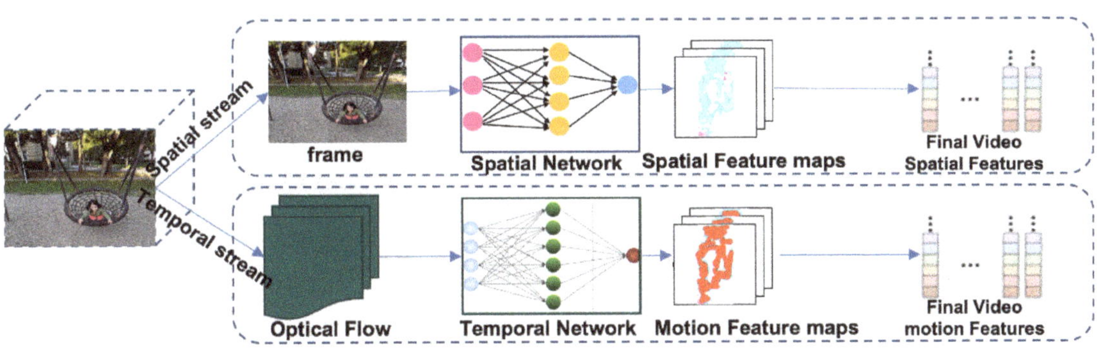

Figure 4. An example of two-stream architecture with optical flow.

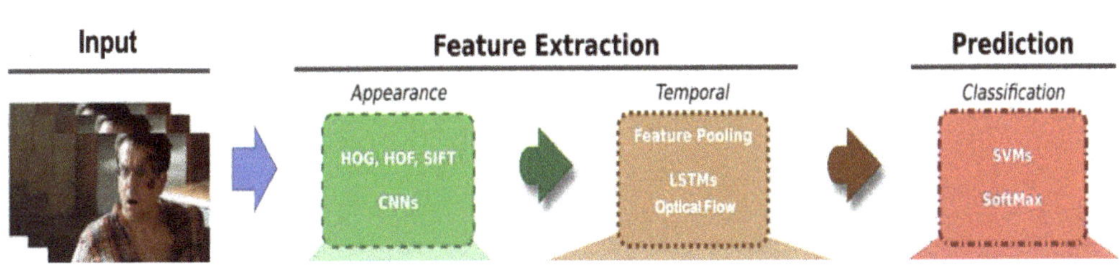

Figure 5. An overview of video classification process.

3.4. Breakthroughs

The breakthroughs in recognition of still-images originated with the introduction of a deep learning model called AlexNet [13]. The same concept of still-image recognition using deep learning is also extended for videos, where individual video frames are collectively processed as images by a deep learning model to predict the contents of a video. The features from individual video frames are extracted, and then, temporal integration of such features into a fixed-size descriptor using pooling is performed. The task is either done using high-dimensional feature encoding [45,46] or through the RNN architectures [47–50]. For un-supervised spatiotemporal feature learning in 3D convolutions, restricted Boltzmann machines [51] and stacked ISA [52] are also studied in parallel. The 3D-CNNs using temporal convolutions to extract temporal features automatically were first proposed by Baccouche et al. [53] and by Ji et al. [54].

3.5. Basic Deep Learning Architectures for Video Classification

The two most widely used deep learning architectures for video classification are convolutional neural network (CNN) and recurrent neural network (RNN). CNNs are mostly used to learn the spatial information from videos, whereas RNNs are used to learn the temporal information from videos, as the main difference between these two architectures is the ability to process temporal information or data that come in sequences. Therefore, both these network architectures are used for completely different purposes in general. However, the nature of video data with the presence of both the spatial and the temporal information demands the use of both these network architectures to accurately process the two-stream information. The architecture of a CNN applies different filters in the convolutional layers to transform the data. RNNs, on the other hand, reuse the activation functions to generate the next output in a series from the other data points in the sequence. However, the use of only 2D-CNNs alone limits the understanding of video to only spatial domain. RNNs, on the other hand, can understand the temporal content of a sequence. Both these basic architectures and their enhanced versions are applied in several studies for the task of video classification.

3.6. Developments in Video Classification over Time

The existing approaches for video classification are categorized based on their working principle in Table 2. The trend observed for the classification of videos from the existing literature is that the recently developed state-of-art deep learning models are outperforming the earlier handcrafted classical approaches. This is mainly due to the availability of large-scale video data for learning deep architectures of neural networks. Besides an improvement in classification performance the recently developed models are mostly self-learned and does not require any manual feature engineering. This added advantage makes them more feasible for use in real applications. However, the better performing recently developed architectures are deeper as compared to the previously developed architectures which brings a compromise on the computational complexity of the deep architectures.

Table 2. Different categories of approaches of video classification.

Categories	Working Principle	References
Hand-crafted approaches	These representations are handcrafted and employ various feature encoding techniques, such as histograms and pyramids.	Spatiotemporal Interest Points (STIPs) [26], iDT [27], SIFT-3D [28], HOG3D [29], Motion Boundary Histogram [30], Cuboids [32], Action-Bank [31], 3D SURF [33], Dynamic-Poselets [34].

Table 2. *Cont.*

Categories	Working Principle	References
2D- CNNs	These are image based models where frame level feature extraction is performed using CNN architecture and classification is performed using state-of-art classification models, for example SVM.	[55]
3D-CNNs	2D image classification extension to 3D for video (For example the Inception 3D (I3D) architecture).	[56]
Spatiotemporal Convolutional Networks	To aggregate the temporal and the spatial information, these methods primarily depend on convolution and pooling.	[54,57,58]
Recurrent Spatial Networks	To represent temporal information in videos, recurrent neural networks such as LSTM or GRU are used.	[47,53,59,60]
Two/multi Stream Networks	In addition to the context frame visuals, these methods use layered optical flow to identify movements.	[50,61–63]
Mixed convolutional models	Models built with the ResNet architecture in mind. They are particularly interested in models that utilize 3D convolution in the bottom or top layers but 2D in the remainder; these are referred to as "mixed convolutional" models. Or the methods based on mixed temporal convolution with different kernel sizes.	[64,65]
Hybrid Approaches	These are models based on integration of CNN and RNN architectures.	[66–68]

Among the initially developed hand-crafted representations, improved Dense Trajectories (iDT) [27] is widely considered the state-of-the-art. Whereas, many recent competitive studies demonstrated that hand-crafted features [35–38], high-level [39,40], and mid-level [41,42] video representations have contributed towards the task of video classification with deep neural networks. The hand-crafted models were among the very early developments of video classification problem. Later, 2D-CNNs were proposed for video classification, where image-based CNN models are used to extract frame level features and based on the frame level CNN features, some state-of-art classification models (for example SVM) are learned to classify videos. These 2D-CNN models do not require any manual feature extraction and these models performed better than the competing hand-crafted approaches. After successful development of 2D-CNN models where features are extracted from frame level, the same concept was extended to propose 3D-CNNs to extract features from videos. The proposed 3D-CNNs are computationally more expensive as compared to the 2D-CNN models. However, these models consider the time variations in feature extraction therefore these 3D-CNN models are believed to perform better as compared to 2D-CNN models for video classification [54,58,69].

The development of 3D-CNN models paved the way for fully automatic video classification models using different deep learning architectures. Among the developments using deep learning architectures, spatiotemporal convolutional networks are approaches based on integration of temporal and spatial information using convolutional networks to perform video classification. To collect temporal and spatial information, these methods primarily rely on convolution and pooling layers. Stack optical flow is used in two/multi-stream networks methods to identify movements in addition to context frame visuals. Recurrent spatial networks use recurrent neural networks (RNN) to model temporal information in videos, such as LSTM or GRU. The ResNet architecture is used to build mixed convolutional models. They are particularly interested in models that utilize 3D convolution in the bottom or top layers but 2D in the remainder; these are referred to as "mixed convolutional" models. These also include methods based on mixed temporal convolution with different kernel sizes. Advanced architectures based on DenseNet have also shown promising results for the video classification task. Some of these notable architectures based on DenseNet include region-based CNN (R-CNN) [70,71], faster R-CNN [72,73], and YOLO [74]. Besides these architectures, there are also hybrid approaches based on the integration of CNN and RNN architectures. A summary of these architectures is provided in Figure 6.

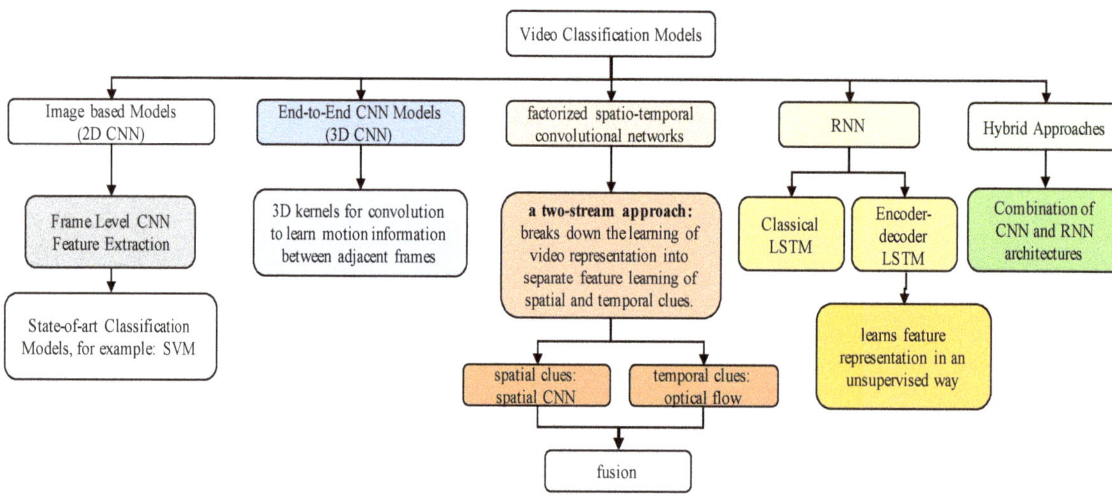

Figure 6. Summary of video classification approaches.

The different deep learning architectures described above employ different fusion strategies. These fusion strategies are either for the fusion of different features extracted from the video or for the fusion of different models used in the architecture. The fusion strategies mainly used for the extracted features are (i) concatenation, (ii) product, (iii) summation, (iv) maximum, and (v) weighted, where the concatenation approach simply combines all the features together, and all the features are used for classification. The product/summation approach performs the product/summation between the features extracted using different strategies and uses the result of product/summation to perform classification. The maximum approach takes the maximum value of the features extracted using different strategies and uses that for classification. The weighted approach gives different weights to different features and performs the classification using the weighted features. Different fusion methods are summarized in Figure 7.

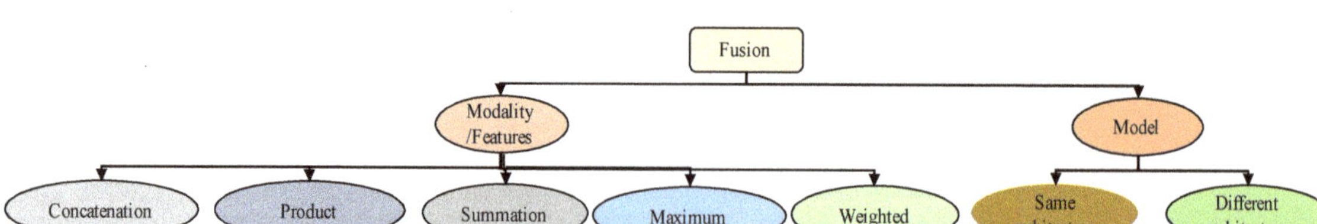

Figure 7. Different Fusion Types.

3.7. Summary of Some Notable Deep Learning Frameworks Developments

A summary of some deep learnings architectures for video classification is provided in Table 3. These studies are summarized based on the architecture, the datasets, the evaluation metrics, the fusion strategy, and the notable findings. The most common architectures for video classification are fundamentally based on the RNN and CNN architectures; classification accuracy is one of the most common evaluation metrics; UCF-101 and Sports-1M datasets are the choice for validation in most cases, multi-class classification problem is considered in almost all cases, SMART blocks outperform 3D convolutions in terms of spatiotemporal feature learning, and average fusion, kernel average fusion, weighted fusion, logistic regression fusion, and MKL fusion are all proven to be inferior compared to the multi-stream multi-class fusion technique. Moreover, a more applied form of classification in videos is to identify/recommend tags or thumbnails in videos, and this specific task is successfully caried out in [75–79].

3.8. Few-Shot Video Classification

FEW-SHOT learning (FSL) has received a great deal of interest in recent years. FSL tries to identify new classes with one or a few labeled samples [80–83]. However, due to most recent work in few-shot learning being centered on image classification, FSL in the video domain is still hardly being explored [84,85]. Some of the notable works done in this domain are discussed below.

A multi-saliency embedding technique was developed by Zhu et al. [85] to encode a variable-length video stream into a fixed-size matrix. Graph neural networks (GNN) were developed by Hu et al. [86] to enhance the video classification model's capacity for discrimination. The local–global link in a distributed representation space was still disregarded nevertheless. To categorize a previously unseen video, Cao et al. [87] introduced a temporal alignment module (TAM) that explicitly took advantage of the temporal ordering information in video data through temporal alignment. To combine the two-stream aspects of videos more effectively, Fu et al. [88] developed a depth-guided adaptive instance-normalization module (DGAdaIN). A C3D encoder was created by Zhang et al. [89] to record close-range action patterns for spatiotemporal video blocks. Few-shot video categorization was addressed by Qi et al. [90] by learning a collection of SlowFast networks enhanced with memory units. To comprehend realistic films of the target classes, Fu et al. [91] presented embodied agent-based one-shot learning, which made use of synthetic videos created in a virtual environment. For the issues of few-shot and zeroshot action recognition, Bishay et al. [92] presented the temporal attentive relation network (TARN), which was trained to compare representations of varying temporal length. By examining local–global linkages and preserving the specifics of properties, Y. Feng et al. [93] recently presented a dual-routing capsule graph neural network (DR-CapsGNN) to address the issue of severely constrained samples in few-shot learning.

Apart from this, contrastive learning has also proved successful in recognizing human actions. Some of the interesting works done in this regard are multi-granularity anchor-contrastive representation learning [94] and X-invariant contrastive augmentation and representation learning [95].

3.9. Geometric Deep Learning

Shape descriptors play a significant role in the description of manifolds for 3D shapes. In general, a global feature descriptor is created by aggregating local descriptors to describe the geometric properties of the entire shape, for example, using the bag-of-features paradigm. A local feature descriptor assigns a vector to each point on the shape in a multi-dimensional descriptor space, representing the local structure of the shape around that point. Most deep learning techniques that deal with 3D shapes essentially use the CNN paradigm. Volumetric 2D multi-view shape representations are applied directly using standard (Euclidean) CNN architectures in neural networks via methods such as [96,97]. These techniques are unsuited for dealing with deformable shapes because the shape descriptors they use are dependent on extrinsic structures that are invariant under Euclidean transformations, as demonstrated in Figure 8a [98], while some other approaches [99–103] create a new framework by adopting the CNN feature extraction pattern to investigate the inherent CNN versions that would enable handling shape deformations by using intrinsic filter structure, as shown in Figure 8b [98]. Geometric deep learning deals with non-Euclidean graph and manifold data. This type of data (irregularly arranged/distributed randomly) is usually used to describe geometric shapes. The purpose of geometric deep learning is to find the underlying patterns in geometric data where the traditional Euclidean distance-based deep learning approaches are not suitable. There are basically two methods available in the literature to apply deep learning on geometric data: (i) extrinsic methods and (ii) intrinsic methods. The filters in extrinsic methods are applied on the 3D surfaces such that it effects the structural deformity due to the extrinsic filter structure. The key weakness of extrinsic approaches [96,97] is that they continue to consider geometric data as Euclidean information. When an object's position or shape changes, the extrinsic data representation fails. Additionally, for these methods to support the challenging-in-practice task of attaining the invariance of shape deformation, complicated models and extensive training are required. The filters in intrinsic approaches are applied on the 3D surfaces without being affected by the structural deformity. Rather than Euclidean realization, intrinsic methods work on the manifold and are isometry-invariant by construction. Some of the works based on intrinsic deep learning include (i) geodesic CNN [99], (ii) anisotropic CNN [100], (iii) mixture model network [101], (iv) structured prediction model [102], (v) localized spectral CNN [103], (vi) PointNet [104], (vii) PointNet++ [105], and (viii) RGA-MLP [106]. The application of geometric deep learning (mostly intrinsic methods) in analyzing videos can help in better understanding from the machine perspective, but it is still an open research problem and needs further investigation. For further details on geometric deep learning, readers are referred to [98,107].

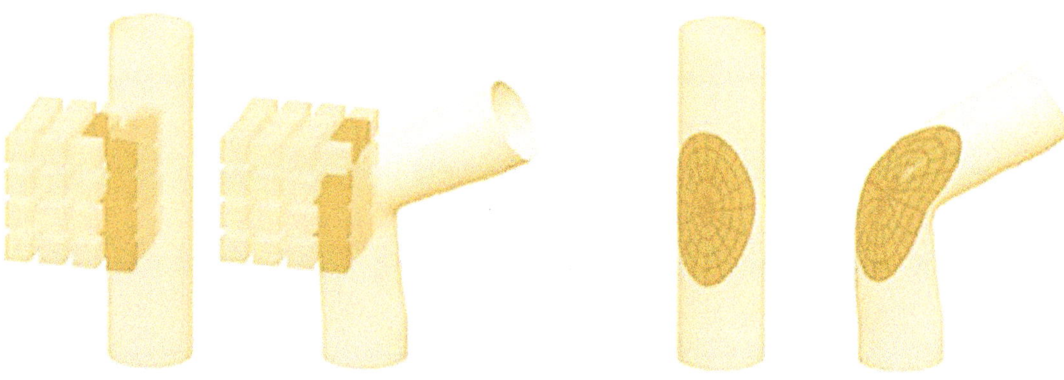

(a) (b)

Figure 8. Illustration of deep learning approaches on geometric data. (**a**) Extrinsic method and (**b**) intrinsic method.

Table 3. Summary and findings of studies based on deep learning models.

Study	Features	Model	Evaluation	Dataset	Problem	Fusion	Findings
[57]	Automatic spatio-temporal features/self-learning. Temporal features captured both locally and globally.	Multiresolution CNN architecture.	By the fraction of test samples that contained at least one of the ground truth labels in the top k predictions.	Sports-1M, UCF-101.	Multi-class	Single frame, Early Fusion, Late Fusion, Slow Fusion.	When compared to a multilayer neural network with rectified linear units followed by a Softmax classifier built using histogram features, the Softmax classifier performed better (both local features such as texton, HOG, cuboids, etc., and global features such as color moments, and hue–saturation).
[108]	Visual (dense trajectory descriptors): A 30-d trajectory shape descriptor, a 96-d HOG descriptor, a 108-d HOF descriptor, and a 108-d MBH descriptor (local visual descriptors). Audio Features: MFCCs and Spectrogram SIFT.	Deep neural network (DNN).	Mean average precision (mAP).	Hollywood2, Columbia Consumer Videos (CCV), and CCV+.	Multi-class	Regularized fusion of multiple features.	Found better than dense trajectory features and classification utilizing the basic early fusion technique.
[109]	Tensor-Train Factorization (global representation for the whole sequence).	Recurrent neural network (RNN).	Classification accuracy.	UCF11, Hollywood2, YouTube Celebrities Face Data.	Multi-class	-	Tensor-Train layer-based RNN such as LSTM and GRU perform better than the plain RNN architectures for video classification.
[110]	Improved Fisher vector (iFV) and explicit feature maps to represent features of conv and fc layers. Long-term temporal information.	A multilayer and multimodal fusion framework of deep neural networks based on fully connected (FC)-RNN.	Classification accuracy.	UCF101, HMDB51.	Multi-class	Multilayer and multimodal fusion framework.	When compared to enhanced dense trajectories, which require a number of handcrafted procedures such as dense point tracking, camera motion estimation, person detection, and so on, the proposed FC-RNN obtained competitive results.
[50]	Convolutional temporal feature pooling architectures (conv pooling, late pooling, slow pooling, local pooling). Global video-level descriptors.	Two CNN architectures (AlexNet and GoogleNet) and LSTM.	By the fraction of test samples that contained at least one of the ground truth labels in the top k predictions.	UCF101, Sports 1 million.	Multi-class	Late fusion	(i) UCF-101 necessitates the utilization of optical flow. (ii) Optical flow is not always beneficial, especially when the videos are captured in the wild, such as Sports-1M. (iii) To make use of optical flow, a more advanced sequence processing architecture such as LSTM is required. (iv) The maximum documented performance is achieved by using LSTMs on both image frames and optical flow for the Sports-1M benchmark.

Table 3. *Cont.*

Study	Features	Model	Evaluation	Dataset	Problem	Fusion	Findings
[111]	Spatiotemporal feature learning: a SMART block and ARTNet for short-term spatiotemporal feature learning with a possibility to explore long-term learning.	ARTNet by integrating the SMART block into the C3D-ResNet18 architecture, where SMART block architecture is composed of appearance branch and relationship branch.	Top-1 and Top-5 accuracy.	Kinetics, UCF101, and HMDB51.	Multi-class	Concatenation and reduction operation.	(i) In terms of spatiotemporal feature learning, SMART blocks outperform 3D convolutions (3D-CNN). (ii) In the case of ARTNet, supplementing RGB input with optical flow improves performance. (iii) The optical flow modality can give additional information. (iv) Optical flow's high computing cost prevents it from being used in real-world systems.
[112]	Spatial, short-term motion and audio clues using CNN. Long-term temporal dynamics. (Multimodal features).	CNNs-LSTM model with multi-stream multi-class fusion process to adaptively determine the optimal fusion weights for generating the final scores of each class.	Classification accuracy.	UCF-101, Columbia Consumer Videos.	Multi-class	Multi-Stream Multi-Class Fusion.	Average fusion, kernel average fusion, weighted fusion, logistic regression fusion, and MKL fusion are all proven to be inferior to the proposed multi-stream multi-class fusion technique.
[113]	Two distinct layers: 1 × 1 × 1 conventional convolutions for channel interaction (but no local interaction) and k × k × k depth-wise convolutions for local spatiotemporal interactions (but not channel interaction). Global spatiotemporal average pooling layer.	Channel-separated convolutional network (CSN). Two models: interaction-preserved channel-separated network (ip-CSN) and interaction-reduced channel-separated network (ir-CSN).	Classification accuracy.	Sports1M and Kinetics.	Multi-class	-	(i) In 3D group convolutional networks, the number of channel interactions has a significant impact on accuracy. (ii) Separating channel interactions from spatiotemporal interactions in 3D convolutions improves accuracy and reduces computing cost. (iii) Three-dimensional channel-separated convolutions offer regularization and avoid overfitting.
[114]	The 3D network is optimized with three loss functions: (i) cross-entropy (CE) loss, (ii) pseudo-CE loss, and (iii) soft CE loss. 2D Image and 3D video model capture short and long visual descriptors.	Semi-supervised learning (VideoSSL) with 3D ResNet-18.	Top-1	UCF101, HMDB51, and Kinetics.	Multi-class	-	(i) For 3D video classification, a direct application of current semi-supervised algorithms (which were initially designed for 2D imagery) cannot yield adequate results. (ii) The accuracy of 3D-CNN models is much improved by a calibrated use of object appearance indicators for semi-supervised learning.

Table 3. *Cont.*

Study	Features	Model	Evaluation	Dataset	Problem	Fusion	Findings
[115]	Modal- and channel-wise attentions.	Expansion-squeeze excitation fusion network	Accuracy, confusion matrix	ETRI-ACTIVITY3D, NUT RGB+D	Multi-class	Multi-modal	(i) Modal-fusion nets (M-Nets) and channel-fusion nets (C-Nets) are capable of capturing the modal and channel-wise dependencies between features in order to improve the discriminative power of features via modal and channel-wise ESEs. (ii) By adding the penalty of the difference between the minimum prediction losses on the single modalities and the prediction loss on the fused modality, multi-modal loss (ML) can further enforce the consistency between the single-modal features and the fused multi-modal features.

4. Benchmark Datasets, Evaluation Metrics, and Comparison of Existing State-of-the-Art for Video Classification

4.1. Benchmark Datasets for Video Classification

There are several benchmark datasets being utilized for classification of videos, AND some of these notable datasets are summarized in Table 4. The details related to these datasets, such as total number of videos contained in the dataset, number of classes present in the dataset, the year of publication of dataset, and the background of videos in the dataset, are included in the summary.

Table 4. Benchmark datasets.

Dataset	# of Videos	# of Classes	Year	Background
KTH	600	6	2004	Static
Weizmann	81	9	2005	Static
Kodak	1358	25	2007	Dynamic
Hollywood	430	8	2008	Dynamic
Hollywood2	1787	12	2009	Dynamic
MCG-WEBV	234,414	15	2009	Dynamic
Olympic Sports	800	16	2010	Dynamic
HMDB51	6766	51	2011	Dynamic
CCV	9317	20	2011	Dynamic
UCF-101	13,320	101	2012	Dynamic
THUMOS-2014	18,394	101	2014	Dynamic
MED-2014 (Dev. set)	31,000	20	2014	Dynamic
Sports-1M	1,133,158	487	2014	Dynamic
ActivityNet	27,901	203	2015	Dynamic
EventNet	95,321	500	2015	Dynamic
MPII Human Pose	20,943	410	2014	Dynamic
FCVID	91,223	239	2015	Dynamic
UCF11	1600	11	2009	Dynamic
YouTube Celebrities Face	1910	47	2008	Dynamic
Kinetics	300,000	400	2017	Dynamic
YouTube-8M	6.1 M	3862	2018	Dynamic
JHMDB	928	21	2011	Dynamic
Something-something	110,000	174	2017	Dynamic

4.2. Performance Evaluation Metrics for Video Classification

The evaluation of video classification models is performed using different performance measures. The most common measures utilized to evaluate the models are accuracy, precision, recall, F1 score, micro F1, and K-fold [8]. Some of the recent studies using these measures are listed in Table 5.

Table 5. Commonly used evaluation metrics for video classification.

Evaluation Metric	Year of Publication	Reference
Accuracy	2020–2021	[116–120]
Precision	2020–2021	[116,118,119]
Recall	2020–2021	[116,118,119]
F1 Score	2020–2021	[116,118,119]
Micro F1	2020	[121,122]
K-Fold	2019	[123]
Top-k	2018,2021	[111,114]

4.3. Comparison of Some Existing Approaches on UCF-101 Dataset

UCF-101 is a benchmark action recognition dataset published by the researchers of University of Central Florida in the year 2012 [124], and the videos in the dataset were collected from YouTube. The total videos in the dataset are 13,320, with 101 action categories. The dataset is challenging because of the uncontrolled environment in the captured videos, and it is widely being used by researchers working on the video classification problem. Therefore, it is easy to compare most of the existing literature based on this dataset. The existing works employing UCF-101 are compared in Table 6, where the methods are arranged in ascending order based on the performance. The results reported in Table 6 are taken from the existing studies in the literature.

Table 6. Comparison of video classification method on UCF-101.

Method	Accuracy
LRCN [48]	82.9
DT + MVSV [125]	83.5
LSTM–Composite [49]	84.3
FSTCN [126]	88.1
C3D [127]	85.2
iDT + HSV [128]	87.9
Two-Stream [61]	88.0
RNN-FV [129]	88.0
LSTM [50]	88.6
MultiSource CNN [130]	89.1
Image-Based [55]	89.6
TDD [35]	90.3
Multilayer and Multimodal Fusion [110]	91.6
Transformation CNN [131]	92.4
Multi-Stream [112]	92.6
Key Volume Mining [132]	92.7
Convolutional Two-Stream [62]	93.5
Temporal Segment Networks [39]	94.2

4.4. Comparison of Different Deep Learning Architectures

In Table 7, some important deep learning architectures are compared in terms of performance and computational requirement. These architectures are the basis of development of different deep learning models for video classification, and from this comparison, an estimation of the requirement of computational cost for each of these architectures can be drawn.

Table 7. Performance comparison of different deep architectures [127].

Architecture Name	Parameters	Error Rate	Depth	Category	Year
LeNet	0.060 M	[dist]MNIST: 0.8 MNIST: 0.95	5	Spatial exploitation	1998
AlexNet	60 M	ImageNet: 16.4	8	Spatial exploitation	2012
ZfNet	60 M	ImageNet: 11.7	8	Spatial exploitation	2014
VGG	138 M	ImageNet: 7.3	19	Spatial exploitation	2014
GoogLeNet	4 M	ImageNet: 6.7	22	Spatial exploitation	2015
Inception-V3	23.6 M	ImageNet: 3.5 multi-crop: 3.58 Single-Crop: 5.6	159	Depth + width	2015
Highway networks	2.3 M	CIFAR-10: 7.76	19	Depth + multi-path	2015
Inception-V4	35 M	ImageNet: 4.01	70	Depth + width	2016
Inception-ResNet	55.8 M	ImageNet: 3.52	572	Depth + width + multi-path	2016

Table 7. *Cont.*

Architecture Name	Parameters	Error Rate	Depth	Category	Year
ResNet	25.6 M 1.7 M	ImageNet: 3.6 CIFAR-10: 6.43	152 110	Depth + multi-path	2016
DelugeNet	20.2 M	CIFAR-10: 3.76 CIFAR-100: 19.02	146	Multi-path	2016
FractalNet	38.6 M	CIFAR-10: 7.27 CIFAR-10 +: 4.60 CIFAR-10 ++: 4.59 CIFAR-100: 28.20 CIFAR-100 +: 22.49 CIFAR100 ++: 21.49	20 40	Multi-path	2016
WideResNet	36.5 M	CIFAR-10: 3.89 CIFAR-100: 18.85	28 –	Width	2016
Xception	22.8 M	ImageNet: 0.055	126	Width	2017
Residual attention neural network	8.6 M	CIFAR-10: 3.90 CIFAR-100: 20.4 ImageNet: 4.8	452	Attention	2017
ResNeXt	68.1 M	CIFAR-10: 3.58 CIFAR-100: 17.31 ImageNet: 4.4	29 - 101	Width	2017
Squeeze and excitation networks	27.5 M	ImageNet: 2.3	152	Feature-map exploitation	2017
DenseNet	25.6 M 25.6 M 15.3 M 15.3 M	CIFAR-10 +: 3.46 CIFAR100 +: 17.18 CIFAR-10: 5.19 CIFAR-100: 19.64	190 190 250 250	Multi-path	2017
PolyNet	92 M	ImageNet: Single: 4.25 Multi: 3.45	– –	Width	2017
PyramidalNet	116.4 M 27.0 M 27.0 M	ImageNet: 4.7 CIFAR-10: 3.48 CIFAR-100: 17.01	200 164 164	Width	2017
Convolutional block attention Module (ResNeXt101 (32 × 4d) + CBAM)	48.96 M	ImageNet: 5.59	101	Attention	2018
Concurrent spatial and channel excitation mechanism	–	MALC: 0.12 Visceral: 0.09	–	Attention	2018
Channel boosted CNN	–	–	–	Channel boosting	2018
Competitive squeeze and excitation network CMPE-SE-WRN-28	36.92 M 36.90 M	CIFAR-10: 3.58 CIFAR-100: 18.47	152 152	Feature-map exploitation	2018

5. Key Findings

From the analysis of the existing literature, the following key findings are drawn for video classification task: (i) The visual description works better than the text and the audio description, and the combination of all modalities can contribute to better performance with an increase in computational cost. (ii) The architectures employing CNN/RNN for feature extraction have the ability to perform better than handcrafted features provided that enough data are available for training. (iii) Tensor-Train layer-based RNN such as LSTM and GRU perform better than the plain RNN architectures for video classification. (iv) It is sometimes necessary to use optical flow for datasets such as UCF-101. (v) It is not always helpful to use optical flow, especially for the case of videos taken from the wild, e.g., Sports-1 M. (vi) It is important to use a sophisticated sequence processing architecture such as LSTM to take advantage of optical flow. (vii) LSTMs, when applied on both the optical flow and the image frames, yield the highest performance measure for the Sports-1M benchmark dataset. (viii) Augmenting optical flow and RGB input helps in improving the performance. (ix) Optical flow modality provides complementary information. (x) The high computational requirement of optical flow limits its use in real-time systems. (xi) Multi-stream multi-class fusion can perform better than average fusion, weighted fusion, kernel average fusion, MKL fusion, and logistic regression fusion on datasets such as UCF-101 and CCV. (xii) In 3D group convolutional networks, the volume of channel interactions plays a vital role in achieving a high accuracy. (xiii) The factorization of 3D convolutions by separating spatiotemporal interactions and channel interactions can lead to an improvement in accuracy and a decrease in the computational cost. (xiv) Further, 3D channel-separated convolutions results in a kind of regularization and prevents overfitting. (xv) Popular frameworks of conventional semi-supervised algorithms (which were originally developed for 2D images) are unable to obtain good results for 3D video categorization. (xvi) For semi-supervised learning, a calibrated employment of the object appearance cues keenly improves the accuracy of the 3D-CNN models.

6. Conclusions

This article reviews deep learning approaches for the task of video classification. Some of the notable studies are summarized in detail, and the key findings in these studies are highlighted. The key findings are reported as an effort to help the research community in developing new deep learning models for video classification.

The latest developments in deep learning models have demonstrated the potential of these approaches for the video classification task. However, most of the existing deep learning architectures for video classification are basically adopted from the favored deep learning architectures in image/speech domain. Therefore, most of the existing architectures remain insufficient to deal with the more complicated nature of video data that contain rich information in the form of spatial, temporal, and acoustic clues. This calls for attention towards the need for a tailored network capable of effectively modeling the spatial, temporal, and acoustic information. Moreover, training CNN/RNN models requires labeled datasets, and acquiring those datasets is usually time-consuming and expensive, and hence, a promising research direction is to utilize the considerable amount of unlabeled video data to derive better video representations.

Furthermore, the deep learning approaches are outperforming other state-of-the-art approaches for video classification. The deep learning Google trend is still growing, and it is still above the trend for some other very well-known machine learning algorithms, as shown in Figure 9a. However, the recent developments in deep learning approaches are still under-evaluated and require further investigations for the video classification task. One such example is geometric deep learning approaches, and the worldwide research interest in this specific topic is shown in Figure 9b, which describes that this topic is still confined to some states of U.S., Europe, and India. Therefore, it has yet to be developed and investigated further. The use of geometric deep learning in extracting rich spatial information from videos can also be a new research direction as a future work for better accuracy in the video classification task.

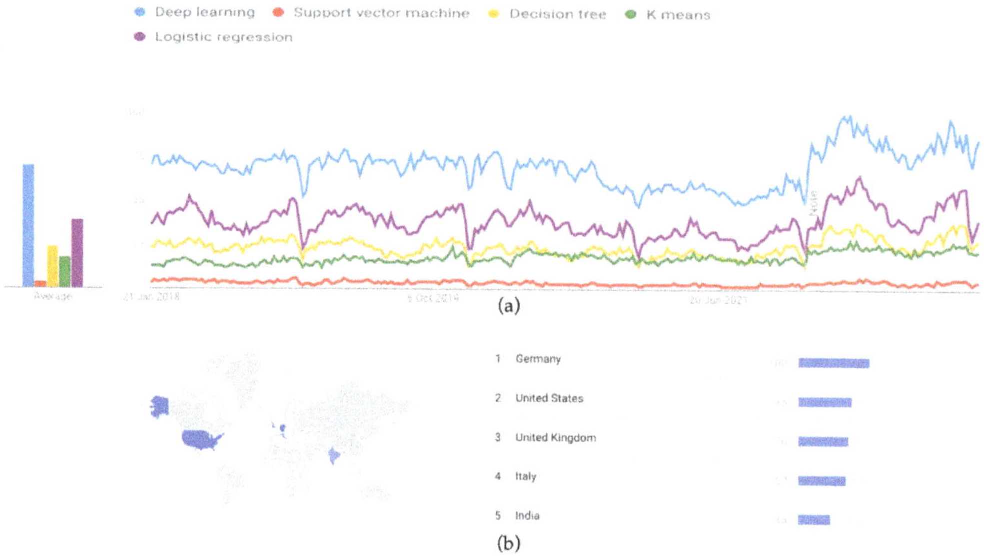

Figure 9. (**a**) Google trend on deep learning vs. some other state-of-the-art methods. (**b**) Worldwide research interest in geometric deep learning.

Author Contributions: Conceptualization, A.u.R. and S.B.B.; methodology, M.A.K.; validation, A.K.; formal analysis, S.B.B.; data curation, A.u.R.; writing—original draft preparation, A.u.R.; writing—review and editing, M.A.K.; visualization, A.u.R.; supervision, S.B.B. All authors have read and agreed to the published version of the manuscript.

Funding: This research received no external funding.

Institutional Review Board Statement: Not applicable.

Informed Consent Statement: Not applicable.

Data Availability Statement: Not applicable.

Acknowledgments: We would like to acknowledge Mälardalen University for supporting the publication charges.

Conflicts of Interest: The authors declare no conflict of interest.

References

1. Samek, W.; Montavon, G.; Lapuschkin, S.; Anders, C.J.; Muller, K.-R. Explaining Deep Neural Networks and Beyond: A Review of Methods and Applications. *Proc. IEEE* **2021**, *109*, 247–278. [CrossRef]
2. Kiranyaz, S.; Avci, O.; Abdeljaber, O.; Ince, T.; Gabbouj, M.; Inman, D.J. 1D convolutional neural networks and applications: A survey. *Mech. Syst. Signal Process.* **2021**, *151*, 107398. [CrossRef]
3. Minallah, N.; Tariq, M.; Aziz, N.; Khan, W.; Rehman, A.; Belhaouari, S.B. On the performance of fusion based planet-scope and Sentinel-2 data for crop classification using inception inspired deep convolutional neural network. *PLoS ONE* **2020**, *15*, e0239746. [CrossRef]
4. Rehman, A.; Bermak, A. Averaging Neural Network Ensembles Model for Quantification of Volatile Organic Compound. In Proceedings of the 2019 15th International Wireless Communications & Mobile Computing Conference (IWCMC), Tangier, Morocco, 24–28 June 2019; pp. 848–852. [CrossRef]
5. Anushya, A. Video Tagging Using Deep Learning: A Survey. *Int. J. Comput. Sci. Mob. Comput.* **2020**, *9*, 49–55.
6. Rani, P.; Kaur, J.; Kaswan, S. Automatic Video Classification: A Review. *EAI Endorsed Trans. Creat. Technol.* **2020**, *7*, 163996. [CrossRef]
7. Li, Y.; Wang, C.; Liu, J. A Systematic Review of Literature on User Behavior in Video Game Live Streaming. *Int. J. Environ. Res. Public Health* **2020**, *17*, 3328. [CrossRef]

8. Islam, M.S.; Sultana, M.S.; Roy, U.K.; al Mahmud, J. A review on Video Classification with Methods, Findings, Performance, Challenges, Limitations and Future Work. *J. Ilm. Tek. Elektro Komput. Dan Inform.* **2021**, *6*, 47. [CrossRef]
9. Ullah, H.A.; Letchmunan, S.; Zia, M.S.; Butt, U.M.; Hassan, F.H. Analysis of Deep Neural Networks for Human Activity Recognition in Videos—A Systematic Literature Review. *IEEE Access* **2021**, *9*, 126366–126387. [CrossRef]
10. Wu, Z.; Yao, T.; Fu, Y.; Jiang, Y.-G. Deep learning for video classification and captioning. In *Frontiers of Multimedia Research*; ACM: New York, NY, USA, 2017; pp. 3–29. [CrossRef]
11. Ren, Q.; Bai, L.; Wang, H.; Deng, Z.; Zhu, X.; Li, H.; Luo, C. A Survey on Video Classification Methods Based on Deep Learning. *DEStech Trans. Comput. Sci. Eng.* **2019**. [CrossRef]
12. LeCun, Y.; Bottou, L.; Bengio, Y.; Haffner, P. Gradient-based Learning Applied to Document Recognition. *Intell. Signal Process.* **2001**, 306–351. [CrossRef]
13. Krizhevsky, A.; Sutskever, I.; Hinton, G.E. ImageNet classification with deep convolutional neural networks. *Adv. Neural Inf. Process. Syst.* **2012**, *2*, 1097–1105. [CrossRef]
14. Simonyan, K.; Zisserman, A. Very deep convolutional networks for large-scale image recognition. In Proceedings of the 3rd International Conference on Learning Representations, ICLR 2015—Conference Track Proceedings, San Diego, CA, USA, 7–9 May 2015.
15. Szegedy, C.; Liu, W.; Jia, Y.; Sermanet, P.; Reed, S.; Anguelov, D.; Anguelov, D.; Erhan, D.; Vanhoucke, V.; Rabinovich, A. Going deeper with convolutions. In Proceedings of the IEEE Computer Society Conference on Computer Vision and Pattern Recognition, Boston, MA, USA, 7–12 June 2015; pp. 1–9. [CrossRef]
16. He, K.; Zhang, X.; Ren, S.; Sun, J. Deep residual learning for image recognition. In Proceedings of the IEEE Computer Society Conference on Computer Vision and Pattern Recognition, Las Vegas, NV, USA, 27–30 June 2016; pp. 770–778. [CrossRef]
17. Huang, G.; Liu, Z.; Van Der Maaten, L.; Weinberger, K.Q. Densely connected convolutional networks. In Proceedings of the 30th IEEE Conference on Computer Vision and Pattern Recognition, CVPR 2017, Honolulu, HI, USA, 21–26 July 2017; pp. 2261–2269. [CrossRef]
18. Khan, A.; Sohail, A.; Zahoora, U.; Qureshi, A.S. A survey of the recent architectures of deep convolutional neural networks. *Artif. Intell. Rev.* **2020**, *53*, 5455–5516. [CrossRef]
19. Ian, G.; Yoshua, B.; Aaron, C. *Deep Learning (Adaptive Computation and Machine Learning Series)*; The MIT Press: Cambridge, MA, USA, 2016.
20. Shah, A.M.; Yan, X.; Shah, S.A.A.; Mamirkulova, G. Mining patient opinion to evaluate the service quality in healthcare: A deep-learning approach. *J. Ambient Intell. Humaniz Comput.* **2020**, *11*, 2925–2942. [CrossRef]
21. De Jong, R.J.; de Wit, J.J.M.; Uysal, F. Classification of human activity using radar and video multimodal learning. *IET Radar Sonar Navig.* **2021**, *15*, 902–914. [CrossRef]
22. Truong, B.T.; Venkatesh, S.; Dorai, C. Automatic genre identification for content-based video categorization. In Proceedings of the International Conference on Pattern Recognition 2000, Barcelona, Spain, 3–7 September 2000; Volume 15, pp. 230–233. [CrossRef]
23. Huang, C.; Fu, T.; Chen, H. Text-based video content classification for online video-sharing sites. *J. Am. Soc. Inf. Sci. Technol.* **2010**, *61*, 891–906. [CrossRef]
24. Lee, K.; Ellis, D.P.W. Audio-based semantic concept classification for consumer video. *IEEE Trans. Audio Speech Lang Process.* **2010**, *18*, 1406–1416. [CrossRef]
25. Liu, Z.; Huang, J.; Wang, Y. Classification TV programs based on audio information using hidden Markov model. In Proceedings of the 1998 IEEE 2nd Workshop on Multimedia Signal Processing, Redondo Beach, CA, USA, 7–9 December 1998; pp. 27–32. [CrossRef]
26. Laptev, I.; Lindeberg, T. Space-time interest points. In Proceedings of the IEEE International Conference on Computer Vision, 2003, Nice, France, 13–16 October 2003; Volume 1, pp. 432–439. [CrossRef]
27. Wang, H.; Schmid, C. Action recognition with improved trajectories. In Proceedings of the IEEE International Conference on Computer Vision, Sydney, Australia, 1–8 December 2013; pp. 3551–3558. [CrossRef]
28. Scovanner, P.; Ali, S.; Shah, M. A 3-dimensional sift descriptor and its application to action recognition. In Proceedings of the ACM International Multimedia Conference and Exhibition, Augsburg, Germany, 25–29 September 2007; pp. 357–360. [CrossRef]
29. Kläser, A.; Marszałek, M.; Schmid, C. A spatio-temporal descriptor based on 3D-gradients. In Proceedings of the BMVC 2008—British Machine Vision Conference 2008, Leeds, UK, September 2008. [CrossRef]
30. Dalal, N.; Triggs, B.; Schmid, C. Human detection using oriented histograms of flow and appearance. In *Lecture Notes in Computer Science (Including Subseries Lecture Notes in Artificial Intelligence and Lecture Notes in Bioinformatics)*; LNCS; Springer: Berlin/Heidelberg, Germany, 2006; Volume 3952, pp. 428–441. [CrossRef]
31. Sadanand, S.; Corso, J.J. Action bank: A high-level representation of activity in video. In Proceedings of the IEEE Computer Society Conference on Computer Vision and Pattern Recognition, Providence, RI, USA, 16–21 June 2012; pp. 1234–1241. [CrossRef]
32. Dollár, P.; Rabaud, V.; Cottrell, G.; Belongie, S. Behavior recognition via sparse spatio-temporal features. In Proceedings of the 2nd Joint IEEE International Workshop on Visual Surveillance and Performance Evaluation of Tracking and Surveillance, Beijing, China, 15–16 October 2005; Volume 2005, pp. 65–72. [CrossRef]
33. Willems, G.; Tuytelaars, T.; Van Gool, L. An efficient dense and scale-invariant spatio-temporal interest point detector. In *Lecture Notes in Computer Science (Including Subseries Lecture Notes in Artificial Intelligence and Lecture Notes in Bioinformatics)*; LNCS; Springer: Berlin/Heidelberg, Germany, 2008; Volume 5303, pp. 650–663. [CrossRef]

34. Wang, L.; Qiao, Y.; Tang, X. Video action detection with relational dynamic-poselets. In *Lecture Notes in Computer Science (Including Subseries Lecture Notes in Artificial Intelligence and Lecture Notes in Bioinformatics)*; LNCS; Springer: Berlin/Heidelberg, Germany, 2014; Volume 8693, pp. 565–580. [CrossRef]
35. Wang, L.; Qiao, Y.; Tang, X. Action recognition with trajectory-pooled deep-convolutional descriptors. In Proceedings of the IEEE Computer Society Conference on Computer Vision and Pattern Recognition, Boston, MA, USA, 7–12 June 2015; pp. 4305–4314. [CrossRef]
36. Kar, A.; Rai, N.; Sikka, K.; Sharma, G. AdaScan: Adaptive scan pooling in deep convolutional neural networks for human action recognition in videos. In Proceedings of the 30th IEEE Conference on Computer Vision and Pattern Recognition, CVPR 2017, Honolulu, HI, USA, 21–26 July 2017; pp. 5699–5708. [CrossRef]
37. Feichtenhofer, C.; Pinz, A.; Wildes, R.P. Spatiotemporal multiplier networks for video action recognition. In Proceedings of the 30th IEEE Conference on Computer Vision and Pattern Recognition, CVPR 2017, Honolulu, HI, USA, 21–26 July 2017; pp. 7445–7454. [CrossRef]
38. Qiu, Z.; Yao, T.; Mei, T. Learning spatio-temporal representation with pseudo-3D residual networks. In Proceedings of the IEEE International Conference on Computer Vision, Venice, Italy, 22–29 October 2017; pp. 5533–5541.
39. Wang, L.; Xiong, Y.; Wang, Z.; Qiao, Y.; Lin, D.; Tang, X.; Van Gool, L. Temporal segment networks: Towards good practices for deep action recognition. In *Lecture Notes in Computer Science (Including Subseries Lecture Notes in Artificial Intelligence and Lecture Notes in Bioinformatics)*; LNCS; Springer: Berlin/Heidelberg, Germany, 2016; Volume 9912, pp. 20–36. [CrossRef]
40. Wang, Y.; Long, M.; Wang, J.; Yu, P.S. Spatiotemporal pyramid network for video action recognition. In Proceedings of the IEEE Conference on Computer Vision and Pattern Recognition, Honolulu, HI, USA, 21–26 July 2017; pp. 2097–2106.
41. Lan, Z.; Zhu, Y.; Hauptmann, A.G.; Newsam, S. Deep Local Video Feature for Action Recognition. In Proceedings of the IEEE Computer Society Conference on Computer Vision and Pattern Recognition Workshops; 2017; pp. 1219–1225. [CrossRef]
42. Duta, I.C.; Ionescu, B.; Aizawa, K.; Sebe, N. Spatio-temporal vector of locally max pooled features for action recognition in videos. In Proceedings of the 30th IEEE Conference on Computer Vision and Pattern Recognition, CVPR 2017, Honolulu, HI, USA, 21–26 July 2017; pp. 3205–3214. [CrossRef]
43. Shen, J.; Huang, Y.; Wen, M.; Zhang, C. Toward an Efficient Deep Pipelined Template-Based Architecture for Accelerating the Entire 2-D and 3-D CNNs on FPGA. *IEEE Trans. Comput.-Aided Des. Integr. Circuits Syst.* **2020**, *39*, 1442–1455. [CrossRef]
44. Duta, I.C.; Nguyen, T.A.; Aizawa, K.; Ionescu, B.; Sebe, N. Boosting VLAD with double assignment using deep features for action recognition in videos. In Proceedings of the International Conference on Pattern Recognition, Cancun, Mexico, 4–8 December 2016; pp. 2210–2215. [CrossRef]
45. Xu, Z.; Yang, Y.; Hauptmann, A.G. A discriminative CNN video representation for event detection. In Proceedings of the IEEE Computer Society Conference on Computer Vision and Pattern Recognition, Boston, MA, USA, 7–12 June 2015; pp. 1798–1807. [CrossRef]
46. Girdhar, R.; Ramanan, D.; Gupta, A.; Sivic, J.; Russell, B. ActionVLAD: Learning spatio-temporal aggregation for action classification. In Proceedings of the IEEE Conference on Computer Vision and Pattern Recognition, Honolulu, HI, USA, 21–26 July 2017; pp. 971–980.
47. Ballas, N.; Yao, L.; Pal, C.; Courville, A. Delving deeper into convolutional networks for learning video representations. In Proceedings of the 4th International Conference on Learning Representations, ICLR 2016—Conference Track Proceedings, San Juan, PR, USA, 2–4 May 2016.
48. Donahue, J.; Anne Hendricks, L.; Guadarrama, S.; Rohrbach, M.; Venugopalan, S.; Saenko, K.; Darrell, T. Long-term recurrent convolutional networks for visual recognition and description. In Proceedings of the IEEE Computer Society Conference on Computer Vision and Pattern Recognition, Boston, MA, USA, 7–12 June 2015; pp. 2625–2634. [CrossRef]
49. Srivastava, N.; Mansimov, E.; Salakhutdinov, R. Unsupervised learning of video representations using LSTMs. In Proceedings of the 32nd International Conference on Machine Learning, ICML 2015, Lille, France, 6–11 July 2015; Volume 1, pp. 843–852.
50. Ng, J.Y.H.; Hausknecht, M.; Vijayanarasimhan, S.; Vinyals, O.; Monga, R.; Toderici, G. Beyond short snippets: Deep networks for video classification. In Proceedings of the IEEE Computer Society Conference on Computer Vision and Pattern Recognition, Boston, MA, USA, 7–12 June 2015; pp. 4694–4702. [CrossRef]
51. Taylor, G.W.; Fergus, R.; LeCun, Y.; Bregler, C. Convolutional learning of spatio-temporal features. In *Lecture Notes in Computer Science (Including Subseries Lecture Notes in Artificial Intelligence and Lecture Notes in Bioinformatics)*; LNCS; Springer: Berlin/Heidelberg, Germany, 2010; Volume 6316, pp. 140–153. [CrossRef]
52. Le, Q.V.; Zou, W.Y.; Yeung, S.Y.; Ng, A.Y. Learning hierarchical invariant spatio-temporal features for action recognition with independent subspace analysis. In Proceedings of the IEEE Computer Society Conference on Computer Vision and Pattern Recognition, Colorado Springs, CO, USA, 20–25 June 2011; pp. 3361–3368. [CrossRef]
53. Baccouche, M.; Mamalet, F.; Wolf, C.; Garcia, C.; Baskurt, A. Sequential deep learning for human action recognition. In *Lecture Notes in Computer Science (Including Subseries Lecture Notes in Artificial Intelligence and Lecture Notes in Bioinformatics)*; LNCS; Springer: Berlin/Heidelberg, Germany, 2011; Volume 7065, pp. 29–39. [CrossRef]
54. Ji, S.; Xu, W.; Yang, M.; Yu, K. 3D Convolutional neural networks for human action recognition. *IEEE Trans. Pattern Anal. Mach. Intell.* **2013**, *35*, 221–231. [CrossRef]
55. Zha, S.; Luisier, F.; Andrews, W.; Srivastava, N.; Salakhutdinov, R. Exploiting Image-trained CNN Architectures for Unconstrained Video Classification. In Proceedings of the BMVC, Swansen, UK, 7–10 September 2015; pp. 60.1–60.13. [CrossRef]

56. Carreira, J.; Zisserman, A. Quo Vadis, action recognition? A new model and the kinetics dataset. In Proceedings of the 30th IEEE Conference on Computer Vision and Pattern Recognition, CVPR 2017, Honolulu, HI, USA, 21–26 July 2017; pp. 4724–4733. [CrossRef]
57. Karpathy, A.; Toderici, G.; Shetty, S.; Leung, T.; Sukthankar, R.; Li, F.F. Large-scale video classification with convolutional neural networks. In Proceedings of the IEEE Computer Society Conference on Computer Vision and Pattern Recognition, Columbus, OH, USA, 23–28 June 2014; pp. 1725–1732. [CrossRef]
58. Tran, D.; Bourdev, L.; Fergus, R.; Torresani, L.; Paluri, M. Learning spatiotemporal features with 3D convolutional networks. In Proceedings of the IEEE International Conference on Computer Vision, Santiago, Chile, 7–13 December 2015; Volume 2015, pp. 4489–4497. [CrossRef]
59. Shu, X.; Tang, J.; Qi, G.-J.; Liu, W.; Yang, J. Hierarchical Long Short-Term Concurrent Memory for Human Interaction Recognition. *IEEE Trans. Pattern Anal. Mach. Intell.* **2021**, *43*, 1110–1118. [CrossRef]
60. Shu, X.; Zhang, L.; Qi, G.-J.; Liu, W.; Tang, J. Spatiotemporal Co-Attention Recurrent Neural Networks for Human-Skeleton Motion Prediction. *IEEE Trans. Pattern Anal. Mach. Intell.* **2022**, *44*, 3300–3315. [CrossRef]
61. Simonyan, K.; Zisserman, A. Two-stream convolutional networks for action recognition in videos. *Adv. Neural Inf. Process. Syst.* **2014**, *1*, 568–576.
62. Feichtenhofer, C.; Pinz, A.; Zisserman, A. Convolutional Two-Stream Network Fusion for Video Action Recognition. In Proceedings of the IEEE Computer Society Conference on Computer Vision and Pattern Recognition, Las Vegas, NV, USA, 27–30 June 2016; 2016; pp. 1933–1941. [CrossRef]
63. Wu, Z.; Jiang, Y.-G.; Wang, X.; Ye, H.; Xue, X.; Wang, J. Fusing Multi-Stream Deep Networks for Video Classification. *arXiv* **2015**, arXiv:1509.06086.
64. Tran, D.; Wang, H.; Torresani, L.; Ray, J.; LeCun, Y.; Paluri, M. A closer look at spatiotemporal convolutions for action recognition. In Proceedings of the IEEE conference on Computer Vision and Pattern Recognition, Salt Lake City, UT, USA, 18–23 June 2018; pp. 6450–6459.
65. Shan, K.; Wang, Y.; Tang, Z.; Chen, Y.; Li, Y. MixTConv: Mixed Temporal Convolutional Kernels for Efficient Action Recognition. In Proceedings of the 25th International Conference on Pattern Recognition (ICPR), Milan, Italy, 10–15 January 2021; pp. 1751–1756. [CrossRef]
66. Wu, Z.; Wang, X.; Jiang, Y.G.; Ye, H.; Xue, X. Modeling spatial-Temporal clues in a hybrid deep learning framework for video classification. In Proceedings of the MM 2015—2015 ACM Multimedia Conference, Brisbane, Australia, 26–30 October 2015; pp. 461–470. [CrossRef]
67. Tanberk, S.; Kilimci, Z.H.; Tukel, D.B.; Uysal, M.; Akyokus, S. A Hybrid Deep Model Using Deep Learning and Dense Optical Flow Approaches for Human Activity Recognition. *IEEE Access* **2020**, *8*, 19799–19809. [CrossRef]
68. Alhersh, T.; Stuckenschmidt, H.; Rehman, A.U.; Belhaouari, S.B. Learning Human Activity From Visual Data Using Deep Learning. *IEEE Access* **2021**, *9*, 106245–106253. [CrossRef]
69. Kopuklu, O.; Kose, N.; Gunduz, A.; Rigoll, G. Resource efficient 3D convolutional neural networks. In Proceedings of the 2019 International Conference on Computer Vision Workshop, ICCVW 2019, Seoul, Korea, 27–28 October 2019; pp. 1910–1919. [CrossRef]
70. Liu, H.; Bhanu, B. Pose-guided R-CNN for jersey number recognition in sports. In Proceedings of the IEEE Computer Society Conference on Computer Vision and Pattern Recognition Workshops, Long Beach, CA, USA, 16–17 June 2019; pp. 2457–2466. [CrossRef]
71. Huang, G.; Bors, A.G. Region-based non-local operation for video classification. In Proceedings of the International Conference on Pattern Recognition, Milan, Italy, 10–15 January 2020; pp. 10010–10017. [CrossRef]
72. Girshick, R. Fast R-CNN. In Proceedings of the IEEE International Conference on Computer Vision, ICCV 2015, Santiago, Chile, 7–13 December 2015; pp. 1440–1448. [CrossRef]
73. Biswas, A.; Jana, A.P.; Mohana; Tejas, S.S. Classification of objects in video records using neural network framework. In Proceedings of the International Conference on Smart Systems and Inventive Technology, ICSSIT 2018, Tirunelveli, India, 13–14 December 2018; pp. 564–569. [CrossRef]
74. Jana, A.P.; Biswas, A.; Mohana. YOLO based detection and classification of objects in video records. In Proceedings of the 2018 3rd IEEE International Conference on Recent Trends in Electronics, Information and Communication Technology, RTEICT 2018, Bangalore, India, 18–19 May 2018; pp. 2448–2452. [CrossRef]
75. Zhou, R.; Xia, D.; Wan, J.; Zhang, S. An intelligent video tag recommendation method for improving video popularity in mobile computing environment. *IEEE Access* **2020**, *8*, 6954–6967. [CrossRef]
76. Khan, U.A.; Martinez-Del-Amor, M.A.; Altowaijri, S.M.; Ahmed, A.; Rahman, A.U.; Sama, N.U.; Haseeb, K.; Islam, N. Movie Tags Prediction and Segmentation Using Deep Learning. *IEEE Access* **2020**, *8*, 6071–6086. [CrossRef]
77. Apostolidis, E.; Adamantidou, E.; Mezaris, V.; Patras, I. Combining adversarial and reinforcement learning for video thumbnail selection. In Proceedings of the ICMR 2021—2021 International Conference on Multimedia Retrieval, Taipei, Taiwan, 21–24 August 2021; pp. 1–9. [CrossRef]
78. Carta, S.; Giuliani, A.; Piano, L.; Podda, A.S.; Recupero, D.R. VSTAR: Visual Semantic Thumbnails and tAgs Revitalization. *Expert Syst. Appl.* **2022**, *193*, 116375. [CrossRef]

79. Yang, Z.; Lin, Z. Interpretable video tag recommendation with multimedia deep learning framework. *Internet Res.* **2022**, *32*, 518–535. [CrossRef]
80. Wang, Y.; Yan, J.; Ye, X.; Jing, Q.; Wang, J.; Geng, Y. Few-Shot Transfer Learning With Attention Mechanism for High-Voltage Circuit Breaker Fault Diagnosis. *IEEE Trans. Ind. Appl.* **2022**, *58*, 3353–3360. [CrossRef]
81. Zhong, C.; Wang, J.; Feng, C.; Zhang, Y.; Sun, J.; Yokota, Y. PICA: Point-wise Instance and Centroid Alignment Based Few-shot Domain Adaptive Object Detection with Loose Annotations. In Proceedings of the 2022 IEEE/CVF Winter Conference on Applications of Computer Vision (WACV), Waikoloa, HI, USA, 4–8 January 2022; pp. 398–407. [CrossRef]
82. Zhang, A.; Liu, F.; Liu, J.; Tang, X.; Gao, F.; Li, D.; Xiao, L. Domain-Adaptive Few-Shot Learning for Hyperspectral Image Classification. *IEEE Geosci. Remote Sens. Lett.* **2022**. [CrossRef]
83. Zhao, A.; Ding, M.; Lu, Z.; Xiang, T.; Niu, Y.; Guan, J.; Wen, J.R. Domain-Adaptive Few-Shot Learning. In Proceedings of the 2021 IEEE Winter Conference on Applications of Computer Vision (WACV), Virtual, 5–9 January 2021; pp. 1389–1398. [CrossRef]
84. Gao, J.; Xu, C. CI-GNN: Building a Category-Instance Graph for Zero-Shot Video Classification. *IEEE Trans. Multimedia* **2020**, *22*, 3088–3100. [CrossRef]
85. Zhu, L.; Yang, Y. Compound Memory Networks for Few-Shot Video Classification. In *Computer Vision—ECCV 2018*; Ferrari, V., Hebert, M., Sminchisescu, C., Weiss, Y., Eds.; Lecture Notes in Computer Science; Springer: Berlin/Heidelberg, Germany, 2018; Volume 11211, pp. 782–797. [CrossRef]
86. Hu, Y.; Gao, J.; Xu, C. Learning Dual-Pooling Graph Neural Networks for Few-Shot Video Classification. *IEEE Trans. Multimedia* **2021**, *23*, 4285–4296. [CrossRef]
87. Cao, K.; Ji, J.; Cao, Z.; Chang, C.-Y.; Niebles, J.C. Few-Shot Video Classification via Temporal Alignment. In Proceedings of the 2020 IEEE/CVF Conference on Computer Vision and Pattern Recognition (CVPR), Seattle, WA, USA, 13–19 June 2020; pp. 10615–10624. [CrossRef]
88. Fu, Y.; Zhang, L.; Wang, J.; Fu, Y.; Jiang, Y.-G. Depth Guided Adaptive Meta-Fusion Network for Few-shot Video Recognition. In Proceedings of the 28th ACM International Conference on Multimedia, Seattle, WA, USA, 12–16 October 2020; pp. 1142–1151. [CrossRef]
89. Zhang, H.; Zhang, L.; Qi, X.; Li, H.; Torr, P.H.S.; Koniusz, P. Few-Shot Action Recognition with Permutation-Invariant Attention. In *Computer Vision—ECCV 2020*; Vedaldi, A., Bischof, H., Brox, T., Frahm, J.M., Eds.; Lecture Notes in Computer Science; Springer: Berlin/Heidelberg, Germany, 2020; Volume 12350, pp. 525–542. [CrossRef]
90. Qi, M.; Qin, J.; Zhen, X.; Huang, D.; Yang, Y.; Luo, J. Few-Shot Ensemble Learning for Video Classification with SlowFast Memory Networks. In Proceedings of the 28th ACM International Conference on Multimedia, Seattle, WA, USA, 12–16 October 2020; pp. 3007–3015. [CrossRef]
91. Fu, Y.; Wang, C.; Fu, Y.; Wang, Y.X.; Bai, C.; Xue, X.; Jiang, Y.G. Embodied One-Shot Video Recognition. In Proceedings of the 27th ACM International Conference on Multimedia, Nice France, 21–25 October 2019; pp. 411–419. [CrossRef]
92. Bishay, M.; Zoumpourlis, G.; Patras, I. Tarn: Temporal attentive relation network for few-shot and zero-shot action recognition. *arXiv* **2019**, arXiv:1907.09021.
93. Feng, Y.; Gao, J.; Xu, C. Learning Dual-Routing Capsule Graph Neural Network for Few-shot Video Classification. *IEEE Trans. Multimedia* **2022**, 1. [CrossRef]
94. Shu, X.; Xu, B.; Zhang, L.; Tang, J. Multi-Granularity Anchor-Contrastive Representation Learning for Semi-Supervised Skeleton-Based Action Recognition. *IEEE Trans. Pattern Anal. Mach. Intell.* **2022**, 1–18. [CrossRef]
95. Xu, B.; Shu, X.; Song, Y. X-Invariant Contrastive Augmentation and Representation Learning for Semi-Supervised Skeleton-Based Action Recognition. *IEEE Trans. Image Process.* **2022**, *31*, 3852–3867. [CrossRef]
96. Wu, Z.; Song, S.; Khosla, A.; Yu, F.; Zhang, L.; Tang, X.; Xiao, J. 3D ShapeNets: A deep representation for volumetric shapes. In Proceedings of the IEEE Computer Society Conference on Computer Vision and Pattern Recognition, Boston, MA, USA, 7–12 June 2015; pp. 1912–1920. [CrossRef]
97. Su, H.; Maji, S.; Kalogerakis, E.; Learned-Miller, E. Multi-view convolutional neural networks for 3D shape recognition. In Proceedings of the IEEE International Conference on Computer Vision, ICCV 2015, Santiago, Chile, 7–13 December 2015; pp. 945–953. [CrossRef]
98. Cao, W.; Yan, Z.; He, Z.; He, Z. A Comprehensive Survey on Geometric Deep Learning. *IEEE Access* **2020**, *8*, 35929–35949. [CrossRef]
99. Masci, J.; Boscaini, D.; Bronstein, M.M.; Vandergheynst, P. Geodesic Convolutional Neural Networks on Riemannian Manifolds. In Proceedings of the IEEE International Conference on Computer Vision, Santiago, Chile, 7–13 December 2015; pp. 832–840. [CrossRef]
100. Boscaini, D.; Masci, J.; Rodolà, E.; Bronstein, M. Learning shape correspondence with anisotropic convolutional neural networks. *Adv. Neural Inf. Process. Syst* **2016**, *29*, 3197–3205.
101. Monti, F.; Boscaini, D.; Masci, J.; Rodolà, E.; Svoboda, J.; Bronstein, M.M. Geometric deep learning on graphs and manifolds using mixture model CNNs. In Proceedings of the 30th IEEE Conference on Computer Vision and Pattern Recognition, CVPR 2017, Honolulu, HI, USA, 21–26 July 2017; pp. 5425–5434. [CrossRef]
102. Litany, O.; Remez, T.; Rodola, E.; Bronstein, A.; Bronstein, M. Deep Functional Maps: Structured Prediction for Dense Shape Correspondence. In Proceedings of the IEEE International Conference on Computer Vision, Venice, Italy, 22–29 October 2017; pp. 5660–5668. [CrossRef]

103. Boscaini, D.; Masci, J.; Melzi, S.; Bronstein, M.M.; Castellani, U.; Vandergheynst, P. Learning class-specific descriptors for deformable shapes using localized spectral convolutional networks. *Eurographics Symp. Geom. Process.* **2015**, *34*, 13–23. [CrossRef]
104. Qi, C.R.; Su, H.; Mo, K.; Guibas, L.J. PointNet: Deep learning on point sets for 3D classification and segmentation. In Proceedings of the 30th IEEE Conference on Computer Vision and Pattern Recognition, CVPR 2017, Honolulu, HI, USA, 21–26 July 2017; pp. 77–85. [CrossRef]
105. Qi, C.R.; Yi, L.; Su, H.; Guibas, L.J. PointNet++: Deep hierarchical feature learning on point sets in a metric space. *Adv. Neural Inf. Process. Syst.* **2017**, *30*, 5100–5109.
106. Li, Y.; Cao, W. An Extended Multilayer Perceptron Model Using Reduced Geometric Algebra. *IEEE Access* **2019**, *7*, 129815–129823. [CrossRef]
107. Bronstein, M.M.; Bruna, J.; LeCun, Y.; Szlam, A.; Vandergheynst, P. Geometric Deep Learning: Going beyond Euclidean data. *IEEE Signal Process. Mag.* **2017**, *34*, 18–42. [CrossRef]
108. Wu, Z.; Jiang, Y.G.; Wang, J.; Pu, J.; Xue, X. Exploring inter-feature and inter-class relationships with deep neural networks for video classification. In Proceedings of the MM 2014—2014 ACM Conference on Multimedia, Orlando, FL, USA, 3–7 November 2014; pp. 167–176. [CrossRef]
109. Yang, Y.; Krompass, D.; Tresp, V. Tensor-train recurrent neural networks for video classification. In Proceedings of the 34th International Conference on Machine Learning, ICML 2017, Sydney, NSW, Australia, 6–11 August 2017; Volume 8, pp. 5929–5938.
110. Yang, X.; Molchanov, P.; Kautz, J. Multilayer and multimodal fusion of deep neural networks for video classification. In Proceedings of the MM 2016—2016 ACM Multimedia Conference, Amsterdam, The Netherlands, 15–29 October 2016; pp. 978–987. [CrossRef]
111. Wang, L.; Li, W.; Li, W.; Van Gool, L. Appearance-and-relation networks for video classification. In Proceedings of the IEEE conference on Computer Vision and Pattern Recognition, Salt Lake City, UT, USA, 18–23 June 2018; pp. 1430–1439.
112. Wu, Z.; Jiang, Y.G.; Wang, X.; Ye, H.; Xue, X. Multi-stream multi-class fusion of deep networks for video classification. In Proceedings of the MM 2016—Proceedings of the 2016 ACM Multimedia Conference, Amsterdam, The Netherlands, 15–19 October 2016; pp. 791–800. [CrossRef]
113. Tran, D.; Wang, H.; Torresani, L.; Feiszli, M. Video classification with channel-separated convolutional networks. In Proceedings of the IEEE/CVF International Conference on Computer Vision, Seoul, Korea, 27 Octobet–2 November 2019; pp. 5552–5561.
114. Jing, L.; Parag, T.; Wu, Z.; Tian, Y.; Wang, H. VideoSSL: Semi-Supervised Learning for Video Classification. In Proceedings of the IEEE/CVF Winter Conference on Applications of Computer Vision, Virtual, 5–9 January 2021; pp. 1110–1119.
115. Shu, X.; Yang, J.; Yan, R.; Song, Y. Expansion-Squeeze-Excitation Fusion Network for Elderly Activity Recognition. *IEEE Trans. Circuits Syst. Video Technol.* **2022**, *32*, 5281–5292. [CrossRef]
116. Li, Z.; Li, R.; Jin, G. Sentiment analysis of danmaku videos based on naïve bayes and sentiment dictionary. *IEEE Access* **2020**, *8*, 75073–75084. [CrossRef]
117. Zhen, M.; Li, S.; Zhou, L.; Shang, J.; Feng, H.; Fang, T.; Quan, L. Learning Discriminative Feature with CRF for Unsupervised Video Object Segmentation. In *Lecture Notes in Computer Science (Including Subseries Lecture Notes in Artificial Intelligence and Lecture Notes in Bioinformatics)*; LNCS; Springer: Berlin/Heidelberg, Germany, 2020; Volume 12372, pp. 445–462. [CrossRef]
118. Ruz, G.A.; Henríquez, P.A.; Mascareño, A. Sentiment analysis of Twitter data during critical events through Bayesian networks classifiers. *Future Gener. Comput. Syst.* **2020**, *106*, 92–104. [CrossRef]
119. Fantinel, R.; Cenedese, A.; Fadel, G. Hybrid Learning Driven by Dynamic Descriptors for Video Classification of Reflective Surfaces. *IEEE Trans. Industr. Inform.* **2021**, *17*, 8102–8111. [CrossRef]
120. Costa, F.F.; Saito, P.T.M.; Bugatti, P.H. Video action classification through graph convolutional networks. In Proceedings of the VISIGRAPP 2021—16th International Joint Conference on Computer Vision, Imaging and Computer Graphics Theory and Applications, Vienna, Austria, 8–10 February 2021; Volume 4, pp. 490–497. [CrossRef]
121. Xu, Q.; Zhu, L.; Dai, T.; Yan, C. Aspect-based sentiment classification with multi-attention network. *Neurocomputing* **2020**, *388*, 135–143. [CrossRef]
122. Bibi, M.; Aziz, W.; Almaraashi, M.; Khan, I.H.; Nadeem, M.S.A.; Habib, N. A Cooperative Binary-Clustering Framework Based on Majority Voting for Twitter Sentiment Analysis. *IEEE Access* **2020**, *8*, 68580–68592. [CrossRef]
123. Sailunaz, K.; Alhajj, R. Emotion and sentiment analysis from Twitter text. *J. Comput. Sci.* **2019**, *36*, 101003. [CrossRef]
124. Soomro, K.; Zamir, A.R.; Shah, M. UCF101: A Dataset of 101 Human Actions Classes From Videos in the Wild. *arXiv* **2012**, arXiv:1212.0402.
125. Cai, Z.; Wang, L.; Peng, X.; Qiao, Y. Multi-view super vector for action recognition. In Proceedings of the IEEE Computer Society Conference on Computer Vision and Pattern Recognition, Columbus, OH, USA, 23–28 June 2014; pp. 596–603. [CrossRef]
126. Sun, L.; Jia, K.; Yeung, D.Y.; Shi, B.E. Human action recognition using factorized spatio-temporal convolutional networks. In Proceedings of the IEEE International Conference on Computer Vision, Santiago, Chile, 7–13 December 2015; pp. 4597–4605. [CrossRef]
127. Tran, D.; Bourdev, L.; Fergus, R.; Torresani, L.; Paluri, M. C3D: Generic Features for Video Analysis. 2015. Available online: https://vlg.cs.dartmouth.edu/c3d/ (accessed on 20 January 2023).
128. Peng, X.; Wang, L.; Wang, X.; Qiao, Y. Bag of visual words and fusion methods for action recognition: Comprehensive study and good practice. *Comput. Vis. Image Underst.* **2016**, *150*, 109–125. [CrossRef]

129. Lev, G.; Sadeh, G.; Klein, B.; Wolf, L. RNN fisher vectors for action recognition and image annotation. In *Lecture Notes in Computer Science (Including Subseries Lecture Notes in Artificial Intelligence and Lecture Notes in Bioinformatics)*; LNCS; Springer: Berlin/Heidelberg, Germany, 2016; Volume 9910, pp. 833–850. [CrossRef]
130. Park, E.; Han, X.; Berg, T.L.; Berg, A.C. Combining multiple sources of knowledge in deep CNNs for action recognition. In Proceedings of the 2016 IEEE Winter Conference on Applications of Computer Vision, WACV 2016, Lake Placid, NY, USA, 7–10 March 2016. [CrossRef]
131. Wang, X.; Farhadi, A.; Gupta, A. Actions ~ Transformations. In Proceedings of the IEEE Computer Society Conference on Computer Vision and Pattern Recognition, Las Vegas, NV, USA, 27–30 June 2016; pp. 2658–2667. [CrossRef]
132. Zhu, W.; Hu, J.; Sun, G.; Cao, X.; Qiao, Y. A Key Volume Mining Deep Framework for Action Recognition. In Proceedings of the IEEE Computer Society Conference on Computer Vision and Pattern Recognition, Las Vegas, NV, USA, 27–30 June 2016; pp. 1991–1999. [CrossRef]

applied sciences

MDPI

Technical Note

Quality Analysis of Unmanned Aerial Vehicle Images Using a Resolution Target

Jin-Hyo Kim [1] and Sang-Min Sung [2,*]

[1] Department of Landscape Architecture, Kyungpook National University, Daegu 41561, Republic of Korea; jhkim85@knu.ac.kr
[2] CCZ Forest Land Management Office, Korea Forest Conservation Association, Daejeon 35262, Republic of Korea
* Correspondence: sungsm@kfca.re.kr; Tel.: +82-10-2574-3519

Abstract: Unmanned aerial vehicle (UAV) photogrammetry is an emerging means of acquiring high-precision rapid spatial information and data because it is cost-effective and highly efficient. However, securing uniform quality in the results of UAV photogrammetry is difficult due to the use of low-cost navigation devices, non-surveying cameras, and rapid changes in shooting locations depending on the aircraft's behavior. In addition, no specific procedures or guidelines exist for performing quantitative quality tests or certification methods on UAV images. Additionally, test tools for UAV image quality assessment only use the ground sample distance (GSD), often resulting in a reduced image quality compared with that of manned aircraft images. In this study, we performed a modulation transfer function (MTF) analysis using a slanted edge target and a GSD analysis to confirm the necessity of MTF analysis in UAV image quality assessments. In this study, we aimed to address this issue by conducting a modulation transfer function (MTF) analysis using a slanted edge target and a ground sample distance (GSD) analysis. This was carried out to confirm the necessity of MTF analysis in evaluating UAV image quality. Furthermore, we analyzed the impact of flight height and mounted sensors on image quality at different study sites.

Keywords: unmanned aerial vehicle (UAV) photogrammetry; ground sample distance (GSD); modulation transfer function (MTF); image quality

Citation: Kim, J.-H.; Sung, S.-M. Quality Analysis of Unmanned Aerial Vehicle Images Using a Resolution Target. *Appl. Sci.* **2024**, *14*, 2154. https://doi.org/10.3390/app14052154

Academic Editors: Zahid Mehmood Jehangiri, Mohsin Shahzad and Uzair Khan

Received: 2 February 2024
Revised: 27 February 2024
Accepted: 28 February 2024
Published: 4 March 2024

1. Introduction

Images obtained using unmanned aerial vehicles (UAVs) are captured at low heights, and thus, have higher resolutions than those captured by manned aircraft and can be acquired anytime and anywhere. Additionally, UAVs are emerging as a means of acquiring high-precision rapid spatial information and data because of their low cost and high efficiency. Images obtained from UAVs are widely used in public and private institutions for surveying civil engineering and construction sites [1,2], estimating the quantity of civil works, analyzing terrain slope, in traffic applications for traffic data collection [3–5], are utilized in agriculture and for the environment [6–8] and coastline detection, and find application in the marine field [9,10] and in studying forest diseases and pests [11]. Therefore, practical applicable operational procedures such as public surveying have been established. However, it is difficult to obtain data with consistent quality and to use UAV images in practical applications because no specific procedures or methods exist for quantitatively testing or certifying the data's quality. This difficulty has been attributed to the use of cheap navigation systems, the unsteadiness of the UAV at the time of image capture, and unfavorable weather conditions. In addition, ground sample distance (GSD) analysis is currently used for assessing the image quality [12].

Several methods are used for testing UAV image quality, including the MTF, edge response, and GSD analyses. Among previous studies on aerial image quality testing, Baer [13] proposed the spatial resolution analysis method using a circular target. The method overcomes the shortcomings of the traditional method that uses edge and slanted edge targets. Wang et al. [14] proposed a method that automatically measures the modulation transfer function (MTF) with a high success rate and acceptable accuracy using the Hough transform for detecting straight lines from manned aircraft or satellite images. Sieberth et al. [15] developed a technique that automatically filters UAV image blurring caused by camera movements induced by strong wind, turbulence, or the operator's sudden movement. The technique enables objective analysis as it automatically detects and removes blurring from UAV images, improves image quality, and reduces time and cost compared to the traditional method based on manual detection by the operator. Orych [16] used the Siemens star to measure spatial resolution in UAV images. The Siemens star facilitates analysis and ensures objectivity in all directions as it is unaffected by flight direction. Additionally, as the Siemens star has a smaller size and smaller dimensions than those of the bar target, which is widely used for manned aircraft images, the Siemens star is an ideal resolution target for the UAV photogrammetry system, which flies at low heights. Likewise, many methods are used for testing UAV image quality, including the MTF, edge response, and GSD analyses.

However, the quality of UAV images is lower than that of manned aircraft images, in some cases because the UAV image quality test tool assesses quality using only the GSD analysis, which, unlike the MTF or edge response analyses, does not consider the contrast levels alongside image resolution. In addition, securing uniform quality in the results of UAV photogrammetry is difficult due to the use of low-cost navigation devices, non-surveying cameras, and rapid changes in shooting locations depending on the aircraft's behavior.

To address this issue, we aimed to investigate the effect of UAV imaging altitude and the performance of mounting sensors on UAV imaging quality. We also aimed to evaluate the necessity of MTF analysis in evaluating UAV imaging quality. To achieve this, we set up inclined corners and bar targets at the shooting site, as shown in Figure 1, and captured images using four types of UAVs and one type of manned aircraft. We conducted shooting using different mounting sensors at the same shooting altitude, and the same mounting sensors were also used at different shooting heights. Subsequently, we generated the final UAV orthoimages and performed both GSD and MTF analyses on the corresponding orthoimages to draw conclusions.

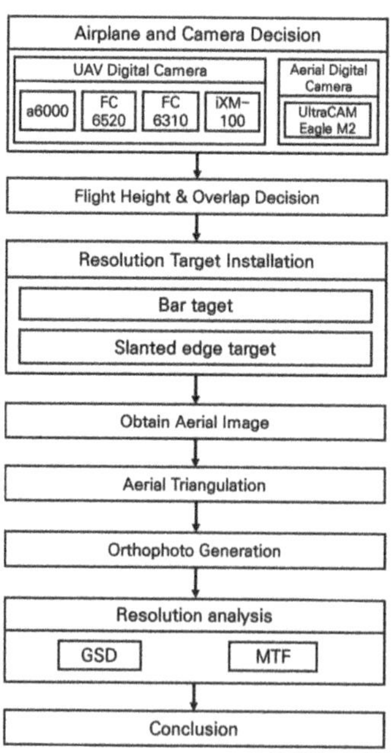

Figure 1. Research flow chart illustrating the experimental setup and methodology.

2. Theoretical Background

2.1. GSD Analysis Using the Bar Target

We analyzed the GSD using a bar target in addition to the MTF to determine the image resolution and contrast. We compared the results to determine the necessity of MTF analysis. The spatial resolution analysis using the bar target is described below. As illustrated in Figure 2, the modulation function in an image can be represented by digital numbers (DNs), which are not continuous for each pixel in the original image [17].

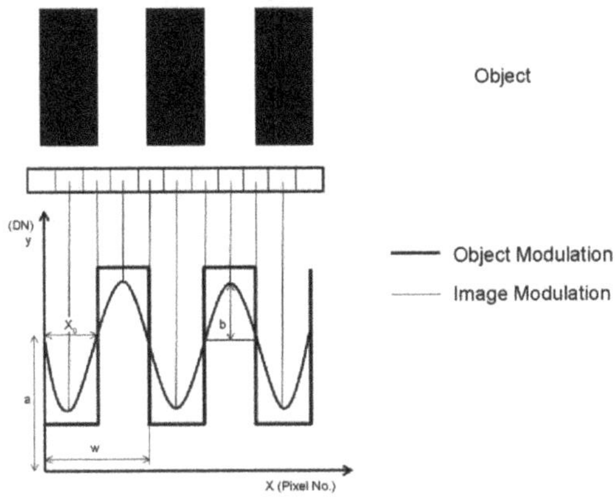

Figure 2. Concept of modulation value analysis.

When a graph of numbers in non-continuous points is fitted by the method of least squares, the curve is expressed as a sine function in periodic form, such as Equation (1):

$$\mathrm{Y} = \mathrm{a} + \mathrm{b}\sin\omega(x - x_0), \quad (1)$$

where a is a coefficient for the digital numbers (DNs) in a pixel from which a curve begins before it moves toward the y-intercept; b represents the amplitude of the sine curve, i.e., the difference between the maximum and minimum values; ω represents the period of the sine function and is related to the image GSD measurement; x represents the pixel sequence; and x_0 is the distance moved in parallel toward the x-axis, causing a phase change that determines the form of the sine function. Hence, an accurate image GSD can be obtained by measuring the size of the black and white lines in the imaged target and dividing the size by the spatial frequency represented in the sine function with the coefficients calculated from Equation (1).

2.2. MTF Analysis

Cameras do not provide images that perfectly represent real objects. object's level of representation is related to the camera's performance; the MTF value, which indicates the object's level of representation, is used for analyzing UAV images. The MTF analysis is based on the camera's lens and performance. The MTF value is expressed as the relative ratio of the actual modulation value of the resolution target to the modulation value of the target in an image. The MTF can be analyzed using a graph based on spatial frequency, which shows how many line pairs (lps) can be included in one pixel when black and white lines form each lp. In the MTF graph, the horizontal axis expresses spatial frequency, and the vertical axis expresses the MTF value [18].

Figure 3 illustrates the DNs extracted from an image with black and white line pairs. In this graph, the modulation value is expressed by Equation (2):

$$\text{Modulation} = \frac{l_{max} - l_{min}}{l_{max} + l_{min}} = \frac{a + b - (a - b)}{a + b + (a - b)} = \frac{2b}{2a} = \frac{a}{b} \quad (2)$$

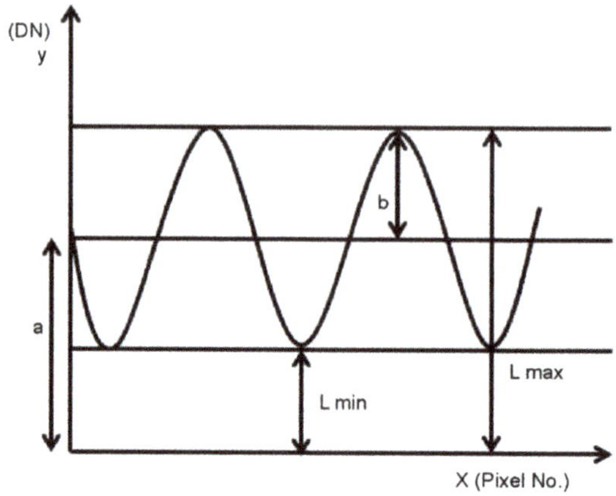

Figure 3. DNs extracted from an image of a pair of black and white lines.

The MTF value is expressed by Equation (3):

$$\mathrm{MTF} = \frac{Modulation_{Image}}{Modulation_{Object}}, \quad (3)$$

where $Modulation_{Image}$ is the modulation value of the image, and $Modulation_{Object}$ is the modulation value extracted from the actual object.

$$\mathrm{MTF} = e^{-2\pi^2\sigma^2{}_{MTF}K^2}, \quad (4)$$

The DNs extracted from UAV images taken by the iXM-100 sensor, as illustrated in Figure 4, were linearized, and an MTF graph was generated using Equation (4), where K represents the spatial frequency equivalent to the vertical axis of the MTF, and $\sigma^2{}_{MTF}$ is the variance of the MTF [19].

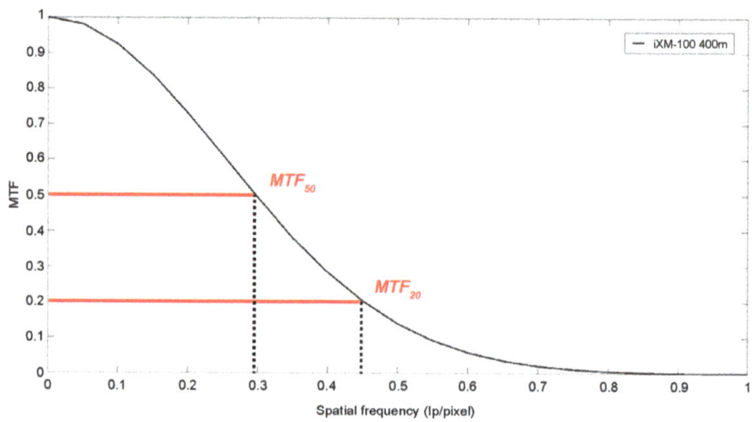

Figure 4. MTF graph for explaining MTF20 and MTF50 (iXM-100 400 m).

We calculated σ_{MTF}, the standard deviation of the MTF, and performed comparisons. We also calculated and compared MTF_{50}, a spatial frequency equivalent to 50% of the MTF graph, and MTF_{20}, a spatial frequency equivalent to 20% of the MTF graph, as illustrated in Figure 4. MTF_{50} is an empirical criterion used in many studies, and it refers to a spatial frequency from which the boundary begins to blur in the operator's eyes. MTF_{20} refers to the minimum spatial frequency at which the boundary is distinguishable by the operator's eyes. It is also an empirical criterion used in many studies.

2.3. MTF Analysis Using the Slanted Edge Target

Figure 5a represents the slanted edge target and the boundary of the target is emphasized with a red dotted line. Figure 5b is the content of extracting DN from the target, and the black dotted line means extracting the DN value at the corresponding location. Figure 5c is the ESF graph generated from the extracted DN, Figure 5d is the PSF graph, and Figure 5e is the finally generated MTF graph.

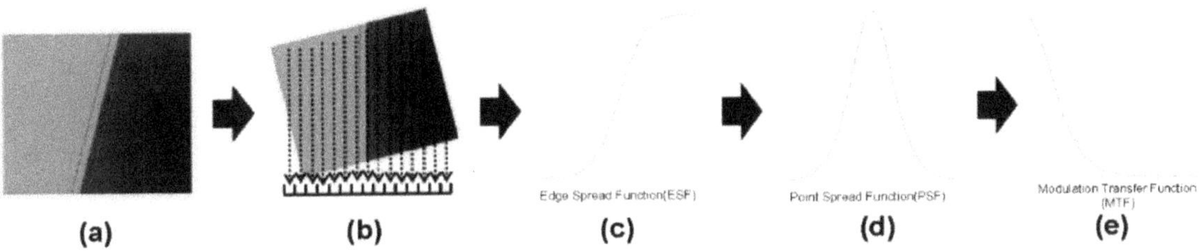

Figure 5. MTF analysis step using slanted edge target. (**a**) represents the slanted edge target and the boundary of the target is emphasized with a red dotted line. (**b**) is the content of extracting DN from the target, and the black dotted line means extracting the DN value at the corresponding location. (**c**) is the ESF graph generated from the extracted DN, (**d**) is the PSF graph, and (**e**) is the finally generated MTF graph.

The first step of the MTF analysis involves using the slanted edge target to find the boundary that is useful for analysis from the slanted edge target, as illustrated in Figure 5a. To find the boundary, a sufficient number of DNs is determined to produce the edge-spread function (ESF) and point-spread function (PSF) graphs stably. If too many DNs are extracted and analyzed, the image noise affects the MTF analysis. Sixteen DNs are typically used, but fifteen to sixteen DNs were used in this study.

In the second step, the mean of the DNs extracted from each line, as illustrated in Figure 5b, is calculated to generate the ESF illustrated in Figure 5c. Unlike the edge target, which is arrayed perpendicularly, the slanted edge target has slanted boundaries and different pixel array angles; hence, the mean DNs calculated from each line obtains the ESF without aliasing.

The most important step in an MTF analysis is detecting the boundary of the slanted edge target and extracting the DNs. If the perpendicular edge target is used, a few DNs are generated, as illustrated in Figure 6a. Figure 6b, however, shows that, if the slanted edge target is used to extract the DNs, multiple scan lines can be used, which makes it possible to extract and analyze an appropriate number of DNs across the boundary. In the case of a vertical edge target, the same set of DN values is generated at any location. However, in the case of a slanted edge target, a set of DN values is generated at different locations; thus, performing the MTF analysis is possible by extracting appropriate DN values across the entire boundary. The slanted edge target has an advantage as it obtains the ESF without aliasing from UAV images using the non-metric digital camera, and it enables a more accurate MTF analysis [20]. In the third step, the PSF graph is generated after the ESF graph is generated and differentiated, as illustrated in Figure 5d. Fitting the graph with the Gaussian function while producing the PSF graph reduces the effect of noise on the MTF value. Finally, the Fourier transform is used to generate the MTF graph from the PSF graph, as illustrated in Figure 5e.

Figure 6. ESF graph of edge target (**a**) and slanted edge target (**b**).

3. Materials and Methods

3.1. Specifications of Resolution Targets

3.1.1. Bar Target

The bar target in this study was developed based on the USAF 1951 test pattern, which is a resolution target used by the US Air Force to test the quality of sensors installed on UAVs, night goggles, and other image devices [21]. The USAF 1951 test pattern consists of 3 bars, and the distance between consecutive bars is fixed as a scale factor. Considering the characteristics of the unmanned aerial image, which has a higher resolution than the manned aerial image, in this study, the size of the bar pattern was successively reduced in 11 steps, as shown in Figure 7. The size was reduced by $1/\sqrt[6]{2}$ (approximately 12%) times in each step; thus, Bar 11 was 15.75 cm wide and 3.15 cm long. Therefore, the shape of a small bar can be visually identified in the image because a high-resolution sensor is mounted [21].

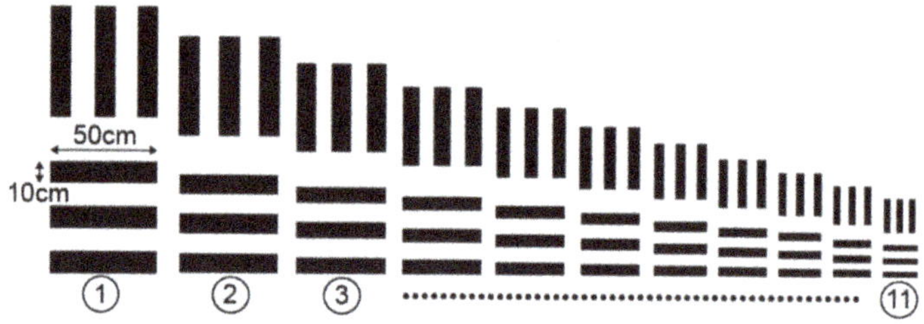

Figure 7. Specification of simple resolution bar target for UAV photogrammetry.

3.1.2. Slanted Edge Target

The image quality verification method using a slanted edge target has been widely used over the past 10 years. It has been adopted by several international standards, including the International Organization for Standardization (ISO). As shown in Figure 8, the angle of the slanted edge of the target was designed to be inclined at 5°, as stipulated in ISO 12233 [22]. In addition, the contrast between the black and white parts of the slanted edge target must be at least 40:1 in ISO 12233. However, the recently revised black and white contrast ratio is as 4:1 [22].

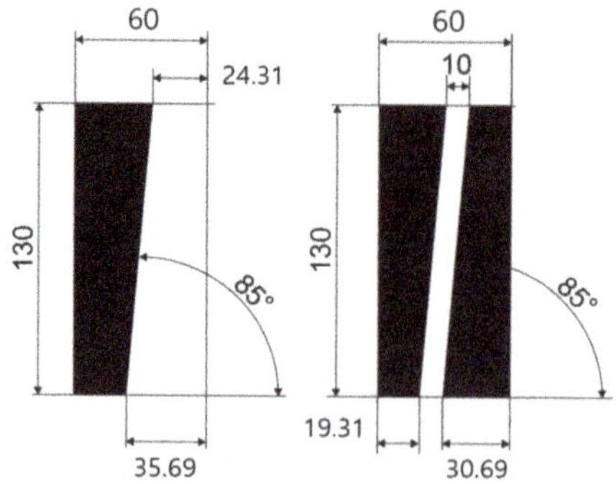

Figure 8. Specifications of the slanted edge target for UAV photogrammetry.

3.2. Resolution Target Installation

A resolution target was used for the UAV photogrammetry to improve the portability, ease of conducting the spatial resolution analysis of UAV photos, and outcomes, and work efficiency. The bar target was divided into 11 sizes; the largest bar was Size 1 (50 cm × 10 cm; width × height), and the size was reduced by $1\sqrt[6]{2}$ (about 12%) at every step to the smallest bar of Size 11 (15.75 cm × 3.15 cm; width × height). The slanted edge target was 60 cm × 130 cm (width × height), and the edge at the center was placed at 5°.

Three locations were selected for the UAV imaging: Miryang, Gyeongsangnam-do; Gimhae, Gyeongsangnam-do; and Beomil-dong, Busan. In Miryang, Gyeongsangnam-do, the a6000 and iXM-100 sensors were used for imaging; in Gimhae, Gyeongsangnam-do, the FC 6250 and FC 6310 sensors were used; and in Beomil-dong, Busan, the UltraCAM Eagle M2 manned aircraft sensor was used. Figure 9 shows the longitude, altitude, and coordinates of the three locations and the camera sensors used for imaging.

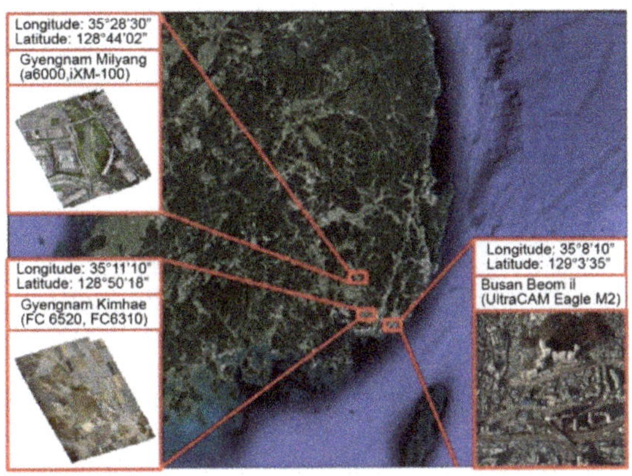

Figure 9. Map of the location of the study area.

3.3. Image Acquisition and Processing

Table 1 lists the UAVs used to acquire the study data; these were: FireFly 6 pro (fixed-wing), Inspire 2, Phantom pro 4, and Matrice 600 (rotary-wing). The table also lists the specifications of the UltraCAM Eagle M2 manned aircraft sensor. The resolution of the camera in each UAV is also listed. In terms of the focal length, pixel size, and CCD sensor size, the iXM-100 sensor included in Matrice 600 had the best performance, followed by the a6000 sensor included in FireFLY 6 PRO, the FC 6520 sensor included in Inspire2, and the FC 6310 sensor included in Phantom pro 4. All unmanned aerial cameras were automatically set to capture the set shot-routed images.

Table 1. Specifications of the UAVs, their cameras, and the manned aircraft sensor [23–27].

UAV Model	FireFLY 6 PRO	Inspire 2	Phantom Pro 4	Matrice 600	Manned Aircraft
Appearance					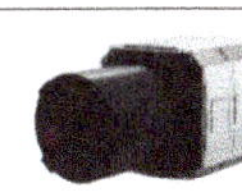
Camera model	a6000	FC 6520	FC 6310	iXM-100	UltraCAM Eagle M2
Focal length	20 mm	15 mm	8.8 mm	35 mm	100 mm
Pixel size	4 × 4 μm	3.28 × 3.28 μm	2.41 × 2.41 μm	3.76 × 3.76 μm	6 × 6 μm
CCD sensor size	6000 × 4000 (24 MP)	5280 × 3956 (21 MP)	5472 × 3648 (20 MP)	11,664 × 8750 (100 MP)	17,310 × 11,310 (193 MP)

To analyze the effects of flight height, camera performance, and imaging conditions on the quality of the UAV photos and outcomes, we designed the flight parameters as illustrated in Table 2. The term "overlap" refers to the degree of route overlap that occurred while capturing the UAV images. In overlap, P is the degree of overlap in the vertical direction of the photograph, and Q is the degree of overlap in the horizontal direction.

Table 2. Flight parameters used for analyzing UAV images.

Camera Model	Flight Height	Overlap	Area	Number of Images	Wind Velocity	Flight Date
a6000	150 m	P = 60% Q = 75%	720 m^2	451	0.9 m/s	19 April 2011
FC 6520	150 m	P = 60% Q = 70%	422 m^2	371	1.9 m/s	18 May 2022
FC 6310	80 m	P = 60% Q = 70%	894 m^2	632	1.9 m/s	18 May 2022
	100 m	P = 60% Q = 70%	894 m^2	556	1.9 m/s	18 May 2022
	150 m	P = 60% Q = 70%	894 m^2	422	1.9 m/s	18 May 2022
iXM-100	150 m	P = 60% Q = 70%	462 m^2	231	1.3 m/s	19 March 2028
	200 m	P = 60% Q = 70%	462 m^2	115	1.3 m/s	19 March 2028
	400 m	P = 60% Q = 70%	462 m^2	52	1.3 m/s	19 March 2028

Metashape (v1.8.2, Agisoft, St. Petersburg, Russia) was used to calculate the results, and all parameters within the software were set to be the same.

4. Results

Based on the resolution target, we analyzed the quality of the images obtained by the UAVs and the manned aircraft. The results were divided into the outcomes of the GSD and MTF analyses as follows:

1. To analyze the effect of camera performance on the quality of the UAV's photos and outcomes, we set the flight height to be almost identical at 150 m, and the overlap at P = 60% and at Q = 70–75%. To compare the camera performance, we indicated the name of each camera model.
2. Using the FC 6310 and iXM-100 sensors, we captured the images at different heights to analyze the effect of flight height on the quality of the UAV's images and outcomes.
3. The GSD and MTF of the manned aircraft images from the UltraCAM Eagle M2 sensor and of the UAV images from the four sensor types were analyzed and compared.

4.1. GSD Analysis

Table 3 presents the results of the GSD analysis using the bar target. The flight height, camera focal length, and pixel size were used to calculate the theoretical GSD, which was compared with the measured GSD. Theoretical GSD can be calculated by multiplying the camera's one-pixel size by the flight altitude and dividing by the focal length. In this context, the measured GSD deviates from the theoretical GSD, owing to errors in the correction values of the camera, the atmospheric conditions during image capture, the unmanned aerial vehicle's dynamics, GPS error values, and other factors.

Table 3. GSD analysis results using bar target.

Camera Model	Flight Height	Bar Target	Theoretical GSD	Measured GSD
a6000	150 m	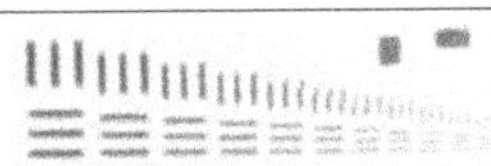	3.0 cm	3.1 cm
FC 6520	150 m		3.3 cm	4.1 cm
	80 m	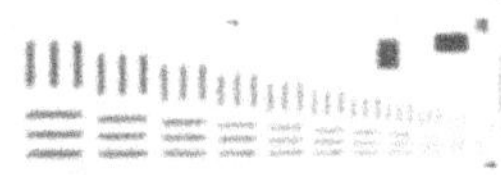	2.2 cm	3.4 cm
FC 6310	100 m		2.7 cm	4.0 cm
	150 m		4.1 cm	5.0 cm
	150 m	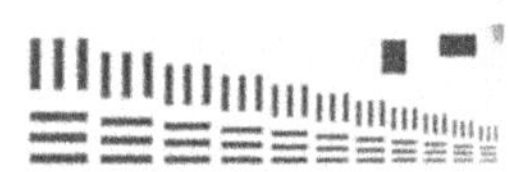	1.6 cm	1.6 cm
iXM-100	200 m		2.1 cm	2.2 cm
	400 m		4.3 cm	4.5 cm
UltraCAM Eagle M2	1000 m		6.6 cm	6.8 cm

Figure 11 displays a graph of the theoretical and measured GSDs using the bar target. The GSDs of the FC 6520 and FC 6310 sensors differed by 18–35% from the theoretical GSD. The GSDs of the iXM-100, a6000, and UltraCAM Eagle M2 sensors, however, only differed slightly, by 0–5%, from their theoretical values. These results suggest that the FC 6520 and FC 6310 sensors were more affected than the other sensors by the factors that reduced the image quality during UAV image capturing. Hence, sensors that differed considerably from the theoretical GSD should be avoided or carefully tested.

Figure 10 is a graph of the GSD results analyzed using bar targets for flight height. For FC 6310 sensors, the GSDs were 3.4 cm (80 m in height), 4.0 cm (100 m in height), and 5.0 cm (150 m in height) as the flight height increased, resulting in poor image quality. The GSDs for iXM-100 sensors decreased to 1.6 cm (150 m in height), 2.2 cm (200 m in height), and 4.5 cm (400 m in height) as the flight height increased. In addition, the GSD for the iXM-100 sensors, which showed the best performance at the same flight height of 150 m, was the best, with a value of 1.6 cm. Subsequently, the GSD values were 3.1 cm for the a6000 sensor, 4.1 cm for the FC 6520 sensor, and 5.0 cm for the FC 6310 sensor.

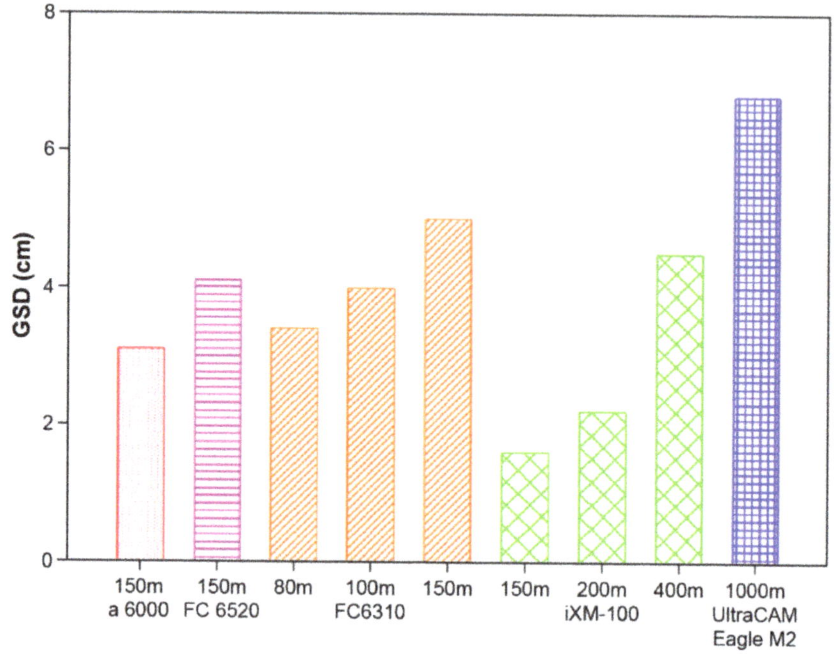

Figure 10. Comparison of the ground sample distance (GSD) analysis results obtained using the bar target.

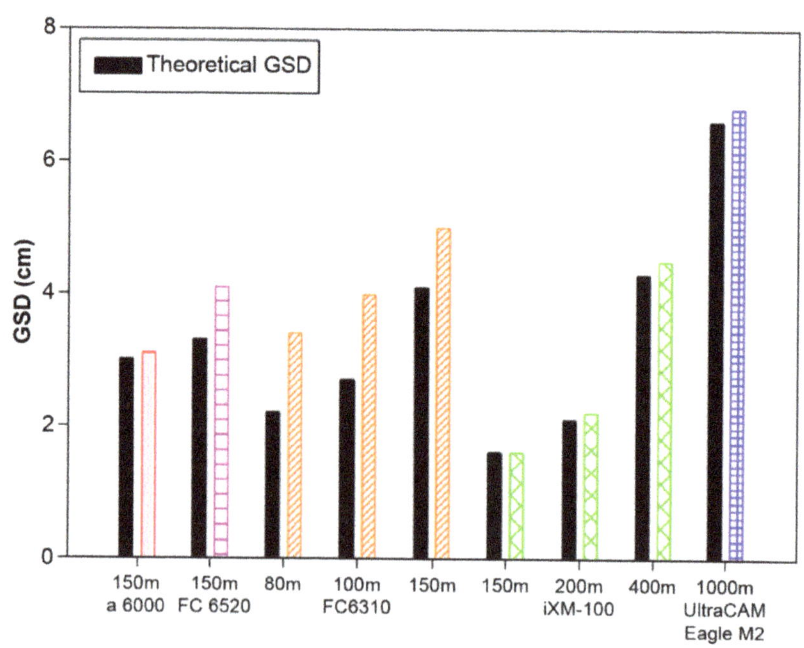

Figure 11. Comparison between the theoretical GSD and measured GSD using bar target.

4.2. MTF Analysis

In Table 4, σ_{MTF} denotes the standard deviation of the MTF; the smaller the value of σ_{MTF}, the clearer the image. MTF_{50} and MTF_{20} were also calculated and compared. Considering the σ_{MTF}, the image quality worsened as the flight height increased in the MTF analysis, similar to those in the GSD, ground resolved distance (GRD), and edge response analyses.

Table 4. MTF analysis results using slanted edge target.

Camera Model	Flight Height	Slanted Edge Target	σ_{MTF}	MTF_{50}	MTF_{20}
a6000	150 m		0.401	0.456	0.694
FC 6520	150 m		0.511	0.381	0.582
FC 6310	80 m		0.443	0.426	0.648
	100 m		0.522	0.336	0.513
	150 m		0.694	0.268	0.408
iXM-100	150 m		0.331	0.545	0.831
	200 m		0.395	0.474	0.722
	400 m		0.635	0.286	0.437
UltraCAM Eagle M2	1000 m		0.715	0.263	0.399

As shown in Figure 12, the iXM-100 sensor maintained a high MTF value as the spatial frequency increased to a height of 150 m and showed significantly better MTF results than those for the other camera sensors. The σ_{MTF} value for the iXM-100 sensor was the lowest at 0.331 (smaller the σ_{MTF} value, better is the image quality). The image quality worsened from top to bottom in the order of the MTF curves of the sensors presented in Figure 12. Specifically, the iXM-100 sensor exhibited the best results at a height of 150 m with σ_{MTF} = 0.331, MTF_{50} = 0.545 lp/pixel, and MTF_{20} = 0.831 lp/pixel. The UltraCAM Eagle M2 manned aircraft sensor exhibited the worst results, with σ_{MTF} = 0.715, MTF_{50} = 0.263 lp/pixel, and MTF_{20} = 0.399 lp/pixel. At the same height of 150 m, the boundary of the black and white lp of 10 cm width began to blur from the GSD values of 3.31 cm for the iXM-100 sensor, 4.58 cm for the a6000 sensor, 3.75 cm for the FC 6520 sensor, and 2.65 cm for the FC 6310 sensor. The boundary of the black and white lp that was 10 cm in width was no longer identifiable above the values of 8.31 cm, 6.98 cm, 5.72 cm, and 4.04 cm for the iXM-100, a6000, FC 6520, and FC 6310 sensors, respectively. Moreover, the boundary of the black and white lp that was 10 cm in width started to blur from the GSD value of 2.63 cm for the UltraCAM Eagle M2 sensor, which exhibited the worst performance, and was no longer identifiable above 3.99 cm.

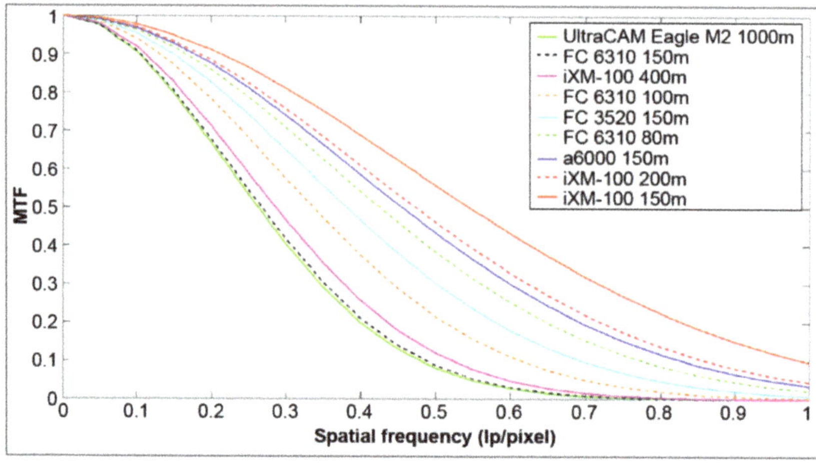

Figure 12. Comparison of the MTF curves obtained from cameras using slanted edge target.

5. Conclusions

In this study, we acquired images from four types of UAVs and one type of manned aircraft to investigate the need for MTF analysis in UAV image quality assessment. We also examined the impact of shooting altitude and sensor performance on the quality of unmanned aerial images. An MTF analysis was conducted using a slanted edge target, while the GSD analysis was performed using a bar target.

First, the trend in σ_{MTF} indicated that the image quality worsened as the flight height increased in both the MTF and GSD analyses. However, the σ_{MTF} values were low for the FC 6520 and FC 6310 sensors, resulting in a slight blurring of the white object. The MTF analysis evaluates both image resolution and contrast; hence, slight blurring has a considerable effect on σ_{MTF}. However, the MTF analysis of the corresponding bar was not possible owing to contrast reduction of the bar target for which the visual resolution analysis was possible. The smallest bar that can be recognized as the bar target, which can be analyzed by visual resolution, determines the GSD of the corresponding image. For example, the FC 6310 sensor, at a shooting altitude of 150m, can analyze up to Bar 5, which is 6.3cm in size due to reduced contrast ratio. However, the GSD results analyzed using Bars 1–5 showed a 5cm result. Similarly, the iXM-100 sensor, at a 400m shooting altitude, can analyze up to Bar 6, which is 5.6cm in size due to reduced contrast ratio. However, the analyzed GSD result was 4.5cm. Therefore, the MTF, which can analyze both the degree of contrast and resolution of the image, was required to verify the quality of the unmanned aerial image, which experienced a deterioration in its quality owing to various factors, such as weather conditions, the use of non-surveying cameras, and low-cost navigation devices. Thus, the MTF analysis was proven to be a more objective and reliable method of analysis than the GSD analysis.

Secondly, we observed a decline in image quality for both the FC 6310 and iXM-100 sensors as shooting altitude increased. Furthermore, when comparing images captured by these sensors at the same altitude of 150 m, it was evident that the GSD and MTF values varied based on the sensor's performance. Consequently, we confirmed that both shooting altitude and sensor performance significantly impact the image quality of UAV images.

Third, despite the UltraCAM Eagle M2 manned aircraft sensor exhibiting the poorest image quality, the results from the FC 6310 sensor were nearly identical at a height of 150 m compared to those from the UltraCAM Eagle M2 sensor. These observations suggest that obtaining high-quality UAV images during UAV photogrammetry is contingent upon the operator accurately determining appropriate camera-sensor parameters, overlapping, and UAV performance before capturing the images, regardless of the number of UAV sensors used.

In this study, we confirmed the necessity of both MTF analysis and GSD analysis for assessing image quality. This was achieved by conducting analyses at various research sites, adjusting the flight height, and using different mounted sensors, which exert the most significant influence on image quality. In future studies, if a test bed with a permanent UAV photogrammetry resolution target can be constructed to analyze the quality of the image under the same conditions, it will be possible to analyze it more quantitatively and objectively, and to accumulate data.

As a result of the analysis, it was determined that MTF analysis, which can analyze the resolution and contrast of images simultaneously, rather than GSD analysis, was a more objective and reliable method. It was found that high-quality unmanned aerial images could be obtained only when workers properly judge the performance of camera sensors, redundancy, and the aircraft's performance.

Author Contributions: Investigation, S.-M.S. and J.-H.K.; Methodology, S.-M.S. and J.-H.K.; Supervision, J.-H.K.; Validation, S.-M.S. and J.-H.K.; Writing—original draft, S.-M.S. and J.-H.K.; Writing—review and editing, S.-M.S. and J.-H.K. All authors have read and agreed to the published version of the manuscript.

Funding: This research was funded by the National Research Foundation of Korea (NRF) under grant number NRF-2018R1D1A1A02085675.

Institutional Review Board Statement: Not applicable.

Informed Consent Statement: Not applicable.

Data Availability Statement: Data are contained within the article.

Conflicts of Interest: The authors declare no conflicts of interest.

References

1. Masita, K.; Hasan, A.; Shongwe, T. Defects Detection on 110 MW AC Wind Farm's Turbine Generator Blades Using Drone-Based Laser and RGB Images with Res-CNN3 Detector. *Appl. Sci.* **2023**, *13*, 13046. [CrossRef]
2. Liu, Y.; Zhou, T.; Xu, J.; Hong, Y.; Pu, Q.; Wen, X. Rotating Target Detection Method of Concrete Bridge Crack Based on YOLO v5. *Appl. Sci.* **2023**, *13*, 11118. [CrossRef]
3. Lu, L.; Dai, F. Accurate road user localization in aerial images captured by unmanned aerial vehicles. *Autom. Constr.* **2024**, *158*, 105257. [CrossRef]
4. Ke, R.; Li, Z.; Tang, J.; Pan, Z.; Wang, Y. Real-Time Traffic Flow Parameter Estimation from UAV Video Based on Ensemble Classifier and Optical Flow. In *IEEE Transactions on Intelligent Transportation Systems*; IEEE: Piscataway, NJ, USA, 2018; pp. 54–64. [CrossRef]
5. Zhao, Y.; Zhou, L.; Wang, X.; Wang, F.; Shi, G. Highway Crack Detection and Classification Using UAV Remote Sensing Images Based on CrackNet and CrackClassification. *Appl. Sci.* **2023**, *13*, 7269. [CrossRef]
6. Ercolini, L.; Grossi, N.; Silvestri, N. A Simple Method to Estimate Weed Control Threshold by Using RGB Images from Drones. *Appl. Sci.* **2022**, *12*, 11935. [CrossRef]
7. Logan, R.D.; Torrey, M.A.; Feijó-Lima, R.; Colman, B.P.; Valett, H.M.; Shaw, J.A. UAV-Based Hyperspectral Imaging for River Algae Pigment Estimation. *Remote Sens.* **2023**, *15*, 3148. [CrossRef]
8. Rajeena, F.P.P.; Ismail, W.N.; Ali, M.A.S. A Metaheuristic Harris Hawks Optimization Algorithm for Weed Detection Using Drone Images. *Appl. Sci.* **2023**, *13*, 7083. [CrossRef]
9. Diruit, W.; Le Bris, A.; Bajjouk, T.; Richier, S.; Helias, M.; Burel, T.; Lennon, M.; Guyot, A.; Ar Gall, E. Seaweed Habitats on the Shore: Characterization through Hyperspectral UAV Imagery and Field Sampling. *Remote Sens.* **2022**, *14*, 3124. [CrossRef]
10. Fabris, M.; Balin, M.; Monego, M. High-Resolution Real-Time Coastline Detection Using GNSS RTK, Optical, and Thermal SfM Photogrammetric Data in the Po River Delta, Italy. *Remote Sens.* **2023**, *15*, 5354. [CrossRef]
11. Domingo, D.; Gómez, C.; Mauro, F.; Houdas, H.; Sangüesa-Barreda, G.; Rodríguez-Puerta, F. Canopy Structural Changes in Black Pine Trees Affected by Pine Processionary Moth Using Drone-Derived Data. *Drones* **2024**, *8*, 75. [CrossRef]
12. Sung, S.M. A Study on Spatial Resolution Analysis Methods of UAV Images. Ph.D. Dissertation, Dong-A University, Busan, Korea, 20 August 2019.
13. Baer, L.R. Circular-edge spatial frequency response test. In *Proceedings Society of Photo-Optical Instrumentation Engineers, Image Quality and System Performance, San Jose, CA, USA, 18 December 2003*; Yoichi Miyake, D., Rene, R., Eds.; SPIE: Bellingham, WA, USA, 2003.
14. Wang, T.; Li, S.; Li, X. An automatic MTF measurement method for remote sensing cameras. In Proceedings of the 2nd IEEE International Conference on Computer Science and Information Technology, Beijing, China, 8–11 August 2009; pp. 245–248.

15. Sieberth, T.; Wackrow, R.; Chandler, J.H. Automatic detection of blurred images in UAV image sets. *ISPRS J. Photogramm. Remote Sens.* **2016**, *122*, 1–16. [CrossRef]
16. Orych, A. Review of methods for determining the spatial resolution of UAV sensors. *Int. Arch. Photogramm. Remote Sens. Spat. Inf. Sci. ISPRS Arch.* **2015**, *XL-1/W4*, 391–395. [CrossRef]
17. Lee, T.Y. Spatial Resolution Analysis of Aerial Digital Camera. Ph.D. Dissertation, Dong-A University, Busan, Korea, 2012; 50p. (In Korean with English abstract).
18. Neumann, A. Verfahren zur Auflösungsmessung Digitaler Kameras. Masters's Thesis, University of Applied Sciences, Cologne, Germany, 2003; 70p.
19. Pedrotti, F.L.; Pedrotti, L.M. *Introduction to Optics*, 3rd ed.; Cambridge University Press: Cambridge, UK, 2017.
20. Crespi, M.; De Vendictis, L. A Procedure for High Resolution Satellite Imagery Quality Assessment. *Sensors* **2009**, *9*, 3289–3313. [CrossRef] [PubMed]
21. Pinkus, A.; Task, H. *Measuring Observers' Visual Acuity Through Night Vision Goggles*; Defense Technical Information Center: Fort Belvoir, VA, USA, 1998.
22. *ISO 12233:2000(E)*; Photography–Electronic Still-Picture Cameras—Resolution Measurements. ISO: Geneve, Switzerland, 2000.
23. Geo-Matching. Available online: https://geo-matching.com/uas-for-mapping-and-3d-modelling/firefly6-pro (accessed on 20 January 2024).
24. Inspire 2. Available online: https://www.dji.com/inspire-2 (accessed on 20 January 2024).
25. Support for Phantom 4 Pro. Available online: https://www.dji.com/phantom-4-pro (accessed on 20 January 2024).
26. MATRICE 600PRO. Available online: https://www.dji.com/matrice600-pro (accessed on 20 January 2024).
27. EagleM2. Available online: https://www.vexcel-imaging.com/EagleM2 (accessed on 20 January 2024).

MDPI AG
Grosspeteranlage 5
4052 Basel
Switzerland
Tel.: +41 61 683 77 34

Applied Sciences Editorial Office
E-mail: applsci@mdpi.com
www.mdpi.com/journal/applsci

www.ingramcontent.com/pod-product-compliance
Lightning Source LLC
Chambersburg PA
CBHW041942260926
51850CB00018B/437
9783725818259